The Broad Spectrum

Studies in the Materials, Techniques, and Conservation of Color on Paper

Edited by
Harriet K. Stratis and Britt Salvesen

Archetype
Publications

First published 2002 by Archetype Publications Ltd.

Archetype Publications Ltd.
6 Fitzroy Square
London W1T 5HJ

www.archetype.co.uk

Tel: 44(207) 380 0800
Fax: 44(207) 380 0500

ISBN 1-873132-57-3

British Library Cataloguing in Publication Data
A catalogue record for this book is available from the British Library.

Typeset in Garamond

Text printed on Garda Matt Art 130 gsm (acid-free)
Jacket printed on Gardagloss Art 135 gsm (acid-free)

Production management and copy editing by Lyn Corson
Designed by Michael J. Claridge
Typesetting and layout by PDQ Digital Media Solutions Ltd., Bungay, Suffolk

Printed and bound in Italy by LEGO, Vicenza

Contents

Preface and Acknowledgments

The Broad Spectrum: The Art and Science of Conserving Colored Media on Paper, a conference held at The Art Institute of Chicago in October 1999, brought together 300 conservators, conservation scientists, curators, art historians, and students from 22 countries. The conference focussed on recent advances in the investigation and conservation of colored media, while embracing a wide range of interests and approaches. Ideas and information – the results of years of art-historical research and technical analysis – were exchanged among distinct but overlapping professional disciplines. In four stimulating days of presentations and discussions, we encountered materials as diverse as pastel, watercolor, distemper, and fluorescent paint; discovered means of examining, conserving, storing, and displaying delicate works on paper; and contemplated the often indistinct line between materials and meanings in works of art.

The principal organizers of the conference, Elizabeth Sobczynski and Harriet Stratis, were joined by the members of a program committee that included David Chandler, Margo McFarland, Cheryl Porter, Gillian Roy, and Joyce Townsend; they were assisted in all aspects of conference organization by Caesar Citraro, Christine Conniff O'Shea, Catherine Hoskins, Julie Houghton, and M. Brigitte Yeh. To all those dedicated colleagues we offer our sincere thanks. We also gratefully acknowledge the financial support provided by The Book and Paper Group of The American Institute for Conservation of Historic and Artistic Works, The International Fine Print Dealers Association, The Fred and Susan Novy Family Foundation, The Print and Drawing Club of The Art Institute of Chicago, Rising and Wyndstone Papers, Graphic Products Corporation, and the Schott Corporation.

As this book of conference proceedings took shape, we benefited greatly from the encouragement and support of James Black at Archetype, whose enthusiasm steadily pushed us forward; Lyn Corson, our production coordinator and copy editor who assured that every detail was in place; Noel and Florence Rothman, who always provided us with a home away from home in which to work when we needed one; and Matthew Glucksberg and Brian Yates, who championed this project from the outset and sustained us throughout. At The Art Institute of Chicago, we thank Caesar Citraro in the Department of Prints and Drawings for helping to develop the cover design and cheering us along the way; Arawa McClendon in the Imaging Department for providing visual materials; Frank Zuccari, Executive Director of Conservation; Suzanne Folds McCullagh, Curator of Earlier Prints and Drawings; and Jay Clarke, Associate Curator of Prints and Drawings.

Finally, we are especially grateful to the authors for providing the compelling and highly diverse series of papers that comprise this volume.

Harriet K. Stratis and Britt Salvesen

Foreword

Agreeing to open *The Broad Spectrum* conference by discussing the relationship between scientific investigation and art history, I realized that I risked preaching to the converted. Yet the task of articulating the connection between two different practices – the technical and scientific on the one hand, and the historical and curatorial on the other – reminded me that some things do bear repeating. The custodians of art are many and the demands on the various professional groups great, with the result that it is all too possible to lose sight of the essential interconnections between the various practices involved in art's custodianship.

The professional expertise required to cover the "field" ranges from those practices and approaches primarily dedicated to art's *physical* body, to those more concerned with what might be termed (for the sake of metaphoric consistency) its *soul*. Practitioners include those who employ knowledge of art's constitution and habits to diagnose its ills, minister to its needs, and ensure its continued well being; they also include those who are dedicated to understanding its spirit, to presenting it in accordance with an interpretation of its trans-material meaning. All carried out in the name of art, these constitute a broad spectrum of tasks – so broad, in fact, that occasionally they appear not so much a continuum of overlapping interests but rather as disjunctive, independent, and even competing bodies of professional expertise.

I have purposely used the oppositional metaphor of body and soul because I find it suggestive of the oppositional dynamic that still lingers in our field and that is inherent in the attitudes toward art that can be subsumed under the term "modernism." With its distinctions of high and low, art and craft, the material and the transcendent, the realm of facts and that of ideas, modernism's legacy can be detected in the ghost of divisiveness that occasionally haunts our discipline. Sometimes it seems that professors in university classrooms, curators in museums, and conservators and conservation scientists in institutional and private laboratories see themselves not simply as differently engaged with different facets of art and its history, but as essentially independent entities pursuing different truths that, depending on the viewpoint, are differently ranked.

In the spring of 1999, the Clark Art Institute in Williamstown, Massachusetts, hosted a symposium devoted to the question of whether it is possible to bridge the "two art histories" – meaning art history as practiced at the university and at the museum. Well attended and sometimes contentious, the symposium was provocative, enlightening, and, I think, ultimately beneficial: it provoked people to reflect on the assumptions they bring to their practice. I recognized, much more keenly than I did when studying art history at Yale University in the early 1970s, that museums can and should assume a unique role as the logical site for bringing together the technical and curatorial, the "science" and the "histories" of art.

Recent changes in the discipline have shifted the focus of art-historical investigation at the university. An emphasis on social history and theory has significantly enriched the field by opening up new ways of considering works of art. But this has also entailed a shift away from the "object" – a term applied with sometimes palpable ambivalence to the physical body of art. Not long ago, students of art history at Yale, Harvard, Oberlin, the Institute of Fine Arts, and other distinguished American institutions associated with museums, typically worked directly with art objects and with those responsible for them. In the 1960s and early 1970s, publications and exhibitions devoted to such artists as Ingres, Degas, Tiepolo, and Goya broke new ground by harnessing the expertise and insights of conservators and art historians such as Marjorie Cohn, Nia Janis, Roy Perkinson, and Eleanor Sayre. Such work exemplified the results – the new insights – that a joint approach can achieve. It was in this climate that I made the decision to pursue a career in museums.

Over the years – first at Yale, then at the National Gallery of Canada, and since 1985 at The Art Institute of Chicago – I have worked closely with colleagues in conservation on every major project I have undertaken. Not because I wanted to make a point, but simply because I realized that by doing so I would see more, understand better, and gain insights I would not have arrived at otherwise. One reason such collaboration seemed natural was that by chance, my first museum positions – initially as a student intern, then as a curator – were in a print department. By tradition print curators must be familiar with the various methods used by artists to make their prints. They require this technical knowledge in order to be able to assess the quality of differing impressions and to understand the stylistic shifts involved when an artist like Dürer moved from making woodcuts to making engravings. Curiously, though, comparable familiarity with media is not traditionally a prerequisite in many other curatorial areas, such as painting. For some reason, many people, and certainly a good portion of our public, believe that they cannot appreciate prints without understanding the differences between the media. Yet lack of technical understanding does not generate a comparable anxiety when they stand before an encrusted pastel that is the result of complex layering and manipulation, or a piece of sculpture by Rodin that exists in different editions, in different media, and in different sizes. Perhaps the lesson is this: what you don't know you don't know won't bother you.

It is difficult for me to imagine how the projects I have been involved with would have fared had I not looked at Degas's pastels with Anne Maheux, at Fantin Latour's paintings with Barbara Ramsay, at Gauguin's graphic work with Peter Kort Zegers, or at the *noirs* and pastels of Odilon Redon with Harriet Stratis. Without the collaborative looking, mock-up experiments, and microscopic study, we could not have fully appreciated how the appearance of pastels by Redon and Degas have altered with time, and we would not have known that seemingly compromised works by Gauguin in fact conform to the artist's intention and methods. Nor could we have responsibly solicited loans of charcoals and pastels without an understanding of these artists' working procedures and use of fixatives.

In my experience, the process of collaboration produces new questions – and new answers – that can radically shift the perspective from which an artist's work is viewed. Of course, collaboration can be seen to challenge cherished notions of professional autonomy, including the self-sufficiency of the practiced connoisseurial "eye" that is traditionally the curator's touchstone. Increasingly, however, we in museums realize that multiple sets of eyes, shaped by different professional experiences, can see further into any question that involves the making and meaning of a work of art. Dialogue sharpens seeing. As we continue our project to publish systematically the Art Institute's permanent collections of drawings and paintings, we, like many of our sister institutions, have built partnerships between curators, conservators, and scientists into our planning. With every new catalogue, attributions change and important discoveries are made.

I find it strange, but also exciting, that so much remains to be discovered in areas that one might think have been mined to exhaustion. *The Broad Spectrum* presents new revelations, and new questions, about such prominent figures as Degas, Cézanne, Turner, Homer, Gauguin, and Hokusai. It also brings less familiar artists and working methods to our attention. While the designations curator, professor, conservator, and scientist are necessary to indicate the different positions we occupy on the broad spectrum of professions devoted to art, it is good to remember that we represent a continuity of curiosity and interest. My own work has, I realize, benefited directly from the efforts made by a number of the contributors to the conference. Similarly the papers presented here may well help some more pieces of the incomplete puzzles we all carry around in our heads drop into place.

Douglas W. Druick
Searle Curator of European Painting and Prince Trust Curator of Prints and Drawings
The Art Institute of Chicago

These remarks opened the conference on October 6, 1999.

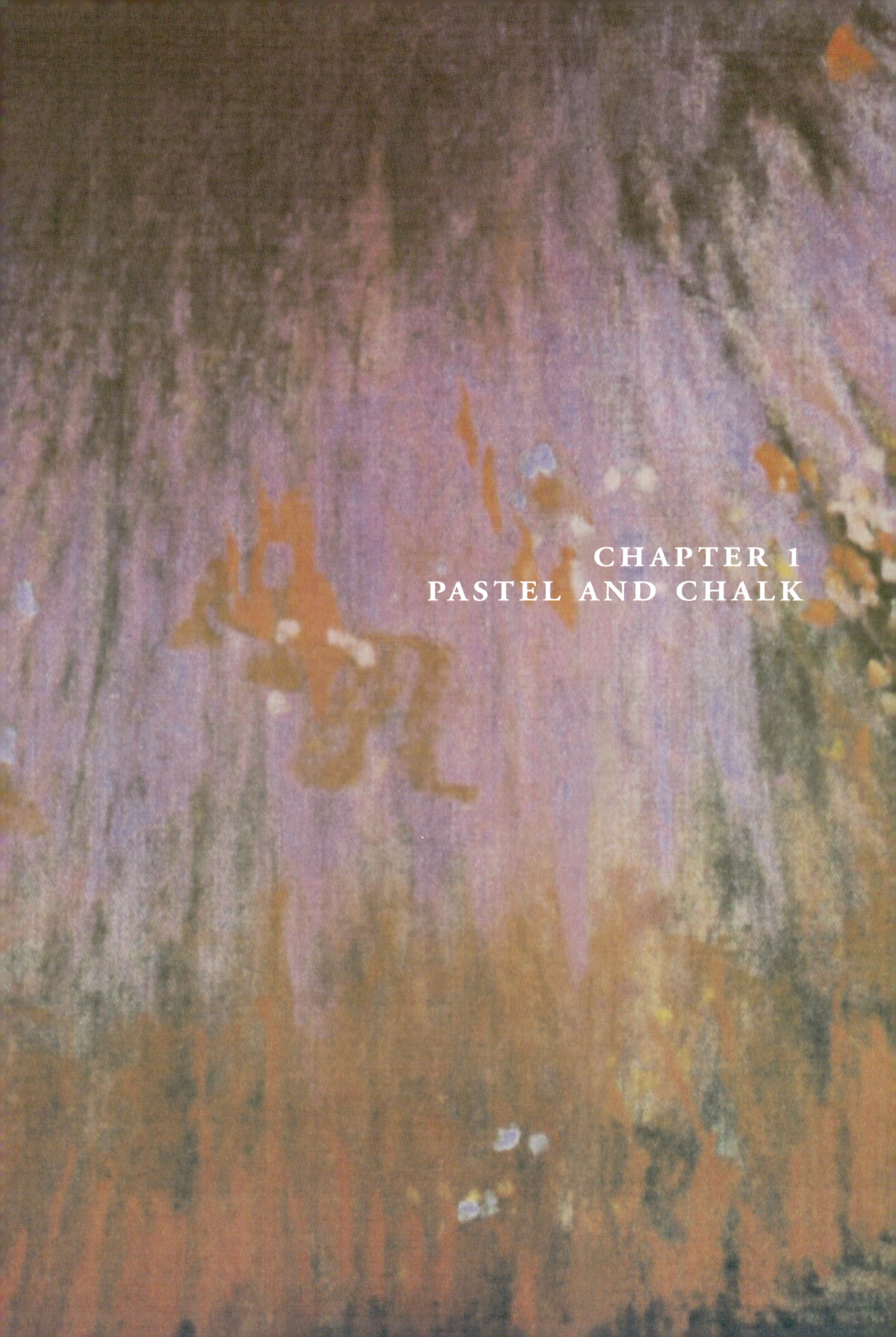

CHAPTER 1
PASTEL AND CHALK

An Aesthetic Overview of the Pastel Palette: 1500–1900

Marjorie Shelley

The appeal of pastel has been its enduring brilliance of color, its matte surface, and velvety richness. These characteristics arise from its particle structure – finely ground powder, randomly distributed in an opaque layer – and the diffuse light reflected from these irregular surfaces (Fig.1.1). Throughout its history the fabrication of this medium has entailed combining pulverized pigment and a white base with a binder, kneading them into a dough or *pastille* (the mass or lump from which the word *pastel* evolves), and, lastly, rolling this into crayons, the name common to these color sticks through the eighteenth century.[1] With a few exceptions, the pigments were those used in painting, and the bases, such as chalk or clay, were determined by the properties of the pigments. They extended the color, thereby producing tints, providing opacity, and giving body and cohesion. The binder, which included a range of gums and sizes, was added only in an amount sufficient to hold the powdered mass together, enabling the dry pastel to be grasped between the fingers without crumbling, yet friable enough to leave a coherent mark when stroked across the support. With few changes in the process and materials, the physical properties of pastel were well adapted to the many aesthetic transformations that occurred over time. As it alternately shifted from a draftsmanly medium into a painterly one, it would serve as the model for, and be modeled on, other art forms. These permutations were accompanied by the changing role of the support, new modes of application, and an ever-expanding palette.

The concept of combining pure pigment with increasing proportions of a white substance in order to produce gradations of color evolved from early Renaissance practices, articulated by Cennino Cennini, for a system of modeling drapery in tempera.[2] It is not

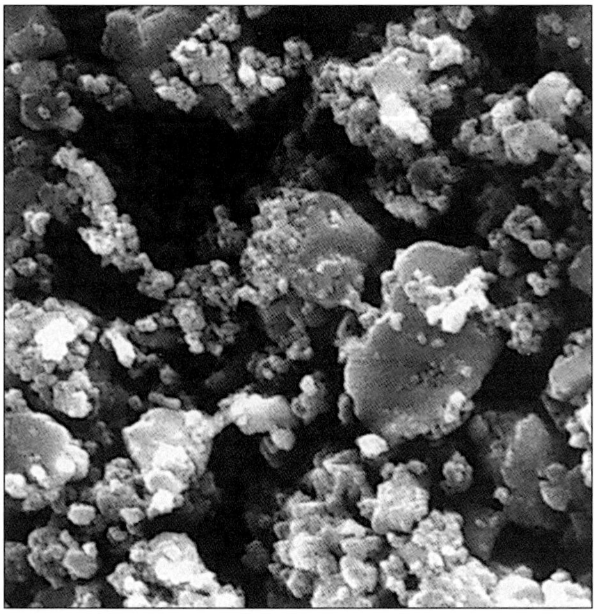

Figure 1.1 Scanning electron micrograph of pastel, 5000x magnification. The irregularly shaped particles of pastel reflect light diffusely, contributing to the characteristic velvety texture of the medium.

known when the idea of transposing the tonal gradations of painting to dry color occurred, but the transition from black and red chalks to a broad palette of powdered tints seems to have taken place only gradually, emerging in an aesthetic atmosphere that placed theoretical precedence on design over color. Leonardo da Vinci was first to record the existence of a material that is associated with pastel, noting in about 1495 his intention to "get from John of Paris [Jean Perreal] the way to draw with dry color."[3] Although he would use such material for chromatic accents, as in the yellow border in the dress of *Isabella d'Este* (Musée du Louvre, Paris), the primary role of dry color for Leonardo, as represented by the *Head of the Virgin* (The Metropolitan Museum of Art, New York), was to assist in creating visual effects seen in nature. With relatively soft and friable materials, Leonardo and his followers incorporated a limited range of hues – ochres, browns, and reds – into their broadly rendered black-chalk drawings, integrating them in order to create relief; the delicate transitions of tone and shadow, or sfumato, whereby both forms and contours were softened and unified with their background.

Seeking to enhance the sense of naturalism, while adhering to the prevailing early sixteenth-century aesthetic that disdained color in drawing, Jean Clouet (d. 1540/41) and his circle, and Hans Holbein (1497/98–1543) employed pastel merely to provide descriptive local accents to their linear black-chalk drawings. Unlike Lombard draftsmanship, color was not integral to the design in the works of French and northern masters. In Holbein's *Mary Zouch* (Windsor Castle), for example, dense and luminous yellow – the use of which was widespread – and salmon pink suggest they are fabricated colors, but applied without the nuances or gradation that would ultimately characterize pastel. The interest of the French court portraitists in coloristic embellishment is demonstrated in the use of the same powdered colors, and the optical enlargement of the palette by overlapping fine strokes of red and black to produce new hues. Though partly determined by aesthetic parameters, their linear precision resulted from the rather unyielding and compact nature of these materials, which were either hard, quarried chalks, or pastels made with bases and binders producing dense compounds. Summary tests utilizing X-ray diffraction conducted at The Metropolitan Museum of Art on yellow samples (earths with small amounts of hematite) from several sixteenth-century drawings revealed the presence of gypsum and the fine clay bentonite, bases that could also serve as binding agents. Under polarizing microscopy, high levels of particle agglomeration were observed, such as those that occur when moisture evaporates from wet pastel. In one of the samples, autofluorescence indicated the presence of a binder; however, its identity could not be ascertained because of the small sample size.[4] These factors suggest that these examples were fabricated crayons, and hence represent a transitional phase in the history of dry color. Whether the use of bases and binders was customary, or if colored natural chalks were commonly employed, or if these yellows were simply calcined, ground, and formed into crayons in the manner described in early treatises demands further investigation.[5]

It is only in the latter part of the sixteenth century, in Italy, in the distinctive compositional and head studies by Jacopo Bassano (1515–1592) and Federico Barocci (1535–1612) that an expansive range of pastel color and varied techniques of application are first observed, pointing to the increasingly painterly and tonal use of this drawing tool. Serving as *ricordi*, or studies for oils, these highly chromatic drawings, enhanced by luminous blue paper, still rely on a framework of black or red chalk. Bassano's more forceful and tactile pastels employ heavily laid-on saturated hues, rubbed colors, and tints. Barocci, using fewer and more delicately applied pigments, stumped the pastel into the sheet, diffusing the color and modeling with it. Although brief statements regarding the use of *pastelli* are found in treatises from the mid-sixteenth century, it is noteworthy that the appearance of works by Bassano and Barocci corresponds to the earliest extant recipe for this medium, implying that its practice had become more widespread. Written in 1574 by Petrus Gregorious of Lyon, the recipe called for various binders and "white lead or chalk to soften the tones and smooth the stroke,"[6] thus referring both to the modification of the hues by the addition of a white substance and to textural adjustments that allowed for the broad handling favored in the pastels of these artists and their followers.

It was in portraiture that pastel would flourish. Correctness of color was valued above design in this genre because it served to convey the verisimilitude that was required for a convincing representation of the face, and as G. P. Lomozzo expressed, the "voice and spirit of the person."[7] It was perhaps because the soft quality of the stroke readily conveyed the varied textures of a sitter's physiognomy and attire that by the seventeenth century pastel came to be associated with this subject matter. This pairing is suggested by Turquet de Mayerne's manuscript entry of 1620, entitled *crayons de toutes couleurs principalement pour visaiges.*[8] By the 1640s the fashion for simple linear crayon portraits as practiced by the followers of Clouet had passed, and the first signs of broadly applied pastel emerged in the works of French and Flemish artists. This change in handling was made possible by adjustments in the fabrication of the crayons. De Mayerne's manuscript, the most detailed of the early technical discussions on pastel, lists two binders (milk and *savon de Venise*) and four bases (alabaster, chalk, white and blue tobacco pipe clays) – choices implying increasing concern with textural properties.[9]

Although drawings incorporating dry color up to the 1660s are not fully chromatic, they nonetheless exhibit variations in depth and transparency as a result of using softer crayons. In portraits by the Lagneau Circle (fl. 1590–after 1650), for example, the powder is manipulated by stumping in order to render the effect of velvet. A similar broadness of application characterizes the pastels by Sir Peter Lely (1618–1680), independent images that were highly regarded in his day and marked the introduction of the medium to England. Executed in black and red chalk on lightly sized paper, *un peu griastre* so that it served as the middle tone,[10] the faces of his sitters were enhanced with passages of delicately modulated pink, blue, white, and yellow pastel. Using a blue support with a harder surface, his contemporary Wallerant Vaillant (1623–1677) combined the precise line of plumbago and ink for the underlying structure with broad areas of white pastel in his *Portrait of a Young Woman* (Fig. 1.2). The technique used here – colored powder applied with the stump or with cotton set in a quill – perhaps evolved in the search for a more expansive means of applying dry pigment. Edward Norgate, in *Miniatura or The Art of Limning* (1648), was the first to describe this method, but, favoring a more linear aesthetic, he dismissed such work as slovenly for its imprecision and large scale.[11] Though presumably little used, this variant on crayon drawing continued to be cited in eighteenth-century treatises and was to persist through the early nineteenth century, as is revealed by Elizabeth Cay's (1770–1831) color box, which contains bottles of powdered pigment and leather stumps (Victoria & Albert Museum, London) and by the remarkably precise portraits produced by James Sharples (1750–1811), a practitioner of this manner.[12]

That these examples are not highly colored must be attributed to a lingering aesthetic preference for restraint, for various sources reveal that a broad range of color was becoming available and that pastel was finding increasing acceptance. In 1662 the Dutch scientist Christiaan Huygens (1629–1695), intrigued by the medium, accompanied Lely to his pastel supplier and purchased 54 crayons.[13] Edward Luttrell (fl. 1680–1724) listed 30 recipes for tints and compound colors in his 1683 manuscript, *An Epitome of Painting*.[14] It was at this time that the first pastel palette, so to speak, was described: Robert Nanteuil's (1623–1678) crayons were chronicled by his protégé as being housed in eight oblong wooden boxes, labeled with the parts of the face for which those colors and half tints were designated.[15] After about 1660 published treatises describing this medium proliferated, and though most originated in England and Holland, those from France included one by the acclaimed champion of colorism, Roger de Piles (1635–1709), written jointly with Jean Baptiste Corneille in 1684.[16] On an official level, the French Académie Royale, founded in 1648, admitted Nicholas Dumonstier as the first pastelist in 1663, and its director, Charles Lebrun (1619–1690), used pastel for preparatory and expressive head studies.

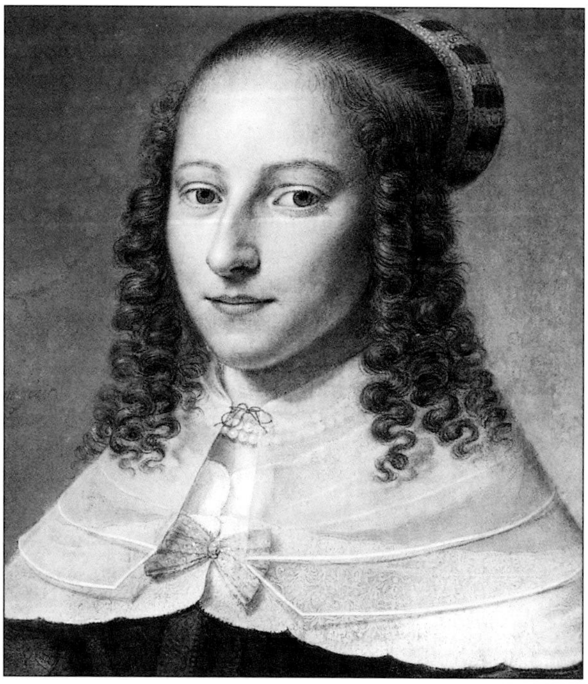

Figure 1.2 Wallerant Vaillant (Dutch, 1623–1677), *Portrait of a Young Woman*, 1648. Graphite, black ink, and gray and white pastel applied with stump on blue paper, 39 x 32.4 cm. The Metropolitan Museum of Art, New York, 1993.199.

Although drawings in pastel would become widespread toward the end of the seventeenth century, the increasing chromatism and painterly handling that emerged after 1650 largely occurred among limners and engravers practicing in England and France. The explicit information on pastel contained in books on miniature painting, and the frequent comparisons of the two techniques in artists' manuals, underscores the link between these sister arts: Norgate, for example, wrote that an artist practicing one would be likely to do both.[17] The support for pastels, as for miniatures, was a card of laminated or pasted paper coated with gouache, the latter now providing a tooth to hold the powdered pigment. Perhaps the earliest examples of this are the pastel-and-chalk portraits on pink prepared paper by Holbein, who was trained as a miniaturist. In the seventeenth century this practice was commonplace, as demonstrated by the pastels by William Faithhorne (1616–1691), with their underlayer of brushed color, or by a toned album page on which Charles Beale (1631–1705) executed a preparatory study (Victoria & Albert Museum, London). Unlike the transparent carnations of limning, the background tone for pastel was muted, echoing the prevailing taste for a dark palette, and also serving as a midpoint for working the lights and shadows in color. In their linearity, however, the austere pastels of this era were modeled on the fine brushwork of these small paintings. Their controlled, precise effects indicate that they were executed with sharpened color sticks the thickness of a quill, and compounded of bases and binders that produced hard and compact crayons.

The link with printmaking also developed at mid-century, notably with Robert Nanteuil, who was best known as an engraver. A pivotal figure in the history of pastel, Nanteuil favored a somber palette of halftones that translated the black-and-white gradations of engraving and anticipates the painterly transitions that characterize eighteenth-century works in this medium. Although he had few competitors, as his student

Domenico Tempesti noted in *Opinions, Maximes et Conseils*, Nanteuil jealously guarded his method of pastel fabrication and thus nothing is known about their composition.[18] The mystery of his nuanced colors was perfunctorily explained at the time as follows: "some pastes for each color must be darker and some less dark, in order to make the lights and shades."[19] Despite his ideas on color application and methodical procedures for executing portraits, Nanteuil's pastels were not often rendered as works in their own right, but rather as disposable models for engraving, a common preparatory role for this medium that is referred to in the diaries of both Samuel Pepys (1633–1703) and John Evelyn (1620–1706).[20]

Especially in England, the new broader handling was related to the invention of mezzotint engraving. Both art forms shared painterly intentions and continuous tonal values, goals served in pastel by softer crayons in more hues – the enlarged palette being credited to miniature painter Edmund Ashfield (fl.1669–1690?).[21] The link between pastel and mezzotint must have been obvious, for in his treatise of 1707, *Het Groot Schilderboek*, Gerard de Lairesse compared the "black art" – in which the rocked copper plate was first covered with charcoal and then drawn on with white chalk (as described by Alexander Browne in *Ars Pictoria* [1669]) – with crayon drawing on dark paper, in which one adds only the heightening.[22] Transposing this concept to a new invention, Edward Luttrell, a mezzotint engraver, used his prepared copper plates for richly colored pastel compositions, relying on their tooth to hold the powder in place and adjusting his paste by decreasing the amount of plaster to improve tenacity (Fig. 1.3).[23] Such works were no longer perceived as drawings colored with pastel, but rather as paintings in dry color modeled with halftones, dominated by the vigorous dark hues currently advocated for oils by Roger de Piles.

Having succeeded in imitating the rich tones of the oil palette, by the end of the seventeenth century the pastel medium had achieved official recognition. In France royal patronage encouraged academic artists such as Joseph Vivien (1657–1734) to use pastel for the production of monumental ceremonial *portraits d'apparat*, composed of multiple sheets of joined paper. The luxury of armor or velvet in Vivien's representations, or in those by Charles-Antoine Coypel (1628–1707), is conveyed with colored highlights and shadows that produce an illusion of texture enhanced by the actual light-reflecting facets of the pastel particles. Following Louis XIV's death, smaller and less formal pastels combining blended passages and directly applied color became highly desirable among private clients. François Lemoyne (1688–1737) and Jean Jouvenet (1644–1717) executed brilliantly colored pastel drawings and sketches on toned paper of subjects derived from large-scale historical canvases and frescos. Benedetto Luti (1666–1724) created independent, idealized heads stumped and hatched in luminous oranges, pinks, and yellows, with delicate effects of light and shadow. These new types of pastel set the stage for the fashionable, graceful portraits that Rosalba Carriera (1675–1757) began producing upon her arrival in Paris in 1720.

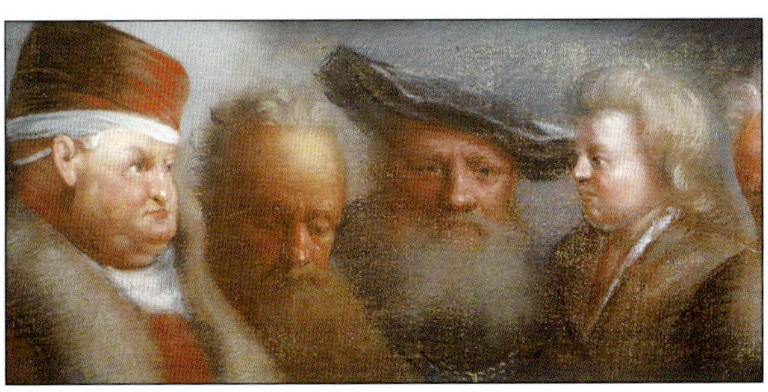

Figure 1.3 Edward Luttrell (English, fl. 1680–1724), *Study of Five Heads after Rembrandt*, 1694 (detail). Pastel on a rocked copper mezzotint plate, 10.5 x 24.9 cm. Yale Center for British Art, New Haven, Conn., B1983.26.

With the materials and techniques of pastel having reached a state of perfection, and the suitability of the medium for portraiture assured, artists and amateurs across Europe and England increasingly adopted it in the first half of the eighteenth century. Widespread interest in all aspects of science and the arts encouraged a proliferation of treatises and encyclopedias expounding on the details of this metier. However, because of the many complications entailed in balancing the ingredients, practitioners were no longer expected to make their own crayons.[24] Though much information remained shrouded in secrecy, or suspect, pastels could now be readily purchased in cities far and wide in a vast range of colors and halftones compounded with an array of binders and bases that enabled fine adjustments in their degree of softness.[25] The latter property was most highly desired, as it conferred a velvety, matte quality unlike any other medium.

So successful were pastelists at producing convincing portraits with these refined materials that they posed a threat to the oil painting establishment and its hierarchy of subject matter. In France, with commissions for historical subjects declining and the privileged classes desiring to record their own images, a tax was imposed in 1747 on portraitists, the consequences of which were largely suffered by the vast number of artists practicing in pastel.[26] The attack was based not only on the recognition that these works would not darken and change in color as did oils, and that sitting for a pastel was less time-consuming, but also on the realization that the pastelist was capable of producing painterly effects.

As in oil painting, fewer than 25 pigments were used to compose the pastel palette; however, the number of tints to be made from them were countless. John Russell (1745–1806), in his *Elements of Painting in Crayons* (1772), suggested that 20 gradations be made of each pigment and compound color, an almost innumerable amount in the end. Despite leaving a treatise, diaries, and a manuscript of recipes (National Art Library, Victoria & Albert Museum, London), Russell's description of his pastel box – like those in other sources – is pitifully brief, and reveals the lack of established practices in this medium. Using trays divided into nine partitions – with black and less-used colors on one side, light tints needing to be kept clean on the other, and between them classes of colors arranged according to gradations in the center units – it was without any overall chromatic scheme.[27] Representations of pastelists at work[28] suggest that unlike the meticulously systematized eighteenth-century oil palette, which was driven by an empirical approach to color, the pastel palette was arranged according to practical concerns. A large number of colors was a necessity, for one had to have the exact tints in advance: mixing on the support or palette was not possible.[29] Maurice Quentin de La Tour (1704–1788) bemoaned this, for unlike the painter who could combine several colors and apply the mixture on the tip of his brush, he required several strokes for the same effect, a technique he claimed held the risk of destroying the freshness of his touch.[30] Yet this great number of pastels, in fact, was not so far afield from the oil painting practices mandated by the Académie from the 1650s throughout the eighteenth century. As John Gage describes, the painter was instructed to premix and lay out all tints prior to working, "limiting the possibility of mixture in the process of painting." Lely had applied 40 tints to his palette, Jean-Baptiste Oudry (1686–1755) even more profusely, advised one to prepare five or six tints for each color and as many mid-tones as possible to "avoid tiring them with the brush." Indeed, with emphasis now on a light-hued palette, a preference that pervaded all aspects of rococo decoration and personal adornment, purity of tone was essential in both oil and pastel.[31]

As indicated by instruction manuals and direct examination, eighteenth-century pastels were often executed on blue paper, a hue that served as a middle tone and as a base for the dead coloring, but which upon completion of the composition was to be entirely concealed.[32] Laid down on canvas, mounted on a strainer, and set in a gilt frame, a pastel of this era had the appearance of an easel painting. A particularly painterly style was initiated by Rosalba Carriera, who worked her soft-focus, delicately stumped pastels in the lighter tones fashionable early in the century. Pastel was adaptable to this taste because of its physical properties: the high percentage of white base, especially in the pale colors, and the diffuse scattering of reflected light from the faceted pastel particles. This optical phenomenon renders a surface matte, suffusing it with bright luminosity, thus heightening the hues and softening the image. Carriera credited her success to using pastels made with tailor's chalk and shells, rather than gum.[33] Indeed, because chalk holds the particles more loosely, pastels made with it would crumble, allowing for tactile, draftsmanly strokes, which would in turn increase the surface area from which light could be reflected (see Fig. 1.4). This effect was exploited in all ranges of the palette, using varied techniques: Jean-Baptiste Perroneau's (1715–1783) veil of green hachures, the rapid spontaneous effects of directly applied lines over stumped color in pastels by La Tour and Jean-Baptiste Greuze (1725–1805), the calligraphic colored strokes over black chalk in portraits by Thomas Gainsborough (1727–1788), or the divided color of Jean Baptiste Simeon Chardin (1699–1779).

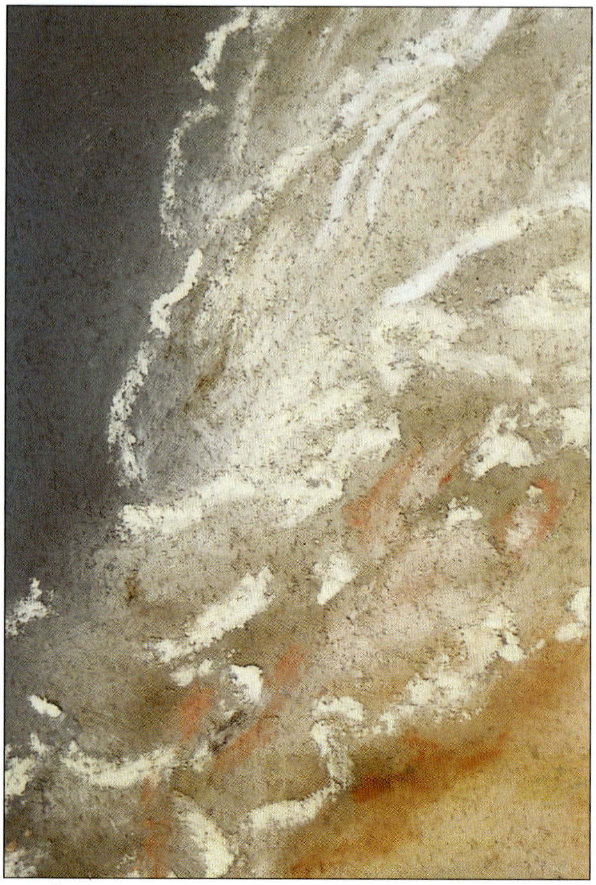

Figure 1.4 John Russell (English, 1745–1806), *Mrs. Shurlock and Her Daughter Ann*, 1801 (detail). Pastel on paper, laid down on canvas, 60.6 x 45.1 cm. The Metropolitan Museum of Art, New York, Gift of Geoffrey Shurlock, 1967. 67.132. The texture of directly applied pastel increases the surface area from which light is reflected.

The vast range of pastels also encouraged another painterly, but precisionist, mode of representation. The uniform flatness of Jean Etienne Liotard's (1702–1789) surfaces betrays that his color is firmly impressed into the support, an effect that he probably achieved by working with heavily bound, compact crayons and with stumps of many sizes for large areas and small details. Although this technique compressed the pastel layer and thus reduced its light-reflecting capacity, Liotard masterfully manipulated his palette to produce crisp, polished forms with subtle chiaroscuro, as exemplified in *Presumed Portrait of Mary Gunning, Countess of Coventry* (Fig. 1.5). According to contemporary manuals, sweetening the colors seamlessly into one another produced effects appearing to be painted with a brush.[34] Among the many practitioners of this manner were John Singleton Copley (1738–1815), with his simplified forms; Hugh Douglas Hamilton (1736–1808), who virtually obscured all traces of his hand; and Anton Raphael Mengs (1728–1779).

Instead of hatching their strokes and dragging the pastel to produce broken, textured effects, they displayed their virtuosity with lateral modeling – deftly blending graduated hues in the most subtle transitions, one into another, to create volume and illusion rather than atmosphere and suggestion.

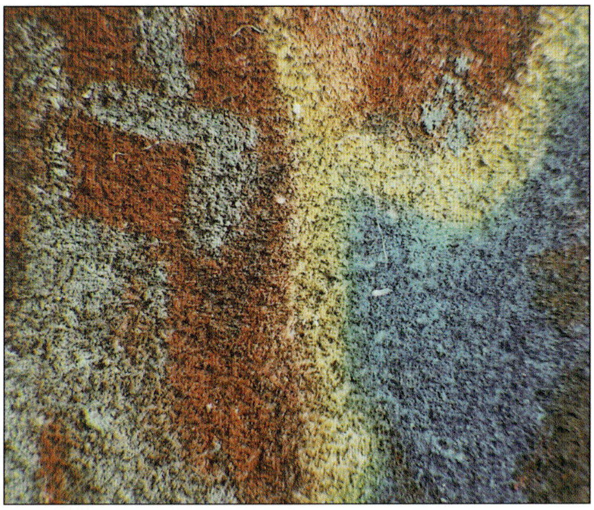

Figure 1.5 Jean Etienne Liotard (Swiss, 1702–1789), *Presumed Portrait of Mary Gunning, Countess of Coventry*, c. 1750. Pastel on parchment, 50.8 x 40.6 cm. New York, private collection. The pastel is deeply impressed into the support by the use of stumps.

Throughout the eighteenth century, the broad range of the pastel palette was augmented with gouache. Made simply from scraped pastel powder mixed with brandy – as in the case of English pastelist Daniel Gardiner (1750–1805)[35] – or similar fast-evaporating mixtures, the two mediums had similar optical and textural properties, owing to the high ratio of pigment to binder and the resulting matte surfaces. First introduced in the mid-seventeenth-century pastels of Edmund Ashfield, though criticized by his student Edward Luttrell for their stiff and hard effects,[36] for Ashfield the concept of painting details of ceremonial attire must have stemmed from using gouache to prepare his supports. In the eighteenth century, La Tour, determined to prove pastel's equality to oil, relied on the precision of the brush for the intricacies of the lace cuffs in his 74-inch-high portrait of *Gabriel Bernard de Rieux* (Figs. 1.6 and 1.7). Pietro Rotari (1707–1762) reversed the proportions of the two mediums, using the pastel for fine details in what were otherwise painted compositions on slightly roughened, solid grounds.[37] The practice of using mixed media would continue through the nineteenth century. Among the many artists incorporating gouache for coloristic and textural effects, Edgar Degas (1834–1917) fully integrated the two as he increasingly turned to opaque mediums.

Denigrated as frivolous, sentimental, and effeminate, pastel fell from favor by the late eighteenth century, as the sobriety and moral purpose of Neoclassicism gained pre-eminence. The medium re-emerged among Romantic and Barbizon artists, notably Eugène Delacroix (1798–1863) and Jean-François Millet (1814–1875), who would exert a profound influence on future generations of Salon and independent artists, and in turn on international practitioners of this medium. Millet's work of the 1860s marks the beginning of pastel's second flowering, which culminated in the pastels of Odilon Redon (1840–1916), Degas, and the vibrant pastels of the Impressionists.

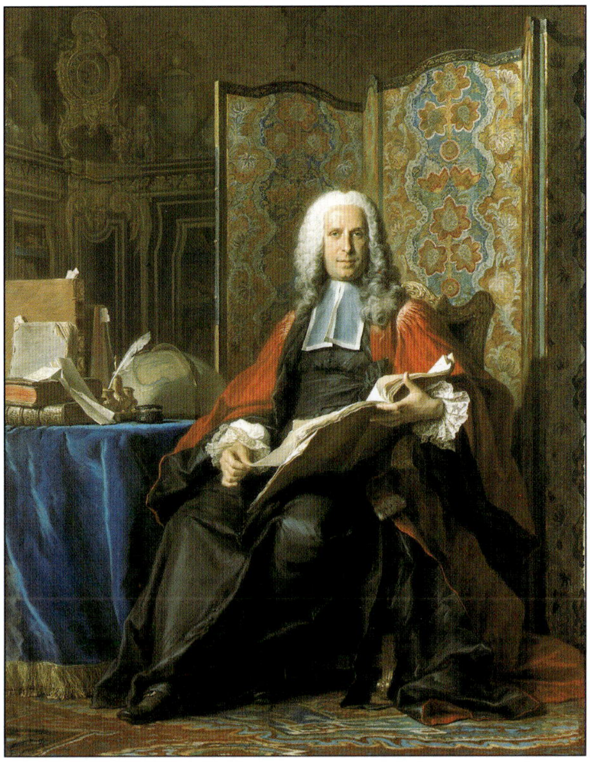

Figure 1.6 Maurice Quentin de La Tour (French, 1704–1788), *Gabriel Bernard de Rieux*, c. 1739–41. Pastel and gouache on paper mounted on canvas, 200.6 x 149.8 cm. The J. Paul Getty Museum, Los Angeles, 94.PC.39.

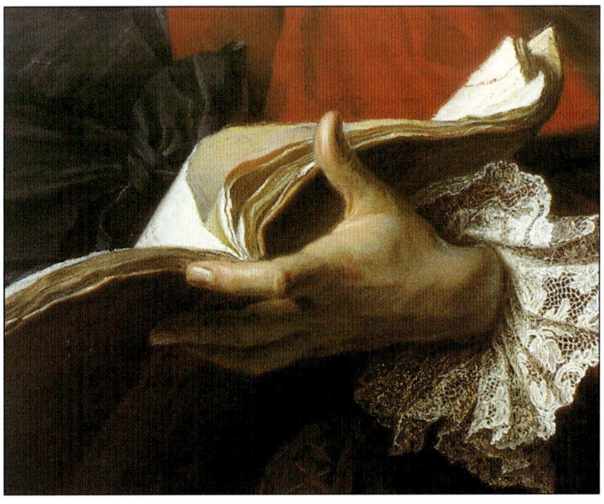

Figure 1.7 Maurice Quentin de La Tour, *Gabriel Bernard de Rieux* (detail). Gouache applied by brush is composed of pastel powder and a binder.

The remarkable technical developments in the

pigment and paper industries in the nineteenth century had an enormous impact on the pastel palette, greatly increasing the variety of materials with which the pastelist worked. More significant, however, were the theoretical transformations taking place at this time that determined how these materials were to be used: the color theories of Michel Eugène Chevreul (1786–1889) and others on chromatic relationships and physiological phenomena, and the new approaches to drawing and painting that began with the reforms in the French Académie in 1863 and were ultimately popularized by Impressionist practices. These changes promoted spontaneity and emphasis on the sketch rather than the polish and finish of academic painting. They also elevated the status of landscape, and with it bright color. Significantly, pastel came into its own in the 1860s and 1870s, the period when vanguard painters challenged the traditional dark and transparent oil palette that they associated with an outmoded system and increasingly expressed the individuality of their artistic perceptions.[38]

The physical and optical properties of pastel made it ideally suited to the new aesthetic. As artists left their dark ateliers and sought sunlight, pastel's portability and lack of a liquid medium were tangible assets; even more significantly, because pastel did not require drying, it could be used to capture transient chromatic effects and would not alter in hue. It was this subject – atmosphere – that Delacroix explored in his pastels and that later became an Impressionist theme, as is reflected in the rising popularity of plein-air landscape in this medium. The post-Romantic emphasis on revealing the working process and conveying a sense of immediacy could be given expression in the draftsmanly quality of directly applied pastel. The traditional painterly modes practiced by academic artists, or even by Degas in his early career (see, for example, *Yves Gobillard* [The Metropolitan Museum of Art, New York]), quickly gave way to new compositional aesthetics, so much so that it could be questioned whether pastel should be called a process of drawing or of painting.[39] The rapidity of the stroke and its high-keyed palette complemented the modern-life subjects depicted, for example, in the Impressionist pastels of Mary Cassatt (1844–1826) and Degas. In the 1880s, when Camille Pissarro (1831–1903) and Pierre-Auguste Renoir (1841–1919) rejected varnish in favor of a matte surface,[40] they turned to opaque media, giving preference to pastel, which lacked a resinous binder, and thus did not yellow. Emulating the effects of pastel in their paintings, and often exhibiting works in both mediums together, these artists would heighten the overall luminosity of their oils with substantial amounts of white pigment to give them the diffuse aspect of dry color.[41] Pastel thus became an expression of this new aesthetic and also a model for painting. Like oil paint, pastel was available in a vast range of brilliant hues, made with both traditional pigments and newly introduced synthetic ones.[42] These premixed colors with their inherently graphic mode of application were readily combined in new complex chromatic interactions in a bright tonal range. Pastels by the Impressionists are characterized by broken and richly textured strokes played against contrasting or complementary underlayers of color, effects allowing for pictorial construction based on color and surface rather than line modeling and chiaroscuro. The broad range of the palette and the diverse ways in which the medium could be handled attracted artists of all aesthetic persuasions. The American Thomas W. Dewing (1851–1938), for example, used a restrained palette in muted gradations that set against dark brown papers for his studies of ethereal young women (see Fig. 1.8); Pierre Puvis de Chavannes (1824–1898) and Ferdinand Khnopff (1858–1921) employed pale colors of equal tonal value to flatten pictorial space, an effect closely linked with the revival of wall decoration.[43] So popular were these soft, diffuse hues that the term "pastel color" was coined to describe them in the last decades of the nineteenth century.[44]

With emphasis on the working process, and thus on drawing, the support assumed a pre-eminent role. Whereas its hue had been concealed in the eighteenth century, the Impressionists and their followers intentionally left it partially exposed to serve as a coloristic element in the composition. They sought new

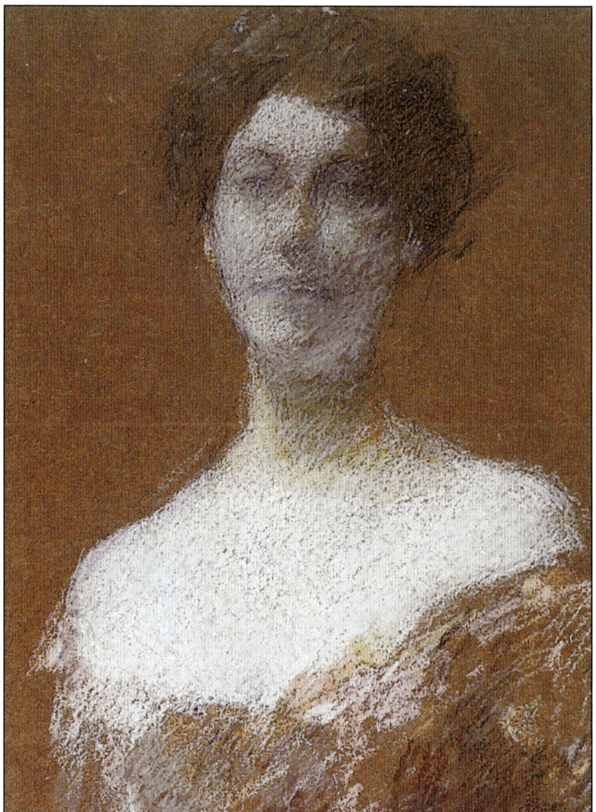

Figure 1.8 Thomas W. Dewing (American, 1851–1938), *Evening Dress,* before 1926 (detail). Pastel on brown wove paper, 36.8 x 27.9 cm. The Metropolitan Museum of Art, New York, 66.157. With a limited number of muted colors, discreet, closely spaced strokes of pastel are combined to produce a blended, volumetric effect.

types of supports, of diverse textures and color, ranging from the dark gouache-coated panels used by William Merritt Chase (1849–1916) to the white laid papers favored by Renoir. Comparable to the white priming left visible in oil painting, these papers were valued for their light enhancement and modification of color. Vibrantly toned supports in every conceivable color, largely produced with aniline or coal tar dyes introduced in the late 1850s, soared in popularity for pastel as the century progressed. These hues allowed for a striking interplay of contrasts and juxtapositions with the opaque hachures and stumping of the medium. However, because of the instability of these dyes and exposure to light, such papers have invariably faded, resulting in an alteration of the intended tonal relationship between the design layer and the support.

Texture, which was obliterated from the eighteenth-century pastel support so as not to interfere with the illusion of the image, was now valued, for the rough grain of the paper broke the stroke, giving it more vitality. Varied supports suited this goal, including the blue-gray commercially primed canvases made for painting often used by Edouard Manet (1832–1883) for his pastels,[45] the coarse wrapping paper retrieved from warehouses in Venice favored by James McNeill Whistler (1834–1903), and the millboard visible in the mixed media works of Henri de Toulouse Lautrec (1864–1901). Abrasive papers coated with dyed pumice were manufactured especially for pastel.[46] These provided a tooth and multiplied the facets from which light could be reflected, yet often failed to hold this fragile powder in place (see Fig. 1.9). Throughout the second half of the nineteenth century, many of the age-old debates on fixing pastels were rekindled. Hoping to retain the inherent brilliance and opacity of pastel, many artists wisely protected their works with glass rather than applying resins and gums to these surfaces. These fixing processes inevitably resulted in color saturation and imparted a gloss, irreparably impairing the distinctive characteristics of this medium.

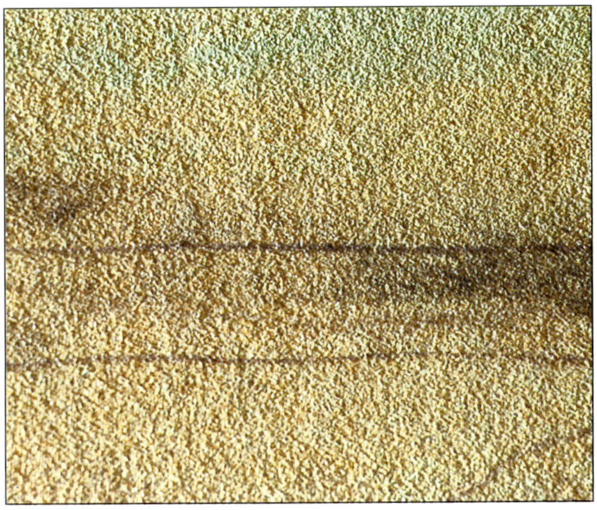

Figure 1.9 John Twachtman (American, 1853–1902), *Landscape*, 1855 (detail). Pastel on dyed pumice-coated paper, 37.5 x 45.7 cm. The Metropolitan Museum of Art, New York, 25.107.2.

The popularity of pastel throughout the second half of the nineteenth century encouraged the publication of numerous treatises on its use. Directed toward the amateur, these manuals typically referred to the chromatic theories of Chevreul, which were described as having inspired the great modern painters. The reader was encouraged not to mix colors but to rely on the purity of the individual sticks of pastel, using them as complements or in contrast to one another and to the support. These ideas reflected the practices of avant-garde pastelists and oil painters, who from about mid-century had rejected modulated values and the multiple halftones that chiaroscuro and spatial depth demanded. Manuals such as J. de la Rochenoire's *Le Pastel Appris Seul Avec Sept Couleurs* (Paris, 1853) recommended a palette limited to seven crayons – a number perhaps based on its auspicious history predating Newton's spectral theories, but prescient in that decades later it would become central to the myth of the Impressionist palette. For the more accomplished pastelist, S. Jozan, in his *Du Pastel, Traité de sa Composition, de sa Fabrication, de son Emploi* (Paris, 1852), advised 150 colors as being sufficient. Yet increasingly, fabrication was becoming exclusively a commercial endeavor. Colormen such as LeFrance and Henri Roche, recipients of commendations at the various world's fairs, would each advertise pastels in more than 1,500 colors in soft and semi-hard textures, for purchase individually or in sets assembled for landscape or portraiture. Toward the end of the century, exhortations directed at the amateur to restrict the number of pastels were common. Reverting to a much earlier model, handbooks including Karl Robert's *Le Pastel, Traité pratique et complet* (Paris, 1890) instructed the reader to outfit the pastel box with a minimum of 400 colors, specifying in greater detail than ever before the range of tints and shades and their organization.[47] How much of this advice was translated into studio practice is difficult to assess. The availability of a vast number of modern and traditional pigments, and the many theories regarding their application, nonetheless, would have a profound impact on this art form, expressed in bolder and more innovative uses of color and the increasing role of the support as a chromatic element in the composition.

Acknowledgment

I am grateful for the dedicated assistance of Luise Trabucchi.

Endnotes

1. The word *pastille* (*pastil*) is cited in many early sources (e.g. Merrifield, Bolognese MS. XVI, p. 346) in reference to the kneaded mass of lapis lazuli and other components (wax, soap, resin) from which the blue pigment ultramarine was extracted. Wm. Sanderson, *Graphice* (London, 1658), pp. 77–78, entitles the entry on making ultramarine "Of Pastills, or Croyons," and the following entry on the colored drawing sticks "To Work in Croyons or Pastills," using these terms interchangeably. The usage of the term *pastil* suggests that the name for the wet mass was also a description of the dry color sticks. The word pastel can also be traced to the name for the blue dye derived from a paste made of the twigs of woad (*isatis tincturia*), suggesting that present English usage derives from processing similarities of these two blue-colored substances.

2. Cennino Cennini, *Il libro dell'arte*, transl. by D. V. Thompson (Dover, 1960), pp. 91–93.

3. Leonardo da Vinci, *Codex Atlanticus* (c. 1495; Ambosiana, Milan, 1483–1518, fol. 247 r-a), in C. Pedretti, *The Literary Works of Leonardo da Vinci* (Berkeley, 1977), Vol. 1, p. 359. While it is possible that Leonardo's term "dry color" was not a reference to pastel, information is equally insufficient to accept it as a reference to fresco, as powdered pigment is the basis of all painting mediums. G. P. Lomazzo, *Trattato dell'arte de la pittura* (1585), transl. by R. Haydocke as *A tracte containing the artes of curious painting* (Oxford, 1598), refers to Leonardo using *pastello* for studies of the heads of Christ and the apostles for *The Last Supper* (1495–97).

4. Christopher McGlinchy (Conservation Scientist, Museum of Modern Art, New York) analyzed the yellow samples from three sixteenth-century drawings (Metropolitan Museum: Milanese Helmet, Domenico Feti, and Clouet school, unaccess.). According to V. Daniels ("The effects of water treatments on paper with applied pastel and powder pigments," *The Paper Conservator* 22 (1998), pp. 29–37), various external factors can cause agglomeration, such as exposure to high levels of humidity. Although it is not possible to determine the relative humidity conditions to which these drawings were subject during their long history, the presence of two bases points to an intentional admixture with the pigment and not a natural occurrence.

5. T. Turquet de Mayerne, *Pictoria Sculptoria and quae subalterarum artium* (1620–32; Lyon, 1967), p. 155, notes that certain earths need no additions: bol, calcined ochre (*ochre double ou triple*), English brown, *terre d'ombre*. Alexander Browne (*Ars Pictoria*, 2nd edn [1675], p. 30) and followers such as William Salmon (*Polygraphice*, 5th edn [London, 1685], p. 6) cite yellow ochre burnt, ground, rolled into sticks, and dipped in linseed oil for use.

6. J. Meder, *Die Handzeichnung, Ihre Technick und Entwicklung* (Vienna, 1919–23), transl. by W. Ames as *The Mastery of Drawing* (New York, 1978), p. 100. Gregorious cites fish glue, gum arabic, and fig juice as binders. For a discussion of the various commentaries on pastel in the sixteenth century see T. McGrath, "Federico Barocci and the history of *Pastelli* in Central Italy," *Apollo* (Nov. 1998), pp. 3–9.

7. G. P. Lomazzo, *Idea del Tiempo della pittura*, ed. and transl. by R. Klein (Florence, 1974), p. 187, in T. McGrath, "*Disegno, Colore*, and the *Disegno Colorito*: The Use and Significance of Color in Italian Renaissance Drawings" (Ph.D. Diss., Harvard University, 1994), p. 58.

8. Mayerne 1967 (as in note 5), p. 117, no. 97.

9. Ibid., p. 117, nos. 154–56.

10. Christiaan Huygens describes Lely's paper in a letter of 13 July 1663; see *Oeuvres complètes de Christiaan Huyghens* (The Hague, 1891), Vol. 4, p. 371.

11. Edward Norgate, *Miniatura or The Art of Limning* (1648/50), ed. by M. Hardie (Oxford, 1919), p. 73. The technique was described by subsequent writers in the seventeenth century, including A. Browne 1675 (as in note 5) who discusses Vaillant, p. 25. In the eighteenth century, the use of the stump was advised for applying costly colors.

12. Sharples's technique was described by J. W. Palmer in *Lippincott's Magazine* (Dec. 1871), cited in K. McC. Knox, *The Sharples, Their Portraits of George Washington and His Contemporaries* (New Haven, Conn., 1930), p. 9.

13. Huygens 1891 (as in note 10), letter of 3 August 1663, Vol. 4, p. 389.

14. Edward Luttrell, *An Epitome of Painting* (1683; MS., Yale Center for British Art, New Haven, Conn.), p. 66.

15. E. Bouvy, *Robert Nanteuil* (Paris, 1924), p. 45.

16. Unlike France, where education was determined by a centralized academic system, a large market existed in England and Holland in didactic artists' manuals for both the gentleman amateur and the pupil who relied on private ateliers, apprenticeship, and written instruction. Roger de Piles and Jean Baptiste Corneille, *Premiers éléments de peinture pratique* (Paris, 1684).

17. Norgate 1919 (as in note 11), p. 72.

18. T. H. Thomas, "The drawings and pastels of Nanteuil," *The Print Quarterly* 4 (1914), p. 356.

19. H. Gautier de Nismes, *L'Art de laver, ou nouvelle maniere de peindre sur le papier* (Paris, 1687), in G. Monnier, *Pastels from the Sixteenth to the Twentieth Century* (Geneva, 1984), p. 109.

20. H. C. Levis, *Extracts from the Diaries and Correspondence of John Evelyn and Samuel Pepys related to Engraving* (London, 1915): Pepys, on 7 November 1666, refers to a pastel that he wished to buy from Faithorne, but was forced to wait until the artist finished using it as a model for his copper plate, p. 65; and Evelyn, on 20 June 1649, reports taking his pastel landscape drawings to France to be engraved, p. 20.

21. Bainbrigg Buckeridge, "An Essay Towards a British School," appended to an English translation of R. de Piles, *The Art of Painting and Lives of the Painters* (London, 1706), biographical entry on Ashfield.

22. A. Browne, quoted in John Chaloner Smith, *British Mezzotinto Portraits* (London, 1878–83), Vol. 1, p. 107; *Het Groot Schilderboek*, 2 vols (Amsterdam, 1707), transl. by G. de Lairesse as *A Treatise on the Art of Painting in All Its Branches* (1738; London, 1817), p. 270; Luttrell 1683 (as in note 14), p. 12, held that in drawing one should apply the heightening first, much as one would have commenced a mezzotint: "you may in this manner draw on any colored paper, as well as white provided you begin with a white cryoon [*sic*] in the lights."

23. Luttrell 1683 (as in note 14), p. 66.

24. Among the numerous treatises and encyclopedias appearing during the eighteenth century were the more widely circulated Watelet and Levesque, *Encyclopédie Methodique* (Padua, 1742); A.-J. Pernety, *Dictionnaire portatif de peinture, sculpture et gravure* (Paris, 1757); R. Dossie, *Handmaid to the Arts* (London, 1764); C. M. P. R. de Chaperon, *Traité de la peinture au pastel* (Paris, 1788); J. Russell, *Elements of Painting in Crayons* (London, 1772); C. de Massoul, *A Treatise on the Art of Painting* (London, 1797).

25. Treatises such as those listed in note 24 cite various cities where pastels could be purchased and the types of shops where they could be found: the 1788 *Traité* notes pigments "chez tous les Epiciers-Droguistes, Marchards des couleurs, or d'Estampes," and provided information on the ingredients and fabrication process so that the artist could make his own pastels or inform himself as to the ingredients.

26. J. Loquin, "La lutte des critiques d'art contre les portraitistes au XVIII siecle." *Archives de l'art français* 7 (1913), pp. 309–20; idem, *La Peinture d'histoire en France de 1747–1785* (Paris, 1912), p. 6. More than 2,500 pastelists were said to have been active in Paris around 1780; A. Bury, *Maurice-Quentin de La Tour* (London, 1971), p. 8.

27. Russell 1772 (as in note 24), p. 40; *Traité* 1788 (as in note 24), p. 26; Watelet and Levesque 1742 (as in note 24), p. 488.

28. Representations of pastelists at work include Arthur Pond's *Rhoda Astley* (National Portrait Gallery, London); Jean Etienne Liotard's *Carolina-Luisa of Hesse Darmstadt* (private collection, Geneva); Alexander Roslin's *Artist and His Wife* (Tamm Collection, Fano, Uppland, Sweden); Pietro Longhi's *The Painter* (Stanford University Museum of Art, Stanford, Calif.); and Rosalba Carriera's *Self-Portrait* (Uffizi, Florence).

29. Watelet and Levesque 1742 (as in note 24), p. 488.

30. La Tour, letter to M. le Marquis of 1 August 1763, in J. J. Guiffrey, "Correspondance inédite de Maurice Quentin de La Tour," *Gazette des beaux-arts* 31 (1885), p. 211.

31. J. Gage, *Color and Culture: Practice and Meaning from Antiquity to Abstraction* (Boston, 1993), pp. 179–80. Premixing tints was recommended in treatises from many centuries, including de Piles 1684 (as in note 16), Lairesse 1817 (as in note 22), W. Hogarth, *The Analysis of Beauty* (London, 1753). Newton's theories were known to artists, and his color circle appeared as an anonymous supplement in Claude Boutet's *Traité de la peinture en mignature* (Paris, 1673), reprinted 30 times during the eighteenth century, but did not appear to have an impact on pastelists; Gage, p. 171. The light-hued palette was theoretically justified by Newton's discoveries in optics according to Algarotti; Francis Haskell, *Patrons and Painters* (New Haven, Conn., 1980), p. 354. On the connection with cosmetics, see Anne Hollander, *Seeing Through Clothes* (Berkeley, 1975); and Joanna M. Kosek, "The heyday of pastels in the eighteenth century," *The Paper Conservator* 22 (1998), pp. 1–9.

32. Many types of supports were used for pastel. For an overview see *Traité* 1788 (as in note 24), pp. 330–39; for marble powder on canvas, see N. Trivas, "Two formulas by Liotard," *Technical Studies in Conservation* 10 (July 1941); for glass powder, oil, and wax on canvas, see Pileur d'Apigny, *Traité des couleurs materielles* (Paris, 1779), pp. 59–60; and for silk papers, see J. Krill, "Silk paper for crayon drawing in the eighteenth century," *IPH Yearbook* (Sept. 1994). The support was intentionally left visible only in preparatory work as the *preparations*, or face sketches, done by La Tour and Russell. For his finished work, La Tour found the blue paper objectionable and would cover it with an egg-yolk mixture rather than a great amount of pastel in order to obscure the hue. Edmond and Jules de Goncourt, *L'Art du XVIIIième siècle* (Paris, 1880; reprint 1943), p. 205. This preparation would probably have also improved the tooth to hold the powdered color.

33. A. Sensier, *Journal de Rosalba Carriera* (Paris, 1865), p. 425.

34. Mr. Lens, *A New and Compleat Drawing-Book* (London, 1751), p. 7.

35. W. T. Whitley, *Artists and Their Friends in England 1701–1799* (London, 1928), p. 278.

36. Luttrell 1683 (as in note 14), p. 32.

37. The surface preparation of this pastel by Rotari appears to be that described by d'Apigny 1779 (as in note 32), pp. 59–60, in which canvas would be imbibed with oil, then coated with glass powder which formed a hard and compact surface. The process is a variant on that invented by M. Pellechet in 1764; Bury 1971 (as in note 26), p. 188.

38. For changes in artists' practices and reforms in the French teaching system, see A. Boime, *The Academy and French Painting in the Nineteenth Century* (New Haven, Conn., 1986); A. Callen, *Techniques of the Impressionists* (London, 1982); for nineteenth-century color theory, see J. C. Taylor, *Nineteenth-Century Theories of Art* (Berkeley and Los Angeles, 1987), pp. 446–79.

39. *Century Magazine* 29 (1884), p. 205.

40. M. Swicklik, "French painting and the use of varnish 1750–1900," *Conservation Research* 41, 2 (1993), p. 167; V. J. Wasiutynski and H. T. Newton, "Absorbent grounds and the matte aesthetic in Post-Impressionist painting," *Painting Techniques: History, Materials and Studio Practice* (Dublin Congress, International Institute for Conservation, London, 1998).

41. For a discussion of the *peinture claire* technique, see D. Bomford, J. Kirby, J. Leighton, and A. Roy, *Art in the Making: Impressionism* (London, 1987).

42. While there appears not to have been research specifically pertaining to pastel colors, writers such as S. Jozan, *Du Pastel, Traité de sa composition, de sa Fabrication, de son Emploi dans la Peinture* (Paris, 1852), p. 25, noted that pastel pigments were the same as used in oil, and based his information on this. Substantial research was carried out during the nineteenth century on these pigments, including traditional colors, new introductions, and issues of stability. Among them are J. F. L. Mérimée, *De la Peinture a l'Huile* (Paris, 1830); G. Field, *Chromatography* (London, 1835); and J. G. Vibert, *The Science of Painting* (Paris, 1891).

43. The aesthetic connection between wall painting and the pastel palette was most clearly envisioned in Gustav Klimt's *Beethovenfries* (1902), executed for the Secession group in Vienna. Made with color sticks bound with casein, gum arabic, and tragacanth, the technique was referred to by the inventor and color theorist Wilhelm Ostwald as *Monumental Pastel* in his booklet by that name published in 1905. M. Koller, "Monumentales pastel: a forgotten invention in wall-painting techniques about 1900," *ICOM Committee for Conservation* (Zagreb, 1978), pp. 1–8.

44. *Daily News* (21 Oct. 1899), p. 7. "The soft wraith-like tints ... are now in fashion again. The modern name for them is 'pastel' ... for these soft, half-faded tones bear the same relation to real colours as pastel do to oil paintings."

45. For commercially primed canvases, see Bomford et al. 1987 (as in note 41), p. 44.

46. J. de Fontenelle and P. Poisson, *Nouveau Manuel Complet du Marchand Papetier* (Paris, 1854), pp. 37–39.

47. Jozan 1852 (as in note 42), p. 21; K. Robert (Georges Meusnier), *Le Pastel, Traité pratique et complet* (Paris, 1890), pp. 26–30.

Distinguishing Between Chalk and Pastel in Early Drawings

Thea Burns

It is difficult to obtain material evidence for the use of pastel by artists in early modern Europe. Without sampling and scientific analysis, viewing with the unaided eye or low-power magnification is insufficient for distinguishing applications of chalk and pastel (Burns 1994). Furthermore, there is a lack of "fit" between the visual and the textual evidence typically cited in twentieth-century studies of pastel. Therefore, historical documentation of the presence and early use of pastel is suspect. Natural red, black, and white chalks were used in drawings beginning in the later fifteenth century in Europe. A few works executed in a slightly broader though still restricted range of friable and other colored media by Leonardo da Vinci's pupils, Boltraffio and Luini, were described by Joseph Meder in his *Die Handzeichnung* as pastels (Meder 1923). By "pastel" Meder meant a work made with cohesive sticks that were fabricated from powdered pigments rather than cut from naturally occurring chalk-like minerals. Based on his interpretation of the visual evidence available to the unaided eye, Meder suggested the introduction of pastel into the artistic repertoire of Western Europe by the first decade of the sixteenth century.

Why have Meder and others identified these early occurrences of colored friable media in drawings as pastel rather than as natural or fabricated chalk? Natural chalks, as Watrous tells us, are mineral substances that are mined from the earth and used for drawing as extracted with only minimal shaping (Watrous 1957). Fabricated chalks and pastels are colored pastes of pigments mixed with a water-soluble binder and, sometimes, with inert clays. Viewing with the unaided eye or low-power stereo-binocular magnification, the identification of chalk and pastel is likely to be made subjectively, on the basis of the number of colors present in a drawing (chalk drawings being thought to feature a more restricted range of colors than pastels) (Burns 1994) and the apparent hardness or softness of the stick used (chalks being deemed harder than pastels) (Leuchs 1829).

In an article published in *The Paper Conservator* in 1994 and in a presentation at the pastel conference at the Tate Gallery in June 1997, I argued that natural chalks occurred in the past and can still be found today in a range of yellow through brown hues, including various reds (Burns 1994, 1997). Particularly notable is the great range in working properties of these natural chalks. To be sure there exist early texts that refer to friable colored sticks for drawing as *pastil*, *pastel*, *pastelli*, and Meder and others have drawn on these literary references to support their arguments for the early appearance of pastel in drawings (Meder 1923; McGrath 1994). Primary written confirmations of the use of pastels and instructions for their fabrication by artists are scarce before the eighteenth century, although a few references exist. As Thomas McGrath explained in his 1994 Harvard University dissertation, "*Disegno, Colore*, and the *Disegno Colorito*: The Use and Significance of Color in Italian Renaissance Drawings," there are both theoretical and practical reasons for this dearth of clear

written references, but those reasons will not be repeated here (McGrath 1994). Suffice it to say that it is a relatively simple matter for an artist to form powdered pigment into a cohesive yet friable stick.

Several late sixteenth-century Italian texts mention *pastelli*, but they require interpretation because we need to recover the contemporary meaning of this uncodified term. The texts cannot simply be taken as providing instructions for the artist's use. Do they give us specific information about artistic practice? There is space here for two brief examples. In his *Disegno*, six dialogues on the merits of painting and sculpture published in 1549 and recently described as a "valuable document on workshop practice" (Quiviger 1996: 142), the sometime painter and writer Anton Francesco Doni (1513–1574) noted the use of "powder upon sheets," which he listed with watercolor, gouache, tempera, and oil as "ways of painting" (Doni 1549: 15a). This mention of powder (*polvere*) need not necessarily allude to fabricated sticks, as McGrath has argued (McGrath 1994, 1998). Rather, Doni may have been referring to the use of naturally occurring chalks, to paper sheets toned with dry pigments (McGrath 1994), or to the use for drawing of dry loose pigment applied by finger, brush, or other tool (de Piles 1684).

A second example may be more familiar to paper conservators. In 1584 the painter Giovan Paolo Lomazzo (1538–1600), Leonardo's successor as a theorist in sixteenth-century Milan, noted Leonardo's use of pastel almost incidentally: "I will not be silent ... about another sure way to color which is called *pastello* which are made with points composed ... of colored powder ... It was done on paper" (Lomazzo 1584). This passage, however, tells us only that Leonardo used fabricated color sticks – "points composed of colored powder," a description that applies to both natural and fabricated color chalks. The main reason to identify this material as pastel is because of Lomazzo's use of the Italian *pastello*, which appears close to our term "pastel."

The term "pastel" evokes for us a range of referents – the tool, the work of art, touches of friable color, full color, and so on. Twentieth-century writers have reiterated the etymology of the word "pastel" from the Italian *pastello* and the Latin *pasta*, both meaning paste. The French *pastel* is also regularly traced back to the Italian word. It is important to clarify to what exactly the sixteenth-century writer of an art treatise was referring when he used the word *pastello*. To my knowledge the pastel literature has not considered that the word *pastel* was used quite widely in various contexts in the early modern period. It was, for example, employed in sixteenth-century French apothecaries' texts to denote a *petit pâte* and a *morceau de pâte*, translatable as a cake or pastille, the form in which pharmacists and others typically dispensed powders – medicinal, confectionery, etc. (La Curne de Sainte Pelaye 1880). Substances to be dispensed were typically crushed, dried in the sun, and formed into *pasteaulx* (Huguet 1925–70: 70).

Dictionaries show that the term *pastel* was also used in early modern France for the herb and dyestuff *guède* (*isatis tinctoria*), which was likewise prepared as a *pastille*. They inform us that this dyestuff was called *pastel*, "a word derived from *pastellus* ... because after having been crushed this plant is made into tablets or balls" (Trévoux 1721). We can trace a similar derivation for the term in English. Particularly interesting is the usage recorded in 1704: "Pastills are Odiferous Tablets ... made up of ... Odorous Bodies with Mucilage of Gum Tragacanth" (Harris 1736). Gum tragacanth was long the traditional binder for pastel. These perfumed tablets often took the form of rolls or rods; they were burnt to deodorize, fumigate, or disinfect. Earlier Edward Phillips defined the English verb "pastillicate" as "a making of anything into the form of a pill or round ball" (Phillips 1658). The precise geometrical shape of the solid formed from powder is nowhere rigidly prescribed – we find that pastilles were shaped into cakes, round lozenges, balls, rolls, rods, and loaves (the French word is *pain*). Therefore I would argue that the term "pastel" was chosen from common early modern usage for fabricated chalks because it characterized the method of their fabrication and their format. It did not, originally, designate a unique artists' material distinguishable from fabricated chalk.

Twentieth-century historians of pastel have found a visual record that has suggested to them that chalks were available initially in a limited color range. For later periods the visual record increased, partly because many more works in colored friable materials survive. Historians have projected backward from the rich later visual record, as well as from an increasingly fuller written record about fabrication and use, to fill in the patchy earlier record and to construct a narrative history (Monnier 1984, 1992). In this way they have come to identify the use of colored friable material in some early drawings as early "pastel." Yet in many cases, pastel rather than fabricated chalk does not necessarily have to be posited. I shall take a skeptical position to make the point that the eventual arrival of pastel as a discrete category of artists' materials is an aesthetic and practical event. Therefore in my argument I have and shall continue to refer to the material under discussion as "chalk" – comprising both natural and fabricated chalk. In my earlier paper (Burns 1994), I identified the existence of ochres ranging in color from yellow through red to brown hues, occurring naturally near the surface of the earth in the appropriate consistency for drawing, with or without minimal shaping.

At the *Carrière des Perchers* in the Nièvre district of France, very pure natural yellow (goethite) and red (hematite) ochres have been exploited in a small way by the same family for 600 years as a sideline to clay extraction for making ceramics (Quieffin 1999). Documents survive which record that from the fifteenth century, ochre from the broader surrounding area has been shipped down the Loire on barges and traded throughout Western Europe (Guillaume 1991). The

deposit at the *Carrière des Perchers*, which consists of a narrow band of hematite between thicker deposits of clay and yellow ochre, lies near the surface and was accessible at the bottom of shallow pits after extraction of the clay. Pieces of red and yellow ochre picked off abandoned heaps today are cohesive enough to handle and to draw with, without shaping, in their natural form. To shape them a fine saw must be used.

McGrath has referred in his dissertation to the use in several early drawings of blue and green pastels that, because of their color, must have been fabricated (McGrath 1994). It is certainly possible that for non-naturally occurring or unobtainable colors, artists resorted to fabricating chalks. McGrath drew our attention to a drawing of the *Head of Christ* by Federico Barocci (Fogg Art Museum, 1986.535) in which the artist undoubtedly used a stick that he had fabricated himself or had fabricated by an assistant. However, despite such evidence, we should not ignore the fact that naturally occurring blue, blue-green, and green earths, though rare, are known and can be found today. In addition, references to the existence of unusual colored earths are found in the primary literature, suggesting that we should keep an open mind about former occurrence and availability. Chaperon, in his 1788 treatise, *Traité de la peinture au pastel*, noted that "there is found in the diocese of Uzès in Languedoc an earth of very fine lemon color. Why has no one put it on the market? Is it so difficult to procure?" (Chaperon 1788: 97). There is no way for us to know whether this earth occurred in a suitable consistency for drawing, but we cannot dismiss the possibility.

Earlier in this century, Meder argued that accurate identification of materials is important for clarity's sake (Meder 1923). Recently, McGrath elaborated, noting that accurate identification of materials may contribute to an understanding of artistic practices, aid in attribution, and enhance our understanding of the relationship of artists' material choices and aesthetic aims (McGrath 1994). I hope to show that by defining the relevant terms on the basis of historical evidence, we can develop a consistent policy for distinguishing between fabricated chalks and pastels – a reasonable policy that will meet present needs and avoid paralysis resulting from a reliance on often-unavailable instrumental analysis.

Primary documentation, both textual and visual, suggests that in the seventeenth century a change took place in the way in which fabricated chalks were used by artists and in artists' expectations of these materials. Portraits by the English painter Edmund Ashfield (active 1660s–1690), for example, include fully colored images, conceived and received as paintings.[1] William Sanderson wrote in his 1658 treatise, *Graphice or the use of the Pen and Pensill*: "Then with your several *Pastills* rub in the *Colours*, & with your fingers-end, sweeten and mixe them together, driving them, one within another, after the manner of *Oyl Painters"* (Sanderson 1658: 78–79). Ashfield's portraits are executed largely in friable dry color on paper supports.

We learn from the correspondence of the Dutch scientist Constantyn Huygens (1596–1687), that the chalk sticks used by Sir Peter Lely (1618–1680) were fabricated by an assistant away from the painter's London studio in sets of colors appropriate for painting faces (Huygens 1890). The recipes for Lely's sticks are preserved in Huygen's letters. Sets of chalk sticks were sometimes fabricated outside the studio in Paris during this period as well (Huygens 1890). However, the French engraver Robert Nanteuil (1623–1678) made his own chalks in his studio in conditions of secrecy that were broken by his pupil Domenico Tempesti only after Nanteuil's death (Piot 1863).

What we might call professional suppliers for chalks first come to our attention in the 1660s; they must have set out to meet the needs of professional portraitists and a growing amateur market. I propose that widespread enthusiasm for colored-chalk portraiture, which could compete with oil painting visually (Sanderson 1658) and was described as easier to master in a practical sense (Boutet 1708; Chaperon 1788), provided a context in which the commercial production of purpose-made sets of fabricated chalks first flourished. These are the origins of the sets we today call pastels.

The French scientist Philippe de la Hire stressed in his 1699 treatise, *Traité de la pratique de la peinture*, that these early sets of pastels were intended to function homogeneously in use (de la Hire 1730). What does this mean? I would like briefly to look at paintings in oil. To achieve unity of effect across the surface of an oil painting, it was varnished. This evened out and unified the surface from irregularities of reflectance due to different pigment particle size and variations in fatness and leanness of paint. Instead of varnishing or fixing early pastels (which posed other problems), uniform handling characteristics were sought in order to achieve overall visual unity of the pastel surface. That expertise in handling the materials needed to make sets of pastels developed by trial and error is clear from the primary source material. Treatise writers of the time agreed that consistency was an important factor, but they were vague about how the reader should judge when the appropriate consistency had been achieved. Roger de Piles, in his 1684 treatise, *Les premiers elemens de la peinture pratique*, wrote that pastels should be neither too hard nor too soft. De la Hire, a few years later, recognized that not all pigments could be treated alike in fabricating pastels if the desired result, a homogenous set of fabricated sticks "of about the consistency of soft Champagne chalk," was to be obtained (de la Hire 1730: 662). He noted that some pigments were not solid enough to hold together on their own in stick form: these benefited from the addition of gum water to the ingredients; others, which needed to be softer, could have some soapy water (*eau de savon*) added.

Almost a century later, in 1788, Chaperon still proceeded by comparison, as had de la Hire. The ideal consistency for pastel was that of a stick fabricated from a burning branch quenched in water (Chaperon 1788).

An increasingly subtle understanding of what was required to fabricate a homogenous set of pastels is reflected in the very detailed instructions provided by Chaperon for each pigment. In one example he tells us:

> to make pastels from Chalk or Troyes white, it was first purified. Reduce one to two pounds to powder and throw this into two to three pints of water. Stir and allow grainy inclusions two to three minutes to settle out. Next, pour the rest into another basin and allow sand inclusions to settle out. Discard the water that has become clear and pour the rest of the water and chalk into several cornets of parchment or paper secured with sealing wax. Fold the top of the cornet over to prevent dust entering. Any residual grainy inclusions will gradually travel to the bottom of the cornet. Pierce the cornets above the sediment to release the clear water. Tie a string around the cornets to separate the grainy sediment from the rest and turn out the latter onto the porphyry where it is finely ground. This material is gathered into small piles with a knife and placed on unsized paper (*papier joseph* or *lombard*), where it is rolled into crayons and allowed to dry.

Chalk by itself, Chaperon noted, is very friable: "If one desires that the crayons be made a little firmer, dissolve a piece of gum arabic, quite white, in several drops of pure water and spread this on the chalk before grinding it."

De Piles, de la Hire, and Chaperon inform us that not all colors in a set needed to be fabricated – natural black, yellow, white, brown, and red chalks could be useful additions to the fabricated sets, though by 1788 Chaperon advised that sanguine crushed well in water provides crayons that are infinitely better than naturally available chalks. By the early nineteenth century, treatise writers recommended that, although natural chalks could be suitable for pastels, it was preferable to prepare all colors artificially. Only then could one be certain of obtaining the appropriate consistency throughout the set (Leuchs 1829).

To obtain the range of colors needed for painting in pastel, the artist had to be supplied with a great number of individual sticks – not just each hue (including mixtures) but also one of each of the whole range of values and intensities. Claude Boutet, in his 1708 treatise, *Traité de la peinture au pastel*, described what made up "a complete assortment ... one needs all the colors both pure and mixed, with the shades of each one, from the lightest to the darkest" (Boutet 1708: 152). The descriptions of Boutet and Chaperon suggest that fabricating whole sets of colored pastels, as opposed to the random stick or two, was a skilled and time-consuming activity. It required an understanding of the working properties of all the individual pigments and a knowledge of how to use additives to adjust them to produce a harmonious working set. Boutet makes this clear in his description of "what one needs to know about the Composition of Pastels":

> You must be particularly careful that the pastels mark easily without needing much pressure, this is because here are colors which are of too hard a consistency being put into pastels ... they must be allied with some other material which is the closest in colour and which is soft; colours ... which are too soft must be soaked in more or less gum water depending on whether the substance is more or less soft; and so as not to be mistaken it will be good to make a little trial before finishing forming the pastels (1708: 156–57).

It is not surprising that de Piles had earlier observed that "pastel crayons are sold in several places in Paris and ... it is more expedient to buy them than to take the trouble to make them" (de Piles 1684: 92).

Constantyn Huygens has left us a personal account of just how difficult it was for the interested amateur to make a satisfactory set of pastels and of the extent to which painting in colored pastels was then a novelty in terms of practice. The information, contained in a series of letters exchanged with his son, Christiaan, in 1663, has not previously been brought fully into the discussion of pastel. Constantyn outlined his experiences and gave us a first-hand insight into the quantity and quality of advice he was able to obtain from artists and from artists' treatises. We shall not here examine the whole of this correspondence; several examples must suffice.

In April 1663 Constantyn asked his son to investigate for him the process of fabricating pastels in London and Paris, where the latter was traveling: "If they use plaster to make the body and what they use to bind the whole ... I wish to know if they grind the colors very fine and small before mixing them with the plaster, because Sanderson in his Art of Painting [1658] says it suffices to break them only, which I do not believe to be good [advice]" (Huygens 1890: 4:332). Constantyn was unable to make his pastels hold together unless he mixed in "something glutinous" (Huygens 1890: 4:340). Through the summer of 1663, even after receiving detailed instructions from Lely's pastel-maker and practical observations and advice from his son, Constantyn continued to experience difficulties. The correct proportions of filler to color, the desirable fineness of the pigment grind, and the reasons for the resulting softness or brittleness of the sticks gave him trouble. The whole experience was frustrating, and Christiaan closed his correspondence on this subject by telling his father: "I have nothing more to send you touching pastel ... A little practice will teach you the rest" (Huygens 1890: 4:396).

Endnote

(All translations are the author's.)

1. Examples are found in the collections of the Ashmolean Museum, Oxford; Yale Center for British Art, New Haven, Conn.; and elsewhere.

References

Boutet, C. 1708. *Traité de la peinture au pastel.* The Hague: Louis & Henri Van Dole.

Burns, T. 1994. "'Chalk or pastel': the use of friable media in early drawings," *The Paper Conservator* 18, pp. 49–56.

Burns, T. 1997. "Identifying chalk and pastel," paper presented at the conference *Rediscovering Pastels*, held at the Tate Gallery, London, 26 June 1997.

Chaperon 1788. *Traité de la peinture au pastel.* Paris: Defer de Maisonneuve.

La Curne de Saint Pelaye 1880. *Dictionnaire de l'ancien langage français depuis son origine jusqu'au siècle de Louis XIV.* Paris: H. Campion.

Doni, A. F. 1549. *Disegno.* Venice.

Guillaume, J.-C. 1991. *Le travail de l'ocre dans l'Auxerrois – 1763–1966: une industrie rurale.* Prècy-sur-Thil: Editions del'Armançon.

Harris, John 1736. *Lexicon technicum (1704–10*). London: Walthoe.

de la Hire, P. 1730. *Traité de la pratique de la peinture*, in *Mémoires de l'Académie Royale des Sciences 1666–1699*, pp. 637–730.

Huguet, E. 1925–70. *Dictionnaire de la langue français du seizième siècle.* Paris: Edouard Champion.

Huygens, C. 1890. *Oeuvres complètes de Christiaan Huygens: Correspondance 1662–1663.* Vol. 4. The Hague: Société Hollandaise des Sciences.

Leuchs, J. C. 1829. *Traité complet des propriétés, de la préparation et de l'emploi des matières tinctoriales et des couleurs.* Paris: DeMalher.

Lomazzo, G. P. 1584. In Ciardi, R. P. (ed.), *Scritti sulle arti.* Florence: Centro Di, 1970.

McGrath, T. 1994. "*Disegno, Colore*, and the *Disegno Colorito*: The Use and Significance of Color in Italian Renaissance Drawings." Ph.D. diss. Harvard University. Ann Arbor, Mich.: U.M.I. Press.

McGrath, T. 1998. "Federico Barocci and the history of *Pastelli* in Central Italy," *Apollo* 148, no. 441, pp. 3–9.

Meder, J. 1923. *Die handzeichnung: ihre technik und entwicklung*, 2nd edn. Vienna: Anton Schroll.

Monnier, G. 1984. *Pastels: from the 16th to the 20th Century.* Geneva: Skira.

Monnier, G. 1992. "Pastel: Its genesis and evolution to the twentieth century," in D. Dethloff (ed.), *Drawing: Masters and Methods.* London: Philip Wilson, pp. 159–76.

Phillips, E. 1658. *A New World of Words.* London: E. Tyler.

de Piles, R. 1684. *Les premiers elemens de la peinture pratique.* Paris: Nicolas Langlois.

Piot, E. 1863. "Robert Nanteuil: opinions, maximes et conseils recueillis par son élève Domenico Tempesti," in *Le cabinet de l'amateur par M. Eugène Piot: années 1861 & 1862.* Paris: Firmin Didot, pp. 247–50.

Quieffin, T. 1999. Personal communication. Les ocres de Puisaye, 58200 Cosne-sur-Loire.

Quiviger, F. 1996. "Doni, Antonio [Anton] Francesco," in J. Turner, *The Dictionary of Art*, Vol. 9. New York: Grove, pp. 142–43.

Sanderson, W. 1658. *Graphice or the use of the Pen and Pensill.* London.

Trévoux. 1721. *Dictionnaire universel françois et latin.* Paris: Trévoux.

Watrous, J. 1957. *The Craft of Old-Master Drawings.* Madison, Wis.: University of Wisconsin Press.

Making Up the Face: Technique and Meaning in the Pastels of Rosalba Carriera

Thea Burns

To most twentieth-century eyes, Rosalba Carriera's portraiture is characterized by artificial, masklike faces, conventional compositions, and standardized techniques of portrayal (Levey 1980) (see Figs. 1.10 and 1.11). They do not capture the "central core of selfhood" that we consider to be the proper subject of a portrait (Brilliant 1991: 67; Eigen 1980–81). It is difficult to accept that those impassive images record the likenesses of individual sitters or register personal qualities or inner lives. Therefore the reputation of Carriera and her pastel oeuvre has suffered: typical assessments dismiss her as "... a charming and secondary talent, essentially feminine" (Wilhelm 1953: 239).

We read that the refined flattery and superficiality of Carriera's conceptions reflected precisely the decadent values of the age that honored her (Fine 1978). This negative view of the Ancien Régime, particularly the period 1715–1740s, does not reflect the enthusiastic reception given Carriera's works by her contemporaries. It first appeared later, in the mid-eighteenth century writings of newly prominent critics who sought to reform art (La Font de Saint Yenne 1747).

Since the French theoretician Roger de Piles (1635–1709) posited the objective of portraiture to be faithful resemblance, our attempt to evaluate Carriera's work in terms of physiognomic likeness seems appropriate (de Piles 1708). Many of her pastels portray real men and women, and her contemporaries appreciated her portraits in terms of their resemblance. Carriera's correspondent, the painter and engraver Antonio dall'Agata, wrote of her pastel *Portrait of Sister Maria Caterina:*[1]

> [This portrait] of the tertiary Domenicana ... agrees well enough with the description made to me some time ago of her life; that she is completely spiritual and does not nourish herself in any other way but for a little hot water; this is expressed in her extreme exhaustion and the splendour of her eyes ... (Sani 1985: 2:565).

De Piles also acknowledged the role of flattery: "Women have told me clearly that they do not esteem painters who make them too closely resembling and that they like better to be shown with less resemblance and more beauty ... I don't doubt that one could not portray them faithfully without displeasing them" (1708: 132).

Recent scholarship has focussed discussion on issues of how much flattery necessarily accompanied Carriera's portrayals versus how much likeness and characterization was retained (Wittkower 1965; Harris and Nochlin 1978; Levey 1980; Sani 1987). Nevertheless, the perceived lack of expressiveness of Carriera's sitters remains disturbing for modern viewers and is one reason why her portrayals are accused of repetitiveness, decorativeness, and an inability to register subtle fluctuations of inner life. It is apparent that these evaluations of Carriera's achievement have been formulated from the perspective of contemporary

expectations of what constitutes likeness in portraiture. Are the terms "decorative," "flattering," and "formulaic" fair and useful assessments of Carriera's portraits or do they impede our understanding of what her peers, who praised them, saw there?

Allegorical allusions rather than physiognomic likeness are another way in which artists have traditionally portrayed a person's identity (Brilliant 1991). Generalizing devices such as classicizing dress, pose, or attributes are reconciled with the physiognomic characteristics to attribute universal, ideal, or other qualities to a sitter (Woodall 1997). Certain of Carriera's pastels retain traditional titles that identify them as classicizing personifications – Diana, Flora, the Elements, the Seasons, and so on (see Fig. 1.11). Undoubtedly some were originally portraits of identified sitters, perhaps lightly disguised as a form of flattery, but they have lost their referentiality with the passage of time. Yet other pastels were imaginative creations from the beginning (Sani 1985).

Attempts have been made to decipher the allegorical meanings behind the visual motifs of Carriera's pastel portraits, but these efforts have proved disappointing (Sani 1987). This is because the iconographic identification is not forceful enough either to advance their interpretation or, more importantly, to explain to us their very great initial popularity. I suggest that iconographers have not provided workable interpretations of Carriera's portraits because their explanations do not relate to the works' pictorial status and are isolated from pictorial context. Rather than functioning denotatively and identifying figures concretely, classicizing features – such as diaphanous garments, flowers, and fruits – contribute instead to "a wash of meaning, a lightly stressed classicism, flattering to the sitter and available to the connoisseur, perhaps, but not required … they enhance the figures but do not fix or limit [them]" (Junkerman 1988: 358).

Iconographers have studied Carriera's paintings as if they are transparent – as if, through the forms of the image, we are given access to an intellectual realm behind the surface. This approach has perhaps been applied, as Svetlana Alpers has suggested in another context, in an attempt to lend the right kind of significance to an otherwise non-signifying art (Alpers 1983). Likeness and allegory are two poles around which twentieth-century scholars have understood Carriera's aims and achievements in portraiture; decoration and formula are pitfalls of their approaches. The real entry to these paintings is not through likeness or allegory and, therefore, we should apply more historically appropriate criteria to Carriera's portraiture.

Carriera's portraits cannot affect viewers today in the same way they affected earlier viewers, for viewing now takes place

> in a mental world that differs fundamentally in its assumptions, values and cultural codes from the world of the Ancien Régime … our inability to get their jokes should tip us off to the difficulty of getting a culture that was fundamentally different from ours (Goulemot 1991: 134, 153–55).

But these works spoke to their intended audience, and if their meaning is obscure today, it may be that we are not asking the right questions. We must attempt to read Carriera's portraits in the context of a now-vanished world.

The degree of likeness required by different societies at different times will vary because views change about what constitutes resemblance (Brilliant 1991). Since the Renaissance, portraiture has sometimes been characterized by yet another function: the need to articulate in the portrayal not only the sitter's likeness but also his or her social identity. Carriera's portraiture was part of a highly developed social system, and we should look to that system to understand the high regard in which her contemporaries held her pastels. Likeness can be an approximation to a social norm that projects, for instance, belonging to a particular social class. Likeness is still a component of the portrait, but it must be negotiated by the skill of the artist as part of a system that poses social and aesthetic demands as well. The private individual is of less interest to society than the public presentation of the sitter.

I shall now suggest how Carriera was able to match up the particular aesthetic of pastel painting with social norms relating to individual appearance. Fashion may result from the desire of privileged classes to distinguish themselves and be distinguishable from the bourgeoisie. By the early eighteenth century, French fashion had established its sovereignty over Europe (Braudel 1982). The elite looked to Paris for information on the care and adornment of the face, the hair, and the body. There, standards to which everyone aspired were set by the most admired of society, the *honnêtes gens.* In summary, the *honnête homme,* motivated by an "elitist impulse," sought to create a "consummately artistic – and of necessity artificial – secondary self designed to exact recognition of superiority" (Stanton 1980: 30). For the elite there was no choice in this, at least initially (Perrot 1984). Sartorial originality was seen as a contemptible attempt to be noticed at all costs (Stanton 1980). Conformity did not render one invisible. Rather one distinguished oneself by one's ability to fashion and display signs of ornamentation within the system; "one must look like other people, but not exactly …" (Lipovetsky 1994: 33–34).

Carriera's preferred portrait format, with the sitter placed close to the picture plane in a neutral space and limited to head and shoulders or upper torso, concentrates attention on the sitter's face. Writers at the time agreed that:

> It is the face that dominates our outward appearance. That is the part of us which pleads, threatens, flatters and expresses our joys and sorrows, and in which others read our thoughts

before our tongue has had time to express them ... (Faret n.d.: 94).[2]

Insight into the enthusiastic reception of Carriera's portraits by her contemporaries is afforded by the corollary to these ideas – that one might manipulate and control one's facial appearance and expression so that only desired messages were conveyed, so that "only those inner feelings will be visible which are the most favorable and advantageous to you" (de Marmet 1662: 166). The face is "asked to operate like the very stuff of art, [it is] malleable matter that can be molded to perfection" (Stanton 1980: 123).

To achieve a fashionable appearance, the face was strengthened by cosmetics. The artifices used were stressed rather than denied (Perrot 1984). It is easy for us to condemn fabricated appearances as arrogant and dishonest, but such moral condemnation was of no concern to the aristocracy of the Ancien Régime, who exploited cosmetics, grooming, and costume to express their social position and essential difference; ignorance in the use of cosmetics betrayed the *nouveau arrivé* (Perrot 1984). At a second toilette (which followed after a first, private one), a woman of the upper class, attended by her servants and her hairdresser, chatting with her priest or lover, easily took five or six hours to fabricate her appearance (Perrot 1984). It was these "works of art" that were re-presented by Carriera in her pastel portraits.

I shall now reconstruct the preferred appearance. The ideal type of woman in this society was the adolescent, "when beauty still seems in the bud"; the ideal feminine face was a refined delicate oval, almost infantile, with dimples on the cheeks, a small round chin, long almond-shaped eyes, and a small, fleshy mouth (Pisetsky 1967) (see Figs. 1.10 and 1.11). Effects of age were replaced by the illusion of eternal youth (Perrot 1984). A certain amount of fleshy padding in the right places, but not obesity, was considered attractive. Plumpness and cosmetics implied leisure, and bespoke rank and fortune. Fabricated appearances demanded hours of preparation and made the toilette into a spectacle, like the portrait sitting.

Carriera exploited the moderate grain of her paper supports to break up the expanse of the background and evoke a sense of space surrounding her sitters. She did this by applying pastel, possibly from the side of the stick, lightly over the textured surface of the paper, so that it deposited preferentially on the raised grain – a technique later called *grainer* (Jombert 1755). A closer look at make-up helps explain the appearance of Carriera's portraiture and underlines the close material and technical links. Carriera would have known from daily experience with cosmetics that in order to achieve the necessary layering and definition, she could not always work with dry powder alone. The first operation of the second toilette was to apply a white mastic of putty-like consistency; called *fard blanc*, it was composed of a toxic mixture of white lead, mastic, and other ingredients.

This foundation layer was spread over the skin so that women displayed perfectly smooth opaque white complexions from forehead to scarcely concealed breasts. If applied with more subtlety, the *fard blanc* layer was of an unequal luminous density – white like alabaster on the forehead and around the lips, a little brownish towards temples and lightly tinted with blue where the blood rose close to the skin. It was particularly important to unify and smooth the face with this foundation layer to efface the ravages of syphilis, smallpox, and other less serious eruptions (Perrot 1984). To evoke the *fard blanc* layer – for example, in the very smooth porcelain surfaces of the face and decolletage of the *Allegory of Painting* (National Gallery of Art, Washington, D.C.) and *Portrait of a Woman* (Fig. 1.10) – Carriera sometimes applied pastel or pigment that had been moistened and mixed to a thick pastelike consistency. Smoothness resulted because the viscous paint flowed together, obliterating the contours of individual brush strokes and filling in the irregularities of the paper grain. Once dry, this layer covered and hid details of the paper surface.

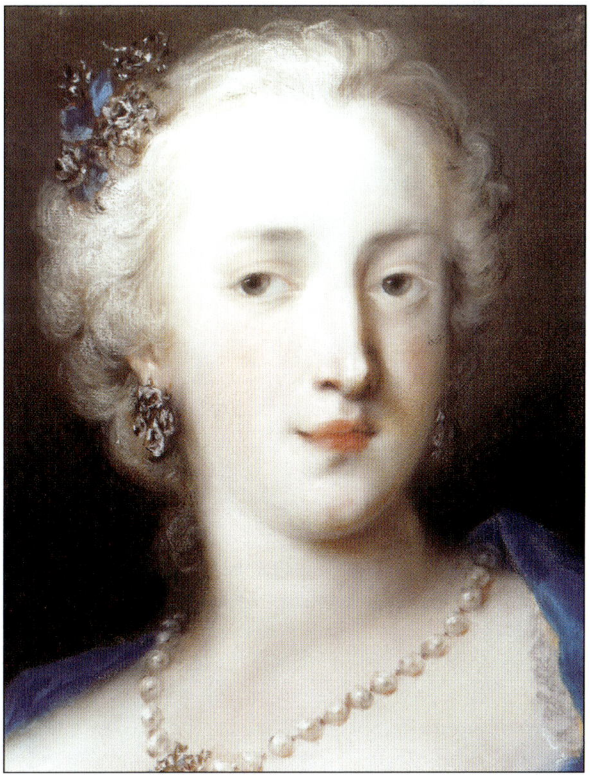

Figure 1.10 Rosalba Carriera (Italian, 1675–1757), *Portrait of a Woman*, c. 1730. Pastel on paper, 31 x 25 cm. Art Gallery of Ontario, Toronto, Purchased as a gift from Mrs. Walter Gordon, 1995, 94/308.

Whether deposited wet or dry, the lean dry pastel layer scattered light and caused the velvety matteness of the surface, known as its *fleur*, for which Carriera's pastels were especially prized. Where dry pastel alone has been deposited, paper fibers may remain raised randomly from the paper surface; pastel particles cling to these fibers and enhance surface matteness by increasing diffusion of incident light from the uneven surface. Although the subtle effects of material manipulation are appreciated with the unaided eye, these effects are best analyzed under the stereo-binocular microscope.

The *fard blanc* layer was an indispensable foundation for setting off the *rouge*, applied next. The preparation of the *fard rouge* was described at the time in terms used also for preparing pigments for painting; the processes of preparation were identical. The material for the *fard rouge* consisted of "a compound of carmine, softened with calcined talc and made into an impalpable powder on the porphyry" (Savary de Bruslon 1742: 3:358). The passion for *rouge* was pursued with a furious intensity in early eighteenth-century France. Violent red was very thickly applied to the cheeks in a circular motion so as to enflame them on "a base of snow," and touched to the lower eyelids, to set the eyes on fire (Perrot 1984). "One does not wish this *rouge* to appear natural, one puts it on to give pleasure to the eyes which see the signs of a drunkeness which promises strayings and enchanting passions" (Lescault 1977: 48). The effect is demonstrated in Carriera's pastel *Portrait of Mlle Clermont* (Musée Condé, Chantilly). Carriera applied dry powdered carmine with brush, finger, or stump and worked it lightly into the paper surface to produce sweetened expanses of matte velvety pastel. Pastel applied dry to a surface and not blended can leave a series of minute grooves, formed in dragging clusters of particles released from the pigment stick (Derow 1988). Sweetening smoothed out this surface and also minimized breaks in the color due to paper texture, evoking the smooth velvety surface and the delicate tonal transitions of flesh.

The eyes were underlined and lids shaded with paint. The eyebrows were plucked and blackened; pomade was applied to make them glisten; the lashes were combed straight, so that each lash would count in the overall effect (Perrot 1984). Lips were heightened with red. Finally a light and circumscribed application of blue picked out one or two veins in the chest and arms to enhance the skin's delicate and aristocratic whiteness (Perrot 1984). Carriera the artist relied on these definite strokes and on contour lines to finish and firm up the image. The artist's *Self-Portrait* (Uffizi, Florence) illustrates her pastel sticks, which are thin yet apparently cohesive and consistent enough to be sharpened to a point for drawing very fine lines.

Carriera, working in pastel, used technically creative ways to create surrogates for flowers, lace, and jewelry, rather than merely imitating them. One example must suffice. The individual pearls of the bracelet worn by Autumn (Fig. 1.11) were made by placing a pearly gray-colored surface layer of wet pastel over a blob of dried white paste. To evoke highlights on pearls, Carriera punctured the center of the gray layer, perhaps with the pointed end of a stump or small brush, to expose the underlying white.

Carriera's pastel portraiture cannot be approached historically in terms of humanistic iconography, naturalistic likeness, or expressive individualism. Produced by a particular society, it is, in effect, part of an

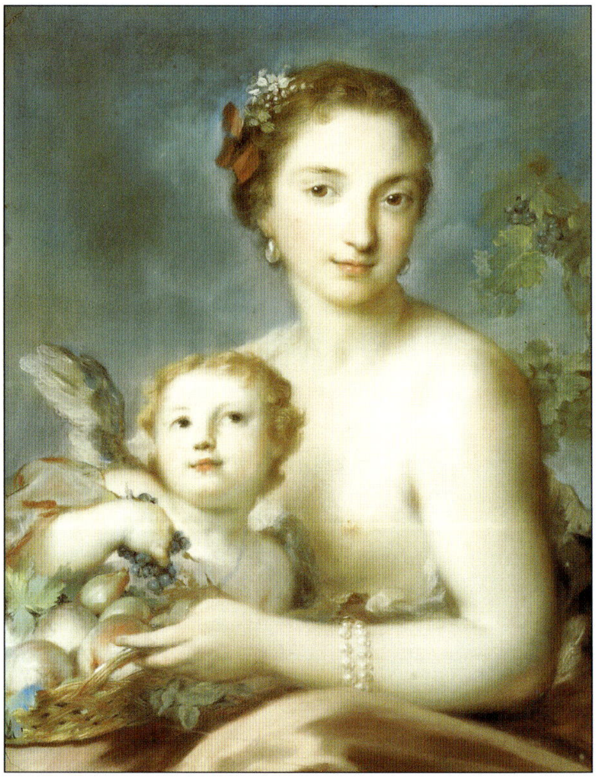

Figure 1.11 Rosalba Carriera, *Autumn*, c. 1730. Pastel on paper, 63.5 x 49.2 cm. The Royal Collection, © 2001, Her Majesty Queen Elizabeth II.

aesthetic of pleasurable illusion best expressed by Roger de Piles. De Piles believed that the value of art was measured in terms of the pleasurable experience it gave the viewer, that sense experience was valuable and pleasurable in itself (Lichtenstein 1993). "If they [paintings] instruct us, so much the better; if they don't, we still have the pleasure of viewing a kind of creation that both amuses and moves us" (de Piles 1708). A painted image supplied absent things, not just their idea, to please and to deceive the eye (de Piles 1708). "Painting is only make up (*fard*), its essence is to deceive, and the greatest deceiver in this art is the greatest painter" (de Piles 1708: 169). The spectator was aware of the illusion and self-indulgently consented to the deception (*se prète à l'illusion*). After an initial resistance (amazement), the spectator was overcome by satisfaction at the effectiveness of the illusion (Lichtenstein 1993: 170; Sani 1985: 2:459). The spectator's attention was, however, necessarily interrupted as it moved between illusion and awareness – from the image of the thing to the thing the image was of (Hobson 1982). The interruption privileged material aspects of the painting and prevented complete absorption by the viewer in the subject of the image (Lichtenstein 1993). The succession and interpenetration of surface and subject made production an integral part of the meaning of the painting.

The original viewers of Carriera's pastels were fascinated by this ingenious deceptive play: play with the impermanence of cosmetics reapplied daily; the re-creation of fragile cosmetic effects in equally fragile pastel; the evanescence – when painted in pastel – of portraits, a genre traditionally intended to provide permanent records of sitters; and, lastly, the permanence of cosmetics when painted in oil paints on canvas. The seductive play with illusion was heightened because identical materials were used to paint (make up) faces and to paint pastels. Lichtenstein recently summarized the pleasure of painted color as pleasure to the eye; although the desire is to touch flesh painted by colorists, "touch is no more real here than the flesh that provokes the desire for it" (1993: 167). The illusion and the viewers' pleasure would dissipate if they came too close, so they are careful not to so as "to keep [their] joy in its effects" (Lichtenstein 1993; de Piles 1677).

Our appreciation of the contribution of Carriera's handling of the pictorial elements of her pastels to the work's meaning is increased when we move interpretation beyond the skillful portrayal of "the sensuous skin of beautiful women" (Freedberg 1971: 77) and the boundaries of the picture plane into the elite culture that surrounded it. The pictorial conventions that Carriera exploited in elaborating her images of beautiful women are indebted to the category of the Venetian sensuous half-length, a mode of depicting women which took on an unprecedented form in sixteenth-century Venice (Junkerman 1988: 18ff.).

Titian's *Flora* (Uffizi, Florence) must serve to summarize the features of a large group of painted women that reproduced the ideal with minor changes. Flora's features may be compared with those of images by Carriera (Junkerman 1988). Characteristics that define the sixteenth-century Venetian group and distinguish it from other contemporary portraiture include the following: a beautiful young woman, displayed in half-length against a dark background, her body lit by a bright light which casts few shadows. Her alabaster skin is pearly soft and delicately heightened with pink; her face is a perfect oval with a bared, wide, high forehead. Her eyes are large and moist with a soft dreamy look. Her hair hangs about her shoulders, loose or simply bound. Her clothing is in a state of *dishabille*. Sometimes her gaze is averted; often she meets the viewer's gaze. Her body is a sensual body and "simply present." The viewer's eye is free to linger on the broad expanses and soft forms of the naked flesh. The figure is placed close to the picture plane and nearly fills the frame; this and the half-length format contribute to the intimacy of focus.

Even Venetian Renaissance female portraits, whose sitters are fully dressed and their toilette perfected, conform to a generic type evoking prevailing ideals of beauty. They can give us some insight into the reception of Carriera's more conventional female portraits. The erotic is avoided and the social aspect stressed by placing the accent on the richness and elegance of jewels and costume. Titian is credited with inventing this type; a well-known version is his *La Bella* (Pitti Palace, Florence). Such images celebrated domestic virtues by positing the reserve and humility appropriate to a wife and mother (Habert 1996).

Claims that Carriera revolutionized pastel have been repeated from her lifetime to the present (Meder 1923; Parker and Pollock 1981; Slatkin 1985); however, until we embed these claims in the appropriate social and technical context, they lack force and concrete meaning and remain, to a large extent, celebrations of her artistic genius and of the value of the work of art. My attempt to recover the conceptual foundations of Carriera's choice and use of materials in her pastels demonstrates the critical need for a technical component to art-historical investigation, provided it serves the process of elucidating aesthetic meaning in a cultural context.

Endnotes

(All translations are the author's except where otherwise indicated.)

1. Sani 1985, 2: 567, note 2, identified the portrait referred to as that of Sister Maria Caterina (pastel, 440 x 350 mm), a version of which is in Venice, Ca'Rezzonico, Museo del Settecento Veneziano (inv. 1790, cl. I).

2. Translated and cited by Stanton 1980, p. 122.

References

Alpers, S. 1983. *The Art of Describing*. Chicago: University of Chicago Press.

Braudel, F. 1982. *Civilization and Capitalism: 15th to 18th Centuries*. New York: Harper and Row.

Brilliant, R. 1991. *Portraiture*. London: Reaktion Books.

Derow, J. 1988. "Gainsborough's varnished watercolor technique," *Master Drawings* 26, 3.

Eigen, M. 1980–81. "On the significance of the face," *Psychoanalytic Review* 67, 4, pp. 427–41.

Faret, N. n.d. *L'honneste homme; ou L'art de plaire à la cour*. Ed. by Maurice Magendie, 1925. Paris: Presses Universitaires de France.

Fine, E. H. 1978. *Women and Art*. Montclair, N.J., and London: Allenhead & Schram/Prior.

La Font de Saint Yenne 1747. *Réflexions sur quelques causes de l'état présent de la peinture en France, avec un examen des principaux ouvrages exposés au Louvre le mois d'août 1746*. The Hague: Jean Neaulme.

Freedberg, S. 1971. *Painting in Italy: 1500–1600*. Harmondsworth and Baltimore: Penguin.

Goulemot, J. M. 1991. *Ces livres qu'on ne lit que d'un main: Lectures et Lecteurs de livres pornographiques au XVIIIe siecle*. Paris: Alinea. Cited by Robert Darnton, 1994. "Sex for Thought," *The New York Review of Books* 41 (22 December 1994).

Habert, J. 1996. *Véronese: une dame vénitienne dite la Belle Nani*. Paris: Réunion des musées nationaux.

Harris, A. S., and L. Nochlin 1978. *Women Artists: 1550–1950*. New York: Alfred A. Knopf.

Hobson, M. 1982. *The Object of Art: The Theory of Illusion in Eighteenth-Century France*. Cambridge: Cambridge University Press.

Jombert, C. A. 1755. *Méthode pour apprendre le dessein*. Paris: Jombert.

Junkerman, A. C. 1988. *"Bellissima Donna": an interdisciplinary study of Venetian sensuous half-length images of the early 16th-century*. Ann Arbor, Mich.: U.M.I. Press.

Lescault, G. 1977. "Banalité sur le maquillage," *Traverses* 7, pp. 48–56.

Levey, M. 1980. *Painting in Eighteenth-Century Venice*, 2nd edn. London: Phaidon.

Lichtenstein, J. 1993. *The Eloquence of Color*. Transl. by E. McVarish. Berkeley, Calif.: University of California Press.

Lipovetsky, G. 1994. *The Empire of Fashion.* Transl. by C. Porter. Princeton, N.J.: Princeton University Press.

de Marmet, M. 1662. *Maximes pour vivre heureusement dans le monde et pour former l'honnête homme.* Paris: Charles de Scery.

Meder, J. 1923. *Die handzeichnung: ihre technik und entwicklung,* 2nd edn. Vienna: Anton Schroll.

Parker, R., and G. Pollock 1981. *Old Mistresses.* London: Pandora.

Perrot, P. 1984. *Le travail des apparences: le corps féminin XVIII–XIXe siècle.* Paris: Seuil.

de Piles, R. 1708. *Cours de peinture par principes.* Reprint 1989. Paris: Gallimard.

de Piles, R. 1677. *Conversations sur la connaissance de la peinture.* Paris: Langlois.

Pisetsky, R. L. 1967. *Storia del costume in Italia.* Milan: Istituto Editoriale Italiano.

Sani, B. (ed.) 1985. *Rosalba Carriera: lettere, diari, frammenti.* Florence: Leo S. Olschki.

Sani, B. 1987. "'La furia francese' di Rosalba Carriera: ses rapports avec Watteau et les artistes français," in F. Moureau and Margaret Morgan Graselli (eds), *Antoine Watteau (1684–1721): le peintre, son temps et sa légende.* Paris and Geneva: Clairefontaine, pp. 73–84.

Sani, B. 1988. *Rosalba Carriera.* Turin: Umberto Allemandi.

Savary de Bruslon, J. 1742. *Dictionnaire universel de commerce.* New edn. Geneva.

Slatkin, W. 1985. *Women Artists in History from Antiquity to the 20th Century.* Englewood Cliffs, N.J.: Princeton University Press.

Stanton, D. C. 1980. *The Aristocrat as Art.* New York: Columbia University Press.

Wilhelm, J. 1953. "Le portrait de Watteau par Rosalba Carriera," *Gazette des beaux-arts* 42, pp. 235–46.

Wittkower, R. 1965. *Art and Architecture in Italy 1600–1750,* 2nd edn. Harmondsworth: Penguin.

Woodall, J. 1997. "Introduction: Facing the subject," in J. Woodall (ed.), *Portraiture: Facing the Subject.* Manchester and New York: Manchester University Press.

Materials, Methods, and Meanings in Edgar Degas's Late Pastels

Richard Kendall

Edgar Degas was arguably the boldest, the most persistent, and the least apologetic pastelist of the nineteenth century, perhaps of the entire modern era. A high proportion of his major works of art were made in pastel on paper, largely supplanting oil on canvas toward the end of his career as his main studio activity. Yet he was also among the least orthodox masters of the medium, regularly – almost systematically – overturning its conventions and extending its boundaries: he avoided the tinted papers and pale palettes associated with the technique, and rarely took steps to preserved the "bloom" so revered by its admirers;[1] he frequently mixed pastel with other materials, combining it with distemper, gouache, watercolor, metallic paint, printing ink, and even oil paint; he applied it to such questionable supports as canvas, tracing paper, silk, wood panel, and still-wet monotype; and he was equally contrary in his use of fixative, emphasizing its role as a creative vehicle rather than a stabilizing process. Although we should not oversimplify the historical uses of pastel, Degas seems to have taken a delight in pushing the medium to creative extremes.[2]

Two further factors complicated Degas's approach to pastel: his public reverence for the tradition he was privately overturning and his attempts to fuse the procedures of pastel with those of painting in oil. Degas's acquaintance with the golden age of pastel began early: he grew up in a home where works by eighteenth-century practitioners hung on the walls, inspiring some rather conventional exercises on colored paper in which he blended muted hues with the stump and added late highlights in the textbook manner.[3] In his middle years, this technique was considerably relaxed, with a wider repertoire of marks, a richer palette, and a more adventurous choice of supports. But even in late life, when his own practices as a pastelist were at their most anarchic, Degas continued to endorse the contemporary taste for the works of Chardin, Liotard, Perroneau, and Quentin de La Tour, telling his niece that a pastel self-portrait by La Tour was "a masterpiece."[4] When he departed from convention, it seems, it was through neither ignorance nor disrespect.

Degas's admiration for the pastelists of the past went hand-in-hand with a resurgence of interest in his later years for the great exponents of oil painting, especially the masters of the Venetian school. He studied their pictures in the Louvre, instructed a pupil to copy a celebrated Mantegna, and cited Titian and Veronese on the subjects of color, underpainting, glazing, and brushwork, himself executing a number of slightly puzzling canvases under their influence.[5] More unexpectedly, he extended some of these preoccupations into the making of his pastels, at times presenting the two techniques as interchangeable. "In oil painting, one should proceed as with pastel," he told the English artist Walter Sickert.[6] What Degas seems to have had in mind was the practice of *layering*, of proceeding in "a series of operations" – as he explained to Paul Valéry – from colored ground to final highlight, a sequence that is common to traditional pastel and oil practice.[7] As so often when studying Degas's technique, therefore, his pastels confront us not only with a means of applying color to a particular surface but with a privately resonant act, a mixture of partly digested history and reckless innovation that pervades every aspect of these pictures.

The pastels of Degas's last decades – effectively those made between 1890 and about 1910 – are the products of this curiously conservative radicalism, in which technical license, homage to the past, and a blurring of the boundaries between media all play their part. Central to their identity is a heightened interplay between medium and image, in which new meanings are created by events on the physical surface and where newly conceived subject matter makes unprecedented demands on technique. This interdependence is one of the key elements of Degas's later work and one that requires a special combination of skills and sympathies from those who deal with it. Art historians need the guidance of conservators and other specialists in assessing the nature of Degas's procedures and to help them unlock the intimate histories of the works in question. Conservators, in their turn, have learned much from the wider pattern of Degas's ideas and from increased scholarly attention to his final decades. Working together, art historians and conservators can arrive at a deeper understanding of some of the most remarkable pastels of our era and perhaps extend the shared territory between our two disciplines. By looking in detail at specific examples of Degas's late pastels, we can enter this rich and complex terrain, where narratives and materials mingle, and where historians and conservators find common cause.

A relatively restrained pastel from the early 1890s, *The Morning Bath* (Fig. 1.12), summarizes the inventiveness of Degas's application of the medium as he approached his 60th year.[8] Selecting an off-white wove paper with a rather coarse, repetitive texture, the artist indulged in a bravura display of parallel strokes, hatchings, smears, powdery swirls, and improvised gestures that vividly evoke the qualities of his subject. In a way that is highly significant for such a classically trained artist, the body of the bather – the most delicately modeled feature in the scene – is not bounded by a firm, old-masterly line but is repeatedly broken by pastel marks, seeming to crumble under the influence of light or blur with movement. At the most immediate level, Degas's handling informs and conditions the nature of his subject; the bather is tremulous because of the way it was drawn, when it might have been reposeful, statuesque, or merely suggestive if rendered with a different touch. Other features conspire to the same end: the hazy background wall, where white pastel was brushed across unfixed green, the scribbled pattern of the curtain, and the unfocussed, unmade bed in the foreground, all of which evoke early morning domestic chaos. Several of these passages go well beyond the constraints of draftsmanship and might well be described as painterly; we know that they are not simply unfinished

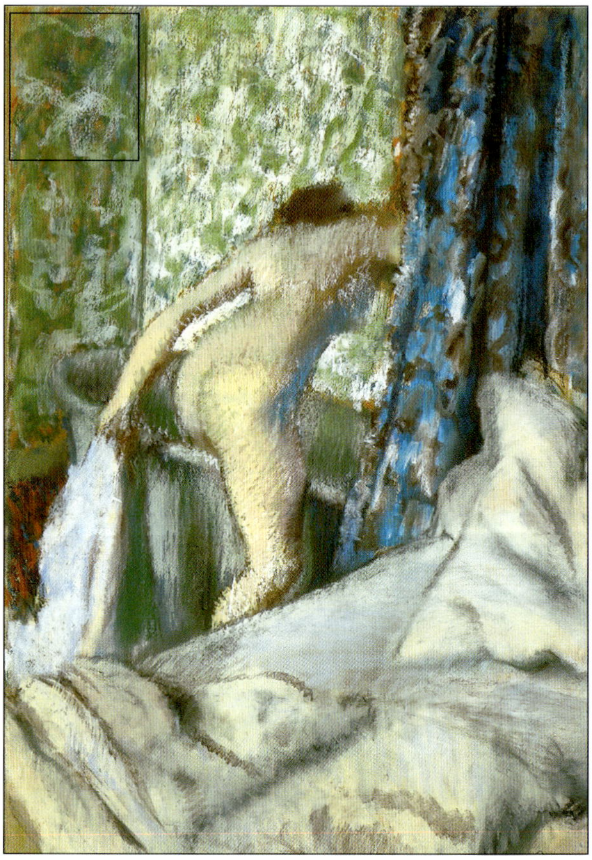

Figure 1.12 Edgar Degas (French, 1834–1917), *The Morning Bath*, c. 1892–95. Pastel on wove paper laid down on board, 70.6 x 43.3 cm. The Art Institute of Chicago, Mr. and Mrs. Potter Palmer Collection, 1922.422.

from the fact that the artist signed the work in this state, probably around 1896, when it was sold to Mrs. Potter Palmer of Chicago.[9]

Equally expressive in *The Morning Bath* is Degas's austere range of color, a harmony of cool whites, greens, and blues that has few precedents in his oeuvre. Crucial to its final impact is a less immediately apparent warmth, most noticeable in the top left of the rectangle, where a base of orange-red pastel can be glimpsed through the final layer of pale green (Fig. 1.13). This fiery substratum once covered the entire upper left area and might be explained in several ways. At its simplest, it could represent Degas's first conception of this part of the image, which he later changed from warm to cold as the composition evolved in order to enhance the unity of the picture. Alternatively, the initial orange may have been laid down as a kind of ground layer or "underpainting," in anticipation of the flourishes of green soon to be "glazed" over it in a rough approximation of the techniques of the masters. A number of works by Degas from this period, in both pastel and oil, show similar preparations in warm browns and Venetian red, apparently resulting from his current interest in writers on technique from Cennino Cennini to Eugène Delacroix.[10] Finally, the fact that the color of the lower layer is the near-complementary of the upper may also

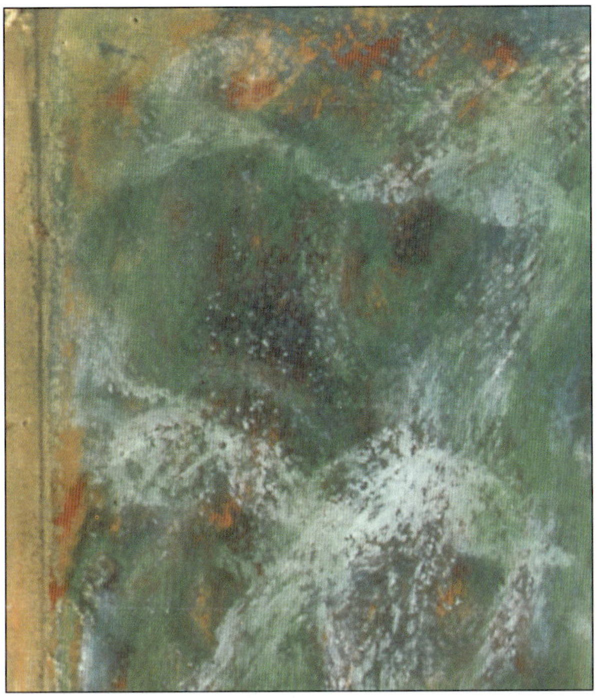

Figure 1.13 Edgar Degas, *The Morning Bath* (detail).

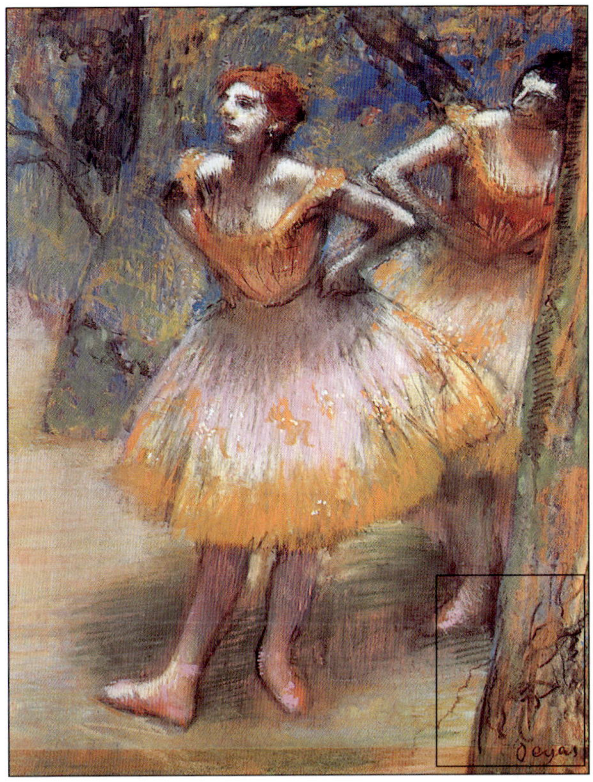

Figure 1.14 Edgar Degas, *Two Dancers*, c. 1893–98. Pastel and charcoal on tracing paper, pieced and laid down on board, 70.5 x 53.6 cm. The Art Institute of Chicago, Amy McCormick Memorial Collection, 1942.458.

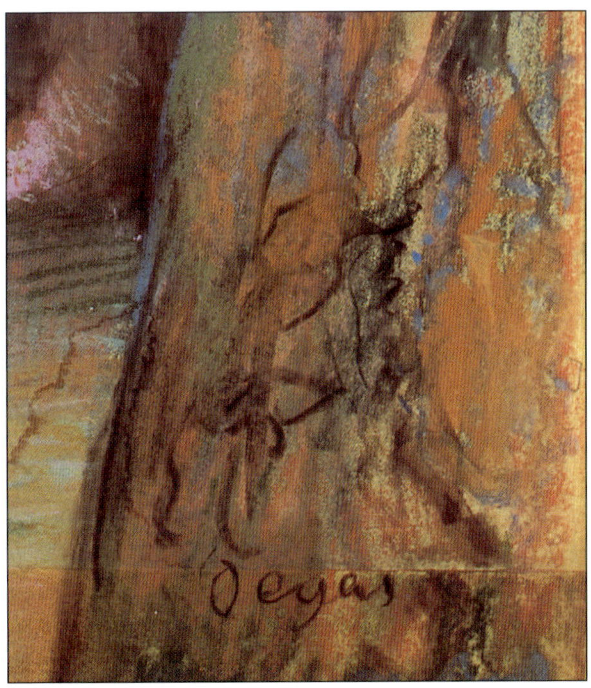

Figure 1.15 Edgar Degas, *Two Dancers* (detail).

point to a more calculated juxtaposition, at a time when Degas was briefly intrigued by the color theories of contemporaries such as Georges Seurat and Camille Pissarro.[11]

Degas followed a comparable, though structurally distinct, approach in the slightly later *Two Dancers* (Fig. 1.14), where complementary hues answer each other across the surface of the work rather than in depth. Here, passages of blue and orange are repeatedly found side by side, as are less-exact pairings of moss green and sugar pink, of yellow and lavender. These colors were applied in a dazzling gamut of strokes, even more varied than those in *The Morning Bath*, ranging from needle-sharp lines to softly modeled, almost rococo shadows, from veils of pastel so light that the paper tint shows through, to violent stabbings and slashings in the right foreground (Fig. 1.15). The result is a world of shimmering, superficial animation, where light sparkles on sequins, plays over theatrical décor, and reflects from painted, powdered flesh. Rarely have medium and subject been so appropriately twinned, exploiting the subtle affinities of pastel with ballet costumes, with the matte distemper used by scene painters and with the brittle glamour of the performance itself.[12]

The subject of *Two Dancers* is one that Degas returned to frequently in his final decades, often repeating this precise configuration of individuals and poses. Variants of the scene exist as charcoal drawings, oil paintings, and pastels in different color harmonies – in pale blue and yellow, for example – all of which emerge from the distinctive mode of serial production he perfected in these years.[13] The use of tracing paper was central to this process, but *Two Dancers* itself is unlikely to have been generated directly in this way since it was drawn on standard wove material.[14] Equally characteristic of such sequences was Degas's strategy of extending his original sheet of paper by adding strips to top, bottom, or sides to vary the final format.[15] In *Two Dancers* an extension of some 8 or 9 cm has been glued to the lower edge, but, atypically, it was not then fully worked with pastel to integrate it with its neighbor, remaining tonally and texturally conspicuous to a striking extent.[16]

The presence of Degas's signature at the precise junction of these two sheets (Fig. 1.15) confirms the finished status of the work, but again seems designed to draw attention toward, rather than away from, the added strip. Such extensions are frequently found in the works of eighteenth-century pastelists, where much effort was traditionally spent in concealing the maneuver and maintaining the integrity of the visual field. Why did Degas, in *Two Dancers*, adopt the opposite strategy? The effect – whether intentional or not – is to emphasize the foreground and the buttress-like stage flat at right, and their role as elements in a two-dimensional design. Simultaneously, these features stress the picture surface and remind us of the papers and paste, the cutting and splicing that goes into the making of any composition, thus insisting on the artificial nature of all creativity. Expressing a similar notion to Stéphane Mallarmé around this time, Degas exclaimed, "Art is falsehood!"[17]

Further encouragement for such speculation comes from *Woman at Her Toilette* (Fig. 1.16), a technically vigorous and spatially challenging study executed around the turn of the century. Degas carried out this pastel on commonplace tracing paper, the support he most favored in his late career. We know from contemporaries that his practice was to take partly completed drawings on tracing paper to the workshop of a certain Père Lézin, who pasted them onto cardboard according to the artist's instructions and usually with considerable accuracy: Degas is said to have regarded Lézin as an "astonishing" craftsman.[18] In this instance a large rectangle of the pale, coffee-colored paper has been extended with a strip of paper at the right-hand side, using the kind of neat "butt" joint that was Lézin's specialty. As in *Two Dancers*, when Degas returned to work on the picture, he seems to have welcomed the resulting anomalies, allowing his bright yellow pastel to heighten, rather than conceal, the fine line of the join along most of its length (Fig. 1.17). More mischievously the same pastel has been brushed across a number of irregular creases at lower right, which are the direct consequence of Lézin's inability – on this occasion – to glue the paper down evenly. Wittily or otherwise, Degas has transformed this area of his bathing scene into a piece of drapery, the creases in the paper becoming folds in a towel or discarded robe, a feature absent from other variants of this design.[19] Here again, process has informed meaning, as real materials rhyme with depicted ones, a piece of fabric with the process of fabrication.

The use of tracing paper and the piecing together of separate sheets has a powerful effect on our reading of Degas's pastels. First, it allows the production of a sequence of near-identical drawings by direct tracing, presenting us with a complete or partial experience of a "series" of works analogous to those of contemporaries such as Claude Monet.[20] In such circumstances our attention is naturally drawn to subtle variations within

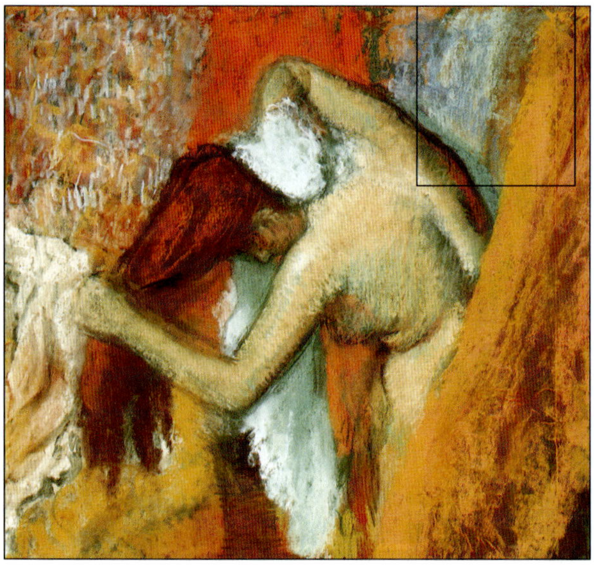

Figure 1.16 Edgar Degas, *Woman at Her Toilette*, c. 1900–05. Pastel and charcoal on tracing paper, pieced and laid down on paper, wrapped around cardboard, 75 x 72.5 cm. The Art Institute of Chicago, Mr. and Mrs. Martin A. Ryerson Collection, 1937.1033.

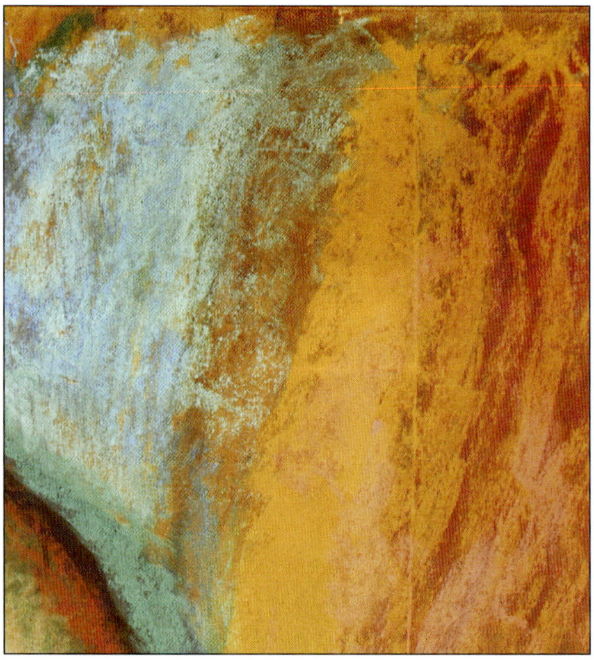

Figure 1.17 Edgar Degas, *Woman at Her Toilette* (detail).

the sequence, to the shifts of color, refinements of surface texture and inflections of mark-making that distinguish individual pictures. Degas seems to have encouraged our preoccupation with the physical membrane of *Woman at Her Toilette* in many ways, drawing his pastel sticks emphatically *across* the picture surface, assembling a brilliant pattern of primary reds, yellow, and blues to delight our eyes, and suppressing perspective and chromatic recession (the vermilion form on the background wall, for example, is also the hottest and most saturated hue in the picture). In comparison with both *The Morning Bath* and *Two Dancers, Woman at Her*

Toilette has a startling flatness, in which the tilted body of the bather and the broad planes of wallpaper and cloth lock decoratively together in a way more often associated with the Nabis or with Henri Matisse. Seen in this way, the heightened joins and undisguised patches of tracing paper take on a new significance, becoming components in a visual syntax that is at once sensuously physical and detached from illusion.

Woman at Her Toilette is one of many unsigned pastels left in Degas's studio at his death that may have been considered unfinished. What is clear, however, is that several changes had already occurred within the work: the red patch above the bather was once blue; her lower body appears to have been indicated before the yellow drapery was added; and part of a chair or couch was formerly visible under the snowstorm of colored strokes at top left. The reason why these passages can still be seen is that they were almost certainly fixed, using the mysterious "ball syringe" observed by Sickert, before Degas began to rework them.[21] Fixing would have prevented the disruption of the initial layer and the contamination of subsequent colors, and – as Anne Maheux has pointed out – may also have roughened the paper surface and thus assisted the adherence of later additions.[22] Denis Rouart, whose 1945 *Degas: à la récherche de sa technique* remains a standard text, noted the effect of "transparency" obtained by this fixing and layering procedure, perceptively linking it with the artist's fascination with historic oil-painting practices.[23]

Degas's habitual use of fixative is widely attested by his contemporaries and by modern conservators, among whom Anne Maheux has published the most complete and detailed account.[24] Although individual works necessarily vary, Rouart's claim that Degas proceeded by "the application of successive layers of pastel in which each one is fixed before it is covered with the next" is now widely accepted.[25] In practice this means that Degas was exploiting fixative as a constructive tool, building up a crust of color in a number of steps that could be either carefully planned or freely improvised. As we have seen in *Woman at Her Toilette*, repeated fixing enabled Degas to obliterate earlier forms, enrich areas of texture, or – in passages such as the woman's body – create the finest of nuances by superimposing latticeworks of charcoal line and modeling, intermediate tones and colored shadows.

In the monumental pastel on tracing paper *The Bathers* (Fig. 1.18), almost every detail of the scene is eloquent of this meshing of Degas's technical originality with the radicalism of his imagery. The composition started life as a single horizontal sheet some 66 cm high, on which charcoal was used to outline the bather group and incorporate at least two figures from other drawings by means of tracing.[26] Considerable labor and refinement went into the orchestration of these charcoal drafts, over which Degas would typically apply fixative to ensure that their forms survived in subsequent phases. Finding echoes in Raphael and Poussin, as well as in Cézanne and Matisse, *The Bathers* already aspired to the proportions of an exhibition-scale canvas and was soon blocked in with appropriately vivid greens, turquoise, Prussian blue, and deep red.[27] As the edges of the sheet show, however, it was only at this stage that Degas decided to add substantial bands of paper to top and bottom, almost certainly fixing it a second time to secure the initial pastel layer, before carrying it across the center of Paris to Lézin's workshop.[28] Once back in his studio with the enlarged sheet, Degas then used pastel again, working over new areas with colors already used elsewhere in the composition. But, as in the foreground of *Two Dancers*, the artist perplexingly failed to complete his task, leaving the mechanisms of his picture open to our inquisitive gaze a century later.

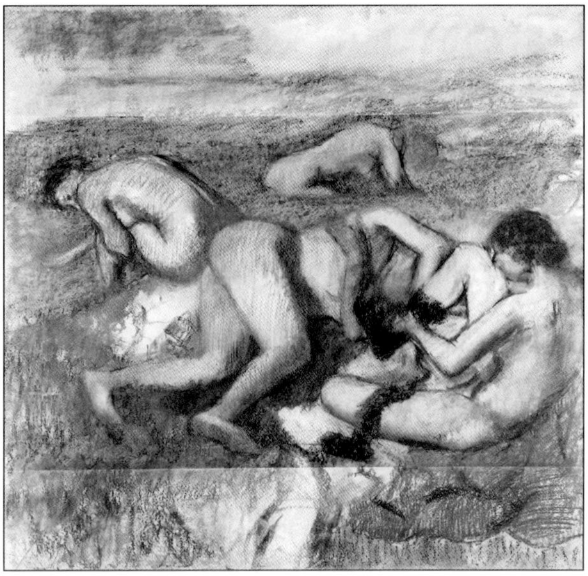

Figure 1.18 Edgar Degas, *The Bathers*, c. 1895–1905. Pastel and charcoal on tracing paper, pieced and laid down on board, 113.4 x 115.7 cm. The Art Institute of Chicago, Gift of Nathan Cummings, 1955.495.

Abandoned just six years before the young Pablo Picasso, a near-neighbor in Montmartre, embarked on his own grandiose study of female nudity, *Les Demoiselles d'Avignon*, Degas's *The Bathers* is one of the great unfinished inventions of his late career, a masterpiece that might have been. But even in its incomplete state – perhaps *especially* in its incomplete state – we can hardly overlook the deep originality of its thematic and material structures. Here is a bold rethinking of space, with forms stacked above each other and volumes in dialogue with the grid-like scaffolding of the pieced paper; here is an insistence on tactility in every register, from delicate hatching to heavy accumulations of pastel; here we encounter an early form of "frottage," where Degas has placed his tracing paper against rough floorboards or plaster, then rubbed pastel over certain areas; and here, in the repeated use of fixative, is the purposeful build-up of dense, tapestry-like color that has little to do with description.

Such qualities are almost meaningless when seen in isolation, whether by the scientist in the laboratory or the iconographer, analyst of style, and social historian in the groves of academe. Where these pastels have been successfully articulated, it has almost invariably been the result of close collaboration or interaction between several disciplines; Paul Valéry, poet and amateur artist, learned much about Degas from Ernest Rouart, also a painter; the invaluable catalogue of the Degas collection in Chicago was written by a paintings curator and a prints and drawings specialist, Richard Brettell and Suzanne McCullagh; work by Harriet Stratis, conservator at the Art Institute, with the curators of several recent Degas exhibitions has greatly extended our knowledge of the artist's practice; and the latest publication on Degas's pastels came from a leading Degas scholar, Jean Sutherland Boggs, and a conservator of works on paper, Anne Maheux.[29] These partnerships surely point the way forward, not just in the study of the meanings of Degas's art but perhaps to the definition of a new, interdisciplinary field where the work of art will once more take center stage.

Endnotes

1. Although some of Degas's completed pastels have unfixed late touches, many other highly developed works appear to have been fixed at the final stage.

2. The combination of convention and extremity in Degas's technique is summarized in A. Maheux and P. Zegers, "Degas pastel research project: a progress report," *The American Institute for Conservation of Historic and Artistic Works; Papers Presented at the Fourteenth Annual Meeting: Chicago, Illinois, 21–25 May 1986* (Chicago, 1986), pp. 66–76.

3. The depiction of a work by Jean-Baptiste Perroneau in a pastel interior by Degas is considered in T. Reff, *Degas: The Artist's Mind* (New York, 1976), pp. 114–15, and in J. S. Boggs (ed.), *Degas* (Paris, Ottawa, and New York, 1988), pp. 155–57. Degas's links with Perroneau are discussed in R. Kendall, *Degas: Beyond Impressionism* (London, 1996), pp. 91–92. One of his earliest developed pastels is *Mme Théodore Gobillard, née Yves Morisot*; P.-A. Lemoisne, *Degas et son oeuvre* (Paris, 1946–49), Vol. 1, no. 214.

4. J. Fevre, *Mon oncle Degas* (Geneva, 1949), p. 70.

5. For Degas's revived interest in these subjects, see Kendall 1996 (as in note 3), pp. 106–23.

6. W. Sickert, *A Free House* (London, 1947), p. 185.

7. P. Valéry, *Degas, Dance, Drawing* (Princeton, N.J., 1960), p. 6.

8. All four pastels discussed in this essay are in the collection of The Art Institute of Chicago. See R. Brettell and S. F. McCullagh, *Degas in The Art Institute of Chicago* (Chicago, 1984).

9. Ibid., pp. 160–62.

10. See Kendall 1996 (as in note 3), ch. 4, esp. pp. 112–19. Degas was far from pedantic in his experiments, sometimes beginning with a cool pastel "ground" and progressing to warmer hues in subsequent applications.

11. Ibid., pp. 101–03.

12. It has rarely been noted that one of Degas's main precursors as a pastelist, Jean-Baptiste Perroneau, also used the medium to represent ballet dancers; see L. Vaillat and P. de Lamay, *J.-B. Perroneau* (Paris, 1923), pp. 38, 213, 215.

13. For Degas's serial processes, see Kendall 1996 (as in note 3), pp. 77–87, 186–88, 254–56.

14. Most commonly, Degas appears to have made these variants by laying a fresh sheet of tracing paper over an existing drawing and tracing it directly, rather than by rubbing the underside of a drawing with charcoal, then retracing the original lines of this drawing onto a sheet beneath.

15. This process is examined in Maheux and Zegers 1986 (as in note 2), pp. 69–71; A. Maheux, *Degas Pastels* (Ottawa, 1988); J. S. Boggs and A. Maheux, *Degas Pastels* (London, 1992), pp. 22–25; and Kendall 1996 (as in note 3), pp. 77–87.

16. In Brettell and McCullagh 1984 (as in note 8), p. 175, it is noted that traces of another signature, corresponding to the first state of the work, have been detected under infrared light at the lower right of the principal sheet. It is possible that the disjunction between the two areas was accentuated by fixative or by subsequent treatment of the picture.

17. Valéry 1960 (as in note 7), p. 84.

18. Cited in Lemoisne 1946–49 (as in note 3), Vol. 1, no. 258. For Lézin, see Kendall 1996 (as in note 3), p. 86.

19. See, for example, Lemoisne 1946–49 (as in note 3), Vol. 3, nos. 1422–27.

20. The evidence from his friends and models that Degas worked on his sequences in a serial fashion is summarized in Kendall 1996 (as in note 3), pp. 103–05.

21. Sickert 1947 (as in note 6), p. 192. A ball syringe may have been similar in appearance to a turkey baster or to a device used in the veterinary sciences, in connection with the practice of horse breeding.

22. Maheux 1988 (as in note 15), p. 33.

23. D. Rouart, *Degas: à la récherche de sa technique* (Paris, 1945), p. 32. English translation by P. C. De Santis, S. L. Fisher, and S. Fletcher, 1988.

24. Maheux 1988 (as in note 15). For accounts by Degas's contemporaries, see Kendall 1996 (as in note 3), pp. 98–99.

25. Rouart 1945 (as in note 23), pp. 34, 66.

26. See Lemoisne 1946–49 (as in note 3), Vol. 3, no. 1080.

27. For historical analogies with this work, see Brettell and McCullagh 1984 (as in note 8), p. 190; Kendall 1996 (as in note 3), p. 278; and R. Kendall, "Degas and Cézanne: Savagery and refinement," in *The Private Collection of Edgar Degas*, ed. by A. Dumas et al. (New York, 1998), p. 206.

28. Lézin's workshop was on rue Guénégaud, on the opposite side of the Seine from Degas's studio on rue Victor Massé.

29. See Valéry 1960 (as in note 7); Brettell and McCullagh 1984 (as in note 8); Boggs 1988 (as in note 3); Kendall 1996 (as in note 3); Boggs and Maheux 1992 (as in note 15).

An Investigation of the Pastels of Giuseppe De Nittis and the Pastel Revival of the Later Nineteenth Century

Anne F. Maheux

Since the peak of its reputation during *l'age d'or du pastel* of the early eighteenth century, the pastel medium has experienced several so-called revivals that have challenged its dubious reputation as an ephemeral medium. In the mid-nineteenth century, pastel continued to be used by artists, typically for portraits, sketches, and studies, but rarely for finished works of art. An examination of the official Salon catalogues of the time signal two acknowledged revivals of interest in the medium: the first occurred in the 1830s and 1840s, and all but dissolved by 1860. A later, more significant episode of popularity was well in progress by the 1880s, and the example of Edgar Degas's work in pastel is usually identified as the driving force responsible for the medium's rejuvenation. However, Degas's adoption of pastel and its presumed influence must be considered within the context of the periodic revivals of interest in the medium in the later nineteenth century, and must also take into account other seemingly extraneous factors. The current study suggests that the monumental pastels by Giuseppe De Nittis (see Fig. 1.19), an Italian artist largely unknown today, were a major proponent in pastel's re-emergence in the 1880s, serving to re-ignite interest in the medium because of their popular subject matter and vital, innovative style.

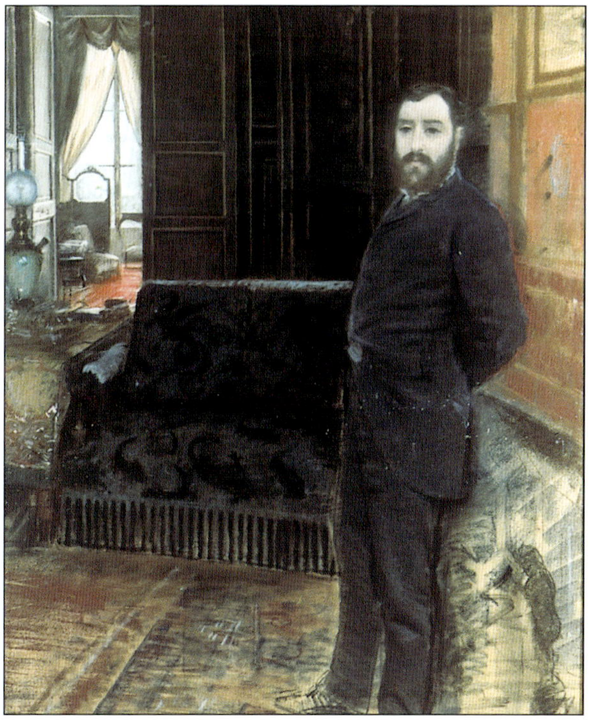

Figure 1.19 Giuseppe De Nittis (Italian, 1843–1884), *Self-Portrait,* 1883. Pastel on prepared fabric, 114 x 88 cm. Museo Civico, Barletta.

De Nittis was born in the eastern seaport town of Barletta in 1846. He attended the School of Fine Arts in Naples as a youth, but soon abandoned formal training to work *en plein air.* After a short but critically successful sojourn in Florence with the Macchiaioli, De Nittis traveled to Paris in 1867, drawn, like other foreign artists, to the city that represented the cutting edge in modern artistic development. Moreover, his first visit to the French capital deliberately coincided with the great Universal Exposition of 1867, which showcased the latest and greatest in technical, mechanical, scientific, and creative innovations. From the outset of his career, De Nittis's greatest ambition was to become assimilated and accepted within Parisian society and its art world, no small feat for a young foreigner. He quickly established contacts with the dealers Goupil and Reitlinger, and with the officially accepted artists Jean-Léon Gérôme and Jean Louis Ernest Meissonier, through whom he gained a foothold in official art circles. These early successes allowed him to settle in Paris, where his marriage in 1869 to the well-connected and charming Léontine Gruville further facilitated his entry into Parisian society.

With the first exhibition of his paintings at the Salon, De Nittis launched his popular standing with the public, critics, and collectors. His Salon entry of 1874, an oil entitled *Che Freddo!*, was a notable success, further establishing his reputation and professional status. Critical reviews cited its significance in opening up a new modern genre for artistic exploration.[1] De Nittis's depictions of the French capital were admired by critics and collectors as faithful reflections of the life, habits, and fashions of contemporary Paris that provided a fundamental method of documenting the spirit of the time in which he lived.

In fact it was because of his secure Salon reputation that Degas asked De Nittis to participate in the first Impressionist exhibition, hoping that his presence would garner more credibility for the Independent group, attract more regular Salon visitors, and secure favorable reviews. The two artists had met several years earlier, probably through the artist Marcellin Desboutin. All of the works exhibited by De Nittis in the Impressionist show, including several extraordinary views of Mount Vesuvius, were executed earlier in Italy, since his more recent successful Salon works were apparently rejected by the other Independents. Although never completely embracing their stylistic methods, De Nittis was able to work in a way that paralleled or extended the Impressionist attitude, while always remaining conscious of his Italian sensibility. Despite the fact that many, including Degas, willingly acknowledged his talent, De Nittis's penchant for self-promotion, and the concessions that he would subsequently make for continued success in a burgeoning art market, would eventually lead to sharp criticism and alienation from contemporary Parisian artists, compatriots, and critics alike at one time or another throughout his short career.

De Nittis reportedly used pastel as early as 1870, although the first pictures known to us were executed around 1878, at roughly the same time that Degas had turned almost exclusively to pastel. Whether or not De Nittis's use of pastel predates Degas's adoption of the medium is a source of great speculation. The reasons for De Nittis's initial attraction to pastel are as unclear as Degas's. However, it has been suggested that De Nittis was searching for a relatively unexploited alternative to oil paint for the realization of full-scale, finished works and was perhaps inspired by the revolutionary treatment of the print medium noted in Japoniste wood engravings.[2] We do know that De Nittis had expressed interest in taking up pastel at least as early as 1876, as Claretie reports in his volume on contemporary French artists published that year: De Nittis "dreams of a new art: pastel. To this exquisite art form of La Tour and Liotard, De Nittis wishes to bring his modern sensibility. He wants to make and exhibit life-size pastels, figures of women and studies of nudes of suppleness and great grace. Pastel appeals to him because he can work more rapidly than with oils, in soft tones of decreasing values."[3] Another critic also recalled meeting De Nittis at about the same time at a pastel store on the rue de La Rochefoucauld, where the artist stood in front of the rows and rows of colors, enthused by the prospect of the great works he would create.[4]

It has been popularly held that De Nittis's attraction to pastel at that time can be attributed to his growing acquaintance with Degas and the close friendship that was developing between the two artists around 1877. In May of that year, after the opening of the third Impressionist exhibition, in which Degas exhibited his pastel-covered prints for the first time, the French artist wrote to Léontine De Nittis, asking why De Nittis had not yet shown him his works, as he had told Degas that he too was making pastels. In response to the critic Ary Renan's later remark that De Nittis owed to Degas his introduction to pastel, Leontine retorted that, to the contrary, her husband had led Degas to the medium.

The widespread shift of interest to pastel at this time was accompanied by a more spontaneous, direct, and informal approach to the medium. While its restored popularity cannot be traced to any one specific reason, several factors may have contributed to its growing use by contemporary artists. Pastel certainly continued to be used for secondary works, and a survey of the objects submitted to the first Impressionist exhibition in 1874 reveals that several artists (Boudin, Degas, Monet, Morisot, and Renoir) showed studies and sketches in pastel. The following year Émile Gavet's collection of the pastels of Jean-François Millet (1814–1875) was featured in an exhibition and sale. Produced mainly after 1865, Millet's pastels were executed on tinted papers in a wide range of colors and subjects that suggested the vast potential that the medium held for contemporary artists. His works were certainly known by other artists and were in fact admired and collected by Degas.

Perhaps the shift of interest to pastel can also be interpreted as a response to the revised view of the relative stability of oil paint. As Douglas Druick and Peter Zegers have pointed out, the solid reputation of oil paint was brought into question in the mid-1870s, with the alarming realization that Edouard Manet's famous *Olympia* (1863; Musée d'Orsay, Paris) had significantly darkened, and that many of the artist's more recent pictures had already become "heavy, opaque and green."[5] Traditional claims to the longevity of the oil medium were challenged with scientific tests on the adverse effect of light on the pigments used by manufacturers and the role played by varnish in these changes. Less complicated in structure than oil paint, pastel could, it appeared, be used with greater certainty.

Also notable is the great amount of attention paid to De Nittis's pastels, particularly those depicting the local character of Parisian life, by the end of the 1870s. De Nittis's early works in pastels were simple landscapes and studies of Paris and London. The first public exhibition of his pastels probably took place in 1878, at the King Street Galleries in London, where he showed, along with oil paintings, a few pastels that illustrated how the chromatic scales of the medium made it particularly suitable for rendering atmospheric effects in the cityscapes of Paris and London.

De Nittis's pastels received a great deal of attention in Paris from the moment he began exhibiting them. On the eve of the opening of the Salon of 1880, he showed his monumental pastels for the first time at *L'Art*, the gallery located on the premises of the journal *La Vie Moderne*, along with a selection of works in other media. The recent establishment of smaller venues like *L'Art* gave artists opportunities to showcase their works in more intimate spaces outside of the constraints of the official Salons. Critics charged De Nittis with being provocative and reckless in opening his own exhibition on the eve of the Salon vernissage, but they nevertheless begrudgingly admitted that it was a triumph. De Nittis also earned the rather backhanded praise of a reviewer who remarked that his pastels surprisingly held their own with oil paintings and watercolors.[6] However, it was the life-size portraits of women – notable for their proficient execution, open, linear style, and not least for their scale – that caught the attention of the press and public. A case in point is *A Woman Seated in a Garden* (1880; private collection, Montecatini Terme), most likely included in the exhibition, which is over 1 m in height.

The following year De Nittis held a private exhibition of pastels at the Cercle de la Place Vendôme, again to much critical acclaim. These works, made up primarily of portraits of women, are again grand in scale, and are ambitious compositions, brilliantly colored and rigorously drawn in a broad manner that departs from more traditional pastel techniques. An 11-page review in the *Gazette des beaux-arts* heralded the exhibition as "amongst the great events of the year – and an important date that will be marked in the annals of art."[7] In the eyes of the critic Lostalot, De Nittis used pastel like the masters: "proceeding with great gashes of the crayon which chop up the canvas in every sense. The juxtaposed colors unite together in harmony to generate a local tone." More discreet praise came from Albert Wolff, the much-feared, much-read, but little-liked critic for *Le Figaro*: "I readily admit that never has pastel been handled with such grand virtuosity. M. De Nittis uses a strength of color and surprises of modeling that oil paint could not exceed."[8] Wolff also pointed out that these strong images happily depart from the artist's usual, more commercial efforts.

De Nittis died of a stroke in August 1884, at the age of 38. His last works were exhibited at the Salon that year. Among them were an oil painting entitled *Luncheon in the Garden* (1883; Civic Museum, Barletta, inv. 861) and the pastel *Autumn* (1883; private collection, Milan), almost 2 m high, both of which seem Impressionist in their *plein-air* theme and technique.

The more general revival of pastel during this period saw younger, lesser-known artists swept up into the action, with increased submissions of pastels to the Salon, and culminating in the foundation in 1885, less than a year after De Nittis's death, of the Société des pastellistes français. Seventeen of De Nittis's works were showcased in the society's inaugural exhibition, and a review in the *Gazette des beaux-arts* mourned De Nittis's death as a great loss to the art of pastel.[9] The vogue for pastel may also have been furthered, as Richard Kendall has pointed out, in publications such as the Goncourt brothers' *L'Art du XVIIIe siècle*, which positioned pastel as a time-honored vehicle that endeavored to prove itself as the vital medium of the new age.[10]

Comparison of De Nittis's pastels with early works by Degas in the same medium suggests interesting parallels as well as points of departure. The two artists undoubtedly saw and admired each other's work; they had occasionally etched together with Desboutin at Cadart's studio in the mid-1870s.[11] One can only conjecture to what degree they might have shared technical innovations. A systematic examination of both artists' pastels executed in the late 1870s and early 1880s reveals their technical origins to be more or less straightforward, though they departed from academic tradition in disparate ways. Although there are isolated instances where De Nittis used laid paper for smaller landscapes and portraits, he favored supports of lightweight, finely woven fabric commercially prepared with a smooth, light gray-beige ground layer, attached to wooden strainers. By contrast, paper of a variety of colors and types was the almost exclusive choice of support for Degas's pastels. The ghostly portrait of his cousin Mathilde (*Woman Seated on a Balcony*, 1872–73, Ordrupgaardsamlingen, Copenhagen, Lemoisne 318), executed on canvas, is a rare exception.

When starting a pastel, De Nittis preferred to block out the forms with colors specific to the intended composition, unlike Degas, who habitually started with a monochromatic base of charcoal or monotype. Scumbling the pastel across the distinct grain of the canvas created textures and highlights, which were either left intact or further refined with smudging to create softened tones. In *Woman in a Garden* (Fig. 1.20), one of the large pastels exhibited in 1881, the figure was initially blocked out in blue strokes and the landscape elements in green. The background was then broadly colored, up to and around the blocked-out sitter, before the figure was completed, with the greatest detail found in the face. The dark colors of the hat and hair were probably executed before the lighter skin tones, so that the falling pastel dust would not disfigure the face. A generous blank area to the left of the figure was quickly tinted with broad, white strokes for the diaphanous scarf wrapped around the neck and fluttering to her side. Final touches to the costume, flower garden, and foliage completed the work.

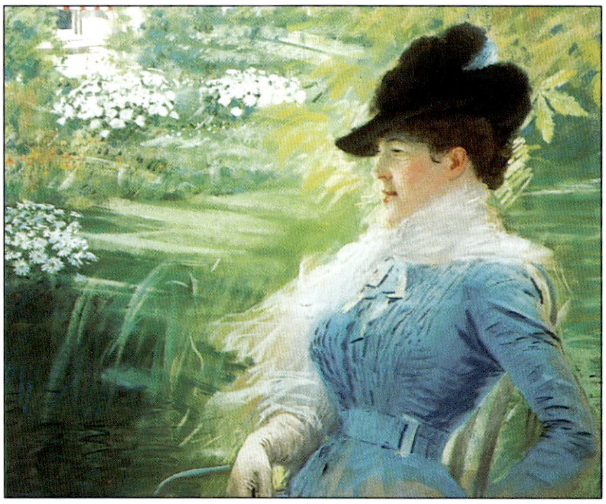

Figure 1.20 Giuseppe De Nittis, *Woman in a Garden*, 1881. Pastel on prepared fabric, 82 x 100 cm. Museo Civico, Barletta.

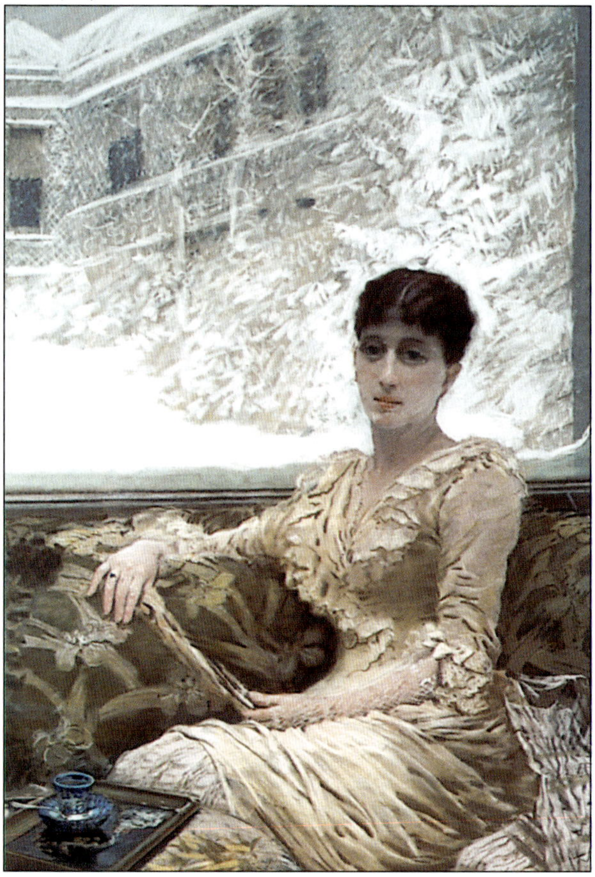

Figure 1.21 Giuseppe De Nittis, *Winter Day (Portrait of Léontine De Nittis)*, 1882. Pastel on prepared fabric, 150 x 89 cm. Museo Civico, Barletta.

Both De Nittis and Degas employed the traditional pastel technique of smudging, or "sweetening," the colors to create seamless passages of color in more highly finished works, like De Nittis's self-portrait of 1883 (Fig. 1.19) and Degas's portrait of his sister Thérèse Morbili (c. 1869; private collection, Lemoisne 255). The smooth, powdery pastel tones are blended to evoke the rich velvet upholstery of the settees in each composition. Both artists used the neutral texture of the exposed supports on which they worked as integral features in their pastels. The particular character of the colored support is easily observed through lightly smudged yet animated pastel strokes that allow the subtle texture of Degas's wove paper and De Nittis's primed canvas to remain visible.

Critics applauded De Nittis's rich portrayals of textures, fabrics, and finishes, which were justly considered unrivaled in their day; some even claimed that they were unrivaled by the confectionery works of Quentin de La Tour and others of the Golden Age. Edmond de Goncourt, in a journal entry of early 1882, said of the portrait of Léontine De Nittis (Fig. 1.21) that it "is the most extraordinary symphony of whites ... I have never seen such vaporous lights; and such a new way of handling the chalk, far removed from traditional techniques."[12] De Nittis's technique departs from the tighter, more traditional handling of the pastel medium to display looser, broader manipulation. He engineered texture, light, and volume with a diverse vocabulary of strokes. In *Bow Window* (Fig. 1.22), this heightened sensitivity and economy of line is achieved with great success: in the texture of the wicker chair, the minute

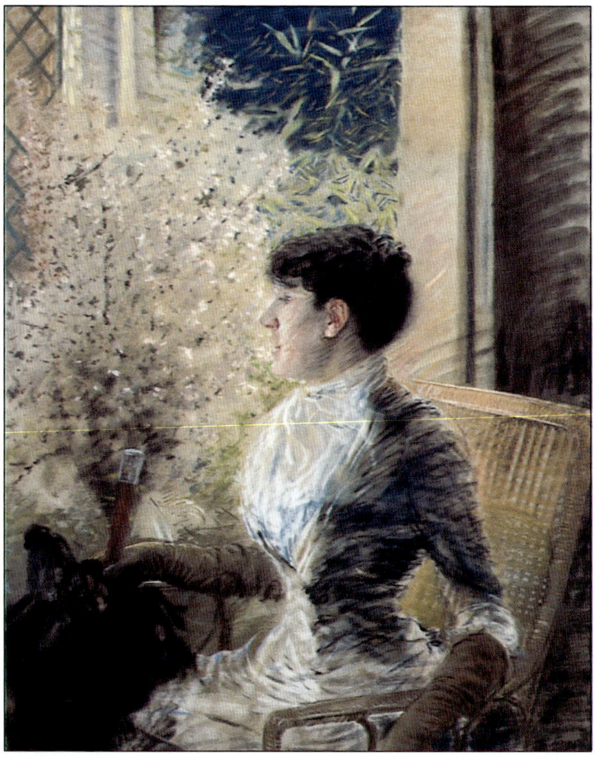

Figure 1.22 Giuseppe De Nittis, *Bow Window*, 1883. Pastel on prepared fabric, 117 x 90 cm. Museo d'Arte Moderna, Ca'Pesaro, Venice.

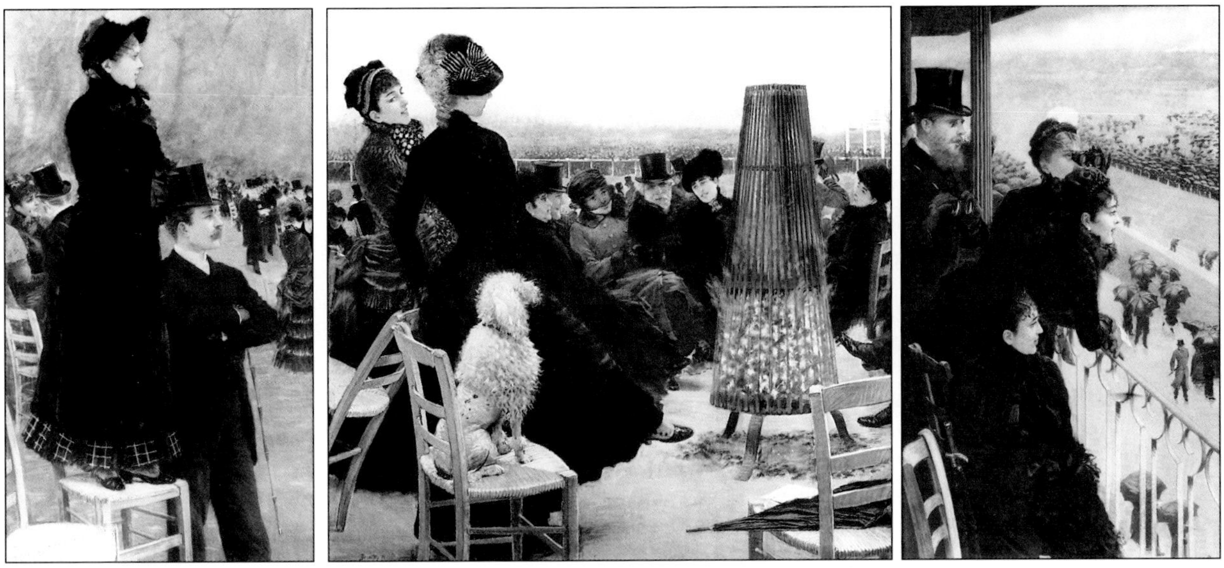

Figure 1.23 Giuseppe De Nittis, *Race Course at Auteuil*, 1881. Pastel on prepared fabric, 191 x 341 cm. Galleria Nazionale d'Arte Moderna, Rome.

petals of white and pink flora, and the translucent scarf at the sitter's neck.

Although De Nittis did not adulterate the pastel with the addition of fixative, there is evidence to suggest that he combined media and worked the pigment into a so-called pastel paste, like Degas did in later years. More often De Nittis sought to diversify the appearance of certain passages by altering the orientation of the crayon, using different facets to vary the strokes, or adjusting the amount of pressure to control the degree of texture picked up from the canvas support or roughness of the paper, sometimes employing "frottage" techniques to create new textures.[13]

Race Course at Auteuil (Fig. 1.23) is by far the most accomplished pastel in De Nittis's oeuvre. A triptych that measures 191 x 341 cm, the composition is ambitious: the central pastel depicts a fashionable group of spectators warming themselves by the glowing coals of a portable fire. It is flanked by two canvases of equal size, each measuring approximately 2 x 1 m. The left canvas shows a lady perched on a chair with a gentleman by her side; the right is a bird's-eye view from the stands of the wet course and the crowd huddled under their umbrellas, resembling, as one reviewer put it, "a bed of mushrooms."[14] Close examination reveals that De Nittis incorporated a thin, fluid paint to block out the figure of the standing woman in the left panel. A discrete drip of brown medium directly beneath her figure at the bottom of the canvas, partially obscured by pastel, betrays its presence. This preliminary paint layer made for a more expedient use of the overlying pastel, which entirely covers it in a lighter, thinner layer than would be necessary if the pigment was applied directly to the uncolored canvas. The light layer of pastel in the sky of each panel, applied in broad strokes, belies a similar treatment, as faint brush strokes are discernible beneath it. A thicker, matte paint is also found in the lower part of the central composition, in the figure of the dog, and along the wooden slats of the chairs. This medium could well be De Nittis's version of pastel paste, accounting for the slightly granular texture of the paint, which is composed of a pastel stick, crushed and ground with water and applied with a brush.

De Nittis's work played an important role in mediating the transition of taste for the public, and his association with Impressionism may have made aspects of that mode more palatable for many French critics and the public at large. The very great extent to which he was associated with the Impressionist movement in the public mind is clearly reflected in the article entitled "The Death of De Nittis and the End of Impressionism" written at the time of the artist's death by Émile Bergerat, director of the journal *La Vie Moderne.* Although his pastels recall the grand portraits of the Golden Age, they are executed in a monumental scale unrivaled in De Nittis's time and undoubtedly signal a movement away from the more traditional aspects of the medium's manipulation. However, the compromises that De Nittis routinely made in both the integrity of his oeuvre and in professional alignment, born of a commercial and personal need to secure his acceptance within his adopted homeland, have relegated him to the ranks of the forgotten second-class artists of late nineteenth-century Paris. Had he lived, De Nittis may have realized the full capacity of his talent as a gifted pastelist who might well have rivaled the innovative genius of Degas.

Acknowledgments

The author gratefully acknowledges the National Endowment for the Arts, the Samuel H. Kress Foundation, and the American Academy in Rome for supporting this project.

Endnotes

1. Bigot, "Le Salon de 1876," *La Revue Politique et Littéraire* (20 May 1876), p. 493, in Caroline Goldberg, "Giuseppe De Nittis: An Italian Perspective on the Parisian Cityscape, 1867–1884" (Ph.D. Diss., New York University, The Institute of Fine Arts, 1995), p. 68.

2. Goldberg 1995 (as in note 1), p. 99.

3. Jules Claretie, *Médaillons et profiles: J. De Nittis. "L'Art et les artistes français contemporains"* (Paris: Charpentier, 1876), p. 415.

4. Paul Mantz, "Les pastels de M.J. de Nittis," *Le Temps* (24 May 1881).

5. Bertall, "L'exposition de M. Manet," *Paris-Journal* (30 Apr. 1876), in Douglas Druick and Peter Zegers, "Scientific realism: 1873–1881," in *Degas* (Paris, New York, and Ottawa, 1988–89), pp. 201–02.

6. Ary Renan, "Joseph De Nittis," *Gazette des beaux-arts* 30 (1884), p. 406.

7. Alfred De Lostalot, "Les pastels de M. De Nittis au Cercle de L'Union Artistique," *Gazette des beaux-arts* (1 Aug. 1881), p. 156.

8. Albert Wolff, "Les pastels de M. De Nittis," *Le Figaro* (23 May 1881).

9. Baron Roger Portalis, "Exposition des pastellistes Français," *Gazette des beaux-arts* (May 1885), p. 438.

10. Richard Kendall, *Degas: Beyond Impressionism* (London: National Gallery Publications, 1996), pp. 94–95.

11. Michel Melot, *The Impressionist Print* (New Haven, Conn.: Yale University Press, 1996), p. 133.

12. Edmond and Jules de Goncourt, *Journal: mémoires de la vie littéraire* (Paris: Robert Lafont, 1889).

13. Close examination suggests that De Nittis used a frottage technique to create the textures in the lower half of the composition of the portrait of his wife (*Ritratto della Signora De Nittis*, pastel on prepared fabric, Museo Civico di Barletta, inv. no. 927).

14. Armand Silvestre, "Expositions des pastels de M. J. De Nittis au Cercle de L'Union Artistique," *L'Estafette* (27 May 1881), pp. 1–2.

James McNeill Whistler: The Color of Line

Margaret F. MacDonald

Works of art come accompanied by piles of documents from owners, dealers, scholars, curators, and conservators. Since each field uses its own language, it is difficult for the uninitiated to understand and assess all this information. Conservation reports are often too technical for the lay person. Records of provenance may be incomplete or deliberately misleading. However, a family "legend" may be the absolute truth, and a picture found in an attic may be a fine and authentic work. In his 1885 *Ten O'Clock* lecture, James McNeill Whistler described scholars:

> sombre of mien, and wise with the wisdom of books, who frequent museums and burrow in crypts; collecting – comparing – compiling – classifying – contradicting.
> Experts these – for whom a date is an accomplishment – a hall-mark, success!
> Careful in scrutiny are they and conscientious of judgement – establishing, with due weight, unimportant reputations – discovering the picture, by the stain on the back [...] speculating, in much writing, upon the great worth of bad work.[1]

Wincing slightly under this century-old onslaught on critics and scholars, I admit that I study the "good, the bad, and the ugly." As author and co-author of two *catalogues raisonnés* on Whistler, I am sometimes offered a "fistful of dollars" to authenticate his work. I know that many people do not believe that anyone can know enough about an artist to be sure of the "authenticity" of a work. Sometimes I can prove beyond reasonable doubt that a particular work is or is not by Whistler, but, more usually, I can only give my considered opinion.

Some art historians query the value of knowledge and connoisseurship, and dispute the concept of "quality," which is sometimes, but by no means always, valuable in distinguishing the work of one artist from another. The works of Whistler and his wife, Beatrice, are a case in point. Drawings that came from Whistler's estate through his sister-in-law to the Hunterian Art Gallery in the University of Glasgow have what might be seen as an impeccable provenance. However, in the light of the knowledge gained from studying the 6,000 letters and manuscripts that accompanied them, it has become clear that certain works assumed on the basis of provenance and "quality" to be by Whistler are by others.

Indeed, when a number of sketches were catalogued in Glasgow long ago, those that are not very good were assumed to be by Beatrice, and those of higher "quality" were assumed to be by James. A caricature of Oscar Wilde (Fig. 1.24), for instance, has often been published as a Whistler despite proof that it was done by Beatrice.[2] When I wrote my *catalogue raisonné* of Whistler's work on paper, I included some 20 of these drawings, and I said that I simply did not know which of them had drawn what.[3] Three years later, after studying Beatrice's largely obscured history and locating more works, she has emerged as a considerable artist and designer with a

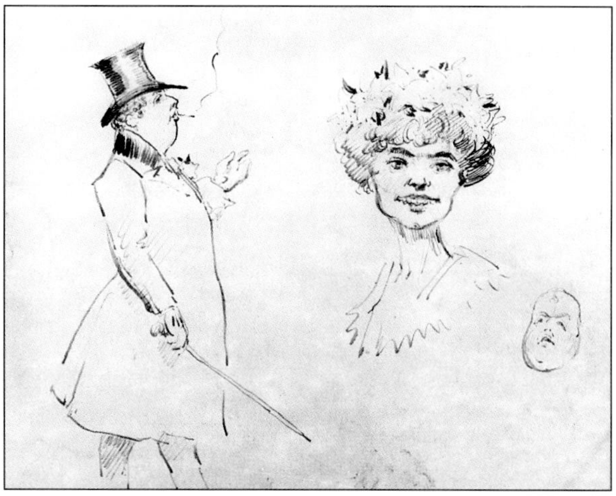

Figure 1.24 Beatrice Whistler, née Philip (British, 1857–1896), *Caricature of Oscar Wilde*, 1882–84. Black ink on paper, 360 x 229 mm. Hunterian Art Gallery, University of Glasgow, GLAHA 46408.

distinctive style. She drew with long, strong, curving, emphatic lines, where Whistler used finer, more wiry, broken lines. All but one of these previously unattributed drawings have therefore been reassigned to her.

A combination of knowledge, judgment of quality, technique, and style is necessary to "authenticate" works. The conservator who restores and conserves the same works must combine this connoisseurship with the scientific and technical skills of the profession. Technical skill alone is not sufficient. A sad case is Whistler's *Study for the Head of Miss Cicely H. Alexander*, which was completed and signed some time later by the artist with his butterfly monogram.[4] Unfortunately a conservator, seeing that the butterfly was a later addition but not recognizing it as that of the artist, cleaned it off!

When I first saw *Harmony in Blue and Gold: The Peacock Room*, it was dark and strange, and I simply did not understand what was going on, since it did not fit with any of the other decorative schemes I had seen. It certainly did not have the passionate vigor and springing lines of the large cartoon, which is in Glasgow. What I did not realize was that generations of over-enthusiastic "restorers" had repainted the room in accord with their own notion of "its former glory." It is only now, after meticulous conservation by a team including Wendy Samet and Joyce Hill Stoner, that Whistler's delicate and expressive brushwork has been revealed.[5]

Technical skill, however, would have prevented some of the grosser mistakes of misdirected conservation. A late pastel by Whistler, *The Captive*, was exhibited in 1898 at the International Society of Painters, Sculptors and Gravers, in London.[6] I sent a photograph of the original catalogue illustration to the current owners, Indiana University Art Museum, and asked if it was the same work. They said it was, and I borrowed it for an exhibition at Knoedler's in New York in 1984. On arrival, the horrible truth was revealed: that it had been heavily restored, each line redrawn firmly, obscuring the lines of the original (Fig. 1.25).[7] This is the worst case I

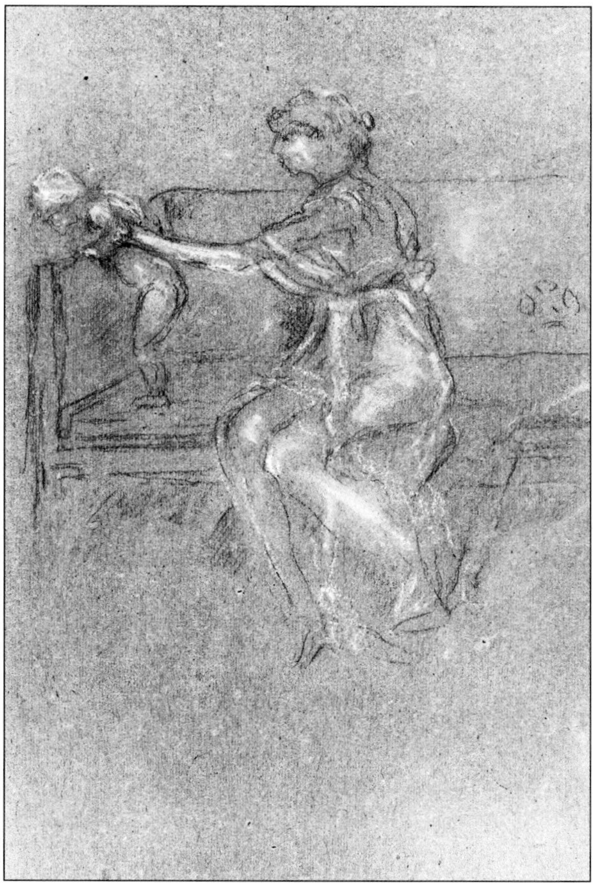

Figure 1.25 James McNeill Whistler (American, 1834–1903), *The Captive*, c. 1898. Pastel and pencil on paper laid down on board, 276 x 177 mm. Indiana University Art Museum, 74.60 (M 1524). The drawing has been extensively reworked by another hand.

have seen, and though I suspect other drawings of having been "strengthened," none have been so completely ruined.

Some works by other artists suffer a worse fate; being altered to make them look more like Whistler's work. This may be limited to adding Whistler's distinctive butterfly monogram to a perfectly good work by someone else. One artist who suffers this fate is Jerome Elwell, who knew Whistler in Venice and was greatly influenced by his work. His monogram, JE and an arrow, is often interpreted as a butterfly and is sometimes altered to be more like one. A drawing that I rejected as a Whistler because of this came back some years later, with an extra butterfly. It appears that Whistler is alive and well, and in this case, living in Florida.[8]

Of course, neither authenticity nor quality depend entirely on signature or condition. For me, it rests more on the shape and color of the line, whether in pen, pencil, chalk, or brush. Whistler's handwriting, signature, and brushwork are equally recognizable.[9] If a brush stroke on a painting known to date from the 1890s is flashy and confident, then it is probably not by Whistler. This was essential knowledge for conservators cleaning the *Portrait of George W. Vanderbilt* in the

National Gallery in Washington[10] – for Whistler was aiming at the most delicate and evanescent of effects. The result was a painting that was never finished. In it Whistler attempted to catch subtle qualities of light and movement, and to reveal the sitter's inner self with delicately probing, thin, soft, and fluttery brush strokes. An unfinished self-portrait, in the University of Glasgow, shows the development of similar brushwork during the painting of his own image.

In drawing, Whistler's line is rather more decisive and always expressive. One drawing of a young girl, known only from a photograph in the Library of Congress, shows a certain uncustomary confusion.[11] "Quality" *does* come into play here, for this drawing is weak and unconvincing in its depiction of space and its lines appear to be limp and meaningless, suggesting that it is not by Whistler. Indeed it is more likely a contemporary fake, possibly by Rosa Corder, an artist who is said to have been the mistress of Whistler's friend C. A. Howell. The fake is based on a sketch by Whistler that Howell owned.[12] Howell was known for removing prints and drawings from the artist's studio, and Corder was his collaborator in producing works that bore the signatures of Fuseli and Rossetti as well as Whistler. Even the slightest of Whistler's drawings have a particular distinctive use of line – and of color in line – that is characteristic to him alone, and this is exactly what this fake lacks. In another example, the splendid pastel *A Nude Figure* in The Art Institute of Chicago, may be compared with a weakly executed and confused pastel, *Reclining Nude*, now in the Fogg Art Museum. The latter was previously attributed to Whistler and is now unattributed.[13] Certainly Whistler was a fine draftsman, and his drawings, watercolors, and pastels show the same technical skill that is seen in his more famous prints and paintings.

Whistler used black and white chalk for his portrait studies on brown paper. The white chalk was more vulnerable to loss, abrasion, and damage than the black, and sometimes it is only under close examination that the highlights and texture provided by the white chalk are still visible. A self-portrait in The Art Institute of Chicago shows this problem: it is a strong drawing, but not as subtle as it originally must have been when the delicate chalk heightening was intact.[14] The original balance of the picture has been irretrievably lost. Although it lurked for years in a box of fakes and forgeries in the Art Institute, in fact it is a fine and interesting drawing by Whistler himself. On the other hand, a drawing depicting Whistler, now in the Chelsea Arts Club (of which Whistler was a member) and published as being by Whistler, is alas not one of his self-portraits. The meaningless lines of the figure negate a certain power in the drawing of the head, and the smudgy overworking of the pastel is quite unlike Whistler's work. I think it may be by his friend and pupil Mortimer Menpes, for it has much in common with other drawings and etchings of Whistler by Menpes.[15]

The brown paper favored by Whistler provided surface texture and color that the artist fully exploited. When Whistler was staying with his patron F. R. Leyland in Liverpool, he asked the French printmaker Auguste Delâtre for more of some brown paper used to wrap up a parcel he had been sent: "C'est juste le papier qu'il me faut pour mes dessins. J'en cherche toujours – mais quoique tous les marchands aient des papier brun, il est rare d'en trouver avec ce beau grain."[16] Through the 1870s Whistler used this sort of rough paper for his working drawings. Two chalk drawings on the recto and verso of the same sheet are typical: *Study for "Arrangement in Black, No. 2: Portrait of Mrs. Louis Huth"* (Fig. 1.26) on one side and *Study for "Symphony in Flesh Color and Pink: Portrait of Mrs. Frances Leyland"* on the reverse.[17] They suggest, as you might expect, variations on the pose and design of the dresses, the color schemes, and harmonies. In fact, the decorative background originally sketched for the oil portrait of Mrs. Huth was transferred in oil to that of Mrs. Leyland.[18]

The line in the drawn studies is quite different from the effect produced in oil. The paintings *Symphony in Flesh Color and Pink: Portrait of Mrs. Frances Leyland* and

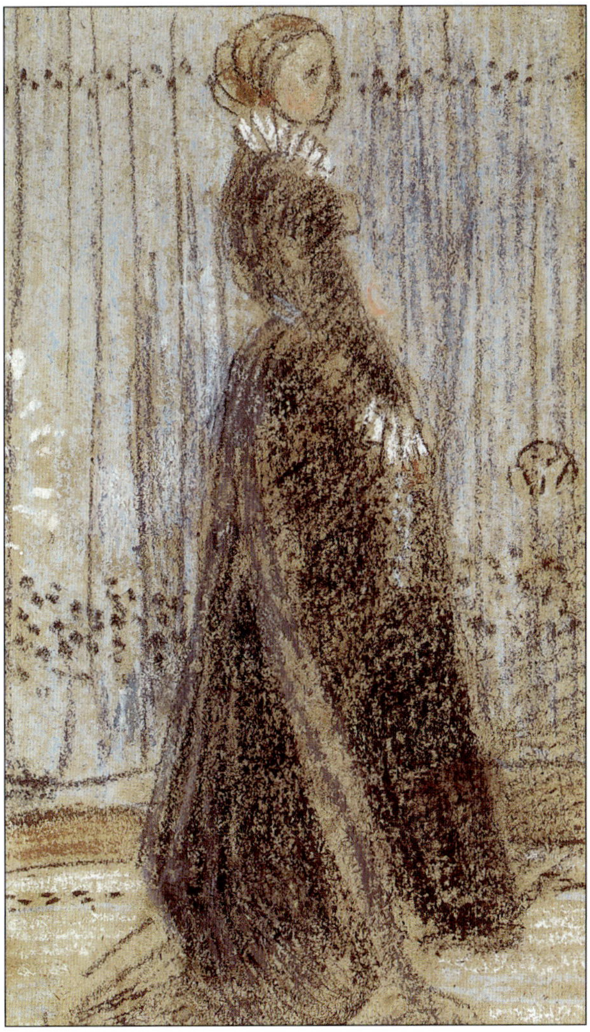

Figure 1.26 James McNeill Whistler, *Study for "Arrangement in Black, No. 2: Portrait of Mrs. Louis Huth,"* c. 1872. Pastel on brown paper, 228 x 122 mm. The Art Institute of Chicago, 1933.211R (M 454).

Arrangement in Black, No. 2: Portrait of Mrs. Louis Huth have soft outlines produced by painting into a wet surface, and reworking and blurring the edges of the brush strokes.[19] The blurring of outlines is less common in Whistler's drawing, but it is apparent at times in both pastels and in prints. The blurred yet hard line of fresh drypoint was often used to soften dramatic outlines, as in Whistler's splendid drypoint portrait of Mrs. Leyland, *The Velvet Dress* (K 105).[20]

The use of a black outline and broken white highlights to suggest a lustrous material was a traditional academic method of drawing, but it was not actually much help in planning a painting. A pastel drawing by its very scale requires a very different approach from a full-length, life-size canvas. There is little suggestion in these pastels that Whistler was using them to anticipate and solve problems on canvas. Brown paper was the place for resolving problems in pastel. As they progress, you can see Whistler evolving a drawing technique to convey movement and color in line rather than paint. A drawing of *Mrs. Leyland Seated* shows the assured quality of his draftsmanship.[21] A minimal amount of highlighting was used on a very fibrous paper to suggest color and texture as well as light. The beauty of the drawing, however, lies in the black-chalk lines, describing with absolute confidence the fine profile, the textures of dress and hat, and the melancholy, pensive pose. This was a drawing finished and signed as a complete work of art in its own right, and possibly exhibited at Whistler's first one-man exhibition in 1874.

By the time Whistler went to Venice in 1879, he had considerable expertise in using chalks. Although he went with a commission from the Fine Art Society to produce a set of etchings, he also intended to draw a series of pastels. From contemporary accounts, as well as from a detailed study of the pastels that have survived (some 80, drawn over 14 months), it is possible to see the development of Whistler's approach to a new subject. He drew famous sites like the Riva degli Schiavoni, as well as the back streets and canals. He used a variety of paper of various colors, from gray-blue to dark brown. *The Beggars – Winter* and *Winter Evening*, for instance, are on a good-quality brown paper of a reddish-purple shade, with a very neat, fine vertical grain, containing small fibers and occasional flecks of wood-stalk.[22] The grain is most visible on one side and is instantly recognizable. However, Whistler sometimes used the less grainy, more irregularly textured side, and this is less distinctive. The paper is of uniform quality, although by now, after a century of wear, individual sheets have darkened or lightened a little (*The Beggars – Winter* has darkened to reddish-brown at the edges). At least 30 Venetian pastels are on this paper, and it was used only in Venice.

Whistler often tried drawing both the outlines and the general background with pastels, but the softness of the pastels applied to the textured paper led to some confusion and lack of definition in these works. He obviously rejoiced in the color of the city and felt a need of color to depict it. He wrote home to his mother:

> After the wet, the colors upon the walls and their reflections in the canals are more gorgeous than ever ... and with sun shining up on the polished marble mingled with rich toned bricks and plaster, this amazing city of palaces becomes really a fairyland, created one would think especially for the painter.[23]

Eventually he evolved a more coherent style, first using broken, wide-spaced lines of hard black chalk to outline the bare essentials of the scene. Then he added color sparingly within these lines. He used a variety of techniques – rubbing, flecking, scumbling, and both thin and broad soft strokes – to vary the effect of color. The colors he had available (over 70), and the shades that he selected, set off by the dark paper, varied over this period.

The artist worked almost entirely on site to produce a most remarkable picture of the city. In one of his finest pastels, *The Zattere: Harmony in Blue and Brown* (Fig. 1.27), alternate dabs of pale pink and orange on the figures and ship are balanced by turquoise and a deeper blue.[24] Touches of white and bone-white, khaki and black, alternate with the more vivid colors. Hazy scumbles of turquoise on sky and sea, and bone-white on the quay, contrast with the firmly applied details. Color

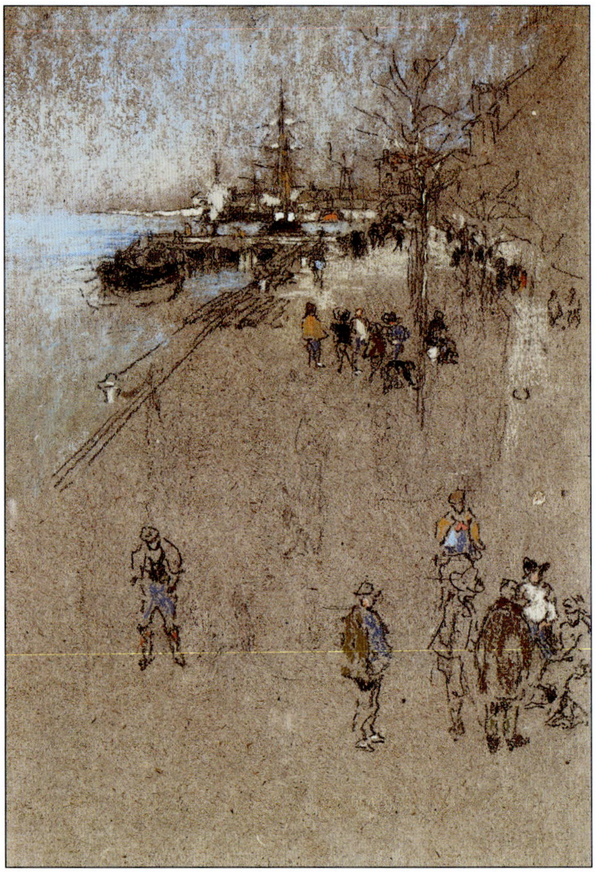

Figure 1.27 James McNeill Whistler, *The Zattere: Harmony in Blue and Brown*, 1880. Pastel and chalk on paper, 300 x 203 mm. Terra Foundation for the Arts, Daniel J. Terra Collection, 1992.162 (M 774); photograph courtesy of Terra Foundation for the Arts, Chicago.

is used decoratively to enliven the surface and as a means of focussing attention on the distant shipping.

When Whistler turned to etching, he evolved new techniques. He started by varying the depth of biting of his etched lines, and added to the variety of effect by combining this with the soft lines of drypoint. He printed on a variety of papers, enjoying the luster of Japanese papers or the patina of old papers with Venetian or Dutch watermarks and noticeable laid lines.[25] He printed the plates with a thin surface tone of ink to produce what are in effect monoprints. The color of ink varies from golden brown through sienna to black, and the ink was always selectively wiped from the plates so that thin films would remain. The conservator is warned that when attempting to wash or bleach such proofs to whiten the paper, much of the subtle tonal quality Whistler sought by varying the plate tone may be lost.

The proofs of one of Whistler's most famous etchings, *Nocturne* (K 184), were printed with the plate cleanly wiped, in black ink (Fig. 1.28). Later prints (most, but not all, printed when he returned to London) were printed with surface tone, which added a luminosity to the surface. The brown ink provides atmosphere and color, giving some impressions of the warmth of sunshine. By the careful manipulation of the ink, a lagoon could look smooth and oily, the sky could be broken by the rays of the setting sun, or San Giorgio could merge into the darkness of a moonlit night.

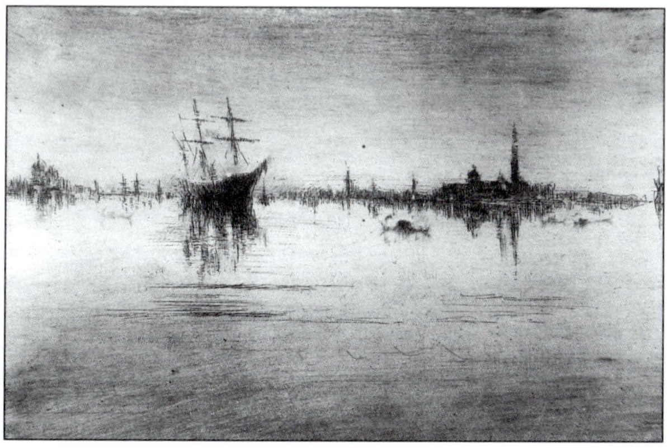

Figure 1.28 James McNeill Whistler, *Nocturne*, 1880. Etching and drypoint on paper, 198 x 294 mm. Hunterian Art Gallery, University of Glasgow (K 184 IV).

Later, in his etchings of Holland, such as *The Square House* (K 404), Whistler gradually steered away from the use of plate tone, although he developed a method of rubbing and softening the line to gain more delicate effects.[26] This, combined with the ephemeral softness of the drypoint line, created a unified surface, with an understated atmospheric harmony. In later pastels he turned from city scenes to concentrate on nude studies, using the skills gained in Venice. Again, the outlines are usually in black chalk, and later in charcoal, as Whistler's appreciation for the nature of the silvery line it produced developed. Moving freely from one medium to another to explore certain themes and subjects, he was inspired by each of them in turn. He experimented, extending the bounds of each medium, and crossed the traditional boundaries that separated, for instance, chalk "studies" from "finished" oil paintings.

In figure subjects such as *The Little Blue Cap* (Fig. 1.29) he used color sparingly in a highly developed combination of line, scumbling, and rubbing.[27] The line itself is neither continuous nor sinuous: it undulates and is broken to create not just an outline but an impression of light and movement, of the space around the figure, of a figure blending with its environment. These effects were closely allied to the artist's experiments in color lithography. In a lithograph such as *Yellow House, Lannion* (Fig. 1.30) of 1893, Whistler attempted to re-create in lithography what he had achieved in pastel, depicting the surface color, pattern, and textures of ancient masonry with colored lines of distinctive shape.[28] He used the grain of lithographic transfer paper as he had used brown paper, to create additional texture, independent of the varying pressure of the crayon. Working with the printer Henri Belfond in Paris, he experimented with various color combinations to produce more muted and harmonious effects. The collapse of his partnership with Belfond, as well as the low success rate for the color lithographs, brought to a halt his experiments with colored lithographic lines.

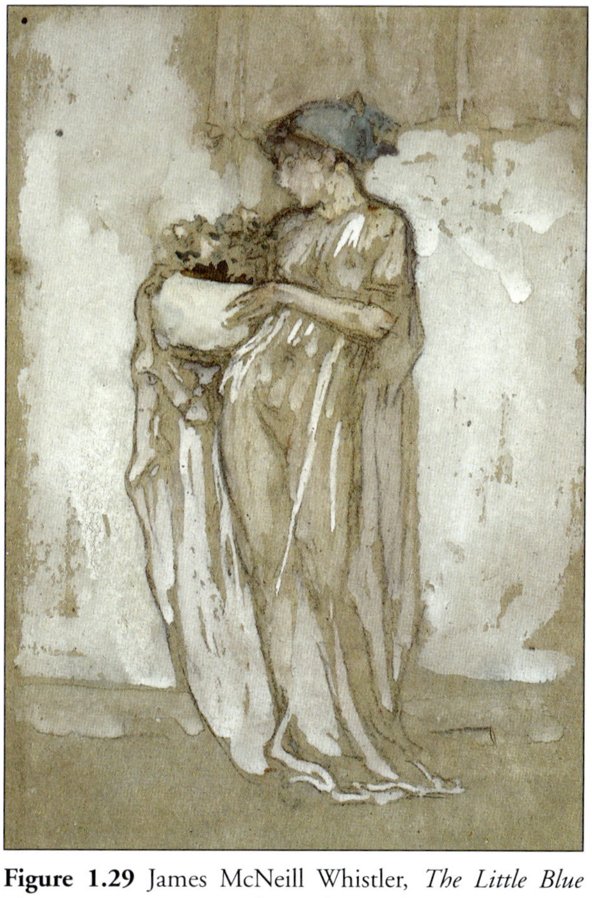

Figure 1.29 James McNeill Whistler, *The Little Blue Cap*, 1893–95. Watercolor and gouache, over charcoal, on brown paper laid down on board, 278 x 182 mm. Promised gift of Dorothy Braude Edinburg to The Art Institute of Chicago, 61.1997 (M 1393).

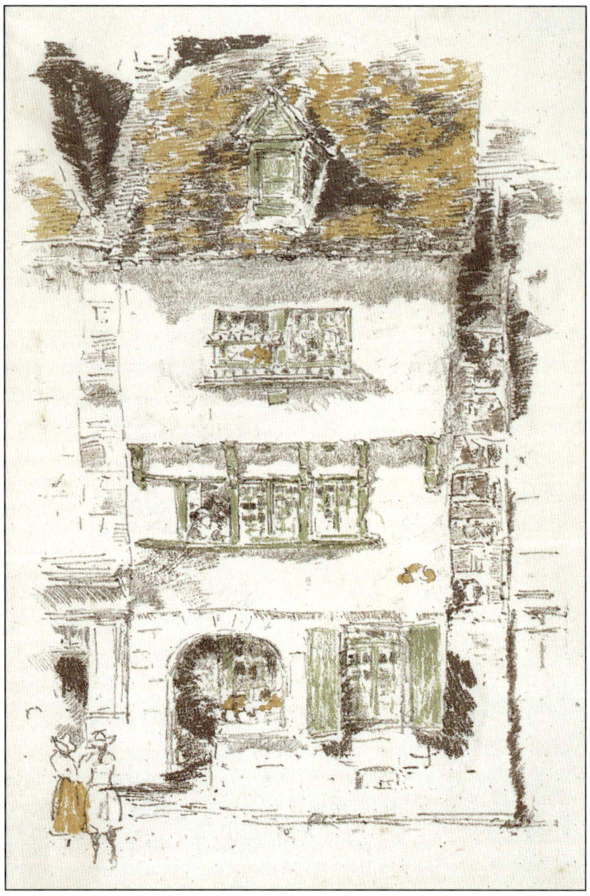

Figure 1.30 James McNeill Whistler, *Yellow House, Lannion*, 1893. Transfer lithograph printed in five colors on paper, 319 x 204 mm. The Art Institute of Chicago, 1927.5849 (C 67).

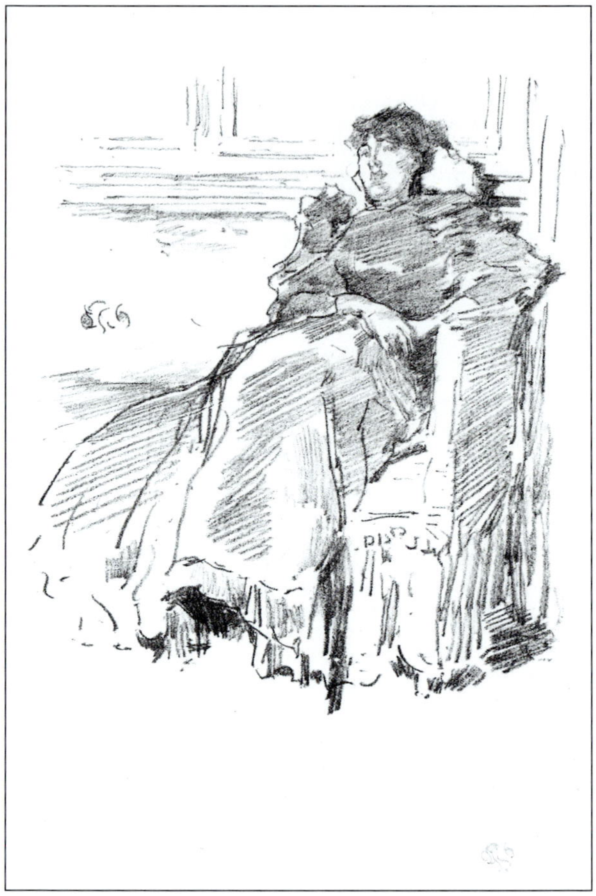

Figure 1.31 James McNeill Whistler, *La Robe rouge*, 1894. Transfer lithograph on paper, 324 x 208 mm. The Art Institute of Chicago, 1927.5821 (C 107).

Instead, he aimed to produce an "impression" of color with printed line alone. He drew a lithograph of his wife, Beatrice, in September 1894 (Fig. 1.31) and wrote to his printer, T. R. Way, "we will call No 1 'La Robe rouge' *the Red dress* – I rather want to see how the color is conveyed! – 'you' will think me quite crazy! – "[29] Areas of freely drawn, closely packed diagonal lines cross the dress within bouncing, wiry, broken outlines. The line is not sharp but softened by the grain of the paper, and this richness of texture conveys a sense of color as well as mood. He felt no need of actual color to illuminate the line. Another lithograph of his wife, *La Belle Dame Paresseuse*, Whistler considered to be "fair and silvery, and beautiful in the quality of blacks! Indeed far and away more like the charcoal drawing itself of the painter – like the burr in a dry point!"[30]

His late drawings – dozens of pencil and pen drawings – such as *An Ajaccio Fisherman* and *At Sea*, in sketchbooks, are extraordinarily vivid.[31] Wiry, broken, hooked lines are combined with short, sharply angular cross-hatching to suggest individual people, places, light, and color. Drawn for himself, they are intensely personal, with a feeling of the excitement of personal discovery on each page. When a critic said to Whistler that his work was "all a matter of taste," Whistler replied, "No, it is all a matter of knowledge!" He was always a seeker after knowledge. A close study of his use of line in drawings and prints shows a serious artist and craftsman continually searching for new ways to resolve old and new problems. The works themselves are wonderful, repaying close examination and revealing to art historian and conservator and connoisseur alike the authentic quality and color of Whistler's expressive line.

Endnotes

1. James McNeill Whistler, "Ten O'Clock Lecture" (1885), reprinted in James McNeill Whistler, *The Gentle Art of Making Enemies* (London, 1890).

2. M. F. MacDonald, *Beatrice Whistler, Artist and Designer* (Glasgow, 1997).

3. M. F. MacDonald, *James McNeill Whistler, Drawings, Pastels and Watercolors* (New Haven, Conn., and London, 1995) (hereafter referred to as M).

4. Private collection: A. McL. Young, M. F. MacDonald, R. Spencer, and H. Miles, 1980, *The Paintings of James McNeill Whistler* (New Haven and London, 1980), no. 128 (hereafter referred to as YMSM).

5. W. Samet, J. H. Stoner, and R. Wolbers, "Approaching the cleaning of Whistler's Peacock Room: Retrieving surface interrelationships in Harmony in Blue and Gold," *Cleaning, Retouching and Coatings, Technology and Practice for Easel Paintings and Polychrome Sculpture* (Brussels Congress, International Institute for Conservation, London, 1990).

6. Indiana University Art Museum, Bloomington (M 1524), reproduced in original and current condition. International Society of Sculptors, Painters and Gravers, *Third Exhibition* (London, 1901), no. 39.

7. M. F. MacDonald, *Notes, Harmonies, Nocturnes,* M. Knoedler and Co., New York, 30 November–27 December 1984, cat. no. 43.

8. Examples of Elwell's work are: *James A. McNeill Whistler [sic], Etude de Venice,* pastel, in Bernard Danenberg Galleries, Inc., *Recent Acquisitions. Important American Paintings,* New York, Winter 1969, cat. no. 32; and *J. A. M. Whistler [sic], Venetian Scene,* in *American Artists Abroad,* Nassau County Museum of Fine Art, 2 June–2 September 1985, pl. 28. I am grateful to Nesta Spink and Ronald Anderson for sharing information on the pastel with forged signatures, which was accompanied by a forged letter of authentication said to be from myself! Both forgeries have not been seen since 1994.

9. An example of 1880s brushwork is seen in *Dordrecht* (M 876); and handwriting on *Trellis* (M 848), both in the Hunterian Art Gallery, University of Glasgow.

10. National Gallery of Art, Washington, D.C. (YMSM 481).

11. Untitled pen drawing, present whereabouts unknown, photograph in Library of Congress, Washington, D.C., Prints and Drawings Department, Pennell-Whistler Collection (see M 717).

12. *Study for "The Blue Girl: Portrait of Miss Elinor Leyland"* (M 717). Rosa Corder (1853–1894) painted copies, portraits, and horses.

13. *A Nude Figure,* chalk and pastel, The Art Institute of Chicago (M 687); *Reclining Nude,* chalk, Winthrop collection, Fogg Art Museum, Harvard University.

14. M 422.

15. T. Cross, *Artists and Bohemians, 100 Years with the Chelsea Arts Club* (London, 1992), frontispiece.

16. Letter from Whistler to A. Delâtre, n.d. [1870/74], Institut Néerlandais, Paris, Delâtre 1972 A779.

17. The Art Institute of Chicago (M 454).

18. Branka Nakanashi, "A symphony re-examined: An unpublished study for Whistler's Portrait of Mrs. Frances Leyland," *The Art Institute of Chicago Museum Studies* 18, 2 (1992), pp. 156–67, 189–91.

19. Frick Collection, New York (YMSM 106); estate of Viscount Cowdray (YMSM 125).

20. E. G. Kennedy, *The Etched Work of Whistler, Illustrated by Reproduction in Collotype of the Different States of the Plates* (New York, 1910) (hereafter referred to as K).

21. Hunterian Art Gallery, University of Glasgow (M 549).

22. Pastels in Freer Gallery of Art, Washington, D.C. (M 727 and M 755, and discussion of paper in M 727).

23. Letter from Whistler to A. M. Whistler, n.d. [Feb./Apr. 1880], Glasgow University Library, Whistler W559; R. H. Getscher, "Whistler and Venice" (Ph.D. diss., Case Western Reserve University, 1970), pp. 117–18; M. F. MacDonald, *Palaces in the Night: Whistler in Venice* (New York and University of California, 2001), pp. 147–48.

24. Terra Foundation for the Arts, Daniel J. Terra Collection, Chicago (M 774).

25. Victoria Bunting, studies carried out in the Freer Gallery of Art, Washington, D.C., conservation department. I am very grateful to Martha Smith for giving me access to her records.

26. J. Heijbroek and M. F. MacDonald, *Whistler and Holland* (Amsterdam, 1997), p. 98.

27. Hunterian Art Gallery (M 1460); The Art Institute of Chicago, gift of Dorothy B. Edinburg (M 1393).

28. *The Lithographs of James McNeill Whistler,* Vol. 1., A Catalogue Raisonné, ed. by H. Stratis and M. Tedeschi (Chicago, 1998), no. 67; hereafter referred to as C.

29. Letter from Whistler to T. R. Way, 23 September 1894, Glasgow University Library, Whistler W1411; see C 123, letter no. 121.

30. Letter from Whistler to Way [14 September 1894], Library of Congress; see C 119, letter no. 118.

31. Private collection (M 1692); National Gallery of Art, Washington D.C. (M 1642).

Blues and Browns and Drabs: The Evolution of Colored Papers

Peter Bower

The color, tone, and texture of any sheet of paper used in a work of art is crucial to the understanding and appreciation of both the work itself and the artist's intentions regarding that work. This paper explores the origin, nature, and importance of the actual color of the paper sheet in the Western papermaking tradition with regard to particular works of art, at the moment of their making and subsequently.[1] It is concerned primarily with colored papers used for drawing in various media, including pastel and chalk, although some reference is made to colored papers used for watercolor.

Any work of art on paper, whatever the media used – graphite, chalk, pastel, ink, watercolor, or gouache – is in a very real sense an intimate object, meant to be seen close to. The actual interplay of the surface and the marks made on it is an integral, often crucial, part of the work's visual effect, although the importance of this does, to some extent, depend upon the concerns and intentions of the individual artist. Subtle qualities of texture, tone, and the actual three-dimensional nature of both the surface and any marks made on it are often ignored when examining a picture, although in a purely physical sense they are the work. Artists choose to work on particular papers for a number of complex reasons: the color of the sheet; the warmth or brightness of tone; the appropriateness of the texture of the finish; the surface strength; the degree of sizing; the dimensions of the sheet; the suitability of the weight and density of the paper for the work in hand. For many artists the paper is not merely a neutral ground carrying the image, but an integral part of that image.

Historically, colored papers have been produced by a variety of means, each of which produce distinctly different results. These include *self-colored* papers, in which the raw materials produce the color in the sheet (e.g. blue rags for blue papers, hemp ropes for brown, etc.) (Fig. 1.32); *dyed in the beater* papers, in which dyes or pigments are added to the rag as it is beaten (Fig. 1.33); *dyed in the vat* papers, in which dyes or pigments are added to the pulp directly in the vat or in a stuff chest; *dyed after making* papers, in which already made white sheets are soaked in color, either by the papermaker or more commonly by colormen or merchants; *coloring by inclusion* papers, in which colored fibers such as wool or silk are added to the base pulp; and *prepared ground* papers, in which the sheet is washed with color, either by a colorman or by the individual artist. Each of these methods produce very different results in terms of the paper's appearance and durability. Papermakers have further complicated the issue by combining two or more of the above methods in the same sheet (Figs. 1.34 and 1.35).

Besides the obvious factors such as the actual color of the raw materials, several other factors of more or less importance can influence the final color of the sheet. These include: the water used in the making; the preparation of the raw materials (i.e. fermentation, cooking, etc.); the method and degree of beating; the method of formation; drying (i.e. airflow and gravity, use

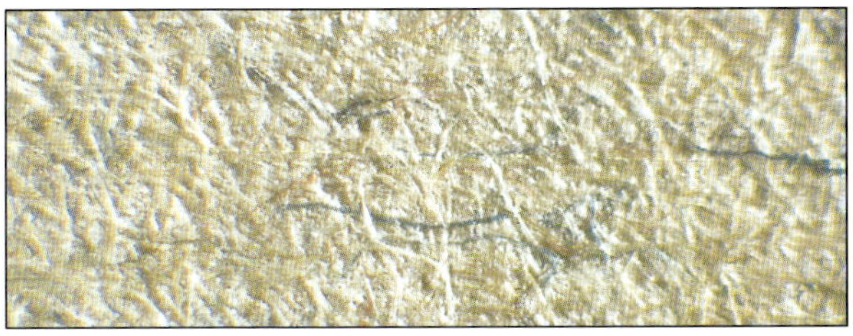

Figure 1.32 A self-colored brown paper, made as a wrapping paper but often used by artists, seen at 30x magnification. The raw materials have produced the color in the sheet: low-grade white linen rag, hemp fiber derived from old rope, and a very small proportion of blue and red rag fiber.

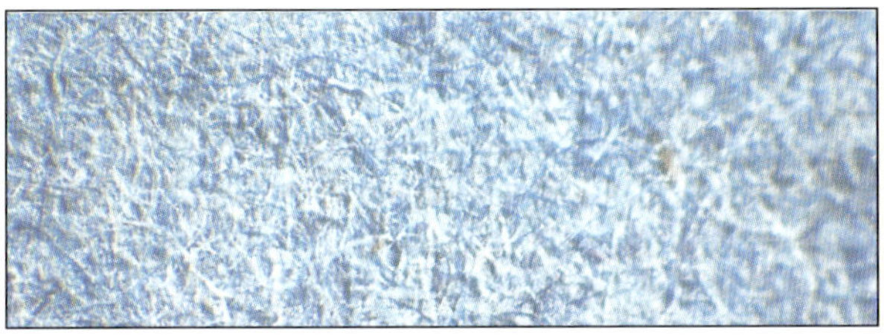

Figure 1.33 A blue paper dyed in the beater where the dye has been added to the rag as it was beaten, seen at 25x magnification.

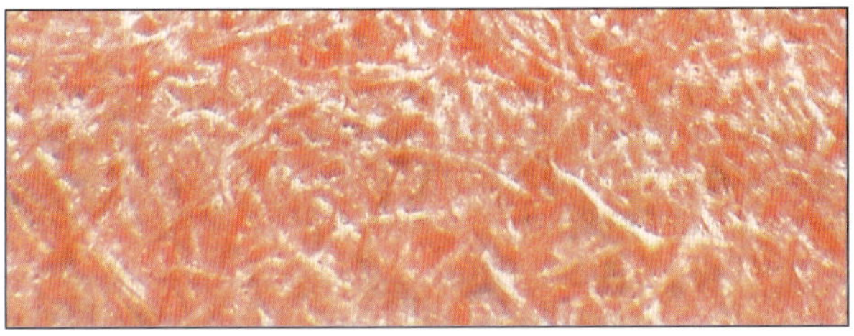

Figure 1.34 The surface of a red paper, seen at 30x magnification. This sheet is a blend of two pulps, one white and the other dyed in the beater. The red dye has bled into the white fiber as well, creating a rich flecked effect.

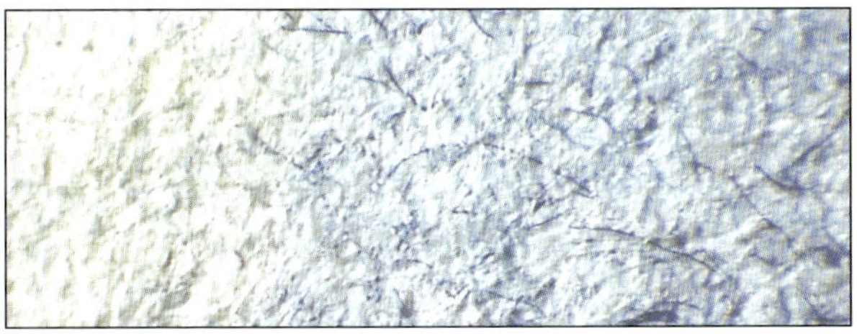

Figure 1.35 The surface of a flecked blue watercolor paper containing blue rags, blue dyed pulp (both dyed with indigo), white rags, and a very small proportion of hemp fiber, seen at 15x magnification. This shows the fugitive nature of indigo, and how the indigo-dyed rag has retained its color better than the dyed pulp.

of heat, etc.); bleaching (whether by sun or by chemicals); sizing; loadings; finishing techniques; and the effect of residues remaining in the sheet.

Other important considerations are the effects of light on the perception of color, in particular the shadows within the surface texture of the sheet that affect the light reflected back to the viewer; the influence of wrapping papers on the design of colored papers used by artists for drawing, watercolor, pastel, and crayon; and the similarities and differences in the manufacture of colored and white paper, particularly in terms of quality control in the mills. It should also be remembered that the pulp is always darker and richer in color than the newly formed wet sheet on the mold, which in turn is darker and richer in color than the dry finished sheet. The complexities of matching colors for the papermaker, particularly in the past, can be very problematic. Dyes behave differently depending on the time of year and even the time of day as temperature and humidity change. Such changes determine the speed and depth of the take-up, or bite, of the dye and can radically alter the intensity of the final color. The old papermakers did not have the benefit of the dedicated environments now found in papermills.

One of the other crucial factors is the origin of the cellulose being colored. For example, although they are all from the same plant, old low-grade linen rags will take dye very differently from new linen cuttings or from raw flax. Each fiber behaves very differently, and matters are further complicated by the amount of washing, cooking, fermentation, and bleaching the fiber has undergone during its preparation for beating. In modern wood pulps, color variations can be remarkable: as one might expect, bleached sulphite and unbleached sulphite, for example, dyed with Xylene Blue A5, produce a greenish turquoise in the unbleached furnish and a rich blue in the bleached furnish.[2]

For most of papermaking history, prior to the advent of chemical dyes, if the raw materials at the papermaker's disposal were not the colors wanted, natural dyestuffs and pigments were the only means of coloring the paper. Natural ochres, smalts, indigo, and very occasionally ultramarine were the main colorants, but a range of various dyewoods was also used. Logwood, *Haematoxylo campeachianum* (sometimes called campechy wood), was used as a source of black, blue, and gray colorings in paper. It also contains a red substance, hematoxyline, which when oxidized produces the brown dye hemateine. Fustic is a yellow dye obtained from the wood of a Caribbean tree, *Cladastris tinctoria.* Catechu, a fast brown dye sometimes called cutch and occasionally also used to prevent mildew, is the dried sap of various species of *Acacia* and *Mimosa.* Annatto is a yellowish-red dye obtained from the pulp surrounding the seeds of the *Bixa orellana* tree. Quercitron, a yellow or brown dye from the North American black oak *Quercus nigra* or the dyer's oak, *Quercus tinctoria,* was usually used to tone blues and grays and deepen blacks. It was used extensively by makers such as George Steart, for his range of colored drawing papers.

The advent of synthetic dyes changed the paper industry. The dyes mostly used since the middle of the nineteenth century are described as Acid, Basic, or Direct. Each of these three types, which should always be added before sizing, has its own uses (Fig. 1.36). Acid

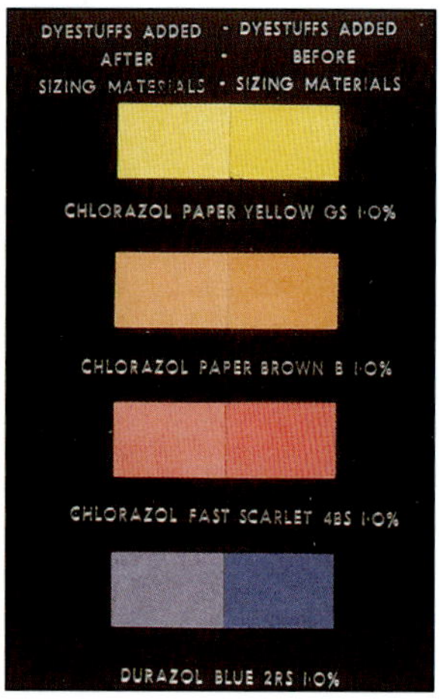

Figure 1.36 Color sampler showing the effects of adding different dyes to the same paper stock before (right) or after (left) sizing. Top: Direct Dye. Second: Basic Dye. Third: Acid Dye. Bottom: Pigments.

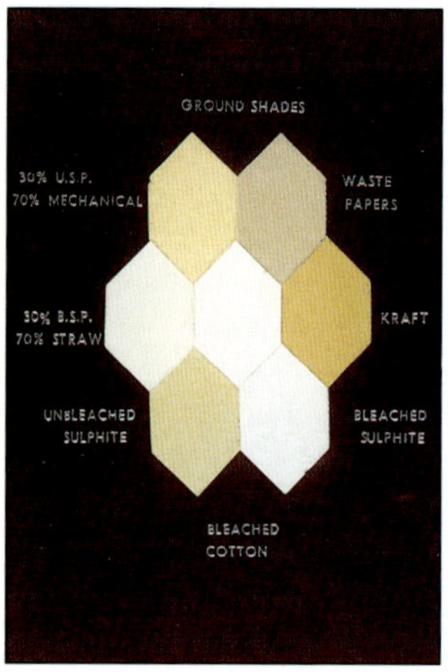

Figure 1.37 Color sampler showing the base color of seven papers made from different furnishes. See Figures 1.38 and 1.39 for how different dyes take on these furnishes.

dyes are adjective, or indirect, dyestuffs that require precipitation by the addition of other agents, such as salts of sodium, potassium, ammonium, or calcium, which act as mordants, to fix the color in the paper. At their best, with hard-sized papers with an excess of alum, acid dyes give relatively brilliant shades but often weak color, and they are comparatively fugitive to light (Figs. 1.37 and 1.39). Basic dyes are also adjective; although some are slightly absorbed directly, most need to be combined with various acids such as chlorides, sulfates, acetates, oxalates, etc. They are most suitable with soft-sized papers. Basic dyes give good strong colors, but are even less light-fast than acid dyes (Figs. 1.37 and 1.38). Direct, or substantive, dyes are directly absorbed by the fibers without the necessity for a mordant. Direct dyes have good light-fast properties but tend to give rather dull or *sad* shades. Consequently they are often *topped* with a small proportion of an acid or basic dyestuff to heighten the color. These dyes are particularly good for unsized papers, since they have an affinity for cellulose fiber that can be further enhanced by heating the pulp and adding a small amount of salt (Figs. 1.37 and 1.38).

Pigments have no real affinity for cellulose and depend on the sizing for their retention. The physical condition of the pigment – its purity, the fineness with which it is ground, and so on – has an important bearing on the finished color. Pigments, being heavier than cellulose fiber, generally sink to the wire side of the sheet during formation, thus producing a greater degree of two-sidedness to the colored sheet than is seen in dyed paper (Figs. 1.37 and 1.39). Most dyes are added at the beater stage, and achieving the desired color is more complex than often realized, depending as it does on a range of variables: the furnish, degree of sizing, order of addition of dyestuff and size, degree of beating, and the presence or not of any fillers. The more well beaten the fiber, the greater the color achieved by dyeing in the beater. This is not a product of the time that the dye is in the beater but of the actual physical condition of the fibers. A smaller amount of dye added at the end of the beating cycle can give a stronger color than more added at the beginning of the cycle. The addition of china clay or titanium dioxide fillers has the effect of weakening but brightening the resulting shade. Fillers, like pigments, can also increase the two-sidedness of the colored sheet.

Many papermakers are cooks rather than scientists. Years ago at Croxley Mill, Hertfordshire (U.K.), where Dickinsons made *Basildon Bond* writing papers, they produced a very particular azure wove paper. I remember standing at the top of the huge hydrapulper as the colorman, Jack, added another small cupful of blended dyestuff to the pulp, perhaps a quarter of a cup at a time. Jack had an old armchair up there and would sit and watch the pulping, adjusting the color by minute degrees as the beat continued. He had an extraordinary record for consistency of color. In fact he was so good at this that when he retired years later, they could not find anyone to match him, and that particular azure paper ceased to be made. The company still produced a blue wove writing paper, but it was no longer called azure: quite simply it did not have the subtlety of the original azure.

Regardless of the care and consideration taken with dyes, it is not just their light-fastness that determines their discoloration. A major factor in the deterioration of colored papers remains the quality of the furnish. The inherent instability of many low-grade furnishes contributes greatly to the discoloration of colored papers (Fig. 1.35). The choice of coloring agent has made many colored papers from earlier centuries very susceptible to light damage. I am sure we have all seen blue and gray papers from the seventeenth and eighteenth centuries that have discolored dramatically, while others have kept their color remarkably well. Indigo, found in so many of these papers and one of the most fugitive colors, actually behaves very differently when present in the sheet as beaten colored rag than when it has been used to dye the pulp in the beater or the vat. Indigo-dyed textiles that have been used and washed many times fade to a basic

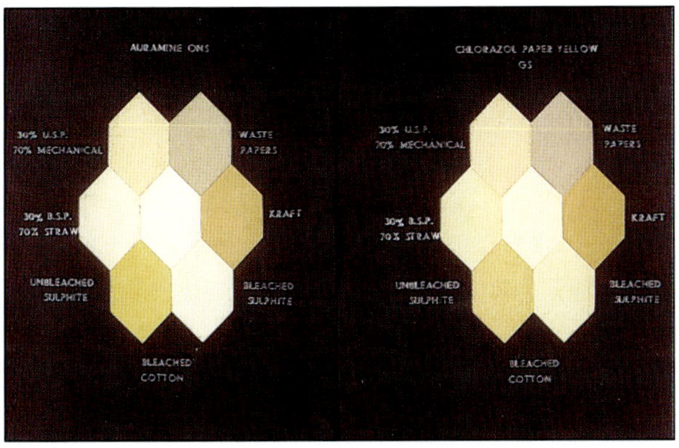

Figure 1.38 Color sampler showing the papers seen in Figure 1.37 dyed with a basic dye (left) and with a direct dye (right).

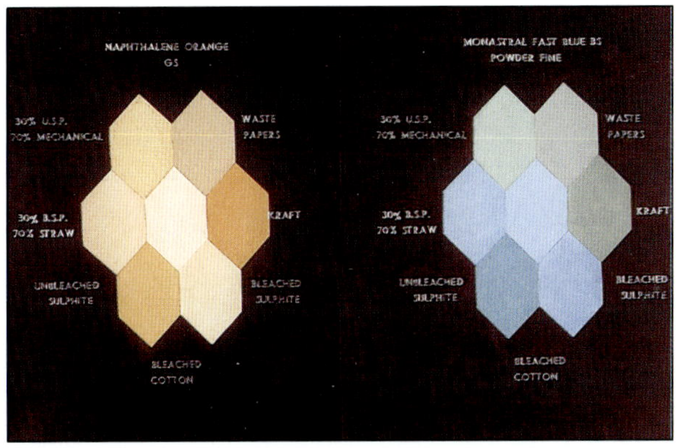

Figure 1.39 Color sampler showing the papers seen in Figure 1.37 dyed with an acid dye (left) and with pigments (right).

stability that is unlikely to get any worse when the discarded textile is turned into paper.

During the eighteenth century, no colored papers were made specifically for artists. Colormen stocked various grades, weights, tones, and colors of a wide range of writing, printing, and wrapping papers, all of which might be used for such work. Most of the papers used were usually either fine writing papers or blue or brown-white wrapping papers. In addition, some artists liked the heavier cartridge papers. The choice of individual papers was very much up to the artist. Many of the so-called "strong wrappings" produced by mills in the eighteenth century in various tones of blue, brown, gray, and buff were well suited to artists' needs. The weight, color, degree of sizing, and surface roughness determined the choice of the individual artist.[3]

The production of colored papers accounted for perhaps half of all paper production in the United Kingdom during much of the eighteenth century. Excise tables show that brown and whited-brown paper easily constituted the largest area of production. Moreover there was a considerable size range in the brown and whited-brown papers, at least toward the end of the eighteenth century. Another paper in use, "cartridge paper," the strong paper designed to carry powder and shot or ball, ranged from a pale brown to close to white and had strength enough to take most of the demands made upon it by artists.[4]

Some papers now look considerably different from when they were first made. There is an unusual laid paper, now a pale buff-gray, that is a single-faced "strong wrapping" paper used by Benjamin West (1728–1820) which was originally probably much closer to white.[5] West used this non-watermarked paper several times, generally working on the felt side, which is marked by a very prominent coarse felt impression.[6] The sheets I examined showed a furnish based on a blend of low-grade white linens with some blue and black rag together with tarred hemp rope. A very small proportion of short red fibers is also present: approximately eight visible in a scan of the whole sheet. The presence of very small amounts of red fiber is more commonly found in British-made, as opposed to continental, wrappings throughout the eighteenth century. The fibers have been very poorly beaten, leading to an irregular formation with considerable process dirt: both iron particles and some tiny specks of wood, probably from the beater roll and its wooden separator bars. Some of the fiber has hardly hydrated at all (in other words, it is free beaten) but much of it is well hydrated (wet beaten), leading to a great variation in opacity and density within the sheet. This is very typical of eighteenth-century British low-grade paper production.

One artist whose choice of paper became increasingly particular throughout his career was Thomas Girtin (1775–1802). Although his painting and drawing techniques changed dramatically throughout his short working life, his use of paper is a particularly vivid example of the importance of the ground in any work of art. His understanding of surface – both visually in terms of what it could contribute to the work and technically in the sense of knowing and using the surface strength of the sheet (the degree of working the surface could take) – became more and more marked as his career developed. He evolved a range of swift wash techniques, with subtle variations of density and tone, that allowed him to draw within the wash itself. The papers he actually used suggest that he took care to seek out those surfaces which suited his intentions, rarely working on just "any old paper," as was the case with some of his contemporaries. Both Girtin and J. M. W. Turner (1775–1851) were very early exponents of the conscious exploration, in both watercolor and pencil, of the inherent strengths and qualities of different papers and paper types.

The 1790s saw the introduction of several new specialist papers designed particularly for watercolor. Girtin used some of these new papers but appears to have ignored others completely, even though they would have been readily available to him. In addition to the white laid and wove Whatman papers and sheets made by Edmeads and Pine, Portals and Bridges, Jean Villedary, and others, he also worked on a range of off-white cartridge papers from several makers, including Austen Stace. But perhaps the most interesting range of papers was the strong wrapping papers, from pale off-white to quite deep buff-grays, from many as yet unidentified sources. Many artists' colormen stocked a range of these papers, specifically for drawing and watercolor, choosing those that they thought most appropriate. Girtin's so-called rough/coarse laids, often referred to as cartridge paper, are not all from the same source; in fact they cover a range of distinctly different papers from a wide variety of different makers.

By the 1820s colored papers had become relatively sophisticated and were being used for an increasing number of purposes. Following the work of Thomas Cobb, George Steart, and others, the range of colors available was also much greater than in earlier years.[7] Both the blue and red woves seen here (Figs. 1.33 and 1.34) were produced by adding dye to the furnish rather than relying on the original color of the rags to give the color of the finished sheet. In English papermaking this practice seems to make its first appearance in the 1790s, but there is evidence from both Italian and French papers that continental makers had been using a similar coloring technique for some papers much earlier.

Merchants and artists' colormen continued to use the same mills for their supplies over long periods of time, but the molds deteriorated and were replaced, methods of pulping changed, and it was never possible for mills to get consistent supplies of rags and other raw materials. All this contributed to variations, sometimes subtle and sometimes quite dramatic, in the appearance and character of different batches of what was nominally the same paper.

European blue paper manufacture dates back to the fourteenth century: a Bolognese statute from 1389, for example, mentions a "royal blue" paper.[8] These blue papers were of varying qualities, being mostly made for the wrapping of trade goods. The first patent in the history of papermaking (issued in 1666 in England) concerned the manufacture of blue paper.[9] During the eighteenth century, a variety of blue papers such as *draft*, *bag*, *fine dark*, and *natural* were all in production in Western Europe.[10]

Some white-paper mills also made small amounts of blue and brown papers, usually for use in wrapping their own papers, but most were made in papermills that specialized only in colored papers. Some blue paper was produced by dyeing white or off-white pulps with dyes such as indigo, logwood, lichen, blueberries, etc., but the bulk of blue paper was made from rags which were themselves blue. Analysis of a very wide range of eighteenth-century blue papers shows that indigo was the most commonly used dye for these rags.

The early years of the eighteenth century saw the advent of a new technology throughout Western European papermaking: the Hollander beater, designed to produce more pulp and at a faster rate than the old stamping engines, also had other advantages. Papermakers soon learned that they could control the beating of individual batches of rags in ways not possible using earlier methods. This allowed the papermaker to develop the potential of a particular rag fiber, or combinations of fibers, to give finished sheets distinct characteristics suitable for very specific tasks. In particular, the length of the fiber and its ability to take up water during beating could now be controlled.

Some of the earliest papers to take advantage of this new development were of course the finest writing and printing papers, but the same evolution can be seen in some of the better blue and brown papers, in particular the French paper known as *Papier Bulle*, sometimes called *Papier Chanvre*. It varies from a warm deep cream to a very deep brownish-buff color and was much liked by many artists in the seventeenth and eighteenth centuries. It was imported from France, mainly from three of the old French papermaking regions: Limousin, Angoumois, and Auvergne. The bulk of French papers imported into England during the eighteenth century arrived from Bordeaux and La Rochelle, but even during the many periods of war with France during the century, some still found their way to England via various Dutch paper factors based in Amsterdam.[11] *Papier Bulle* corresponds to the English paper types variously called *rope browns, hemp browns*,[12] or *whited browns*.[13] Later, by extension, the term also came to be applied to some unbleached papers of a warm cream color, made from a blend of hemp and linen fiber. In France, under various decrees issued by the *Conseil du Roi* of 1739 and 1741, which regulated the manufacture of paper in France, each sheet should contain, besides the watermark, a countermark giving the maker's name, the area of France, and the quality of paper being made. But in England these papers were almost never watermarked before c. 1800.

Among the early nineteenth-century proliferation of new, colored all-purpose papers, one maker stands out. George Steart's development of colored watercolor and drawing papers paved the way for makers throughout Europe. For 30 years at De Montalt Mill, near Bath (U.K.), he produced a range that both satisfied and challenged the demands made by a great variety of different artists. J. M. W. Turner (1775–1851), John Varley (1778–1842), Richard Parkes Bonington (1802–1828), Peter de Wint (1784–1849), David Cox (1783–1859), John Sell Cotman (1782–1842), and John Constable (1776–1837) all worked on Steart's papers. Visually they owe a lot to the appearance of eighteenth-century wrapping papers, but with the addition of the best papermaking developments of the 1790s to the 1820s.[14] Steart's drawing papers were very popular with practicing artists, rivaled only by the two Whatman papers and the range of papers called Creswick. All these papers are essentially simple papers: pure linen, gelatin-sized, and made with great care and attention – precisely the attributes that Steart brought to his endeavors.

Many of the changes in papermaking were "invisible." Technical innovations and growing skill among papermakers as their increasing specializations drove them to produce more and better papers produced some profound changes in the ways sheets behaved in use. However, many of these changes were not always obvious to the uninitiated, although artists such as

Turner, Girtin, and Cotman show a deep understanding of the possibilities of these new surfaces. Some papers became so specialized as to lose much of their potential. A later generation, in particular David Cox the Younger (1809–1885) and Samuel Palmer (1805–1881), would move away from these use-specific watercolor papers, both artists in their different ways becoming increasingly particular about their papers. Palmer used extraordinary preparations to get the surfaces to his required state before working, and later in his career arranged to have London boards put together for him by Newman. Cox's two favorite papers were not actually drawing papers – one was a pure linen map paper, obtained from his brother-in-law, who purchased paper for the Ordnance Survey, and the other a strong hemp/linen wrapping paper, designed for wrapping reams of paper.

On Steart's retirement his range of blues, grays, buffs, sand, and other colors were no longer available. Colormen, such as Winsor & Newton, began to sell versions of the most popular papers made by other makers. One of the artists who benefited from these copies of Steart's papers was Turner's friend and contemporary Clarkson Stanfield (1793–1869), whose group of eight watercolors and chalk drawings of the January 1839 eruption of Vesuvius on just such a paper were originally attributed to Turner.[15] Although this paper bears great visual resemblance to Steart's blue woves, closer examination shows several differences: Steart's papers contain blends of blues, whites, and other colored linen rag, including very small amounts of red, yellow, brown, and black fibers in some batches. The paper Stanfield used was made from a blend of two different blue rags and one white rag, which were mostly cotton rather than linen. This particular example resembles the version produced for Winsor & Newton.

In complete contrast to the ever-popular flecked blue papers are those rich deep blue papers that first made their appearance on the continent in the late eighteenth century. Much deeper in tone than most blue papers, some of them have kept their color remarkably well. This blue paper was made by dyeing the pulp directly in the beater rather than relying on the original color of the rags. A particularly fine example of this paper is the group of 20 approximately quarter sheets found in use by Turner. Some are worked in pencil and some in color. All of the sheets, which are very opaque with a rather wild look-through, contain either half the watermark or half the countermark. Each complete watermark reads: RICHARD BOCKING / & / COMPAGNIE with a BASLE CROZIER on a SHIELD. Although the use of the Basle Crozier (Baselstab) might suggest a Swiss origin for this paper (and indeed variations in this mark had been in common usage by many Basle papermakers from the 1580s onward[16]), over time the use of this mark spread to many German papermills, including the Kautenbach mill run by Richard Bocking (1794–1858), following his father and uncle, Hans and Adolf Bocking, who had founded the mill in 1781.[17]

It is highly likely that Turner acquired these sheets at Trarbach, only 4 km north of the Kautenbach mill. Perhaps some local resident, seeing the blue papers he was working on anyway, told him of the local mill's blue paper or gave him some. Although most of Turner's 1839 blue paper Mosel pictures depict towns and villages from the river, a few works show that he explored inland as well. Turner spent some time at Trarbach, and drew the *Entrance to Trarbach from Bernkastel*,[18] a depiction of the way south out of Trarbach that would have taken him to the Kautenbach mill. A second work, *Trarbach from the South*,[19] is worked from a viewpoint farther up the hill, south along the track to Bernkastel.

Turner, like so many artists, chose his colored papers very carefully. The choice available to him and his contemporaries was far greater than for artists working in earlier centuries. Colored papers had evolved out of the primarily utilitarian and economic use of available raw materials for simple wrapping papers. Those earlier papers were blue and brown and drab simply because the blend of materials used produced those colors. Without chlorine or its descendants, the rags would bleach no further, and without the chemical dyes developed in the nineteenth century, few other colors were easily possible. The last 150 years have seen an increasingly sophisticated range of colored papers made available, but in many ways we are spoiled for choice. So many of the colored papers produced now are designed so specifically for particular functions, such as business stationery, cover papers, etc., that, unlike the old wrapping papers, they are not versatile enough to satisfy the demands that artists might make on them. As a result, artists often still find themselves unable to find just the colored paper that they want.

Acknowledgments

I would like to thank Camilla Baskomb, Sally Bower, Robin Hamlyn, Richard Hills, Annie Lyles, Barry McKay, Antoinette Owen, Jacob Simon, Rod Tidnam, John van Oosterom, and Barry Watson for their advice, assistance, and practical help with different aspects of the research behind this paper.

Endnotes

1. I will be concentrating on British and Western European papermaking practice. The coloring of Asian and Islamic papers is a complex and very different subject. One recent and important study is P. J. Gibbs and K. R. Seddon, *Berberine and Huangbo* (London, 1998).

2. Over the years the specialist literature has produced a range of very instructive publications and technical circulars: e.g. Sandoz, *Acid and Basic Dyestuffs on Paper* (Basle, 1951), or ICI, *Some Variables in Paper Dyeing* (London, n.d.).

3. Trade catalogues provide considerable information about papers available to artists: e.g. Reeves and Sons, *Trade Catalogue* (London, 1878), pp. 91–92; Winsor & Newton, *Trade Catalogue* (London, 1851), offered both "Turnbull's Crayon Boards" as well as "Tinted Drawing or Crayon Papers"; Reeves, *Trade Catalogue* (1856), also offered "Tinted Drawing and Crayon Papers."

4. Public Record Office documents PRO Customs 3/27–80 and Customs 17/7–22 describe the papers available between 1782 and 1793.

5. Benjamin West, *Study for the Figure of Leonidas*, Victoria & Albert Museum, London, Dyce 776.

6. One of the more unusual aspects of this paper is a series of irregular secondary "chain" lines visible intermittently in various parts of the sheet. Many of these are not parallel to the actual chain lines, but are at varying angles to them.

7. Thomas Cobb operated North Newington Mill, Oxfordshire, in partnership with John Pain and William Judd, from 1792 until sometime between 1812 and 1816. He was responsible for several innovative patents regarding colored paper between 1796 and 1812.

8. Gasparinetti, "Ein Altes Statut von Bologna über die Herstellung und den Handel von Papier," *Papiergeschichte* 6, 3 (1956), p. 46.

9. Charles Hildeyerde, Patent No. 147, 16 February 1666: *Making blue paper used by sugar bakers.*

10. Made from blue colored rags rather than other pulps dyed by the papermakers.

11. See W. A. Churchill, *Watermarks in Paper in Holland, France, England, Etc. in the 17th and 18th Centuries* (Amsterdam, 1935), pp. 22–27, for a discussion of Dutch merchant and factoring activity in France.

12. Thin brown wrapping papers were originally made from old rope stock, hemp, and hemp waste, and consequently are of high quality and strength.

13. Whited browns were made by blending old rope stock with various qualities of white linen rags.

14. Until recently papers made by Bally Ellen and Steart were not generally recognized because most of them are extraordinarily opaque and the watermarks are rarely visible in ordinary transmitted light, even on a light box. Most of the B E & S watermarks, found in the blue, buff, and brown papers that Turner used, are only visible using a fiber optic light source or by X-ray. See P. Bower, *Turner's Later Papers* (London, 1999), pp. 94–97, for further information on Steart's working practices. See also P. Bower, "The evolution and development of 'drawing papers' and the effect of this development on watercolour artists 1750–1850," *The Oxford Papers*, Studies in British Paper History (London 1996), Vol. 1, pp. 63–74, for a discussion of some of the developments in artists' paper usage during this period.

15. A. Scott-Elliott, *Catalogue of the Oppé Collection* (unpublished), p. 238. The three watercolors in Wilton (1979), nos. 1039–41, were attributed to Clarkson Stanfield by Eric Shanes in a review of Wilton (1979), *Turner Studies* 1, 1 (1981), p. 47.

16. See W. Fr. Tschudin, *The Ancient Paper Mills of Basle and their Marks* (MCPHI, Hilversum, 1958), Vol. 7, for the history of this mark.

17. Johann Hönl, "Die Papiermühle in Kautenbach," *Kreisjahrbuch* (Bernkastel-Vittlich, 1982), pp. 144–47.

18. Tate Gallery, TB CCXXI G, gouache and watercolor on B E & S blue paper and TB CCXC 22r, pencil.

19. Tate Gallery, TB CCXX P, gouache and watercolor on B E & S blue paper.

Housing a Pastel Collection at the National Archives of Canada

Part I: Pick It Up . . . and Move It Out

Gilbert L. Gignac and Gregory J. Hill

This paper provides an overview of the initiatives taken between 1994 and 1997 by the National Archives of Canada to develop two practical methods to prepare its pastel collection for its imminent move into a new facility. The National Archives of Canada was created in 1872, and since 1967 occupied quarters shared with the National Library of Canada at 395 Wellington Street in downtown Ottawa. Due to a lack of space, most of its collections found themselves housed in several inferior satellite buildings.

The room that was adapted to house the pastel and other art collections was on the third floor at 395 Wellington Street, and it exemplified the overall inadequacies of this facility to provide good storage and operational space to preserve the art collections: space was too confined and insufficient, ceilings were too low, shelving and cabinets were cumbersome and inadequate. Originally, environmental conditions in this space were relatively stable, but within 12 years they were destabilized when HVAC systems were replaced. Extreme fluctuations in humidity levels resulted, with instances of 80%–90% relative humidity (RH) in summer and 10%–20% RH in winter. Environmental conditions were unstable and continued to deteriorate. Flooding from broken pipes and drain back-ups became routine.

It was conceded by all that the singular act that would most enhance the overall preservation of the collections, including the pastel collection, would be to provide them with ideal housing conditions in adequate space. In 1992 the Archives started to build a new storage facility in the nearby town of Gatineau, Québec, across the Ottawa River from its central site in downtown Ottawa. The new Gatineau Preservation Centre was officially opened on 5 June 1997. Succinctly put, this new facility consists of a concrete slab that supports an inner three-story concrete block of 48 vaults whose entire top surface houses new conservation and restoration laboratories. The whole is covered by a single spanned steel roof and wrapped in an outer glass shell.

Besides its essential role as keeper of the records of the Canadian government since 1872, the National Archives of Canada has consistently collected private archives and works of documentary art in all media: oil paintings, watercolors, prints, drawings, posters, miniatures, caricatures, sketchbooks, albums, illustrated books, and pastels. The presence of such a vast array of artworks in the traditional archival milieu is unusual. This collection of 260,000 works constitutes the largest collection of works of art on paper in Canada and one of the largest in the Americas. These images are collected by the National Archives by virtue of the pertinence and relevance of their subject matter, regardless of their aesthetic qualities. In some way, the work must document the history of Canada. The collection, built on iconography, falls into three general categories: portraits, places, and "the business of life" of Canadians.

In Canada pastel as an art form has not yet been studied in depth. Scholars are still in the very earliest stages of staking out the parameters of the field. Incidental notes are scattered here and there in monographs on individual artists or in collection surveys. The art of pastel in Canada began with portraiture in the late eighteenth and early nineteenth centuries; landscapes appeared by the mid-nineteenth century, and pastels relating to "the business of life" emerged at the very end of the nineteenth and into the twentieth century.[1] The earliest pastels by a Canadian artist in the National Archives collection are three portraits by Québec artist Louis Dulongpré (1759–1843).[2] The earliest pastel by a non-Canadian artist in the National Archives collection is a small portrait of the eighteenth-century cartographer Frederick W. Desbarres (1721–1824), drawn in 1785 in London by English artist James Peachey (active 1773–1797).[3]

Our largest pastel, entitled *Chief Crowfoot Addressing the Marquis of Lorne (Governor General of Canada, 1878–1883): Pow-wow at Blackfoot Crossing, Alberta, Sept. 10, 1881*, was drawn in 1887 by English artist Sir Sydney Prior Hall (1842–1922). Measuring 146.8 x 238.5 cm (57 x 94 in.), it is likely the largest nineteenth-century pastel on a single sheet of paper in any Canadian collection and possibly one of the largest in the Americas. Because of its fragile plate glass and its considerable weight, this work, which was ready to move as it was, was transported vertically, alone in a separate convoy, and taken directly to be hung on a temporary rack where it awaited to be re-hung in its permanent, immobile location.

As the construction of our new building evolved, so did intensive preparations to safely move the art collections, among them our eighteenth-, nineteenth-, and twentieth-century pastels. Integrated departmental planning meetings were held at all levels, and complex schedules were drawn up. The work ahead was a monumental collaborative effort. The needs of collections, as they related to protection and packing, had been identified by the custodian and set forth in 32 individual projects, of which the pastel project was only one. Funds were identified and secured for the additional supplies and equipment needed to facilitate the preparation of collections. Custodians of various collections drew up very precise move specifications for a contract to be given to experienced art movers, who were involved in planning meetings, including walkabouts. Because of seasonal extremes in the Ottawa climate, all art moves were to be carried out during late spring and early fall. A year before the move, conservation laboratories closed to allow furniture, equipment, and supplies to be transferred to the new laboratories in Gatineau. In the interim, over the course of two years from 1995 to 1997, paper conservators became part of move teams whose function was to examine all works of art, including the pastels, and improve their housing, thus ensuring their "move readiness" for shipment to their new home.

The pastels fell into two distinct groups: those in original frames and those that remained unframed. The slow rise of our new building unfortunately coincided with a massive downsizing of the Canadian civil service, and the cutbacks had a significant impact on the National Archives move initiatives. When National Archives exhibitions services were drastically reduced, we lost the important services of the carpenters on whom we depended to help maintain the framed collections. During this crisis, the fine-art custodian seized the opportunity to define a preparation project for framed works of art on paper and secured the collaboration of one of the master carpenters who was being redeployed. It was proposed that the carpenter team up temporarily with a paper conservator and the custodian to carry out the essential work of improving the state of the collection of framed works of art on paper.

A small temporary workshop was set up near the paper lab. Basic procedures were adopted to prepare the framed works on paper, intended not only to comply with move requirements but also to guarantee the objects' long-term preservation during future consultation, photography, exhibition, or loan. The procedures to properly prepare each framed pastel were analyzed and discussed by the project team: custodian, conservator, and carpenter formulated a customized approach for each work. It was decided that with repair, stabilization, consolidation, and build-up in mind, this custom approach would address all the structural needs of the frames, such as frame joints, surface finish, glazing, spacers, new backing boards, and new secondary supports for the works of art they were to contain.

Following the removal of backing boards, works were carefully extracted from their frames and examined for treatment. Approximately 20% of the framed pastels required a small amount of repair and consolidation, mostly along the edges. Japanese tissue and wheat starch paste were used to carry out the essential paper repairs and hinging of drawings onto four-ply rag backing boards once treatment was complete. Where appropriate, new window mats were cut to fit the works.

Many works were still nestled in their original frames, and the intent was to carefully respect their stylistic integrity during the necessary consolidation and build-up. Initially, each frame was examined for structural problems, often resulting in re-gluing of joints, local repairs, and refinishing when absolutely necessary (see Fig. 1.40). General cleaning of the frame and glass followed. When possible, the original glass was reused, and when not possible, ⅛ in. tempered glass was installed. Spacers incorporated into the frames were constructed of ¼ in. thick kiln-dried wood, lightly tinted to match the frame. When a pastel was framed without a mat, spacers were lined with a layer of four-ply rag board, thereby separating wood from the drawing. New backings of acid-free corrugated board were cut for the backs of the frames.

The total thickness of the package that included glass, spacer, object, mat, and backing board was determined to assess the degree of build-up required for the verso of each frame. Generally, the back had to be

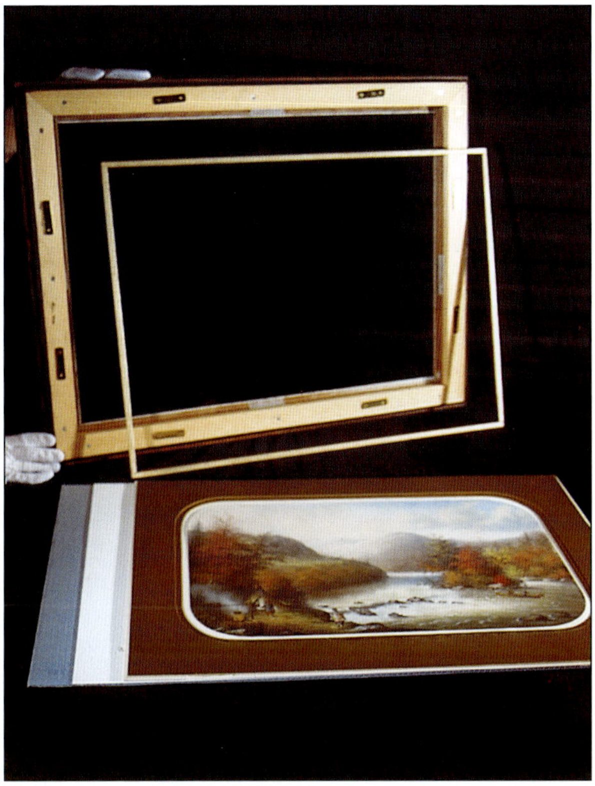

Figure 1.40 The verso of a reconditioned frame showing the spacer, backing boards, pastel, and original mat for Alfred Worsley Holdstock (Canadian, 1820–1901), *Gatineau River, Québec*, 1871. Pastel on paper, 353 x 532 mm. National Archives of Canada.

built up from a half to a full inch. Although the resulting changes were practically unnoticeable, these build-ups caused most of the frames to almost double in thickness. Color-matching minimized the contrast between the new addition and the original frame, as did a slight angling of the addition. The backs of the frames were also built up using clear, kiln-dried wood. After the work on the frames was completed, they were left to aerate for about three weeks before being used, to allow for off-gassing of the volatile components of paints and finishes.

Some unframed pastels were rehoused in recycled frames from a collection of unused period frames stylistically matched up by the custodian. This sometimes entailed cutting them down to fit the artwork and building up the back to accommodate new glass, spacers, mats, and backing boards. Once completed, these framed works of art on paper were ready to be wrapped and moved. In total, 72 works were framed or reframed in this intensive six-month project.

When preparing for the move, the National Archives had to face the astonishing fact that in the last 20 years, the acquisition of pastel and charcoal drawings had increased dramatically, entering the collection by the dozen, or even by the hundred. For example, in 1994 the National Archives acquired 126 large-format twentieth-century pastels by Alma Duncan (b. 1917), and, in that same year, a collection of 23 large-format charcoal drawings by Lauren Harris (1885–1970). Most arrived unframed, and their temporary housing in map folders, which were kept in plan file cabinets or boxes, unfortunately ended up becoming permanent.

Prior to the move, the art custodian had requested the rehousing of a rare collection of unmatted late nineteenth-century charcoal landscape drawings of the Pacific Northwest by Edwin Sandys (active 1880–1890). The request stipulated that the extremely friable surfaces of the drawings be protected without contact with the housing materials, and that the housing would minimize handling during future use. Furthermore, the housings were to be lightweight, given the size of some of the objects, and not too bulky, given the number of objects requiring housing.

Conservators in the prints and drawings laboratory immediately set about creating sink mats, consisting of a solid bottom support, a middle window cut wide, and a stiff lid (see Fig. 1.41). The bottom supports and some of the lids were fashioned from acid-free, corrugated board. This material is reputedly acid-free with a 3% calcium carbonate buffer and is available from several conservation supply houses. Cold extraction pH testing produced an average pH of 8.0 to 9.0. As an added feature, the corrugations of two layers of board placed perpendicular to each other gave greater dimensional stability. In previous experimental models, four- and eight-ply acid-free rag boards were tried as a suitable mat, but lids of this thinner board gradually acquired an undesirable sag across wider window spans and eventually touched and rubbed the friable media of the drawing surface. This newly devised protective sink mat proved dimensionally stable, strong, light, and versatile.

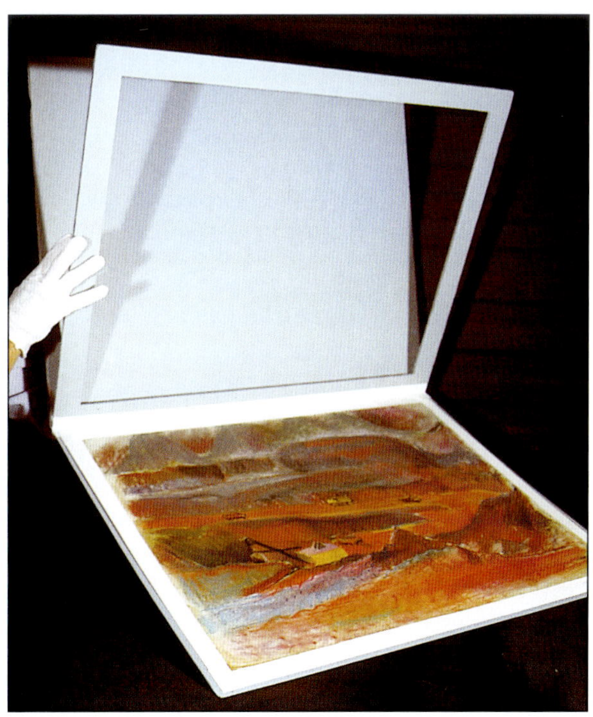

Figure 1.41 Sink-mat housing for the unframed pastel by Alma Duncan (Canadian, b. 1917), *Giant Machines Look Small*, 1956. Pastel on paper, 505 x 657 mm. National Archives of Canada.

The sink-mat model developed for the Sandys charcoal drawings was appropriated for all 250 unframed pastels. Each drawing was examined and treated for structural damages. Only essential tear repair was completed prior to the construction of the sink mat. Following this, the pastels were hinged with Japanese tissue and starch paste onto a secondary support of two-ply, acid-free card. The mats were custom-cut to each item, with windows very slightly overlapping the edge of the primary support. The minimum width of the window mat around the object was approximately 5 cm, allowing a minimum of 4 cm of two-ply card on all sides of the drawing. The secondary support plus object was then set into the mat using Mylar® or paper corners. This allowed the drawing, mounted on the two-ply card, to be removed with ease from the sink mat for photography or for placement into exhibition matting and framing when required, without having to unhinge and re-hinge the actual drawing every time.

A total of 253 such mats was constructed to suit the various sizes of pastel drawings and drawings in other friable media. The outside dimensions of the mats were standardized to fit into drawers and boxes. The mats were easily handled and bar-coded on the top front edge and safely stacked in boxes and drawers without fear of damage to drawing surfaces. Three such sink-matted drawings can be accommodated in Solander® boxes or other flat corrugated cardboard storage boxes. Once all the pastels were repaired, mounted, matted, and safely housed in storage boxes, they were move-ready.

In 1996, after many years and a long winter of anticipation, the new facility was finally ready and so were the collections. The collections of unframed friable drawings were well protected in sink mats stored in containers of consistent size that could be picked up, placed horizontally on carts, and moved out. But framed works of art on paper, also appropriately housed, were moved quite differently. In the past, due to an acute shortage of space, framed pastels were kept in an upright vertical position, which is rightly discouraged because of the risk of loss of friable media. Although good conservation practice recommends the storage and movement of pastels in a horizontal position when possible, our custodian and conservator concluded that this collection of framed pastels would be more safely moved in the vertical position.

This unorthodox conclusion was the result of considering numerous factors. The staff was confident that the pastels had been securely mounted within frames that were structurally sound. The glass of each framed work was carefully taped. The fragility and brittleness of the early nineteenth-century period glass was carefully weighed against the risk of breakage if shipped horizontally. Each frame was double-wrapped in bubble pack with a layer of fluted polypropylene (Coroplast®) protecting the front and back. The frames were to travel in foam-padded carts that were equipped with soft pneumatic tires, and the truck had the required air suspension drive. The journey time between sites was approximately ten minutes along on a well-paved route and the weather was perfect. Also, staff had confidence in the experienced, specialized art mover.

The move team followed a precise plan and schedule. The meticulously prepared and well-packed collections were carefully loaded into carts and rolled onto the truck. Because of the low door at the loading dock at Wellington Street, the specified truck could not enter the loading bay. A ramp extension was installed from dock to truck, allowing carts to remain level during loading. A total of 83 large Solander® boxes, 40 plan file drawers, and 200 framed items was moved. When the truck arrived in Gatineau, staff received, unpacked, horizontally shelved, and scanned all of these pastels and other framed works on paper into their new permanent locations within a few hours (see Fig. 1.42).

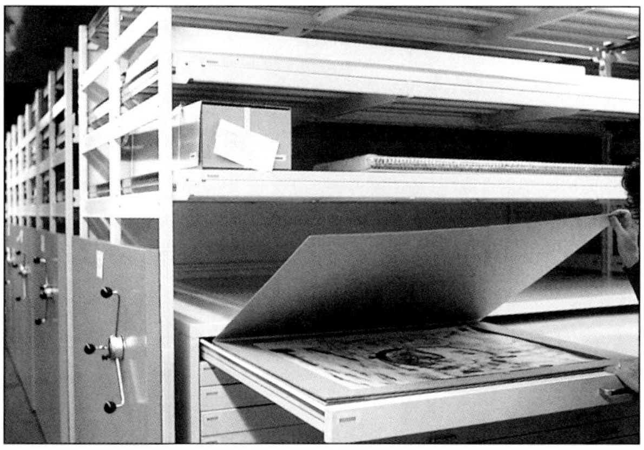

Figure 1.42 Interior of an art vault following the move.

In hindsight, it is clear to us that planning simultaneously for the collection's rehousing and for the subsequent move was key to the success of this major collaborative undertaking. It also became evident that our early, fully integrated planning meetings, which included archivist, curator, custodian, and conservator, overseeing documentation and physical control, proved most critical to the greater success of this preparation and move project. From the outset, understanding the needs of the collections and the concerns of each team member expedited the entire process. Although professional conservators experienced a fair amount of frustration in being removed from their laboratories, their invaluable, enthusiastic, and direct involvement in the preparation of the collection greatly enhanced our preparedness for the move. Move planning with specialized art movers increased our confidence that our specifications and standards would be adhered to.

Not only were we able to prepare for the immediate move of the pastel collections but we also seized the unexpected opportunity to achieve improved long-term housing. This move provided a unique opportunity to re-examine the housing of every single pastel. In developing housing methods, we learned that simple, practical, and effective procedures had universal application to a broad array of pastels from various schools and periods in every conceivable format. Today, when handling our pastel collection for its various uses, the two successful methods applied to their rehousing have proven to be flawless.

Acknowledgments

The authors would like to acknowledge Anthony Bravo, Gaston Tremblay, Wanda McWilliams, and Exhibit Transportation Services, Canadian Conservation Institute.

Endnotes

1. A few very early pastel portraits found in other Canadian collections have been associated with two eighteenth-century Québec artists, Louis de Heer (1760–before1808) and François Beaucourt (1740–1790). See *Le Musée du Québec, 500 oeuvres choisies* (Québec: Musée du Québec, 1983). The earliest known and substantiated pastel artist in early Canada is François Baillairgé (1759–1830), who was also the first Canadian-born artist to study in Paris (1779–81), at the Académie Royale de Peinture et de Sculpture. See D. Karel et al., *François Baillairgé et son oeuvre (1759–1830)* (Québec: Musée du Québec, 1975).

2. R. Derome et al., *Dulongpré: A Closer Look* (Montreal: McCord Museum of Canadian History, 1988).

3. One of the more recent and substantial studies in the field of early pastel in Canada was carried out by Mary Allodi, curator emeritus at the Canadian Department of the Royal Ontario Museum in Toronto. In the course of her research for a major exhibition on the German-Canadian artist William Berczy (1744–1813), which was held at the National Gallery of Canada in 1992, she was able to re-attribute a large number of small pastel profile portraits in Canadian collections to the American pastelist Gerrit Shipper (1775–c. 1835). See M. Allodi et al., *Berczy* (Ottawa: National Gallery of Canada, 1991).

Housing a Pastel Collection at the National Archives of Canada

Part II: God Is in the Details: The Conservation Treatment of Early Canadian Pastels

Maria Bedynski and
Gregory J. Hill

Demands for loan and in-house exhibition, and the anticipated move of the National Archives collections to a new building, prompted the conservation treatment of three pastel portraits by Louis Dulongpré (1759–1843) and a pastel and gouache landscape by William Armstrong (1822–1914), two of Canada's most significant nineteenth-century painters. The extremely fragile and vulnerable condition of these pastels precluded their travel and exhibition, and initiated close examination, scientific analysis of materials, and art-historical research.

Dulongpré was born in 1759 in St. Denis, France.[1] It is believed that he studied art in Paris, but by 1778 he was in America with the French Navy supporting the American patriots during their War of Independence. Following the war, he came to Canada, arriving in Montreal in 1784. In 1787 he married Marguerite Campeau, whom he later depicted with their daughters in one of his finest pastel works, *Les Saisons*.[2] Dulongpré's pastel portraits were greatly influenced by the popular pastel portrait paintings in the rococo style of pre-revolutionary France. Surely surrounded by these paintings in his youth in France, he no doubt encountered others that had been brought to Canada by artists returning from Europe.[3] The sophistication and lightness of French rococo, however, is tempered in Dulongpré's pastels by simplicity and lack of pretension. His portraits are warm and witty, sometimes sad, but always very human. Consistently restricted to head and shoulders and lacking distracting accessories, the sitters are set off by plain but usually richly colored backgrounds. Dulongpré's works often resemble the lively pastel portraits painted by American folk artists. His sitters included prominent members of the community – clergy, important citizens, and seigneurs – with whom he socialized.[4]

In 1832 the Dulongprés left Montreal for Saint-Hyacinthe, in Québec, to be closer to their friends, the prominent Papineau and Dessaule families. Dulongpré died ill and poor, at the age of 84 in 1843. His obituary states that he painted more than 4,000 portraits in pastel and in oil.[5] The speed of the pastel technique, its ease of application, and its relatively low cost partially explain this large number. The extreme fragility of the medium and the fact that Dulongpré rarely signed his work may explain why so few of them survive. Today we can account for a mere 1% of his oeuvre. The National Archives of Canada has seven portraits by Dulongpré in its collection, three in pastel and four in oil. Five of the portraits depict members of the one of Canada's most prominent French families, the Papineaus.

Louis-Joseph Papineau became one of Québec's most influential politicians and the leader of the French-Canadian cause in the Rebellion of 1837. In Dulongpré's 1796 portrait (Fig. 1.43), Papineau is depicted as a 10-year-old boy dressed in a student uniform of the Séminaire du Québec. The pastel was brought to the conservation laboratory in 1996 to be reunited with its original frame, which had been stored separately since a previous conservation treatment in 1979. The unfixed pastel measures 378 x 300 mm and was executed on brown laid paper. It was originally solidly mounted on a piece of linen fabric that was stretched around a wooden strainer. During the previous treatment, however, the work was removed from its frame and strainer, separated from the linen support, and hinged onto four-ply rag board. The original strainer and linen backing were subsequently lost.

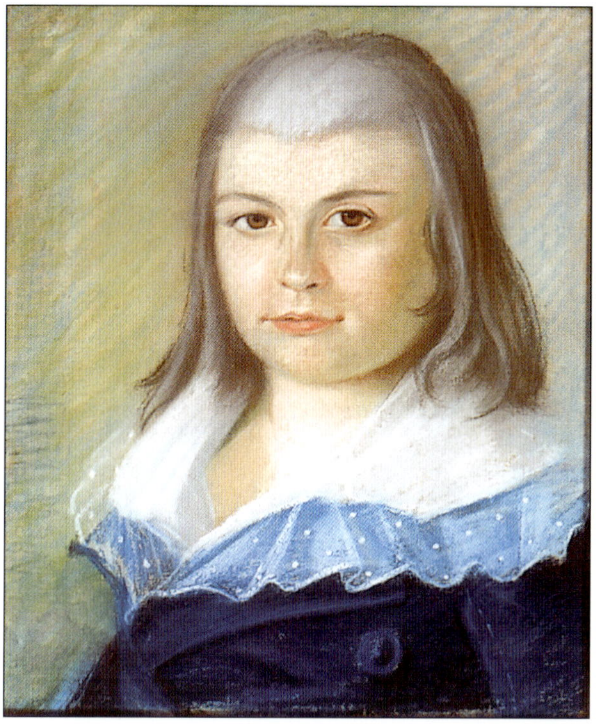

Figure 1.43 Louis Dulongpré (Canadian, 1759–1843), *Portrait of Louis-Joseph Papineau as a Child*, 1796. Pastel on paper, 378 x 300 mm. National Archives of Canada.

The primary support was extremely brittle and discolored with multiple small tears and losses along the edges. Some severely skinned areas on the verso were almost transparent, and had been patched with a heavy white Japanese paper and water-soluble adhesive, causing an overall distortion of the primary support. A thick, dark residue of protein adhesive was present along the top edge. The sheet had also suffered water damage, and prominent tide lines were visible along the top, left, and bottom edges of the sheet; the worst staining appeared around the lower left corner. The pastel layer had suffered abrasions and substantial pigment loss, especially in the top right quadrant and around the forehead. The image was also obscured by surface dust. Overall, the work was in an extremely fragile state, in dire need of treatment prior to replacing it in its original frame.

After removal from the rag board support, the pastel was laid face down to facilitate removal of the Japanese paper hinges and paper patches adhered to the verso. A pre-stretched piece of silicone release paper served as its temporary support. The hinges, old repairs, and crusty pieces of protein-glue residue were softened with a 2% methyl-cellulose poultice and then removed mechanically with a spatula. Fragments of the original cloth backing were found underneath the paper patches. They too were removed from the verso of the painting and later analyzed.

The colors were tested for their sensitivity to water, ethanol, and a number of solutions with different proportions of water and ethanol. Exposure to water slightly darkened the light blue pigment. There was, however, no visible color change when the pigments were treated with ethanol and water/ethanol solutions. Tests indicated that a 50:50 solution of ethanol and water was the optimal choice for the removal of water stains. To facilitate penetration of the solvent, the painting was first '"relaxed" by placing it on a humidified sheet of Gore-Tex® under the dome of a suction table for about 30 minutes. It was then placed face-up on a sheet of blotting paper on the suction table, which was set to operate at 50 millibars of pressure. The water and ethanol solution was delivered locally to the tide lines of the stains as a fine mist, using an ultrasonic mister.[6] The surface was slightly wetted and allowed to dry slowly under the dome. This process was repeated several times until no stain residue was visible on the blotter.

Once the stains had been reduced as much as possible, the portrait was repaired. The tears around the perimeter were joined with a Japanese *kozo* tissue, and small losses were filled using tinted Japanese paper. The pastel's four edges were extended with a 1 in. wide strip of Japanese paper of similar color, texture, and weight to the original sheet of brown laid paper. These extended edges would later be wrapped around a new secondary support panel. All repairs were applied with rice starch paste, which prior to use, was placed on blotting paper to absorb excess moisture. The repaired pastel was then '"relaxed" again using a humidified sheet of Gore-Tex® under the dome of the suction table and left to dry slowly with light pressure applied to its extended edges. Powdered pastel pigments and a fine brush were used to tone down several serious abrasions and to minimize what remained of the water stains. Retouching was minimal; no attempt was made to inpaint areas of heavy pigment loss.

The medium's sensitivity to water and the primary support's tendency to stain easily when dampened precluded the use of an aqueous lining system. Following literature review, consultation with a paper conservator at the Canadian Conservation Institute (CCI),[7] and a number of tests and mock-ups, an ultra-light, machine-made, Japanese *kozo* tissue was chosen as the lining material, and Rhoplex N-580, an emulsion of a pressure-sensitive acrylic polymer, was selected as the adhesive. A 1992 CCI report on adhesives confirmed that Rhoplex N-580 is one of the more stable and neutral acrylic adhesives and is usually used for pressure-sensitive applications.[8]

A sheet of the *kozo* tissue was cut slightly larger than the dimensions of the pastel with its extended edges. It was then placed on a piece of Teflon® cloth, brush-coated with Rhoplex N-580 diluted with distilled water (1 part adhesive to 3 parts water), and left to air-dry for several hours. When peeled off the Teflon®, the lining's underside displayed a thin, uniform coating of the tacky adhesive. It was transferred adhesive-side up to a non-stick polyester reemay sheet. The pastel was placed face-up on the lining material, and the two pieces were transferred as a unit onto the suction table. The suction was slowly increased to the maximum pressure of 30 millibars. After 30 seconds at maximum pressure, the suction was gradually reduced to zero. The lining tissue appeared to adhere thoroughly across the entire pastel verso. Although it is extremely light, it provides substantial support. In addition, the acrylic adhesive does not impregnate the paper fibers, so the lining is reversible and can be removed mechanically or with the application of acetone, methanol, or hexane.

The wooden strainer onto which the pastel was originally mounted has been lost, and it was therefore decided that a replica of the original strainer would not be created. Instead, a supporting panel of museum mounting board would be used. Three layers of four-ply museum mounting board were covered with a cushioning layer of perma-life paper to provide added support and dimensional stability. The restored pastel was placed face-up on the panel. Its extended edges were wrapped around the panel and adhered to the verso with Lascaux 360-HV adhesive. Two original manuscript labels attached to the verso of the frame were restored, lined with a Japanese tissue, and attached to a new backing board. The original frame was modified – its thickness was doubled – to accommodate new glass, new spacers, the restored pastel painting attached to its panel support, and a backboard. A sheet of Plexiglas seals the frame on the back, so the manuscript labels remain visible.

The *Portrait of Marie-Anne Cherrier* (Fig. 1.44) and the *Portrait of Jean-Marie Cherrier* (Fig. 1.45) depict the maternal aunt and uncle of Louis-Joseph Papineau. The two portraits arrived in the conservation laboratory in 1993 to be examined prior to being displayed in an in-house exhibition. Their condition was not as poor as the pastel described above, but the pastel surfaces were more friable. In contrast to the conservation treatment of the portrait of Louis-Joseph Papineau, the work on these pastels concentrated mainly on surface cleaning of the pastel layer, and on modification of their original frames. Both pastels were executed on brown laid paper supports measuring 380 x 300 mm. They were solidly mounted onto linen fabric that was stretched around and adhered to the verso of a wooden strainer. The fabric was secured with flat-head tacks around the sides of the strainer. Both works were framed in what appeared to be their original, black-painted wooden frames, with ill-fitting original glass that was uneven in thickness and severely curved. This resulted in gaps between the glass and frame, which allowed for a heavy deposit of dust and dirt on the surface of the paintings. The lack of a backing board and a paper dust seal on the verso of the frames also encouraged the accumulation of dirt.[9]

The design layer of the portrait of Jean-Marie Cherrier had sustained considerable loss of pigment, especially from the hair and forehead areas, and from all four corners. The paper support and fabric were slightly water-stained along the right edge, though the effect on

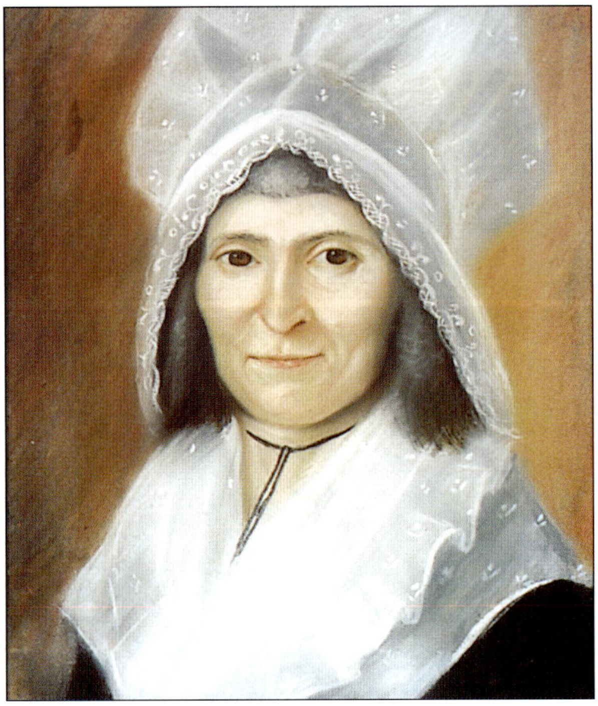

Figure 1.44 Louis Dulongpré, *Portrait of Marie-Anne Cherrier*, c. 1795. Pastel on paper, 380 x 300 mm. National Archives of Canada.

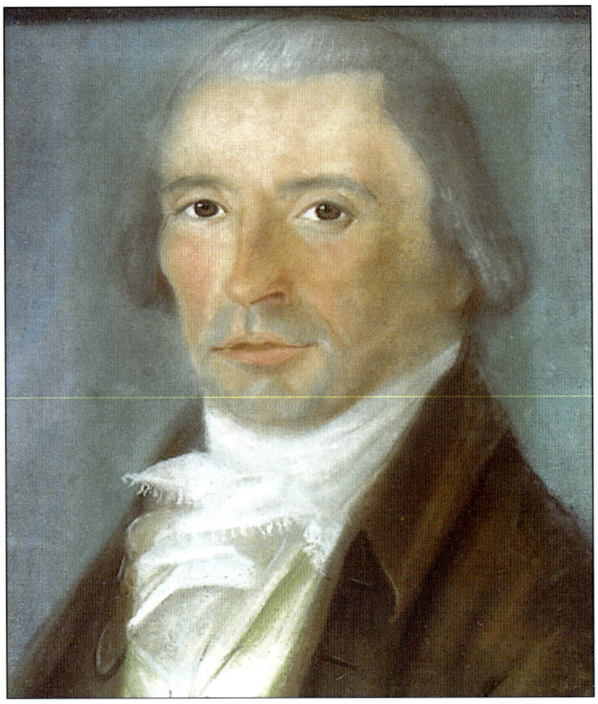

Figure 1.45 Louis Dulongpré, *Portrait of Jean-Marie Cherrier*, c. 1795. Pastel on paper, 380 x 300 mm. National Archives of Canada.

the pastel layer was minimal. In the portrait of Marie-Anne Cherrier, there was a small loss of the paper support at the top left corner, and the perimeter of the sheet was lifting. Both the fabric and primary support were water-stained. Fortunately, here too the pastel layer was only slightly affected.

The heavy layers of dust on the pastels' surfaces were reduced greatly by working under a microscope with a mini-vacuum device. A fine Japanese brush was used to dust the cloth on the versos of both works. The lifting edges of the primary support in the portrait of Marie-Anne Cherrier were re-adhered with a rice starch paste. Small abrasions in the design layer of the portrait of Jean-Marie Cherrier were minimally retouched with powdered pastel pigments. No attempt was made to inpaint areas of heavy loss.

The sheets of glass that glazed the portraits were of poor quality: greenish in color, of uneven thickness, and containing air bubbles. Both sheets were also severely curved (that of Jean-Marie Cherrier more than that of Marie-Anne Cherrier), causing slight optical distortions of the images. This type of glass was made by blowing spheres, which were subsequently cut into cylinders and opened flat while still malleable. It was more readily available and much less expensive than the superior plate glass variety, which was produced by pouring molten glass onto iron tables.[10] Although neither sheet of glass used for the Cherrier portraits fits properly into its frame and though both lack the properties that are considered desirable for glazing pastels, it was felt that the glass bore important artifactual information and should not be replaced with modern glazing. Conservators and curators also felt that in a strange way these sheets suited and complemented the portraits. Their odd shapes were accommodated by inserting them between two layers of specially designed spacers. Museum mounting board strips cut to the shape and thickness of the glass and covered with black felt paper were used to seal the frames tightly and protect the pastel paintings from future dirt and dust accumulation (see Fig. 1.46).

The frames then had to be modified. The frame thickness was doubled in the style of the original; the additions were painted black with an acrylic paint and a Plexiglas sheet was added to seal the backs. Fragments of newsprint from the verso of the original frame of the portrait of Marie-Anne Cherrier were consolidated and then removed, washed, deacidified, lined with a Japanese tissue, and stored separately from the pastel.

The nature and history of pastel portraiture in early Canada remains largely unexplored, and Louis Dulongpré's oeuvre has been little studied. The dismantling and examination required to treat the pastels provided an ideal opportunity for the material analysis of the three portraits. The CCI's Analytical Research Services Laboratory therefore carried out pigment, adhesive, and fiber analysis. The pigment and adhesive samples were first analyzed by Fourier transform infrared (FTIR) spectroscopy, using the diamond anvil microsample cell and the infrared microscope accessory

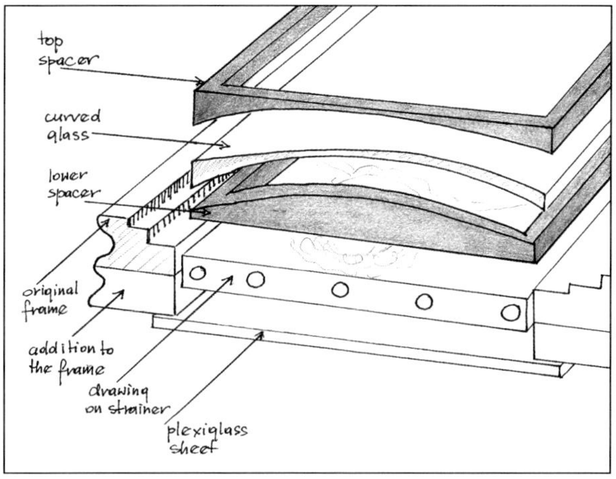

Figure 1.46 Cross section of spacers designed and constructed for the irregular sheets of glass glazing the pastel portraits of Marie-Ann Cherrier and Jean-Marie Cherrier.

as required. Pigment samples were then analyzed by X-ray microanalysis (scanning electron microscopy). Several pigment samples were also analyzed by X-ray diffraction in order to identify crystalline components. Finally, certain pigment and fiber samples from the portrait of Louis-Joseph Papineau were examined by polarized light microscopy.

The pastel portrait of Louis-Joseph Papineau was found to be drawn on a brown laid rag paper composed of well-beaten linen and cotton fibers. The paper had been attached to the fabric support with a collagen-type protein adhesive such as bone glue or rabbit-skin glue. The adhesive attaching the portraits of Marie-Anne and Jean-Marie Cherrier to the fabric supports was a mixture of starch and a collagen-type protein. Similar pigments were used in all three pastel portraits. Typical samples of the pastels consist of calcium carbonate, often containing a minor amount of gypsum, and a mixture of pigments. Red and yellow iron-containing pigments such as ochres and umbers, Prussian blue, bone black, charcoal black, and vermilion were used in a variety of mixtures to provide brown, blue, black, red, and green colors. All the pigments identified in the pastel samples were in common use at the end of the eighteenth century.[11] Examination of pigment particles by polarized light microscopy revealed that the particles of Prussian blue in the portrait of Louis-Joseph Papineau resembled those of ultramarine rather than the modern blue pigment.[12]

Pigments such as Prussian blue, vermilion, ivory black, red ochre, and yellow ochre were advertised for sale in Québec's newspapers as early as 1765. France, and then England (from 1763), were the main exporters of art supplies to Canada at that time. Painting materials were also brought to Canada by artists coming from the United States. However, the geographical location and economy forced painters in Québec to fabricate their materials from local natural resources. In a February 1827 lecture to the Québec Literary and Historical Society, its secretary and treasurer William Green discussed the native Canadian pigments derived from local earths, clays, and plants.[13] Further historical research and scientific, stylistic, and iconographic analysis of materials and techniques will be required to assist in a better understanding, dating, and attribution of the insufficiently recognized pastels of Louis Dulongpré.

Sunset on Thunder Bay: Silver Islet Seen from the Main Land (Fig. 1.47) represents one of very few known works in pastel by William Armstrong and certainly one of his most technically accomplished. In 1851, at the age of 30, Armstrong emigrated to Canada from Ireland as an apprentice civil engineer. Having previously studied art in Dublin, he immediately began to paint the landscape, concentrating on the Great Lakes. Armstrong is known for his documentary pictures of Canada, executed mainly in watercolor, and his works were reproduced in many of the illustrated papers of the day. In 1995 this pastel was requested for loan by the Thunder Bay Art Gallery for an Armstrong retrospective. The loan was significant because it was in this region, on the west side of Lake Superior, that the image was painted. The general feeling at the Archives, however, was to refuse the request as the work had traveled extensively in the past. As a step in loan consideration, the work was sent to conservation for evaluation. This revealed that the artist worked on a light brown, medium-weight laid paper and left the pastel unfixed. Microscopic examination confirmed that in the lower foreground and the left side of the composition, Armstrong brush-applied gouache directly on top of the pastel to delineate fine details; the pastel layer over which the gouache was applied is generally thinner and less built-up than in other areas. Microscopic examination also revealed that the bond between the gouache and the pastel was precarious: several areas exhibited active flaking (see Fig. 1.48).

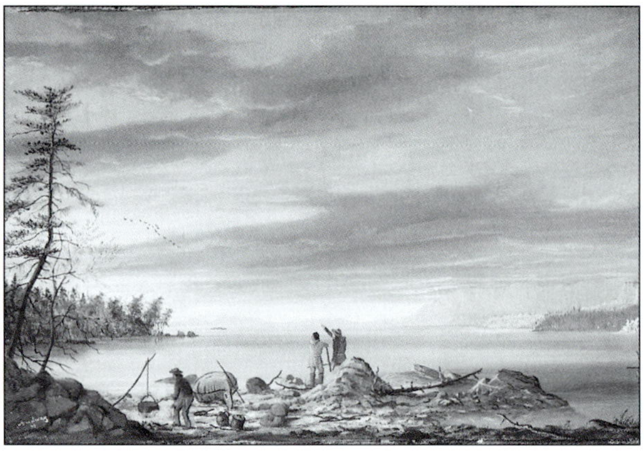

Figure 1.47 William Armstrong (Canadian, 1822–1914), *Sunset on Thunder Bay: Silver Islet Seen from the Main Land*, 1869. Pastel and gouache on paper, 362 x 616 mm. National Archives of Canada.

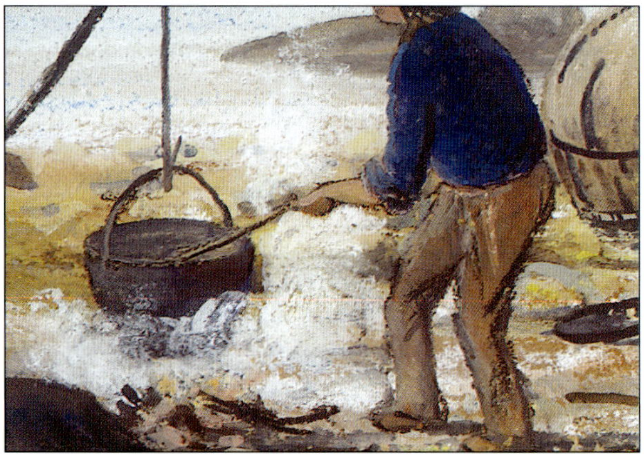

Figure 1.48 William Armstrong, *Sunset on Thunder Bay: Silver Islet Seen from the Main Land* (detail of flaking gouache from the foreground).

Because an ultrasonic mister had been used to consolidate gouache just prior to this evaluation, the conservation staff was keen to test its effectiveness in this application. The custodial, conservation, and curatorial team decided to reserve their final decision on the loan to Thunder Bay until treatment was completed. Then-recent international retrospective exhibitions such as *Degas* (1988) and *Odilon Redon: Prince of Dreams 1840–1916* (1994) were examples in which pastels had successfully traveled.[14] Consequently, staff felt that if the results of the treatment were favorable, loan of the piece was possible.

A method of using the ultrasonic mister for consolidation was developed in consultation with Anne Maheux at the National Gallery of Canada and Sherry Guild at the CCI, both of whom had used the mister successfully for similar applications. The conservator commenced by testing the technique on mock-ups of gouache over pastel loaned by the National Gallery of Canada. Conservation concerns were essentially twofold: that the technique should not alter the appearance of the media (such as color saturation and surface characteristics), and, secondly, that the technique be effective in consolidating the gouache. Testing allowed the conservator to determine the most effective and least invasive consolidant and to develop some facility with the equipment and technique.

Following testing of several different concentrations of gelatin and methyl cellulose, the consolidant of choice was found to be a 1% solution of 100 bloom gelatin in water. The standard mister set up follows CCI guidelines.[15] The treatment was conducted on the suction table to contain the mist and keep the support flat during treatment and drying. Four successive applications of consolidant resulted in no visible change and were sufficient to achieve adhesion. The mist's volume and rate of delivery were controlled by varying the aperture of the aerosolizer control valve, situated on the Nalgene bottle containing the consolidant. The treatment was carried out under the microscope and its effectiveness determined by a very gentle prodding of the flaking area with a fine brush. After all affected areas had been treated successfully, the decision was made to proceed with the loan.

A method for framing and transporting this fragile work was then devised. The framing system had to be suitable for exhibition and for long-term preservation and access needs. A suitable period frame of low historic and monetary value and minimally ornamented was selected and sized to fit the work. The verso of the frame was built up to accommodate the glass, spacer, and backing board. Both the spacer and the addition to the back of the frame were toned to match the frame. The wooden, ¼ in. spacer was lined with four-ply rag board on the edge in contact with the painting, and the work was hinged onto a secondary support of four-ply rag board. Because it was decided not to mat the work, the frame was sized to leave a 0.5 cm border around the work.

The choice of glazing is a critical factor when transporting pastels, since it is considered essential that they travel horizontally to ensure that gravity and vibration do not displace friable media. Old, brittle glass is more likely to break as a result of significant vibration and shocks, and it is therefore important to select a glass that is strong and less likely to cause damage if it were to break. Laminated glass, the interlayer of which holds pieces together if broken, was used in this case.[16] The work would travel approximately 1,500 km by truck with an experienced art handler. A large, heavy crate was constructed to accommodate it lying horizontally, together with several other framed, non-pastel items being loaned to the exhibition. The work was loaned and returned without incident.

The pastel collection at the National Archives of Canada is and always will be an actively used one. Previous storage systems often did little to protect the fragile surface of pastels, and previous treatments frequently altered pastels with little acknowledgment of their original presentation, the intent of the artist, or the aesthetics of the period. Today, less invasive treatments of pastels that respect their integrity without compromising their stability have taken precedence. Adapting period frames to allow for the safe housing of pastels, either for long-term storage or for short-term loan, has allowed us to present some of these works in accordance with original sensibilities.

The rehousing project discussed in the previous paper by Gignac and Hill (p. 49), and the treatments and considerations for travel discussed in this paper have allowed the National Archives to successfully fulfill two of its primary mandates – preservation and access. Housing this collection, whether for a move to a new facility, for loan and exhibition, or to respect the aesthetic sensibilities of the period and the artist, required an integrated approach. The experience has proven to us that involving all parties concerned – conservators, custodians, archivists, circulation staff, physical control staff, and technicians – in the early planning stages of these projects was essential to their successful realization.

Acknowledgments

The authors acknowledge with gratitude the help of Claude Belleau, Conservator at the Musée du Québec; Jean-Noél Dion, Archivist at the Séminaire du Saint-Hyacinthe; Francine Gauthier, Conservator at the Centre de conservation du Québec; Sherry Guild, Conservator at the Canadian Conservation Institute; Anne Maheux, Conservator at the National Gallery of Canada; and Suzanne Saint-Amour-Colas at Societé du patrimoine religieux du diocése de Saint-Hyacinthe for their assistance in the art-historical research, helpful discussions, and advice. Special thanks are extended to Alison Bullock for her invaluable help in editing this paper.

Endnotes

1. R. Derome et al., *Dulongpré. A Closer Look* (Montreal: McCord Museum of Canadian History, 1989); J. R. Harper, *Painting in Canada: A History* (Toronto: University of Toronto Press, 1994), p. 56.

2. His son, Louis-Joseph, Jr., became a painter himself and helped his father on larger projects. He is also believed to have made copies of some of his father's work. From 1787, Dulongpré worked as a teacher of music and dance in a school for girls and boys that he established in Montreal. Concurrently, as co-founder and theatrical manager of the Théâtre de Société, he returned to painting in the form of stage sets. Deriving little income from these pursuits, Dulongpré changed careers in 1792 and took up painting. He went to the United States to complete his training and stayed, mainly in Baltimore, from June 1793 until March 1794. Upon his return he advertised in the *Québec Gazette* (10 Apr. 1874) that he had "lately arrived from the Colonies, where he has improved in the Art of Drawing under the best Academicians and would paint in Miniature and Crayons."

There were very few professional painters in Canada at that time. After 1759 there were British military artists present, but their favorite subject was landscape and their chosen medium was usually watercolor. During the summer months, itinerant artists from America journeyed through Québec City and the countryside, many supporting themselves by painting portraits. Local contemporaries of Dulongpré included François Baillairgé, who painted in Québec City; and François Malepart de Beaucourt, John Ramage, and Louis-Chretien de Heer of Montreal. All these men were capable artists, and all made portraits in pastels; however, Louis Dulongpré used the medium most.

3. R. S. Levenson, "Materials and techniques of painters in Québec City, 1760–1850," *The Journal of Canadian Art History* 7, 1 (1983), p. 24.

4. Dulongpré supplemented his income by occasionally painting religious pictures for parish churches and convents, and by doing architectural work and gilding. He may also have been one of Canada's first restorers, as he was known to touch up canvases from the Desjardins' collection, which was shipped to Québec from Paris by the Abbot Phillip-Jean-Louis in 1816; ibid., p. 27.

5. Derome et al. 1989 (as in note 1), p. 11.

6. For the effects of water treatments on pastels on the suction table, see V. Daniels, "The effects of water treatments on paper with applied pastel or powder pigment," *The Paper Conservator* 22 (1998), pp. 29–38; J. Kosek, "The porosity of pastels and the effect of water treatment on the suction table: A preliminary investigation," *The Paper Conservator* 14 (1990), pp. 17–23; M. K. Weidner and S. Zachary, "The system moisture chamber / suction table / ultrasonic humidifier / air filter," in H. D. Burgess (ed.), *Conservation of Historic and Artistic Works on Paper, Symposium* 88 (Ottawa: Canadian Conservation Institute, 1994), pp. 109–15.

7. S. Guild, personal communication, Canadian Conservation Institute, 1996.

8. J. Dawn et al., "Adhesive testing at the Canadian Conservation Institute – An evaluation of selected poly-vinyl-acetate and acrylic adhesives," in *Environment and Deterioration Research Report No. 1603* (Ottawa: Canadian Conservation Institute, 1992).

9. Two paper labels with manuscript inscriptions in black ink were adhered to the verso of the fabric of the portrait of Jean-Marie Cherrier, and an old newsprint clipping featuring a reproduction of a portrait of Benjamin-Hyacinth-Martin Cherrier had been glued horizontally below the paper labels. A paper label with a manuscript inscription in black ink was attached to the verso of the fabric of the portrait of Marie-Anne Cherrier. Residues of embrittled and discolored newsprint remained adhered to the back of the strainer and frame.

10. This latter method was developed in Europe around 1700 but was rare in Canada in the eighteenth century and even into the early nineteenth century. See T. Burns, "The historic framing and presentation of European pastel portraits in the early eighteenth century," in N. Bell (ed.), *Historic Framing and Presentation of Watercolours, Drawings and Prints* (Worcester: Institute of Paper Conservation, 1996), pp. 10–20; J. Simon, "The production, framing and care of English pastel portraits in the eighteenth century," *The Paper Conservator* 22 (1998), pp. 10–21.

11. Canadian Conservation Institute Analytical Report ARS 3519. 1996. Analysis of Three Pastel Portraits by Louis Dulongpré. 6 June 1996.

12. The same observations have been made regarding samples of original paint taken from the eighteenth-century Independence Hall in Philadelphia; samples from works by J. M. W. Turner; and samples of original blue paint of the sculpted décor of the Ursuline Chapel in Québec City, painted between 1726 and 1736 (see CCI Analytical Report ARS 3519, 6 June 1996). See also F. S. Welsh, "Particle characteristic of Prussian blue in a historical oil paint," *Journal of the American Institute for Conservation* 27, 2 (1988), pp. 55–64; J. H. Townsend, "The materials of J. M. W. Turner: Pigments," *Studies in Conservation* 38, 4 (1993), pp. 231–55; and C. Payer et al., "The interior décor of the Ursuline Chapel in Québec City: Research in conservation," in Dorge and Howlett (eds), *Painted Wood: History and Conservation* (Malibu: Getty Conservation Institute, 1998).

13. Levenson 1983 (as in note 3), p. 11.

14. *Degas* was presented at Galeries Nationales du Grand Palais, Paris (February–May 1988); National Gallery of Canada, Ottawa (June–August 1988); The Metropolitan Museum of Art, New York (September 1988–January 1989). *Odilon Redon, Prince of Dreams 1840–1916* was presented at The Art Institute of Chicago (July–September 1994); Van Gogh Museum, Amsterdam (October 1994–January 1995); Royal Academy of Arts, London (February–May 1995).

15. C. Dignard, R. Douglas, S. Guild, A. Maheux, and W. McWilliams, "Ultrasonic misting. Part 2, treatment applications," *Journal of the American Institute for Conservation* 36 (1997), pp. 127–41.

16. Considerable work has been done on this topic by The Art Institute of Chicago, as reported by Harriet Stratis, "Pastels and charcoal drawings in travelling exhibitions: Guidelines for their preparation, transport and installation," presentation to the Book and Paper Group, American Institute for Conservation Annual Conference, Arlington, Va., 1–7 June 1998.

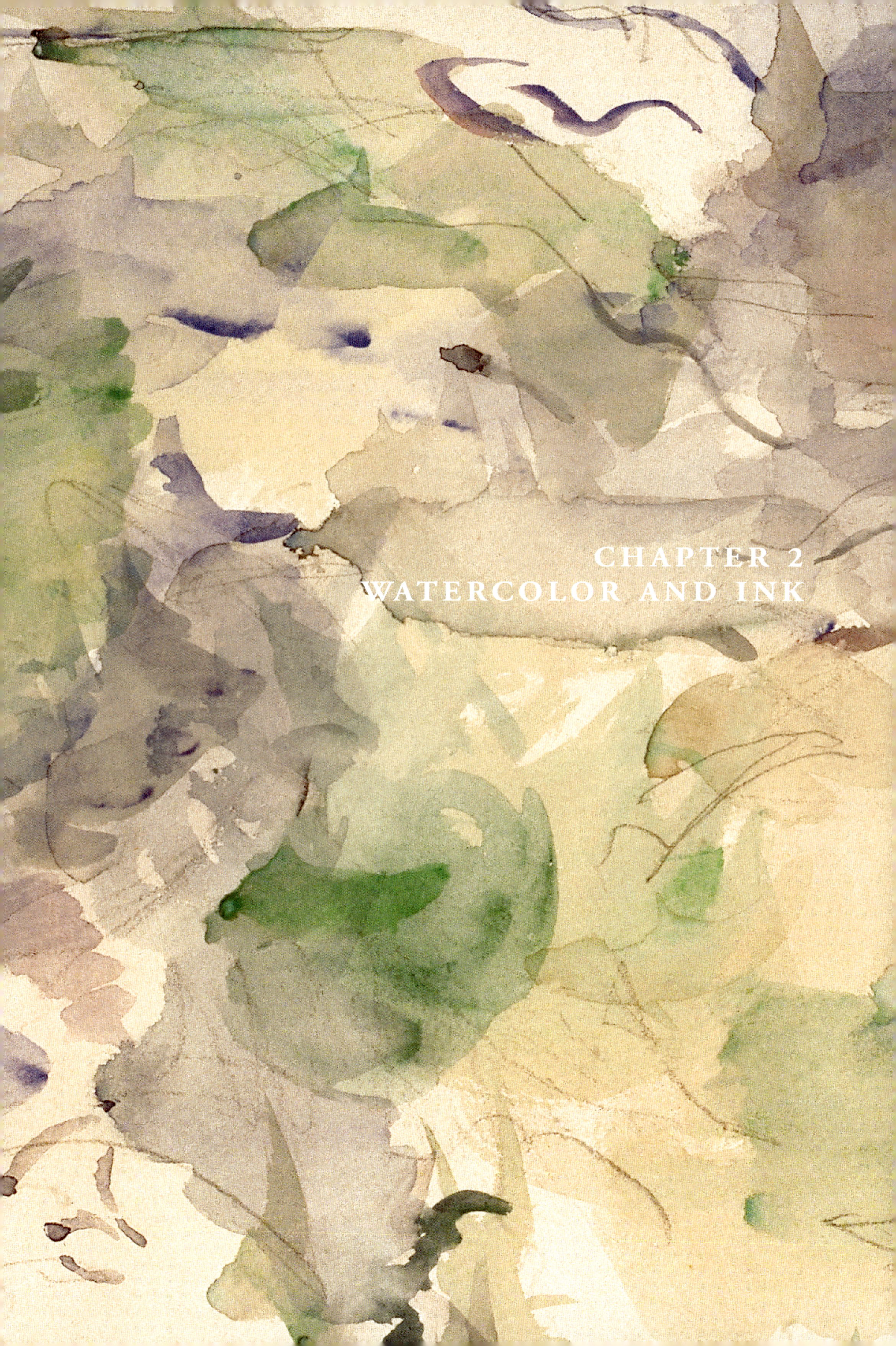

CHAPTER 2
WATERCOLOR AND INK

Cause and Effect in the Historical Development of Watercolor

Marjorie B. Cohn

I last thought seriously about the history of watercolor in the historical past – 1977. Since then, I have left conservation and become a print curator, and so instead of revisiting *Wash and Gouache,*[1] I shall summarize the history of watercolor by focusing on the work of a great printmaker, Albrecht Dürer (1471–1528).

At the beginning of the twenty-first century, watercolor as a genre of art implies certain physical properties and even a certain style. One thinks of purity of color, transparency of the pigment layer over white paper, and the potential for freedom in handling. These qualities are perceptible in even the earliest works that we, with the wisdom of hindsight, describe as watercolors: works by Albrecht Dürer. By his lifetime, there was, of course, a vast corpus of images made with water-based painting systems using opaque pigments applied with brushes, or transparent dyes applied with pens to papyrus, parchment, and cloth, as well as paper. I shall follow the majority of the consciously historical literature on watercolor in excluding these from the definition of the medium.

A consideration of Dürer's watercolors represents a useful beginning for more than chronological reasons.[2] His watercolors are of three distinct types, as defined by differences in technique, and these three modes remain a constant in the use of the medium to this day. In his first mode transparent and opaque media are applied with fine, discrete strokes; in his second mode he uses broad applications of wash to fill in pen outlines; and in his third mode he turns to the use of fully transparent washes. Among his most famous works in any medium are his natural-history studies in what I call his first mode of watercolor. In these, the pigment is usually applied in fine strokes, or at least in discrete touches, which carry so little medium that the paint dried rapidly and the sequence of strokes could have been relatively continuous. Among them I refer to *Large Piece of Turf* (Fig. 2.1) and *Young Hare* of 1502 (Albertina, Vienna). It is the artistic concept that is new here; contemporary illuminators would have recognized Dürer's materials and technique.[3]

It was this kind of watercolor, descriptive of natural, usually living, things, to which the earliest manual writers addressed themselves. Instructions are found up until the nineteenth century itemizing the pigments specific to the species or substances to be represented.[4] Even in the nineteenth century, after the British school of watercolor had perfected a transparent wash style, the same precision of touch and willingness to mingle opaque and transparent pigments that we see in Dürer's work in his first mode was maintained. William Henry Hunt (1790–1864) specialized in natural objects, although he suffused nature's objects with Victorian sentiment.

This transparent/opaque play reached its epitome in the watercolors of John Ruskin (1819–1900) and in his instructional manual, *Elements of Drawing.* I do not think it accidental, or simply a response to Winsor & Newton's aggressive marketing of Chinese white – the

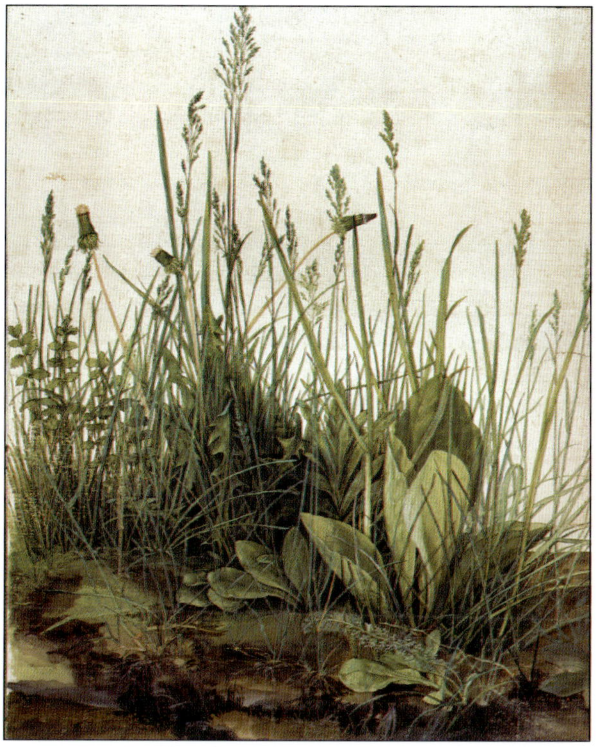

Figure 2.1. Albrecht Dürer (German, 1471–1528), *Large Piece of Turf*, 1503. Watercolor on ivory laid paper. Albertina, Vienna, 3075 D54.

firm's proprietary brand of zinc carbonate body color – that Ruskin recommended a blend of opaque and transparent (though he did indeed specify the use of Chinese white). He acknowledged that the mixture of color with opaque white was considered "illegitimate" by his contemporaries, among whom the transparent wash style reigned supreme; but for him, "body-colour is, in a sketch, infinitely liker nature than transparent colour."[5] Note the word "nature."

In his manual Ruskin specified 24 colors by their brand names, as well as Chinese white. By the nineteenth century, these were all easily found prepared for use, but earlier manuals usually give recipes. In *The Excellency of the Pen and Pencil...* (London, 1668),[6] we are told how to refine pigments from raw materials and also how to grind and gum them, and perhaps add sugar candy for humectation. The need for these instructions was obviated by the rise of the commercial colorman.

The geographical progress of the art-supplies trade, based no doubt on the increasing demands of national art cultures as well as capitalism, can be inferred from a French manual and its English translation. The 1711 edition of *Traité de mignature*[7] does tell its reader how to make colors, but it also recommends a Parisian colorman who could supply them ready-made. The 1733 translation, *The School of Miniature*, does not have a comparable London colorman to recommend. By the end of the eighteenth century, however, *The Artist's Assistant* reported to its British readers that "Most of these colours may be had ready prepared in shells [or] at the colour shops in powders."[8] The powders would still have to be ground and gummed by the artist, for which instructions are given, but there are no instructions for making pigments from raw materials.[9]

An interesting way to interpret this progression is as a reflection of the rapid spread of imperialism abroad and industrialization at home, which was chronologically coincident with the development of watercolor. This was no accident: gentlemen and military officers (often the same thing, of course) frequently underwent training in drawing and watercolor painting in order to fit them as observers of nature, topography, and culture in colonial possessions. These painters were not professional artists, and they would have had neither masters to teach them the chemistry and craft of color manufacture nor apprentices to do the dirty work. At home and especially abroad, they depended on a cadre of commercial suppliers. Colormen in turn benefited from access to new sources of raw materials secured by military men and traders.

Indian yellow turned up on European palettes only in the late eighteenth century. By the late nineteenth century, Indian yellow had become so generic a nomenclature that its specific hue required simulation by cadmium-based pigments when they became more readily available by the end of that century. Indigo, also largely imported from India though known in Europe from medieval times, did not really displace woad, the native equivalent blue, until the eighteenth century. As an aside, I will concede that the name "Chinese white" doubtless derives not from British rule of Hong Kong but from Winsor & Newton's promotion of the paint's porcelain-like whiteness, hardness, and fine texture.

But to continue my musings on the geopolitical economy of pigments: Indian red, a lake that had been made since antiquity from the Old World's kermes insect, was displaced, or at least supplemented, by carmine or Spanish lake, made from the cochineal insect found in Mexico after the Spanish conquest. Indian lake is somewhat cooler in hue than Spanish lake, but the two were often confused or conflated. By the mid-nineteenth century, however, instead of cochineal, Mrs. Merrifield would recommend madder, processed industrially in Europe rather than by natives in colonial outposts.[10] A transparent crimson derived from plant roots rather than insects, it had been used as a dye for centuries but was only lately recognized as a watercolor pigment. Industrial chemistry's synthetic alizarin was just around the corner.

With these transparent reds in mind, let us return to Dürer, and what I propose as his second mode of watercolor, as exemplified in *Madonna with Animals in a Landscape* (Fig. 2.2). Its technical characteristics are self-evident: this is a pen drawing filled in with appropriate, though schematic, colors. Note how transparent the washes are in contrast to the natural-history studies of his first watercolor mode. However, Dürer did use his second mode for natural-history specimens such as *Costume Study of Three Lithuanians* of 1521 (Louvre, Paris) – that is how I would describe his attitude toward these Lithuanians in native dress – and also for landscapes such as the undated *Covered Footbridge by the Hallertor in Nuremberg* (Albertina, Vienna), subject categories that are more comfortably associated with what I define as his first and third modes of watercolor painting. The second mode continued to be used by artists for original drawings, notably in the seventeenth century by Adriaen van Ostade (1610–1685) and in the nineteenth century by Jean Auguste Dominique Ingres (1780–1867).

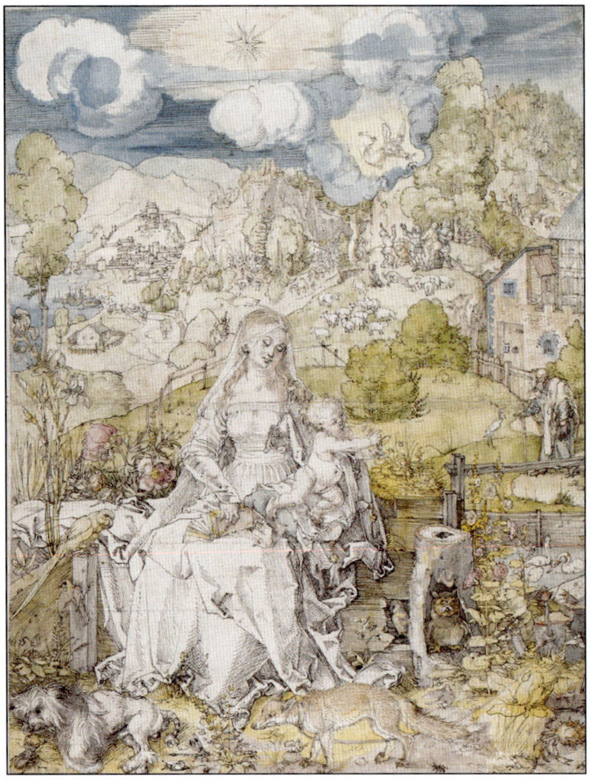

Figure 2.2. Albrecht Dürer, *Madonna with Animals in a Landscape*, 1503. Watercolor and pen and black ink on white laid paper. Albertina, Vienna, 3066 D50.

To return to Durer's *Madonna with Animals in a Landscape*, the historical antecedent for this mode of watercolor is clearly recognizable: it is simply a unique equivalent of the hand-colored multiple, the innovative fifteenth-century cut-rate alternative to manuscript illumination. To be sure, the watercolored print had its own sort of status, as a publisher could offer the same print uncolored even more cheaply. Hand-coloring of prints has continued right up to the present day as a means to vitalize their surfaces and also perhaps to enhance their market appeal. Even today, as in the lithographs of Howard Hodgkin (b. 1932), the traditional separation of the role of printmaker and colorist is maintained, though (as in the case of the Dürer woodcuts) it is still the artist-printmaker and not the painter-assistant who gets top billing.

I do not mean to denigrate the potential of print coloring for great achievements. The mature British style of watercolor permeated the print-coloring trade at its most ambitious and expensive level, as evidenced in the work of John James Audubon (1785–1851). In his work the prevailing transparent watercolor mode takes precedence over traditional natural-history painting, although Chinese white appears for occasional highlights even on Audubons. In raking-light views of his numerous studies of birds, one often sees that the darker and richer colors are heavily gummed to provide details of the plumage. This standard means to intensify coloring in natural-history illustration and portrait miniatures was, by the 1830s (the decade of the Audubon publication by Havell), also recommended in watercolor manuals for landscape foregrounds and foliage. But as with Dürer and Hodgkin, Audubon gets the credit, not the colorist.

Interestingly, A.-M. Perrot's *Manuel du coloriste*,[11] a nineteenth-century manual that describes in detail shop organization and materials and techniques for print coloring, continues to give instructions comparable to those in seventeenth-century manuals for the manufacture of pigments. But this section of the book precedes the Introduction, is not included in the Table of Contents, and is paginated in roman rather than arabic numerals. It is as if it were of use to only a minority of readers, perhaps those whose commercial operations were so large that they warranted pigment production.

In coloring prints, and especially maps, where boundary demarcations, internal detail, and textual information are of the essence, the overlying watercolor must be transparent. Rosamond Harley has noted that the eighteenth-century artists' manuals concerned with print coloring feature organic pigments such as red lakes, which are likely to be more transparent than mineral pigments.[12] Appreciation for the transparency and also the relatively light weight of organic pigments would become even more subtle, however, once watercolors were fully released from their association with print and map coloring in the development of the British landscape painting school of the late eighteenth and early nineteenth centuries. But of course I want to begin my discussion of this high point of historical watercolor art three centuries earlier, again with Dürer's *Madonna with Animals in a Landscape.*

Note the sky, where the billowing clouds are brushed on without an outline. It is a more artistic rendition of atmosphere than the washed bands that denote "sky" in numerous sixteenth-century German woodcuts, but the intention to render thin air thinly is the same. Dürer did not have the paper and paints that would contribute to the exquisitely transparent skies of nineteenth-century watercolors such as those by Winslow Homer (1836–1910). But he was sensitive enough to the problem not to anticipate the recommendation of the eighteenth-century French miniature painting manual cited earlier: to paint skies with a mixture of ultramarine and white, that is, with two opaque mineral colors that would not wash well.[13]

Gradually, less-than-solid forms became liberated from pen outlines. In the seventeenth-century watercolor drawing *Landscape with a Bridge and a Tower* (Fig. 2.3) by Hendrick Baretnsz. Avercamp (1585–1634), earth and architecture are described with pen and ink, but

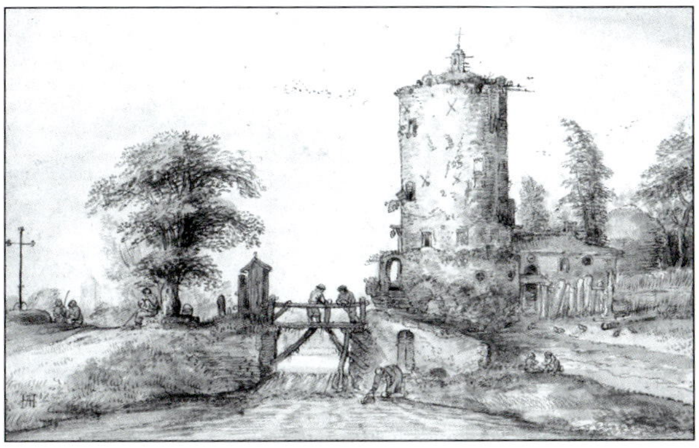

Figure 2.3. Hendrick Baretnsz. Avercamp (Dutch, 1585–1634), *Landscape with a Bridge and a Tower*, c. 1615–20. Brown ink, watercolor, and black chalk on off-white laid paper, 13.5 x 21.6 cm. Courtesy of the Fogg Art Museum, Harvard University Art Museums, Loan from Maida and George Abrams, 0025.1998.24. (Photo credit: Allan Macintyre.)

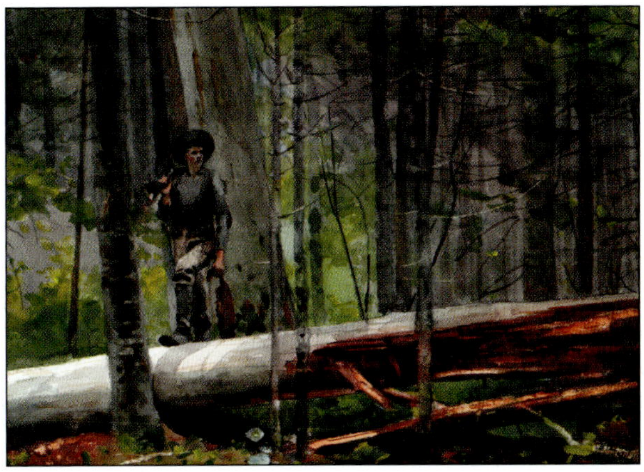

Figure 2.4. Winslow Homer (American, 1836–1910), *Hunter in the Adirondacks*, 1892. Watercolor over graphite on white paper, 35.3 x 50.7 cm. Courtesy of the Fogg Art Museum, Harvard University Art Museums, Anonymous Gift, 1939.2.30. (Photo credit: Allan Macintyre.)

brushed-on foliage, clothes, pen-drawn tree trunks, and water flow freely in watercolor. Additionally, you can see that warm or cool colors harnessed to stiff or soft drawing implements play a major conceptual role, a scheme that would be maintained by artists' manuals addressed to topographic draftsmen.

By the later nineteenth century, watercolor landscape painters would be completely free from the strictures both of pen outline and conceptualized coloring, even for the most weighty foreground elements. They would learn to incorporate, even into their rendering of foliage, the full spectrum of color and range of opacity and transparency available in the watercolorist's palette, where modern chemistry was multiplying the colors. There is no better example than Winslow Homer's *Hunter in the Adirondacks* (Fig. 2.4). Admire the contrast of opaque cadmium yellow and transparent mixed grays and blues to represent the reflection of filtered sunshine in deep shade. Note how the volume of the shattered log that defines the baseline of the composition is made up of overlapping transparent washes of brown and blue, and how its broken core was first hacked down to the paper's white by the artist and then slathered over with sienna.

More than 450 years of progress in watercolor materials and technique separate Homer from Dürer, but even if Dürer's landscapes are assessed according to the standards of a fully mature art form, those that he painted in what I define as his third mode are astonishing. Referring only to the brushwork, the artist easily passed from fully liquid washes to modeling with a dry brush. In the view of *Innsbruck from the North* (Fig. 2.5), look at his virtuoso rendering of reflections in water; and in his view of a sunset in *Weir-House* of c. 1498 (British Museum, London) note his seamless integration of the finest linear details, including white gouache highlights, over the broad and transparent stretch of water. Scholars date these from the very beginning of Dürer's career, in the last years of the fifteenth century. Works that seem incredibly progressive in fact were painted a half dozen or more years before the nature studies and the *Madonna with Animals in a Landscape* discussed above.

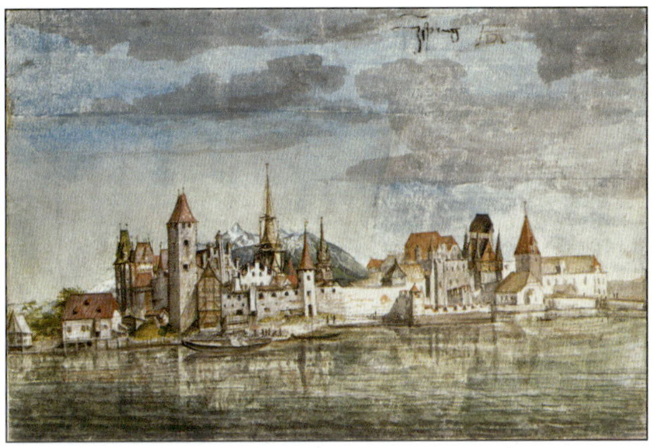

Figure 2.5 Albrecht Dürer, *Innsbruck from the North*, c. 1495. Watercolor on ivory laid paper. Albertina, Vienna, 3056 D36.

Although Dürer's polychromatic, freely brushed technique seems to have sprung full-blown, the later recovery of watercolor to this level passed through orderly stages, entirely predictable from the precedent of print coloring. The first stage not only preserved the convention of the outline, it also approximated printed crosshatching or aquatint shading with so-called dead coloring. In painting with dead coloring, values are established by modeling in gray, and transparent colored washes are then overlaid to establish the desired hues.

In the development of watercolor style, "neutral tint" – a compound of a warm and a cool color – supplanted gray made from diluted black. It gives a livelier cast to the finished work, but artists were understandably cautious about its components. It was well understood that the multiplication of pigment mixtures would lead to a "dingy and opaque" effect in a style where every artistic effort and technical component was geared toward brilliance and transparency.[14]

After the artist no longer prepared his own pigments, manual writers recommended appropriate mixtures, ones in which the chemical compounds were compatible and would blend with predictable results. One British landscape artist and teacher, William Payne (1760–after 1830), so popularized his personal blend of neutral tint that it joined the ranks of named colors. Payne's gray was originally a mixture of crimson lake and indigo, with perhaps an addition of lamp black or raw sienna. By the early twentieth century, it had retained its name but changed a component, with synthetic alizarin substituting for the earlier organic lake.[15]

Generally, the recommendation was to avoid mixed colors, and the transparent wash style, in which the dead coloring stage was eliminated and the artist laid down washes of pure hues, triumphed – somewhat more than 400 years after Dürer had painted his landscapes. I do not mean to imply that this style was the result of a desire to sidestep a certain technical problem. It was simply one facet of the grander Romantic style, the result of a complex weave of intellectual, aesthetic, and cultural developments.

And the romantic paradox – whether history produces the transformative man of genius, or the man of genius transforms history – even raises its head in watercolor. How much of the development of the medium's full technical bravura can be attributed to J. M. W. Turner (1775–1851), who arrived on the scene at the moment when commercial resources and public expectations had been fully prepared for his achievement?

Artists' manuals begin to describe the wash style of landscape representation – indeed, any sort of landscape painting – only in the later eighteenth century. I believe this hand-in-hand development was yet another concomitant of the rise of the colorman, who provided the new wove paper and packaged paints that fostered broad and transparent handling. His rise had a symbiotic relationship with the widening population of trained amateur painters, who depended upon the assistance of professional instructors and manual writers, as well as upon chemists and paint, paper, and brush manufacturers. The spate of nineteenth-century manuals published by art-supply firms such as Winsor & Newton was the natural outcome.

I might add that despite (or perhaps because) all this development in Britain was firmly based in that nation's pioneering industrialization – with its appetite for raw materials and its burgeoning middle-class wealth – the processes by which fine-art supplies were produced were increasingly distrusted. Once imperial industrialization of the production of artists' pigments was complete, manufacturers hastened to reassure customers that suspicions of a loss of quality that had formerly been guaranteed by craftsmen were unwarranted. Hence an 1889 Winsor & Newton advertisement depicts galley slaves putting finishing touches on watercolor pigments by hand in the Watercolor Room at Windsor & Newton's factory.

In any case, it was within Turner's lifetime that watercolor first became capable of the effects that define it to this day. An anonymous, late eighteenth-century British representation of the Globe Theatre (British Museum, London), purportedly dating from Shakespeare's lifetime, has been adjudged a forgery not only because it gets the theater wrong, it gets the rendering wrong. A transparent, wet-into-wet watercolor technique simply did not exist at the turn of the seventeenth century.

In the long term, no one could be assured of watercolors' appearance, whether in physical or optical mixture (that is, in superimposed transparent layers) if the conditions of viewing or care were inimical to the artist's paints, which may have been chosen in ignorance, the artist not having made them himself. And now there were new problems. For example, Mrs. Merrifield observed that cobalt blue would change its appearance in artificial light.[16] But by the mid-nineteenth century, when she issued this warning, who would know if cobalt blue was a component of a mixture?

Most important, of course, was the distortion of hue in a mixed color as a result of fading of one component. Let's look again at Dürer. His *Linden Tree on a Bastion* (Fig. 2.6), an outdoor natural-history study whose technique causes it to fall in my first category, makes it clear that he did have an adequate green at his disposal. Its relative opacity suggests a mineral pigment, which was entirely suitable to the dry and opaque handling of the artist's first mode. The washes in Dürer's broadly worked third-mode landscapes are far more transparent, and I wonder if their dominant, and unnatural, yellow tones were not once green, made with a light-resistant yellow pigment mixed with a fugitive blue. The alternative course of fading can be perceived in many cases, where foliage that originally must have been green has now faded to blue. Hooker's green, made from gamboge and indigo, became unpopular because of this liability.[17]

These few instances of one component of a mixed tint fading are my way of referencing an entire branch of historical pigment studies, that of the permanence of colors. Over the centuries, concern for permanence depended upon the mode of painting as well as the pigments available at the time. In the seventeenth and eighteenth centuries, prior to the developed practice of exhibiting framed watercolors but also prior to the banishment of opaque white from the watercolorist's palette, it was the chemical make-up of the pigment in whites that was distrusted, specifically the lead component that might blacken.[18] Other suspect colors included opaque vermilion, also known to have a tendency to blacken. Yet even before the development of the fully transparent wash style, there was recognition of fading associated with colors derived from plant sources. The same manual that cautioned against lead white remarked that blue from the "Cornbottle flower" was "an excellent Colour that will not fade for several Years."[19]

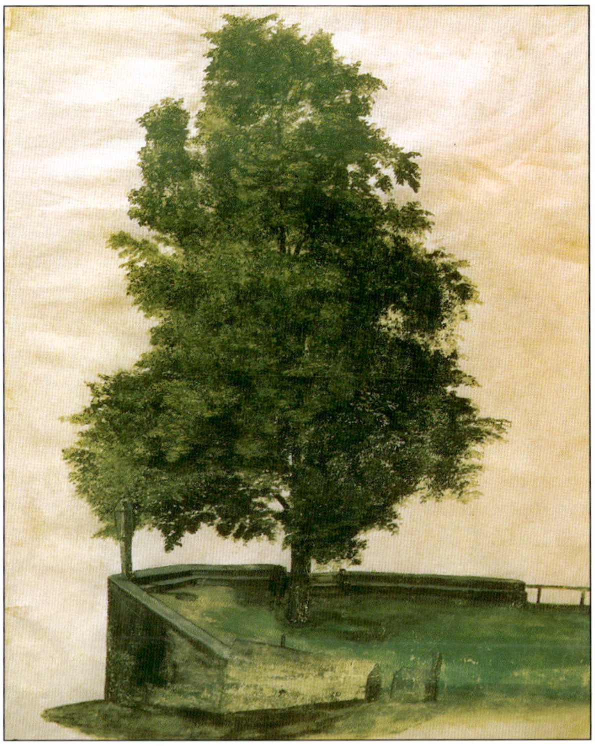

Figure 2.6 Albrecht Dürer, *Linden Tree on a Bastion*, c. 1494. Watercolor and gouache on cream laid paper. Museum Boijmans Van Beuningen, Rotterdam, MB1958/T28.

I have discussed in *Wash and Gouache* the tribulations of green in the artist's palette, which extended even to moral condemnation,[20] and I won't repeat myself beyond indicating that the issue became increasingly contentious as landscape painting in transparent watercolors developed. Let me here simply refer to the commonsensical opinion of George Field, a chemist and colorman obsessed with the spectrum who published in 1835 the first comprehensive and scientific study of the deterioration of pigments.[21] Field disallowed organic yellows and greens, and also organic blues (except indigo), because of their fugitive nature, but he pragmatically observed that mixed tints would only be as strong as their weaker component and suggested that greens should be made from components that faded at the same rate.[22]

Developments in industrial chemistry made Field's study obsolete almost immediately, to the dismay of artists who relied on his recommendations.[23] Uncertainty ultimately led to the disrepute of watercolor as a permanent medium, ironically just as science and industry began to collaborate on the development of truly permanent synthetic pigments. With the status of an official government inquiry, Russell and Abney repeated Field's work. Their 1888 report, replete with spectrographic analyses, makes a graphic point of the transformation of hue in pigment mixtures of unequal permanence that Field would have appreciated.[24] But by the end of the nineteenth century, watercolor lost its newly won institutional prestige and retreated to the status of a sketch medium not far above its beginnings.

It would perhaps be useful symbolism to end my very compressed summary of the historical development of watercolor with a scientific diagram, but I would rather end with Dürer's *Pond in the Forest* of c. 1496 (British Museum, London). His view of a forest pool at sunset is unfinished. I think of this watercolor, painted in the fifteenth century, as symbolic of the unfinished business to come.

Endnotes

1. M. B. Cohn, *Wash and Gouache: A Study of the Development of the Materials of Watercolor* (Cambridge, Mass.: Center for Conservation and Technical Studies, Fogg Art Museum, 1977).

2. See F. Piel, *Albrecht Dürer, Aquarelle und Zeichnung* (Cologne: Dumont, 1983) pp. 56–58, in which the author traces the roots of Dürer's innovations in watercolor to late medieval transformations of visual culture and painters' techniques, the function of light in art, and the private use of images. Dürer also worked with white body color and dark inks on colored grounds, a technique by definition excluded from this paper.

3. There are a few other European watercolors of this sort as early or even earlier than Dürer's works, notably a number by Pisanello of dogs and other animals.

4. See, for example [C. Bowles], *The Art of Painting in Water-Colours* (London: the author, 1786), in which instructions for 74 species of flowers are given. In 1774, Bowles also published an illustrated manual specific to flower painting "containing sixty plates of beautiful flowers … To which is added, an accurate description of their colours, with instructions for drawing and painting them according to nature," to quote the title page.

5. J. Ruskin, *Elements of Drawing* (1st edn, 1857; New York: Merrill & Baker, n.d.), pp. 141 and 143. The emphasis on nature is warranted, as Ruskin justified his polemics for Turner and the Pre-Raphaelites by their purported truth to nature.

6. I cite the editions that I have consulted; these may not, of course, be the earliest editions of these works.

7. *Traité de mignature, pour apprendre aisément à peindre sans maître, avec le secret de faire les plus belles couleurs …* (Paris: Christophe Ballard, 1711), p. 14. The author, traditionally thought to be Claude Boutet, has recently been identified as Christophe Ballard; R. D. Harley, *Artists' Pigments c. 1600–1835: A Study in English Documentary Sources* (London: Butterworths, 1970), p. 21.

8. The quotation is from the 1814 edition of this manual, *The Artist's Assistant in Drawing, Perspective, Etching, Engraving, Mezzotinto-Scraping, Painting on Glass, in Crayons, Water-Colours, and on Silks and Satins* (London: Henry Mozley, 1814), pp. 36–37, written by Carrington Bowles (1724–1793). I have been unable to check earlier editions to learn when this passage first appears. The earliest edition I have traced bibliographically, the third, dates from 1772.

9. Interestingly, in the late nineteenth century, a kind of reverse watercolor instruction manual for naturalists – R. Ridgway, *Nomenclature of Colors for Naturalists* (Boston: Little, Brown, 1886) – lists traditional color names, which imply the earlier centuries' artists' familiarity with their ingredients, together with their latter-day realization in every conceivable contemporary pigment further identified by its commercial supplier. It even includes as a red *either* Winsor & Newton's madder carmine *or* Schoenfeld's deep madder lake, which presumably were the same hue. Rather than prescribing what color to use for what bird drawing (Ridgway's manual was primarily directed toward ornithologists), the manual instead gives actual color samples and the color mixtures that produced them, so that the naturalist could compare his feathered specimen to the tables and describe it according to names of hues which were defined by their widespread familiarity through watercolor painting. In the chart of yellows, for example, the samples are referred to by traditional pigment names. Orpiment orange (or more properly realgar) would once have been made from the mineral arsenic disulfide, gamboge from a natural

resin extruded from a tree, and Indian yellow from the condensed urine of cows fed on mango leaves. But the hand-painted sample of each of these is composed of mineral colors, with their principal ingredients artificial cadmium pigments introduced in the 1840s.

10. Mrs. [M.] Merrifield, *Practical Directions for Portrait-Painting in Water-Colours*, 2nd edn (London: Winsor & Newton, 1851), p. 18.

11. The full title of the book (by A.-M. Perrot, "reviewed and augmented" by E. Blanchard, published in 1834 by Roret in Paris) makes explicit the scope of the hand-colored print industry: *Manuel du coloriste pour l'enlumiure, le lavis et la retouche des gravures, images, lithographes, planches d'histoire naturelle, cartes géographiques et plans topographiques.* In the organization of the shop, where the colorists were each assigned a single hue, it is recommended (pp. 19–20) that the most skilled worker be given the most prominent and extensive color, not because her work will be done well (this is implicit) but because she will be faster and can finish a large amount of coloring in the same amount of time that less skilled workers applying less significant colors will require for their portions of the work, thus not slowing the rate of collective production of finished prints.

12. Harley 1970 (as in note 7), pp. 21–22.

13. *Traité* (as in note 7), p. 20.

14. M. Turner, *The Young Ladies' Assistant in Drawing and Painting* (Cincinnati: Corey and Fairbank, 1833), p. 26.

15. Harley 1970 (as in note 7), p. 152; *A Few Notes on the Composition and Permanence of Artists' Colours* (London: Winsor & Newton, 1938), p. 9.

16. Merrifield 1851 (as in note 10), p. 20.

17. J. Winter, "Gamboge," in E. W. Fitzhugh (ed.), *Artist's Pigments: A Handbook of Their History and Characteristics* (Washington, D.C.: National Gallery of Art, 1986), Vol. 3, pp. 144–45.

18. *The Art of Drawing and Painting in Water-Colours* (Dublin: J. Potts, 1763), p. 56.

19. Ibid., p. 77.

20. Cohn 1977 (as in note 1), pp. 42–43.

21. Every paper on watercolor should contain at least one Field quotation: *"the Chromatic System, like the universe, is an absolute unity comprehending a relative infinity:* – a perfect system!"; G. Field, *Chromatics, or, an Essay on the Analogy and Harmony of Colours* (London: the author, 1817), p. 27.

22. G. Field, *Chromatography* (London: Charles Tilt, 1835), p. 128.

23. William Holman Hunt cited in Cohn 1977 (as in note 1), pp. 63–64.

24. W. Russell and W. Abney, *Report ... on the Action of Light on Water Colours* (London: H. M. Stationery Office, 1888).

Observations on the Functions of Watercolors and Pastels in Italian Renaissance Drawings

Thomas McGrath

In Renaissance Italy, as today, a draftsman's choice of medium was inspired to a significant degree by the type or function of the drawing. When intricate detail was necessary, as in scientific or anatomical studies, artists frequently drew in ink applied with a sharp quill pen. A concern with the distribution of light and shade might encourage the use of charcoal or chalk on toned paper. In those instances where descriptive color was desired, artists had essentially two options: watercolors and colored chalks. An artist's decision to employ one rather than the other, however, as well as the manner in which he handled his medium of choice, depended on a number of factors, including subject, aesthetic aims, and, in the case of preparatory studies, the medium of the final work. Watercolors and pastels in sixteenth-century Italy were associated with different sets of tasks, and in this paper I will consider some of the most popular ways each was used.

Both watercolors and colored chalks were known and employed in a number of different centers in Italy by the first decade of the sixteenth century.[1] Watercolors, however, had a longer history. Watercolor (and gouache, a more opaque water-based medium) appeared as early as the late trecento in the modelbook designs of Michelino da Besozzo and Giovannino de' Grassi, and by the early 1400s Cennino Cennini provided recipes for colored washes and grounds.[2] Although Cennini did not really discuss drawings in which different colors were used for descriptive purposes, he provided all of the necessary technical information for making polychrome images on paper.

Watercolors as a preparatory tool

The late fourteenth-century modelbooks of Michelino and Giovannino served as repositories of motifs, providing formulaic solutions to visual problems. The designs contained within them were not preparatory in the strict sense because they seem to have been made independent of any specific project. Preparatory drawings, however, did occasionally feature watercolors. While there are a few scattered instances of preparatory drawings with watercolors from the fifteenth century, it was not until the sixteenth century that artists began to use watercolors in preparatory drawings with any consistency or regularity. One of the first artists to make watercolors an integral part of his preparatory process was Giovanni da Udine (1487–1564), the chief painter of birds, animals, and grotesque decoration in Raphael's workshop. Vasari tells us that Raphael assigned Giovanni the task of painting the musical instruments in the *Saint Cecilia* altarpiece, begun around 1513, and he later executed much of the fruit and foliage decoration in the Villa Farnesina and the Vatican *logge*.[3]

More than six watercolor drawings can be securely attributed to Giovanni da Udine, and many of them, like the *Parrot* (Fig. 2.7), depict birds, both at rest and in flight. According to Vasari, Giovanni decorated a hall in the Vatican with many images of colorful parrots, for which the live parrots in Pope Leo X's menagerie served as models.[4] Most of these drawings seem to have been made in preparation for decorative fresco projects like those in Raphael's Vatican *logge*, although only one sheet exhibits motifs that correspond precisely.[5] In many of these drawings, one can detect indications of a loose, preliminary sketch in red chalk, and most exhibit a remarkable freedom and spontaneity that suggest they were made as studies rather than copies.

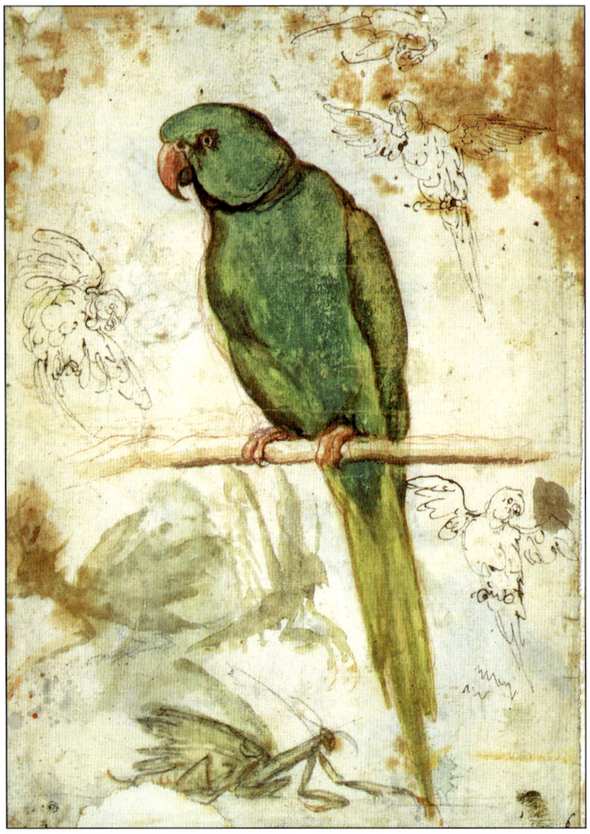

Figure 2.7 Giovanni da Udine (Italian, 1487–1564), *Parrot*. Red chalk; pen and brown ink; yellow, green, red-orange, and black-gray watercolors, 29.8 x 20.8 cm. Nationalmuseum, Stockholm, inv. no. 384/1863.

Like most watercolor drawings that survive from sixteenth-century Italy, those by Giovanni da Udine relate not to panel paintings but to frescoes. This is not surprising when one realizes, first of all, that the effects that can be achieved with watercolors have little in common with those exhibited by oils. The effect of glazing, for example, is relatively easy to accomplish with oils but much more difficult with watercolors, for at least three reasons. First, the refractive indices of the two media are very different; second, the brushing of one layer of watercolor over another will tend to create more of a mixing of pigments than a superimposition, since water from the wet paint will partially dissolve the dry; and third, saturation levels in watercolors are difficult to control, yielding gradations of tone that often appear less subtle than those attainable with oils.

Fresco painting, on the other hand, has several important features in common with watercolor drawings. In each technique there is a partial absorption of the medium into the support that affects the way colors can be handled.[6] Mistakes are difficult to correct, and a certain virtuosity is required. Additionally, the actual composition of the paints is very similar; in both the pigment is suspended in a water-based vehicle. If untempered, the same watercolor used to draw on paper can also be used to paint on wet plaster. There is a remarkable similarity, for example, between Giovanni da Udine's watercolor drawings of plants and animals, such as the *Parrot*, and similar motifs he painted in fresco in the Vatican *logge*. While the use of watercolors to study color effects for frescoes never became a routine practice for most artists, it was more common in some artistic centers than it was in others.[7] In Emilia, for example, a number of painters took advantage of the medium, including Bedoli, Parmigianino, Niccolo del Abate, and Dosso Dossi.[8]

Some of the same qualities that made watercolors unsuited for studying oil paintings made them an effective tool for working out large decorative schemes for walls and ceilings, particularly those that included *grotteschi*. Such decorative schemes relied not on subtle manipulations of tone or virtuoso effects like glazing, but rather on broad and often intricate arrangements of bright, unmodulated hues. There are over a dozen surviving colored drawings from sixteenth-century artists that feature grotesque decoration of some sort, and all utilize watercolor or gouache rather than colored chalk.[9] These drawings, such as the *Ceiling Design* by an artist in the circle of Vasari (1511–1574) (Fig. 2.8), feature a restricted number of identical hues repeated and carefully arranged within a complex decorative scheme.

The frequent use of watercolors in drawings for *grotteschi* was motivated in part by the fundamental importance of color in this decorative genre. Nero's Domus Aurea in Rome, the primary example of antique wall painting that inspired grotesque decoration, was praised for its bold color and fantastic invention, and virtually every writer who discussed grotesques delighted in the unusual and inventive combinations of bright hues.[10] The tendency of watercolors to produce bold, unmodulated areas of color may have lessened their effectiveness as a preparatory tool for figural compositions in panel paintings, but it actually worked to the medium's advantage in drawings for grotesque decoration.

Watercolors in prospectus drawings

In the drawings considered so far, watercolors were used largely, although not exclusively, as a preparatory tool. However, Renaissance artists frequently employed watercolors for prospectus drawings, which were not connected to the preparatory process per se. Prospectus drawings, sometimes called *modelli*, were submitted by an artist to his patron as a means of communicating ideas

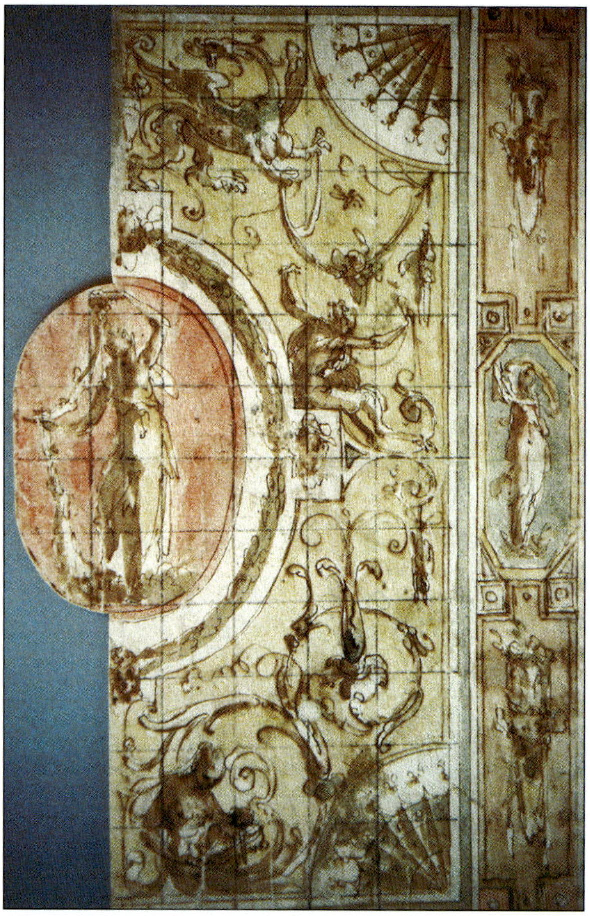

Figure 2.8 *Ceiling Design* by an artist in the circle of Vasari (Italian, 1511–1574). Pen and brown ink, brown wash; pink, yellow, green and blue watercolors; squared in black chalk. Staatliche Graphische Sammlung, Munich, inv. no. 41711.

and securing approval for a proposed design. The practice of adding color to prospectus drawings probably derived in part from the tradition of adding touches of color to *sinopia* underdrawings for frescoes. Red and yellow washes appear, for example, in a number of the fourteenth-century *sinopie* in the Camposanto, Pisa,[11] as well as in a late fifteenth-century *sinopia* showing the *Virgin and Child Enthroned with Saints*, originally executed on the left wall of the Cappella della Confraternità di San Luca in Ss. Annunziata, Florence, which features haloes tinted yellow and parts of draperies toned red.[12] Such coloring had no real preparatory function. The artist, after all, hardly needed a reminder that haloes were golden or that the Virgin's inner robe was crimson. Although the coloring in these *sinopie* was not elaborate, and was of little use as a set of instructions for an assistant, it imbued the images with a certain pictorial quality and helped the patron better envision the proposed work. It also contributed to a tradition whereby color in prospectus drawings was acceptable and perhaps in some cases expected.

In order to examine the ways in which watercolors functioned in prospectus drawings, it should be understood that prospectuses were not regarded as final proposals that could not be altered. This is suggested by Ghirlandaio's 1485 contract for the *Adoration of the Magi* altarpiece, made for the Ospedale degli Innocenti in Florence.[13] Although the contract required Ghirlandaio to submit a (monochrome) prospectus drawing, he was allowed to make changes in the design of the fresco, provided that his changes did not violate the manner and composition ("modo e composizione") of the drawing, and did not meet with the disapproval of the monk overseeing the project. Hannelore Glasser, in her study of artists' contracts in the early Renaissance, concluded that a prospectus drawing was usually regarded as an idea or generic scheme rather than a model in the strict sense of the word.[14] Given this mutable relationship between prospectus and final work, an artist could indicate a particular color scheme in his drawing even before the design had been finalized, knowing that it would be scrutinized by the patron, who might request modifications.

Perino del Vaga's (1500/01–1547) *Design for an Altar Wall* (Fig. 2.9) helps illustrate the advantages of using watercolors in prospectus drawings, as well as some of the potential problems. The sheet's function as a prospectus is perhaps most clearly suggested by the addition of the hastily sketched figure in the lower right, almost certainly a donor, who probably demanded that his likeness be included after Perino had already submitted his initial idea for the project. Watercolors certainly provide a more vivid sense of the proposed fresco's appearance than would a drawing in pen and ink alone. The transparency of the medium, however, encourages a rather high-keyed palette, and modeling has been simplified. While there is some attempt within the canopy to achieve a smooth modulation from light to dark, shading is usually restricted to two or perhaps three gradations of tone. This is most apparent in the yellow and pink draperies of the female saint on the right.

These limitations of the watercolor medium were even more evident in a well-documented prospectus drawing that unfortunately no longer survives: Giulio Romano's *Coronation of the Virgin* for the semidome over the main altar of Santa Maria della Steccata in Parma. In an agreement of 14 March 1540, Giulio Romano explicitly promised to submit "un desegno colorito con acquarelle in carta."[15] The confraternity of the Steccata liked Giulio's proposal and hired Michelangelo Anselmi to execute it. Giulio's drawing, however, was not intended simply, or even primarily, as a set of instructions for another artist. In a letter of 12 September 1542, Giulio noted that he had made the colored drawing "so it could be carefully examined before being made into the final work."[16] He went on to note that "it would be an easy matter to correct any deficiencies; or, if the drawing were not found satisfactory, it could be abandoned and not executed."

Although the confraternity of the Steccata instructed Anselmi to proceed with Giulio's proposal, they disapproved of the final result, complaining that the colors appeared too pale and desaturated at the center of

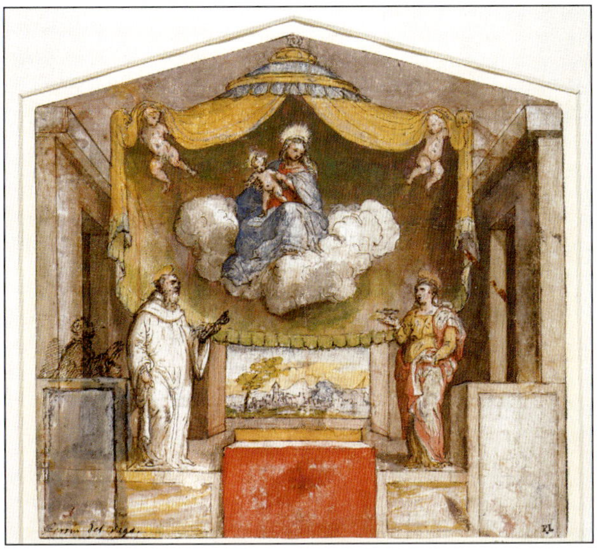

Figure 2.9 Perino del Vaga (Italian, 1500/01–1547), *Design for an Altar Wall*. Pen and brown ink; red chalk; gray, purple, red, yellow, green, and blue watercolors, 19.4 x 20 cm. British Museum, London, inv. no. 1885.5.9.42, reproduced by permission of the Trustees of the British Museum.

the composition, making many figures difficult to read.[17] The confraternity went so far as to demand, unsuccessfully, that Giulio come to Parma and repaint the fresco. While there are a number of possible explanations for the confraternity's unfavorable reaction, an important factor was probably the format Giulio had used to communicate his vision of the *Coronation*'s colors. It would have been nearly impossible, after all, for Giulio to use in his small drawing the variety of pigments in the proper degree of saturation required for the large fresco. It is quite possible that Anselmi translated the watercolor drawing too literally, failing to render the subtle gradations of tone necessary for the ambitious light and color effects Giulio desired.

This limitation of the watercolor medium was generally recognized, I think, because I know of no watercolor drawings for figural compositions that clearly served as guides or *modelli* for assistants. While the addition of watercolors made drawings more attractive to patrons, they did not necessarily make them more useful or legible to members of the workshop. Instructions regarding color could have been, and often were, conveyed much more easily and with a greater degree of precision through verbal notations. Words could identify many different visual effects and identify specific pigments. Such notations appear in Lorenzo Lotto's *Landscape with the Camp of the Assyrians*, now in a private collection, which seems to have functioned as a *modello* for an assistant.[18] In the drawing, which records a design for an intarsia panel, the word *nocte* indicates a dark, night sky, *biaco* a white tent, and *veste de laca* the precise hue of a woman's dress. *Laca* or *lacca* referred not only to varnish, but also to lake pigments, and specifically to the most common lake pigment, red.[19] While watercolor or gouache could convey a more vivid and immediate impression of a proposed work, it could also obscure certain aspects of a drawing and make them difficult to read. This was a concern of Giovan Antonio Dosio, who noted in a letter of 1574 that he had deliberately refrained from using "acquerello" in one of his architectural drawings in order to ensure that the measurements he had inscribed on the sheet would remain visible.[20] In those cases where precision and legibility were valued – as in a *modello* for an assistant – verbal notations had distinct advantages over watercolor or gouache.

Colored chalks and *pastelli*

On certain occasions the desire to add descriptive color to drawings encouraged artists to turn not to watercolors, but rather to fabricated chalks, or *pastelli*. *Pastelli* – made by mixing a dry earth pigment with some sort of binding agent (such as gum arabic, glue, or plaster of Paris) to form a paste that was rolled into sticks and set aside to dry – were often used in combination with natural chalks, which were mined directly from the earth and fashioned into sticks with a knife. Natural chalks can be virtually impossible to distinguish from *pastelli* without the aid of laboratory analysis, and very few Renaissance drawings in colored chalks have been properly analyzed.[21] Yet natural chalks other than black, white, and red were then, as they are today, extremely uncommon in Italy, and they receive, to the best of my knowledge, no mention in period treatises. *Pastelli*, on the other hand, were discussed or mentioned by several sixteenth-century writers, including Anton Francesco Doni in 1549, Benvenuto Cellini in the 1560s, Giovan Paolo Lomazzo in 1584, and Giambattista Armenini in 1586.[22] The rarity of naturally occurring colored chalks is underscored by Vasari's (mistaken) belief that red chalk was not indigenous to Italy and had to be imported from Germany.[23] If an artist wanted to draw with colored chalks, it was far easier for him to fabricate *pastelli* than to locate deposits of natural chalk in the specific hue desired. Lomazzo noted in his 1584 *Trattato dell'arte* that making *pastelli* was so simple anyone could do it; he reported that Leonardo had actually done so as early as the mid-1490s, in order to make studies for the heads of the apostles in his *Last Supper*.[24]

The *pastello* studies for Leonardo's *Last Supper* mentioned by Lomazzo have not survived; there are, however, several drawings by Leonardo's contemporaries and at least one drawing by Leonardo himself that help substantiate Lomazzo's claims. Although predominately in black chalk, with touches of red on the face, Leonardo's large *Portrait of Isabella d'Este* of about 1500 exhibits some yellow chalk on the costume, whose vibrancy suggests *pastello*.[25] A greater range of colored chalks, both natural and fabricated, appears in drawings by Leonardo's followers. Giovanni Antonio Boltraffio, for example, combined black, white, brown, yellow, and red chalks in his *Study for Saint Barbara* from about 1502,[26] and Bernardino Luini embellished his Albertina

Portrait of a Woman with black, brown, yellow, and red chalks.[27]

The use of colored chalks and *pastelli* for head studies

These drawings, like the vast majority of surviving drawings in colored chalks from the sixteenth century (with the notable exception of those by Jacopo Bassano beginning in the 1560s), are portraits or head studies. When studying the human head in color on paper, artists turned to chalks much more frequently than they turned to watercolor. This phenomenon predates the introduction of *pastelli.* Antonio Pollaiuolo's *Head of Fides*,[28] made around 1470, and Luca Signorelli's *Head of Saint John the Baptist*,[29] from about 1484, both take advantage of natural red and black chalks to suggest the tonalities of flesh, determine the optimal placement of red highlights, and add a sense of three-dimensionality. The use of *pastelli* in sixteenth-century drawings of heads and faces, then, should be seen more as a continuation of an established tradition than a complete innovation.

Proper coloring was fundamental to a successful depiction of the human head, and its importance encouraged artists to study color effects on paper. In any depiction of a human being, one of the primary objectives was making the subject appear vibrant and alive, and color was a crucial factor in this regard. According to Lodovico Dolce in 1557, "coloring is so important and compelling that, when the painter produces a good imitation of the tones and softness of flesh ... he makes his paintings seem alive, to the point where breath is the only thing missing in them."[30] Dolce continued, "the main problem of coloring resides, then, in the imitation of flesh, and involves diversifying the tones and achieving softness."[31] Armenini added complexity to the problem when he observed that different faces exhibit different colors: "One must consider the variation of the tints which differ according to the sex, age, and quality of the persons," he wrote in 1586.[32] Color was equally useful, if not indispensable, for conveying an individual's personality and state of mind. Lomazzo, writing in 1585, suggested that color was expressive of a person's "voice and spirit, because in representing the complexions, he paints in their faces melancholy, anger, happiness and fear."[33] Anton Francesco Doni likewise believed that the color of a person's face revealed "the depths of his heart and mind," arguing that a person's complexion betrayed his profession.[34] Color, then, carried a wealth of information that was indispensable to an artist intent on depicting a flesh-and-blood individual.

The various color accents that described complexion and added vivacity to the face, however, had to be reconciled with the modeling of the head as a three-dimensional form. The face was not regarded as an object uniform in hue and tone, like a blue or red drapery, but rather a relief surface exhibiting important tonal variations that had to be accurately rendered if the figure were to appear convincingly alive. In the aforementioned drawings by Pollaiuolo and Signorelli, for example, we can see the artists wrestling with this problem. Both works are preparatory to known paintings; both are full-sized cartoons, pricked for transfer; and both carefully indicate the optimal placement of red highlights that give human flesh a sense of suppleness and vitality. For Signorelli, who executed his drawing in black, white, and red chalks, the tonalities of Saint John's face demanded particular attention, since in the altarpiece, now in Perugia's Museo del Duomo, Saint John would be juxtaposed with the old and leathery-skinned Saint Onuphrius. Importantly, it was colored chalks rather than watercolors that offered the degree of control necessary to achieve a convincing rendering of the human head as a three-dimensional object, with all of its subtle gradations of tone.

Colored chalks, both natural and fabricated, continued to be used in studies of the human head throughout the sixteenth century.[35] *Pastelli* appear in a number of such drawings by followers of Leonardo, including Boltraffio and Luini.[36] They also appear in Parmigianino's *Head of a Boy*[37] made during the artist's years in Rome during the mid-1520s, suggesting that this practice was not confined to Lombardy.

Colored chalks and *pastelli* in preparatory drawings

In the second half of the cinquecento, Federico Barocci continued this tradition of studying heads and faces with *pastelli* and colored chalks. His use of *pastelli*, however, was much more systematic than that of earlier artists, and the medium became an integral part of his preparatory process.[38] For some projects, such as the *Martyrdom of San Vitale* altarpiece in the Brera, as many as 25 studies in *pastelli* survive. To a remarkable degree, Barocci was able to approximate with *pastelli* the color effects that would appear in his works on canvas and panel. His command of the medium allowed him to layer one hue over another to produce striking chromatic effects sympathetic to those in an oil painting, which could not really be achieved with watercolors. Barocci might, for example, superimpose red over yellow ochre, as in the *Head of a Bearded Man*, to describe the appearance of a figure's russet-orange beard,[39] and red and white were often combined on blue paper to produce vibrant flesh tones.

In other instances, however, Barocci seems to have used one fabricated chalk of a specific hue rather than combine two chalks on the surface of the paper. This happens frequently with flesh tones. In the *Study for the Head of Christ*, for example, the stray lines in the lower left margin, where Barocci tested his chalks, indicate that the color of the face was largely achieved with one very pale pink *pastello*, which was sometimes darkened with brownish red and lightened with white.[40] Given that the

ground color in Barocci's drawings often anticipated the ground color in his paintings, as Marcia Hall has demonstrated, it is tempting to speculate that the artist followed a system whereby many of his *pastelli* corresponded with specific pigments or pigment mixtures that he used on his canvases and panels.[41]

While Barocci was integrating *pastelli* into his drawings for oil paintings, one of his contemporaries, Giovanni de' Vecchi (1536–1615), continued to rely on watercolors as a preparatory tool. The quantity and nature of Giovanni's commissions clearly indicate that he was one of the leading artists in Rome in the second half of the sixteenth century. However, he has received little scholarly attention.[42] This is somewhat understandable when one realizes that many of his most important works have been lost and that virtually nothing he painted before his frescoes in the Palazzo Farnese at Caprarola in 1574, when he was about 38 years old, survives. Unfortunately, there are fewer than fifty extant drawings attributed to Giovanni, although at least ten feature watercolors. While his design for the decoration of an apse exhibits a precision and attention to the architectural setting that suggests a prospectus drawing,[43] other watercolors, like the *Visitation* (Fig. 2.10), contain corrections that identify them as studies in which the artist explored possibilities of design and color. Most of Giovanni's watercolors, moreover, appear to be preparatory not to oil paintings, but to frescoes, the medium in which Giovanni excelled. It is interesting to note that at the end of the sixteenth century, and even at the beginning of the seventeenth, the preparatory practices of Barocci and Giovanni de' Vecchi suggest a continuation of a convention we observed earlier: when used as a preparatory tool, watercolors were still associated primarily with frescoes and *pastelli* with oils.

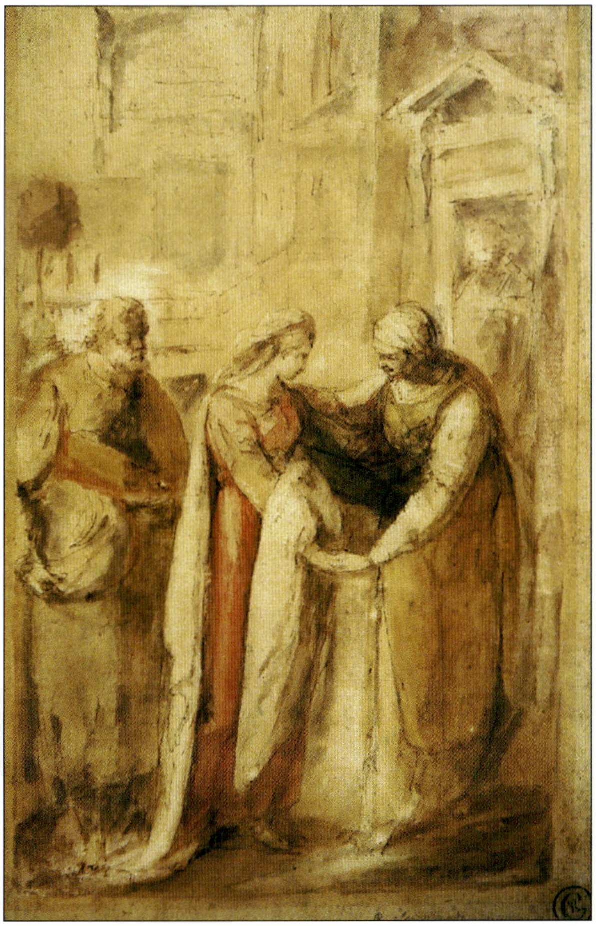

Figure 2.10 Giovanni de' Vecchi (Italian, 1536–1615), *Visitation*. Pen and brown ink over black chalk, with pale brown, gray, pale purple, olive and rose-red watercolors. British Museum, London, inv. no. 1965.10.9.3, reproduced by permission of the Trustees of the British Museum.

Both watercolors and *pastelli* afforded Italian Renaissance artists a means of introducing color into their drawings. Watercolors and *pastelli*, however, were regarded as two distinct tools, best suited to different tasks. Sometimes the physical properties of a medium largely determined a particular function, as in the use of watercolors to study fresco paintings. The practice of embellishing prospectus drawings with watercolors, on the other hand, must have been encouraged more by the expectations of the patron and the force of tradition. In head studies, a preference for pastels arose partly because they allowed for smooth and subtle gradations of tone, but it was the value placed on color as a bearer of meaning in depictions of the human face that prompted artists to turn to a colored medium in the first place. In other words, the various functions of watercolors and pastels depended not only on the preferences of artists and the various properties of the two media, but also on the attitudes and expectations of sixteenth-century audiences and patrons.

Endnotes

(All translations are the author's unless otherwise indicated.)

1. For more on the history of colored media in Renaissance drawings, see J. Meder, *The Mastery of Drawing*, 2 vols, transl. by W. Ames (New York, 1978), passim, but in particular Vol. 1, pp. 85–113; and J. Watrous, *The Craft of Old Master Drawings* (Madison, Wis., 1957), pp. 83–86 and 91–129. More recent discussions of colored drawing media in the Renaissance include M. Cohn, *Wash and Gouache*, exh. cat. (Cambridge, Mass.: Fogg Art Museum, 1977); T. Burns, "Chalk or pastel?: The use of coloured media in early drawings," *The Paper Conservator* 18 (1994), pp. 49–56; D. DeGrazia, *Correggio and His Legacy*, exh. cat. (Washington, D.C.: National Gallery of Art, 1984), pp. 36–39; E. Olszewski, *The Draftsman's Eye*, exh. cat. (Cleveland: Cleveland Museum of Art, 1981), pp. 17–18; and A. Petrioli Tofani, "Materials and techniques," in G. C. Sciolla (ed.), *Drawing: Forms, Techniques, Meanings* (Turin, 1991), pp. 187 ff. See also T. McGrath, "Color in Italian Renaissance drawings: Reconciling theory and practice in Central Italy and Venice," *Apollo* 146, no. 429 (1997), pp. 22–30; and idem, "Federico Barocci and the early history of *pastelli* in Central Italy," *Apollo* 148, no. 441 (1998), pp. 3–9.

2. On modelbooks, see R. W. Scheller, *A Survey of Medieval Modelbooks* (Haarlem, 1963). For Cennini's discussion of colored washes, see C. Cennini, *Il libro dell'arte*, transl. by D. V. Thompson, Jr., as *The Craftsman's Handbook* (New Haven, Conn., 1933), pp. 9–10.

3. G. Vasari, *Le vite de' più eccellenti pittori, scultori ed architettori* …, 7 vols, ed. by G. Milanesi (Florence, 1878–85), Vol. 6, p. 549 ff. For color reproductions of both the drawings and the frescoes from the *logge*, see N. Dacos and C. Furlan, *Giovanni da Udine 1487–1561* (Udine [?], 1987), passim.

4. Vasari 1878–85 (as in note 3), Vol. 6, p. 555.

5. This is the sheet containing sketches of flowers and fruits in Vienna, Albertina, inv. no. 1518, SV 94. For more on the drawing, see V. Birke, *Die Italienische Zeichnungen des Albertina* (Munich, 1991), no. 54.

6. For an excellent explanation of fresco technique, see V. Malaguzzi-Valerj, "Ancient fresco technique in the light of scientific examination," in W. J. Young (ed.), *The Application of Science in Examination of Works of Art* (Boston, 1973), pp. 104–06.

7. For a discussion of some important regional variations in the production and use of colored drawings in Renaissance Italy, see McGrath 1997 (as in note 1).

8. For more on colored drawings by these artists, and for a catalogue of Italian Renaissance drawings that feature descriptive color, see my doctoral dissertation, "*Disegno, Colore*, and the *Disegno Colorito*" (Ph.D. diss., Harvard University, 1994).

9. The colored grotesque drawings with which I am familiar include: Giovanni da Udine workshop, Florence, Uffizi, inv. no. 78 ORN; Marco da Faenza, Florence, Uffizi, inv. no. 981 E; Marco da Faenza, Florence, Uffizi, 4089 S; Perino del Vaga, Turin, Biblioteca Reale, inv. no. 15739 D.C., verso; Perino del Vaga, Toronto, Collection of Marvin Gelber; Perino del Vaga, Paris, Louvre, inv. nos. 600 and 601; Giulio Campi, Milan, Ambrosiana, Cod. F 262 INF. no. 21; Taddeo Zuccaro workshop, Naples, Capidimonte, inv. no. 115.

10. For example Vasari 1878–85 (as in note 3), Vol. 6, p. 551; and G. B. Armenini, *De veri precetti della pittura* (originally Milan, 1586), ed. by M. Gorreri (Turin, 1988), p. 220.

11. On the Camposanto *sinopie*, see B. Dodge, "The role of the sinopia in the Traini Cycle in the Camposanto, Pisa," in Henk van Os and J. R. J. Asperen de Boer (eds), *La Pittura ne. XIV e XV secolo. Il Contributo dell' analisi tecnica all storia dell'arte* (Bologna, 1979), pp. 125–53. While Dodge does not explicitly discuss the role of *sinopie* as prospectuses, she does note that those finished with red and yellow appear "presentational in nature" (p. 127).

12. This *sinopia*, presumably rejected, is now on display in the chapel's anteroom. See U. Baldini in *The Great Age of Fresco*, exh. cat. (New York: The Metropolitan Museum of Art, 1968), pp. 218–19. Although Baldini asserts that "the halos were added later in yellow," there is no evidence to suggest that they are not original.

13. For Ghirlandaio's 1485 agreement with the Ospedale degli Innocenti, Florence, see M. Baxandall, *Painting and Experience in Fifteenth-Century Italy* (Oxford, 1988), p. 6.

14. H. Glasser, "Artists Contracts in the Early Renaissance" (Ph.D. diss., Columbia University, 1965; Ann Arbor: University Microfilms, 1970), p. 121.

15. L. Testi, *Santa Maria della Steccata in Parma* (Florence, 1922), doc. VII, pp. 268–69. For a summary of the documentation surrounding Giulio's work for the Steccata, see F. Hartt, *Giulio Romano* (New Haven, Conn., 1958), pp. 247–50.

16. In his letter to the confraternity of 12 September 1542, Giulio noted: "qual disegno si poteva esaminare benissimo p[ri]ma che se mettessi a op[er]a; se serria corretto in tutto q[uan]to q[ue]llo fossi mancato facil[men]te o vero sel detto disegno no' fossi satisfatto si poteva lassar di no' metterlo a op[er]a" (ibid., doc. XIX, p. 278).

17. Unfortunately, a detailed written record of the confraternity's criticisms of Anselmi's fresco does not exist; rather than sending a letter to Giulio in Mantua, the confraternity dispatched one of its own members, who explained the situation to Giulio and tried to get him to come to Parma. The envoy's brief account of his trip to Mantua is published in ibid., doc. XVIII, p. 277. The confraternity's criticisms are easily inferred from Giulio's response (ibid., doc. XIX, pp. 277–79).

18. Pen and brown wash with white gouache; 31.5 x 24.5 cm. See N. Turner, *The Study of Italian Drawings. The Contribution of Philip Pouncey* (London, 1994), pp. 52–54, cat. 57.

19. On red lakes see H. Schweppe and H. Roosen-Runge, "Carmine – cochineal carmine and kermes carmine," in R. L. Feller (ed.), *Artists' Pigments: A Handbook of Their History and Characteristics* (Washington, D.C., 1986), Vol. 1, pp. 255–83. The authors observe that "lacca" or "lacha" was known by the Middle Ages as a red pigment for painting wood or walls (p. 261). See also M. Hall, *Color and Meaning. Practice and Theory in Renaissance Painting* (Cambridge, 1992), pp. 15, 32, and 54.

20. G. Bottari and S. Ticozzi, *Raccolta di lettere sulla pittura, scultura ed architettura ...* (Milan, 1822–25), Vol. 3, p. 301.

21. The difficulty of distinguishing between natural and fabricated chalks with the naked eye has been noted by T. Burns, "Rosalba Carriera and the early history of pastel painting," in S. Fairbrass (ed.), *The Institute of Paper Conservation. Conference Papers Manchester 1992* (Manchester, 1992), p. 36; and D. Mayer and P. Vandiver, "Red chalk: Historical and technical perspectives," in W. Strauss and T. Feller (eds), *Drawings Defined* (New York, 1987), pp. 174–75.

22. A. F. Doni, *Disegno* (Venice, 1549; facsimile edition, ed. by Mario Pepe, Milan, 1970), p. 15a; B. Cellini, *Opere*, ed. by G. G. Ferrero (Turin, 1971), pp. 807–08; G. F. Lomazzo, *Trattato dell'arte della pittura* (originally Milan, 1584), in *Scritti sulle arti*, Vol. 2, ed. by R. Ciardi (Florence, 1974), p. 170; and Armenini 1988 (as in note 10), pp. 64 and 133.

23. Vasari 1878–85 (as in note 3), Vol. 1, p. 74.

24. Lomazzo 1974 (as in note 22): "Non tacerò anco d'un altro certo modo di colorare che si dice a pastello, il quale si fa con punte composte particolarmente in polvere di colori che tutti si possono comporre. Il che si fa in carta, e molto fu usato da Leonardo Vinci, il qual fece le teste di Cristo e de gl'apostoli a questo modo, eccellenti e miracolose, in carta. Ma quanto è dificile il colorire in questo nuovo modo, tanto è egli facile a guastarsi." Leonardo's use of *pastelli* was also implied by Leonardo himself in the Codex Atlanticus, 247a; see J. P. Richter, *The Literary Works of Leonardo da Vinci* (New York, 1939), Vol. 2, p. 349, where he referred to "il modo di colorire a secco." It is further worth noting that Armenini 1988 (as in note 10) makes several references to *pastelli* in his *De' veri precetti della pittura* (e.g., pp. 64 and 133).

25. Paris, Louvre, inv. no. M.I. 753; black, red, and yellow chalks on paper pricked for transfer; 63 x 46 cm.

26. Milan, Ambrosiana, inv. no. Cod. F 290 inf. 8; 53.3 x 40.5 cm.

27. Vienna, Albertina, inv. no. B 405; 41.4 x 28.4 cm.

28. Florence, Uffizi, inv. no. 14506 F.

29. Stockholm, Nationalmuseum, inv. no. 9.

30. Lodovico Dolce, *L'Aretino* (1557), ed. and transl. by M. Roskill, in *Dolce's "Aretino" and Venetian Art Theory of the Cinquecento* (New York, 1968), pp. 152–53.

31. Ibid., pp. 154–55.

32. Armenini 1988 (as in note 10), pp. 128–29; the translation is E. J. Olszewski's; G. B. Armenini, *On the True Precepts of the Art of Painting*, ed. and transl. by E. J. Olszewski (New York, 1977), p. 178.

33. Gian Paolo Lomazzo, *Idea del tempio della pittura*, 2 vols, ed. by Robert Klein (Florence, 1974), Vol. 1, p. 187.

34. Doni 1970 (as in note 22), p. 29.

35. On the use of pastels in Italian Renaissance drawings, see McGrath 1998 (as in note 1).

36. For example, Boltraffio's *Head of Saint Barbara* of c. 1502 (Milan, Ambrosiana, Cod. F 290 inf. 7; metalpoint and black, yellow, brown, red, and white chalks on tan prepared paper; 53.3 x 40.5 cm); or Luini's *Portrait of a Woman* of c. 1520 (Vienna, Albertina, inv. no. B 405; black, yellow, red, and brown chalks with some traces of white chalk; 41.4 x 28.4 cm).

37. Paris, Ecole des Beaux-Arts, inv. no. 234; black, red, yellow, brown, and white chalks on faded blue paper; 29.1 x 19.1 cm.

38. The most complete discussions of Barocci's handling of *pastelli* include E. Pillsbury and L. S. Richards, *The Graphic Art of Federico Barocci* (New Haven, Conn., 1978); S. Folds McCullagh, "Serendipity in a Solander box: A recently discovered pastel and chalk drawing by Federico Barocci," *The Art Institute of Chicago Museum Studies* 17, no. 1 (1991), pp. 53–65, 93–94; and A. Emiliani, *Federico Barocci*, 2 vols (Bologna, 1985), passim.

39. Washington, D.C., National Gallery of Art, Woodner Collection, inv. no. 1991.182.16; black, purple, orange, blue, yellow, brown, aqua, and red chalks on brown-gray (faded blue) paper; 38 x 26.5 cm.

40. Cambridge, Mass., Harvard University Art Museums, inv. no. 1986.535. The drawing relates to the Urbino *Last Supper* from the 1590s.

41. Hall 1992 (as in note 19), p. 196.

42. For more on Giovanni de' Vecchi and his drawings, see R. Roli, "Giovanni de' Vecchi," *Arte antica e moderna* 29 and 31–32 (1965), pp. 45 ff. and 324 ff.; P. Pouncey and J. Gere, *Italian Drawings in the ... British Museum: Artists Working in Rome* (London, 1983), pp. 176–78; J. Heideman, "St. Catherine of Siena's life and thought. A fresco cycle by Giovanni de' Vecchi," *Arte Cristiana* 77 (1989), pp. 451–64; J. Heideman, "Giovanni de' Vecchi's fresco cycle and its commissioners in the Rosary Chapel in Santa Maria sopra Minerva in Rome," in P. van Kessel (ed.), *Power of Imagery: Essays on Rome, Italy and Imagination* (Rome, 1993), pp. 149–62, 297–304; and P. Tosini, "Rivedendo Giovanni de' Vecchi: nuovi dipinti, documenti e precisazioni," *Storia dell' arte* 83 (1994), pp. 303–47.

43. *Design for an Apse*, Oxford, Christ Church, inv. no. 1342.

The History and Technology of Renaissance and Baroque Hand-Colored Prints

Susan Dackerman and Thomas Primeau

During the Renaissance and Baroque periods, the application by brush of washes, body color, and precious pigments to engravings, etchings, and woodcuts was a common aspect of print production. Now considered anomalies in the history of the graphic arts, painted prints were highly prized in their day. Although skillfully colored impressions of prints by exemplary artists such as Albrecht Dürer, Pieter Brueghel, and Hendrick Goltzius are known, this aspect of print production has not been addressed adequately in the art-historical literature. Most recently, Jan Van der Stock's *Printing Images in Antwerp* presents enlightening archival materials related to the practice of coloring prints in the Low Countries.[1] Municipal documents indicate that those artists engaged in coloring prints from the fifteenth to the seventeenth centuries usually were skilled professionals, obligated to belong to their city's painters' guild, the Guild of Saint Luke. Guild membership was dictated by the use of specific materials and tools; therefore, those who used paint and paintbrushes were obligated to belong to the Guild of Saint Luke.[2] The evidence suggests that young artists, before establishing their reputations, were most commonly involved in coloring prints. In fact, the first biographer of Netherlandish artist Carel van Mander stated in his 1604 *Het Schilderboek* that the well-known painter Quentin Matsys began his career by painting prints.[3]

Archival documents also reveal much about the marketing and consumption of hand-painted prints in the sixteenth and seventeenth centuries. Colored prints were often a standard offering of print publishers and sellers.[4] More expensive than their black-and-white counterparts, they could be purchased in a variety of formats depending on the use intended by the patron. Colored prints were bound into albums, mounted on linen for wall display,[5] or sold individually to be pasted into books and bibles, or tacked to walls. The application of pigments also varied from very extensive modeling, to the addition of broad, flat areas of color, to the application of an overall tone to the paper. Sixteenth-century sales records indicate that various levels of coloring quality were available. For instance, the account books of Christopher Plantin record the prices of books of colored engravings by Peeter Baltens. The most expensive are described as "beautifully colored" while the less expensive are called "simply colored."[6] The use of costly pigments such as gold also increased the price of a colored print.[7]

Unfortunately, because of their everyday uses and the fragility of paint on paper, many of the hand-colored prints produced in the fifteenth, sixteenth, and seventeenth centuries have deteriorated or been destroyed. Only those safeguarded in albums or print cabinets survive in good condition, leaving a somewhat limited number of objects for modern scholarly investigation. Hand-painted prints have gone unstudied in the scholarly literature for a number of other reasons as well. The most significant is the prevailing aesthetic of fine-art prints as black-and-white objects. In nineteenth-century France, distinctions between fine art and commercial or journalistic art were formalized, and color was designated an aspect of commercial printmaking. Colored prints from earlier times were regarded as belonging to the realm of "popular" art rather than to the tradition of fine art. Print purists – curators, collectors, and art historians – considered hand-colored impressions of prints, even those by artists as important as Dürer and Rembrandt, to be adulterated objects. Hand-colored prints continue to be neglected by scholars because they are so difficult to categorize. As both painted and printed images on paper, they blur the traditional distinctions among the different media of printmaking, drawing, and painting. Since scholars tend to limit their specialization to a single medium, objects that share characteristics of three different media tend to fall through the cracks.

Uncertainty about the vintage of the coloring has also hindered the art-historical assessment of these objects. For the last two centuries, print scholars have underestimated the number of prints that were painted during the Renaissance and Baroque periods, assuming it to be a practice undertaken by amateur painters and collectors sometime over the course of a print's history. They knew that nineteenth-century dealers and collectors often "enhanced" worn or damaged impressions of prints to increase their appeal or marketability. Informal discussions with book and print dealers indicate that the practice of hand-coloring antique prints is common even today.

Because doubts exist about the "authenticity" of the coloring on many prints, determining the relative date of the print colorists' materials is a cornerstone of this study. As an introduction to the type of technical research we will be undertaking to study hand-colored prints from American and European collections, we first examined a group of prints from the Baltimore Museum of Art for information about the pigments and binders used in the paints. Working with conservation scientist Janice Carlson at the Winterthur Museum Analytical Laboratory, we were able to perform energy dispersive X-ray fluorescence (XRF) analysis in order to identify the elemental composition of the pigments on several hand-colored sixteenth-century prints. In addition to using this non-invasive methodology, Janice Carlson also performed Fourier transform infrared (FTIR) microanalysis on samples taken from the prints. The data derived from this analysis have provided us with a better understanding of the materials used to color prints and demonstrate some of the problems that have complicated the art-historical view of these objects.

To begin this study, we selected three hand-colored sixteenth-century prints. Each of the objects is elaborately decorated but is distinguished by the range of colors used and the manner in which paints were applied. *The Descent of the Holy Spirit* (Fig. 2.11), a woodcut illustration by Hans Schäuffelein (1480/85–1537/40) from a 1514 evangelical text published in Basel, features a style of coloring that can be seen on many prints, particularly those found in early printed books. A student of Dürer, Schäuffelein was a prominent designer of woodcuts in Augsburg in the first decades of the sixteenth century. Because the practice of coloring prints evolved in part from the tradition of hand-coloring book illustrations, it is important for us to begin our study by looking at an example from an early printed book. The colors on this leaf are bold, applied in broad patches with little attempt at modeling or shading within fields of color. Also evident is the extensive use of a rich, transparent red paint that is often heavily applied and shows evidence of cracking and cupping.

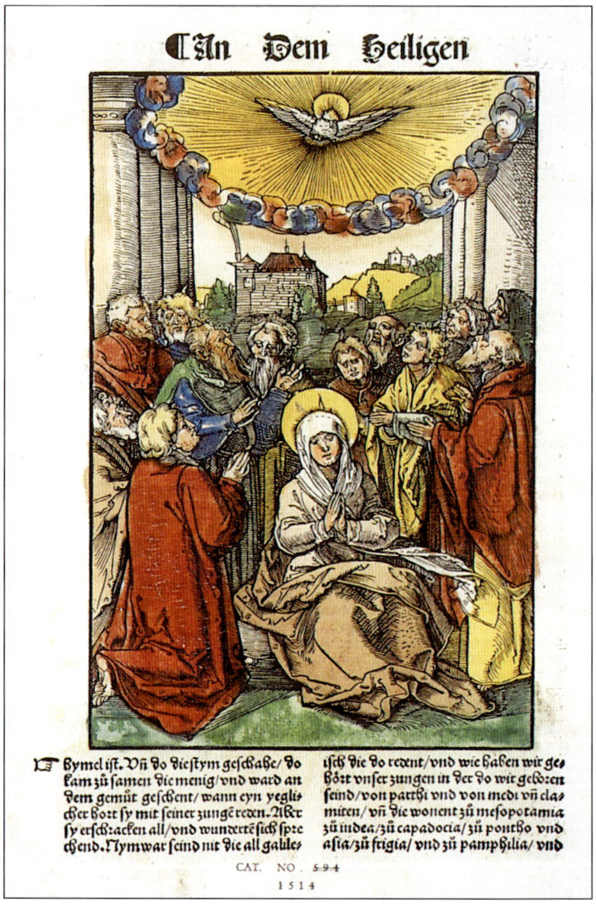

Figure 2.11 Hans Schäuffelein (German, 1480/85–1537/40), *The Descent of the Holy Spirit*, 1514. Hand-colored woodcut with letterpress. The Baltimore Museum of Art, Blanche Adler Memorial Fund, BMA 1946.124a.

Unfortunately, instead of learning about the nature of materials used by sixteenth-century artists, we discovered that the coloring on the Schäuffelein contains pigments that were not in use until the nineteenth century. XRF analysis (Table 2.1) revealed that the blue paint showed the presence of copper and lead, suggesting the use of azurite ($2CuCO_3.Cu(OH)_2$) and lead white ($2PbCO_3.Pb(OH)_2$), both well-known early pigments. But we also found that the color contains cobalt and zinc, components of cobalt blue ($CoO.Al_2O_3$) and zinc white (ZnO), two pigments that were not available until the early 1800s.[8] It should be noted that the presence of cobalt does not always indicate the use of a modern

Sample #	Area	Color	Elements – Major	Elements – Minor	Probable pigment(s)
1	Paper/mat	–	–	–	–
2	Robe	Red			Organic red
3	Sleeve	Blue	Zn, Cu, Pb, Co		Lead white, zinc white, cobalt blue, copper blue or green
4	Landscape	Green	Cu	Zn, Ca, Pb, Cr, Fe	Copper green, zinc white, lead white, chromium green
5	Sky	Yellow			Organic yellow
6	Halo	Yellow			Organic yellow

Table 2.1 XRF analysis of the Schäuffelein print (carbon target, 45 kV, 0.5 mA, 150 sec).

pigment. The blue pigment smalt, widely used since the sixteenth century, is a combination of the oxides of silicon, potassium, arsenic, and cobalt.[9] However, if this pigment were present, the XRF spectrum would also have indicated the presence of arsenic.

Likewise, the presence of copper in an area of green paint indicates the use of a traditional copper green pigment, possibly malachite ($CuCO_3.Cu(OH)_2$), but the presence of a small amount of chromium suggests that the area also contains chromium oxide green (CrO_3), another nineteenth-century pigment.[10] Zinc white was also found in this area. The mixture of traditional and modern pigments found in these two areas of color suggests that the print may have been colored at an early date and then later retouched with modern pigments. Microscopic examination of the paints, however, was inconclusive in determining whether the modern paints were applied over the older paints.

XRF testing of the yellow and red paints revealed no inorganic components. The use of organic colors is not unusual; many early texts related to the preparation of painting materials provide an extensive list of them. Sixteenth-century documents related to the coloring of prints specifically refer to the use of organic colors such as sap green.[11] FTIR analysis of a sample of the red paint indicated only the presence of a proteinaceous material, likely gelatin.

The Alliance with White Russia (Fig. 2.12) is a woodcut by Hans Burgkmair (1473–1531) from *Der Weisskunig*, an idealized biography of Emperor Maximilian's life. Like Schäuffelein, Burgkmair was active in Augsburg in the early sixteenth century. We were anxious to test the pigments on this particular woodcut because of Burgkmair's deep-rooted interest in the addition of color to printed images, and also because he is suspected of painting other of his woodcuts.[12] Unfortunately, this print was not one of them. This print is painted with bright, transparent colors that were carefully and evenly applied within the lines of the printed design. A few touches of gold paint were applied to the crowns and robes of the figures. Throughout the print, layers of glaze, possibly gum arabic, were selectively applied to provide deeper saturation of the colors and a glossiness that accents the gemlike appearance of the colors.

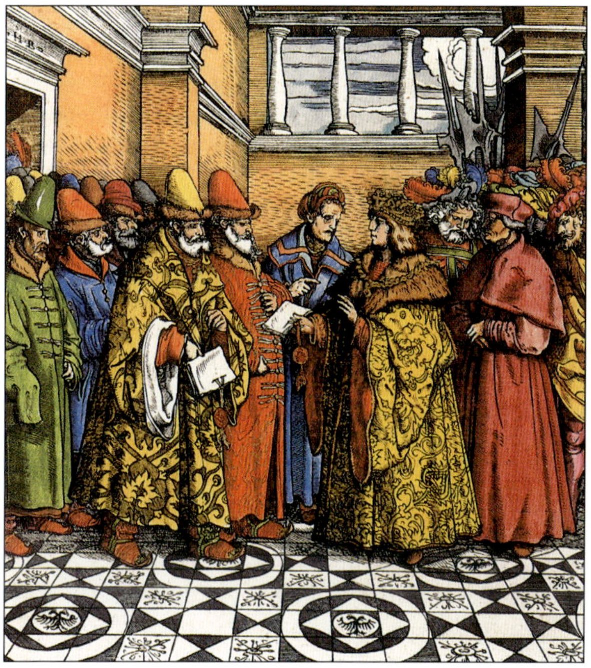

Figure 2.12 Hans Burgkmair (German, 1473–1531), *The Alliance with White Russia*, 1514. Hand-colored woodcut. The Baltimore Museum of Art, Blanche Adler Memorial Fund and Board of Trustees Fund, BMA 1998.104.1.

Microscopic examination of the colors on the Burgkmair shows a high degree of uniformity and an even dispersion of pigment particles, suggesting the use of modern, machine-ground pigments. The suspicion that the print features modern coloring was supported by elemental analysis, which revealed that all of the color areas tested contain traces of zinc and barium, major components of a white pigment mixture known as lithopone ($BaSO_4.ZnS$) (Table 2.2). Lithopone was first produced and patented in England around 1874 as Orr's white.[13] Another indication that the print was colored in the nineteenth century or later is the green in this palette: it contains chromium, indicating the likely use of viridian ($Cr_2O_3.2H_2O$), a bright, transparent chromium oxide green that was patented in 1859.[14]

After the disappointment of discovering the dubious nature of the coloring on these two prints, we were relieved by the results of our examination of the third

Sample #	Area	Color	Elements – Major	Elements – Minor	Probable pigment(s)
1	Paper edge	–	–	–	–
2	Hat	Orange	Zn, Hg, Fe, Ba		Vermilion, lithopone
3	Robe	Green	Zn	Ba, Ca, Fe, Cr, Pb	Viridian (chrome oxide green), lithopone

Table 2.2 XRF analysis of the Burgkmair print (carbon target, 45 kV, 0.4 mA, 150 sec).

print, *The Triumph of Patience* (Fig. 2.13), a 1559 engraving by Dirck Coornhert (1519–1590) after Maerten van Heemskerk. Coornhert was a northern Netherlandish artist who made engravings for the prominent Antwerp publisher Hieronymus Cock. From Cock's sales records, we know that he sold both colored and uncolored impressions of the prints he published.[15] This print shows the personification of Patience, triumphantly enthroned on a processional cart drawn by Hope and Desire. The figure of Fortune is bound behind the cart as the procession travels along a windy coastline. In the sea beyond, a sinking ship demonstrates the outcome of Fortune's vicissitude.

The Coornhert print is a good example of a beautifully and skillfully colored print. This dramatic sheet was elaborately painted by a very skilled hand that was successful in enhancing the print without obscuring its masterfully engraved graphic matrix. This was achieved by using bright colors and a combination of transparent and opaque paints. The colorist did not simply apply colors in broad patches defined by the elements of the printed design in the "paint by numbers" approach seen in many hand-colored prints. Rather, a more painterly handling of the colors can be observed in the careful blending that accents the modeling of the figures and drapery. Opaque body colors were layered with transparent paints to create a heightened sense of surface texture; an opaque white was thinly applied in areas to produce a scumbling effect that softens the hair and feathers. The use of bright colors and silver and gold highlights lend to the print a sense of transcendence appropriate for the allegory depicted.

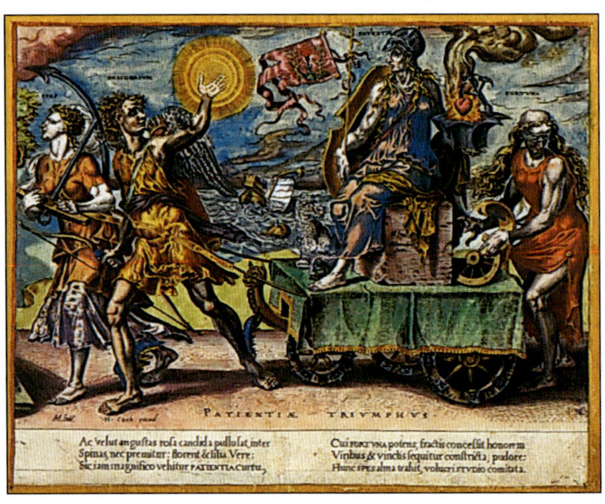

Figure 2.13 Dirck Coornhert (Dutch, 1519–1590) after Maerten van Heemskerk, *The Triumph of Patience*, 1559. Hand-colored engraving. The Baltimore Museum of Art, Blanche Adler Memorial Fund, BMA 1998.006.

The pigments used on the Coornhert have the coarse particle appearance and irregular dispersion associated with early, hand-ground paints. The results of XRF and FTIR analysis have provided identification of a palette that is consistent with a sixteenth-century attribution (Table 2.3, opposite). As the richness of the coloring on this print suggests, the colorist used a wide variety of pigments, sometimes applied in pure form but more often blended with other pigments to create a variety of effects. We can look at the different shades of blue on this print as an example of how the colorist worked. The light sky blue of the background is pure azurite applied as a thin wash. For the deeper blue tones, as seen on the wings at the feet of the figure of Desire and on the robe of the figure of Patience, azurite was mixed with smalt and lead white to create an opaque paint that was applied thickly. FTIR analysis of a sample taken of the blue at the area of the wing confirmed the use of azurite and lead white in this area and also found evidence of calcite and bone white. The presence of polysacharide in the sample indicates a gum arabic binder.

The brilliant orange and red paints throughout the print were, when studied with XRF, found to be combinations of vermilion (HgS) and red lead (Pb_3O_4). FTIR also detected the presence of brown ochre. A sample from the red robe of Fortune again indicated a gum arabic binder, but it also showed the presence of a proteinaceous material such as gelatin.

XRF analysis of the white paints revealed that a lead white was applied to the print. FTIR examination of the same area was helpful in confirming the use of basic lead carbonate, and it also indicated that the lead white was mixed with calcite. This was possibly done to adjust the covering power of the lead white paint. Other pigments identified on the Coornhert palette include: malachite, yellow ochre, organic yellow, and purple, and metallic silver and gold. The use of a variety of pigments to create a complex coloring system suggests that the enterprise of painting prints involved the talents of skilled artisans who possessed a sophisticated knowledge of pigments and their manipulation. The palette found here includes some of the more expensive pigments of the day, such as azurite, silver, and gold,[16] demonstrating that there was an affluent class of patrons for hand-colored prints in northern Europe at this time.

Since beginning this project, we have corresponded with other researchers who have conducted analytical studies on hand-colored prints. The results of their

Sample #	Area	Color	Elements – Major	Elements – Minor	Probable pigment(s)
1	Paper/mat	–	–	–	–
2	Sun	Gold	Au	Pb	Gold, lead white
3	Sky	Blue	Cu	–	Copper blue
4	Cloud	Magenta	Pb	Hg	Red+/or white lead, vermilion
5	Rose	Red	Cu (sky)	Pb, Hg	Red+/or white lead, vermilion
6	Helmet	Silver (black)	Pb, Ag	Zn	Silver, lead white
7	Anvil	Silver (black)	Pb, Ag		Silver, lead white
8	Robe	Blue	Cu, As, Co, Fe, Ca	Pb, Bi	Copper blue, smalt, lead white, calcite or gypsum, bismuth white
9	Robe	Orange	Hg, Pb		Vermilion, red+/or white lead
10	Border	Yellow	Fe, Cu (sky)	Au (stripe), Pb	Yellow ochre, lead white
11	Cloth	Yellow		Pb (trace)	Organic yellow
12	Cloth	Green	Cu	Pb	Copper green
13	Background	Lt. green	Cu	Pb	Copper green
14	Wing	Blue	Cu, Pb		Copper blue, lead white
15	Sash	Purple	Cu, Pb		Lead white, organic purple
16	Sail	White	Pb	Cu	Lead white
17	Lines	Black	Pb	Cu, Ag	Lead white, silver

Table 2.3 XRF analysis of the Coornhert print (carbon target, 45 kV, 0.4 mA, 150 sec).

investigations have provided useful comparative information. A similar group of pigments to those found on the Coornhert were found on a hand-colored Dürer woodcut at the Museum of Fine Arts in Boston during a 1986 XRF study of that print.[17] The results of this investigation, conducted by Annette Mannick, indicate that the Dürer palette includes malachite, azurite, smalt, vermilion, lead white, lead black, ochre, umber, and silver and gold.

Another technical study that should be noted is an examination of a hand-colored engraving attributed to Jacob Matham after Hendrick Goltzius.[18] The print, showing a personification of Pride, is in the collection of the Harvard University Art Museums. This study, conducted by Aparna Tanden and Craigen Bowen, examined samples from the print using a scanning electron microscope with an energy dispersive spectrometer (SEM-EDS). They identified the presence of azurite, malachite, lead white, and earth pigments.

The painting of prints in the sixteenth and seventeenth centuries was a complicated enterprise. Municipal and legal documents from the period indicate that efforts were made to regulate the practice of painting prints. It is hoped that this investigation of print colorists' materials will not only help us to identify prints with coloring contemporary to their production, but also provide a better understanding of the practices of printmakers and publishers in sixteenth-century Northern Europe. Technical and scientific analysis, however, can only demonstrate that the pigments and binders of painting materials are characteristic of types used in the fifteenth to the seventeenth centuries, not that they unquestionably were applied at the time of the prints' production. Still, the data gained from technical analysis can provide insights into print colorists' methods when viewed in conjunction with archival documentation.

This paper is an introduction to a more detailed study that will culminate in the exhibition and catalogue *Painted Prints: Renaissance and Baroque Hand-Colored Engravings, Etchings, and Woodcuts* to be presented at the Baltimore Museum of Art in 2002, and the Saint Louis Art Museum in 2003. Because the questions surrounding these prints are technical, as well as stylistic and historical, the work has been undertaken as a joint curatorial and conservation project.

Acknowledgments

The authors would like to thank Janice Carlson, Senior Research Scientist at the Winterthur Museum Analytical Laboratory, for providing XRF and FTIR analysis. We are also grateful to our colleagues at the Baltimore Museum of Art.

Endnotes

1. J. Van der Stock, *Printing Images in Antwerp: The Introduction of Printmaking in a City* (Rotterdam, 1998).

2. Ibid., p. 38.

3. Carel Van Mander, *Het Shilder-Boeck waerin voor eerst de leerlustighe den groudt der edel vry shilder const in verscheyden deeln wort voorg hedraghen* (Haarlem, 1604).

4. Van der Stock 1998 (as in note 1), p. 133.

5. L. Voet, *The Golden Compasses: A History and Evaluation of the Printing and Publishing Activities of the Officina Plantiniana at Antwerp* (Amsterdam, 1969–72), p. 243.

6. S. Kostyshyn, "Door tsoecken men vindt: A reintroduction to the Life and Work of Peeter Baltens Alias Custodis of Antwerp (1527–1584)" (Ph.D. diss., Case Western Reserve University, Cleveland, Ohio, 1994), p. 383.

7. Van der Stock 1998 (as in note 1), p. 118.

8. R. J. Gettens and G. L. Stout, *Painting Material: A Short Encyclopaedia* (New York, 1969).

9. Bruno Mulenthaler and Jean Thissen, "Smalt," in A. Roy (ed.), *Artists' Pigments: A Handbook of Their History and Characteristics* (Washington, D.C.: National Gallery of Art, 1993), Vol. 2, p. 115.

10. Gettens and Stout 1969 (as in note 8), p. 107.

11. Van der Stock 1998 (as in note 1), p. 93.

12. C. Dodgson, "Rare woodcuts in the Ashmolean Museum, Oxford – II," *Burlington Magazine* 39 (1921), pp. 70–74.

13. F. W. Weber, *Artists' Pigments* (New York, 1923), p. 83.

14. Gettens and Stout 1969 (as in note 8), p. 174.

15. J. Van der Stock, "The impact of the prints of Pieter Brueghel the Elder," in D. Freedberg (ed.), *The Prints of Pieter Brueghel the Elder* (Tokyo, 1989), p. 94.

16. A. Burmester and C. Krekel, "The relationship between Albrecht Dürer's palette and fifteenth/sixteenth-century pharmacy price lists: The use of azurite and ultramarine," in *Painting Techniques; History, Materials and Studio Practice: Contributions to the Dublin Congress of the International Institute for Conservation of Historic and Artistic Works, 7–11 September 1998* (London, 1998), pp. 101–05.

17. A. Mannick, "Analytical Study of Dürer's *Christ Bearing the Cross*" (unpublished typescript, 1986).

18. A. Tanden and C. Bowen, "Technical Study of Two Dutch Sixteenth Century Hand-Colored Engravings" (unpublished typescript, 1997).

The Analysis of Watercolor Materials, in Particular Turner's Watercolors at the Tate Gallery (1790s to 1840s)

Joyce H. Townsend

It is not always possible to use non-destructive techniques for the analysis of the materials found in works of art on paper: considerations of physical security, environmental conditions in analytical laboratories, and staff, transport, and insurance costs may make it impractical for large and/or valuable artworks to be moved for analysis. Current analytical techniques for very small samples from pastel and related materials have been reviewed recently (Townsend 1998), and will be discussed briefly here as they relate to watercolors. Watercolors are less likely than pastels to be in too fragile a condition for safe transport, yet they pose similar problems when sampling is contemplated. The decision to remove even a tiny, sub-milligram sample from a small artwork in excellent condition can never be taken lightly.

Focussed studies on the works of one artist offer the best results, through information-gathering from a number of works in combination with limited sampling from a small selection of these. Those chosen for sampling are either the most damaged, the largest, the ones that have naturally ragged or torn edges, or those where a good deal of paint runs right out to the edge (see Fig. 2.14). Specks of paint that have landed on the reverse of images, or on the facing pages of sketchbooks, etc., also offer possibilities for sampling. The examination of a single work, on the contrary, may reveal a great deal about paint application, but lead the analyst to conclude that nothing could possibly be removed from such a pristine image!

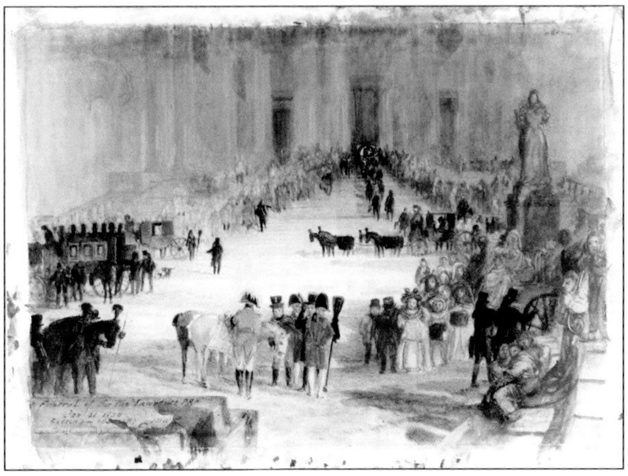

Figure 2.14 J. M. W. Turner (British, 1775–1851), *Funeral of Sir Thomas Lawrence: A Sketch from Memory*, 1830. Watercolor and gouache on white paper, 61.6 x 82.5 cm. Tate D25467.

A detailed study of one artist may also involve the search for palettes and other painting materials that might still be extant (see Fig. 2.15), their surviving materials archives, research into the colormen (the suppliers of artists' materials) whom he or she used, and the making of reconstructions (Zegers, this volume, pp. 138–44). Reconstructions can be used for accelerated aging experiments, and then as a source of material for developing analytical methods, trying out novel cleaning methods, or gaining an understanding of the effects that interventive conservation treatments may have had on the original appearance of the artworks. They can also be used to test whether an experimental paint formulation made from the artist's recipes would actually work, or whether the technique requires changes such as the use of diluents, different tools to apply the paint, a warm or cool studio to influence working temperature, or working with the support horizontal so the paint does not run off. These studies utilize the conservator's skills very well: a good example is a study of Degas's pastel and coloring methods (Fletcher and Desantis 1989), which led to conclusions as to which proteinaceous media he could have used in practice.

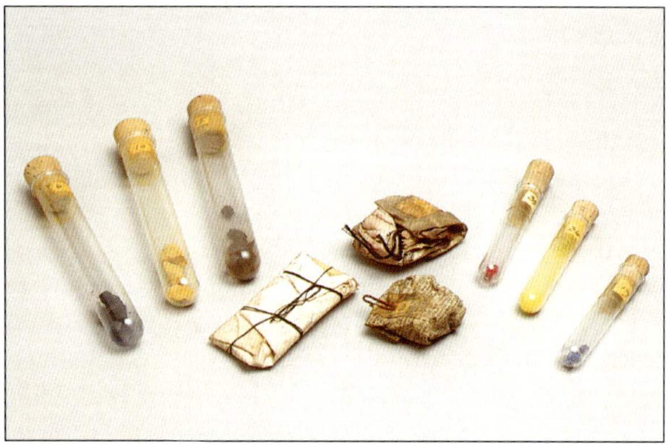

Figure 2.15 Selection of dry pigments found in Turner's studio after his death. Tate.

A study of the oil paintings and watercolors of J. M. W. Turner (1775–1851) illustrates this integrated approach (Townsend 1991). About 200 watercolors were examined, and small samples or splashed material were removed from a small fraction of these. Without the known existence of watercolor palettes used by Turner, the study could not have expanded, as will be described below. Conclusions were greatly enhanced by comparison with the paintings, and to surviving oil paint palettes (which were more numerous than watercolor palettes).

Analysis of watercolor materials

The discussion in this section applies to Turner's materials, that is to say, to watercolor materials in use up to 1850. Many new pigments came into use between 1850 and the present, which has made recognition of the materials in more modern works far more challenging.

Rather few pigments, and no watercolor media, can be recognized by viewing the watercolor through a stereo microscope, typically at 7- to 40-times magnification. A very few organic pigments, such as madder on an aluminum-based substrate (Townsend 1993a), can be recognized by their ultraviolet fluorescence, and lead white and zinc white may be distinguished from all the other white pigments. Infrared imaging spectroscopy (Casini et al. 1999) and false-color infrared photography (Moon et al. 1992) allow a few more to be recognized, and so may X-radiography (Townsend 1992). Large paintings could be sampled more easily than works on paper in the Turner study, and materials found in the oils could then be sought deliberately in watercolor: emerald green, vermilion, and lead white proved to show up clearly in X-radiographs of mock-ups and genuine watercolors; cadmium yellow is completely absent from his palette, so non-fluorescent, pale yellow inorganic pigments in the watercolors could be inferred to be chrome yellow; the coarsely ground particles of natural ultramarine that Turner favored are recognizable at 40-times magnification. Optical grays and greens, found more frequently than pure black or green pigments, could be seen as well (Fig. 2.16).

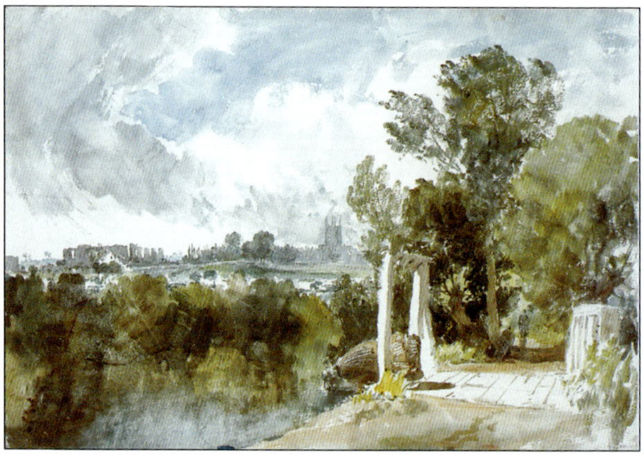

Figure 2.16 J. M. W. Turner, *River Scene, perhaps at Benson, near Wallingford*, c. 1805. Watercolor and pencil on white paper, 25.9 x 37.1 cm. Tate D05917.

Analysis carried out at that time involved sampling rather than travel to equipment that could have been used for non-destructive examination. This was the curator's preference for valuable works such as Turner's. The non-destructive technique of X-ray fluorescence spectroscopy (XRF), carried out in another laboratory, would have yielded some information on materials, although Turner's complex use of thin layers and the small scale of his brushwork would always have made interpretation difficult from an analyzed spot diameter of 2–3 mm. Nonetheless, experiments with mock-ups of very thin washes show that it could have worked in some cases. The analytical method of choice for inorganic pigments would now be Raman microscopy (Burgio 2001). This can be carried out non-destructively on the artwork itself or on tiny samples removed from works that cannot travel. It also yields some information on organic materials (Davey et al. 1994), and the same sample could then have been used for Fourier transform infrared (FTIR) microscopic identification of the medium type and of organic pigments as well. Possible identifications would have included: gum, protein, beeswax or spermaceti wax (the latter a product of the whaling industry, used for candles in the nineteenth century), and drying oil, although FTIR cannot identify specific types within these classes.

At the time a program that made the best use of in-house and/or easily available techniques for pigment identification involved removing tiny samples for optical microscopy in transmitted light, dividing them into one sample for energy-dispersive X-ray analysis in the scanning electron microscope (EDX–SEM), and another for FTIR microscopy (the latter sample was later made into a pigment dispersion). FTIR microscopy gave minimal information on media: gums for the watercolors and a drying oil for the inks used on engravings. Proteinaceous size on Turner's papers, presumed to be glue, was also identified by FTIR microscopy. A method of analysis that gave more detailed results for the medium would have been welcome.

Turner's watercolor techniques and materials

The rather few descriptions of Turner at work and examination of his watercolors from all periods of his life suggest that he often began by working on wet, indeed soaked, paper. After arranging several papers for soaking, he could return to the first one, which had partially dried, developing the others in their turn as they reached a satisfactory dryness. Then he worked on more localized areas, and finally he created highlights by various means, as will be described below. Figures 2.16 and 2.17 illustrate these processes.

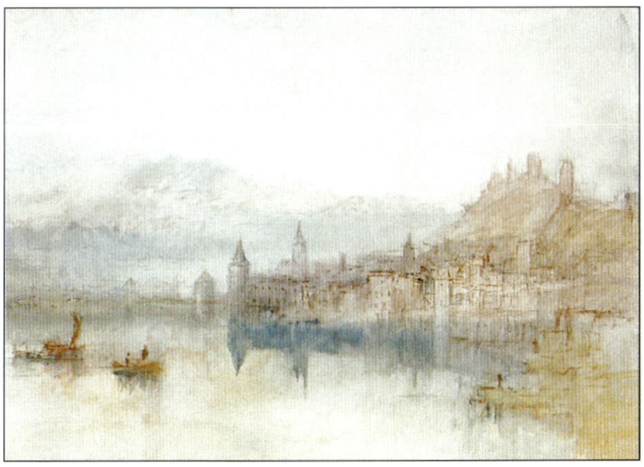

Figure 2.17 J. M. W. Turner, *Lucerne from the Lake: Sample Study*, c. 1844–45. Pencil and watercolor on white paper, 23.3 x 32.7 cm. Tate D36252.

Bower (1990, 1998) has discussed the qualities that Turner sought in paper, one of which was the strength to withstand all the working while the fibers were wet – achieved by careful selection of the right amount of gelatin sizing on the surface. There is a parallel in Turner's oil paintings: during the lifetime of his father, who worked as his studio assistant, he often used a priming of lead white and whole egg to give a high absorbency; later he purchased canvases with absorbent primings from colormen (Townsend 1994). The oil paintings always have white primings, with the exception of a few very early works with red primings. The majority of the watercolors have a support of white paper, with no preparation layer. Turner experimented in the early decades of his career with blue, red, green, mauve, and grayish-brown washed papers, but he never used them regularly. However, one material of Turner's that is unique to the watercolor medium is his use of blue papers in the 1820s and 1830s: the source and coloring of these papers have been discussed recently (Winner 1999). When these papers have not faded due to light exposure, they are a pale sky-blue color, and Turner worked with them by using thick washes of lead white gouache mixed with red and yellow ochres for buildings and foreground details, transparent washes for skies, and reserving areas for water in the foreground (see Fig. 2.18).

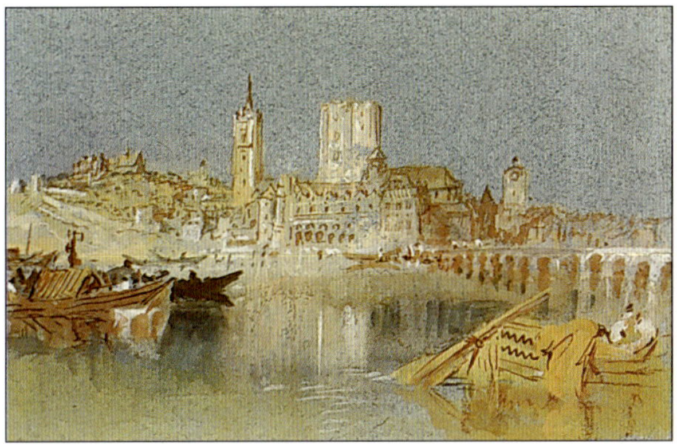

Figure 2.18 J. M. W. Turner, *Beaugency from the South-East*, c. 1826–30. Watercolor and gouache, partly worked in pen, on blue paper, 12.4 x 18.3 cm. Tate D24661.

In fact every aspect of Turner's oil painting technique was influenced by his early and then parallel experiences in the watercolor medium. Analysis of materials from fifty-odd paintings and five palettes, as well as the watercolors mentioned above revealed that he used the same pigments in both media throughout his life. He also used many new industrial products, as has been detailed elsewhere (Townsend 1993a): manufactured iron oxides in brilliant orange and scarlet shades in watercolors from the 1790s, and thereafter in paintings; cobalt blue in the 1800s in oil and in the 1810s in watercolor; chrome yellow in the 1810s in both media; chrome orange in the 1820s in both media; emerald green in the late 1820s in oil and in the early 1830s in watercolor. His palette therefore expanded continuously, and only a few traditional pigments such as orpiment and azurite dropped out of it over the decades. Zinc white (available in an opaque and eminently useful form after 1834) and synthetic ultramarine (produced beginning in 1826–27) are notably absent from both media. Opaque pigments such as vermilion, natural ultramarine, chrome yellows and oranges, and emerald green are notably present in his watercolors. So too are ochres, siennas, and umbers, particularly in brown and fairly "monochrome" watercolor studies for engravings made in the 1810s and 1820s (Townsend 1996). They were used by Turner more regularly than traditional watercolor materials such as bistre and sepia. Indeed, his innovative use of opaque and brightly colored pigments in watercolor medium, often within a few years of their first manufacture, greatly influenced both his contemporaries and later artists.

The reasons for Turner's use of predominantly inorganic pigments such as these were technical as well as aesthetic. He could have achieved some of his surface effects through the use of organic pigments such as red and yellow lakes, bistre, sepia, and the like, or mixtures thereof. Yet not all: he used coarsely ground pigments (ranging in size from 1–2 micrometers for the manufactured iron oxides to over 20 micrometers for natural ultramarine) in preference to fine ones so that he could wash out highlights later, using a brushload of clean water. The organic pigments have particles of submicrometer size, which bind too firmly to gelatin-sized paper to allow for complete washing-out of a highlight. The coarser pigments facilitated the use of "tonal scale practice" described by Shanes (1997a, b), a method of working up an image tonally from light to dark without having to use a separate brush for tonal variations that, in Turner's case, can be counted by the dozen. It involved successively darkening the wash on the palette by adding more pigment and applying that color everywhere it would be needed before darkening the wash further on the palette and repeating the process. This method enabled Turner to work quickly, and adapted well to his known habit of working on several sheets at once. It also presupposes that the artist could visualize the final image.

Turner did nearly all this tonal calculation in his head, and rapidly: in oil he did not produce detailed pencil sketches, oil studies, or small-scale versions of large canvases. Instead, he worked up transparent washes on a white, absorbent priming, gradually adding white pigment to make the image more opaque, before proceeding to apply localized washes of paint of varying gloss and transparency on top (Townsend 1993b). Possibly it was the idea of an even better future image that made Turner finally stop working on the canvas or the sheet in hand, rather than complete satisfaction with it. The Turner Bequest at the Tate Gallery, London, includes about 200 unfinished oils, to some of which Turner returned years later. He also added to "finished" oils that were unsold and therefore accessible in his studio. The finished watercolors, by contrast, often went out to be engraved and were inaccessible until they were returned to Turner for later sale.

In the watercolor medium, Turner had many other ways of creating highlights, some of which owe their origin to his lifelong production of engravings, and hence his skills with engraving tools and familiarity with working tonally, from dark to light. Sometimes he left substantial reserves of white paper as lights (see Fig. 2.16). Localized highlights were created after all of the tonal scale washes had dried, by washing off color with a wet brush; rubbing it off with stale bread of varying age and hardness (a method described by his contemporaries, but not of course made evident by surface examination); or scratching at the paper with the end of a brush, sharpened fingernail (this was also described by contemporaries), or an engraving tool (see Fig. 2.19). Localized, very wet washes could also be applied within such highlights to leave a hard edge as they dried (see Fig. 2.20), or the wet washes could be textured with various brushes or with an engraver's rocker. Some of these effects have been illustrated elsewhere (Townsend 1996). In short, his handling of watercolor was innovative and masterly.

Turner's watercolors are mostly in excellent condition today, except for those that have been exposed to too much light. However, the thicker washes have a tendency to crack and flake today, which might be related to gum type, and to whether or not he used sugar or honey to plasticize the paint film.

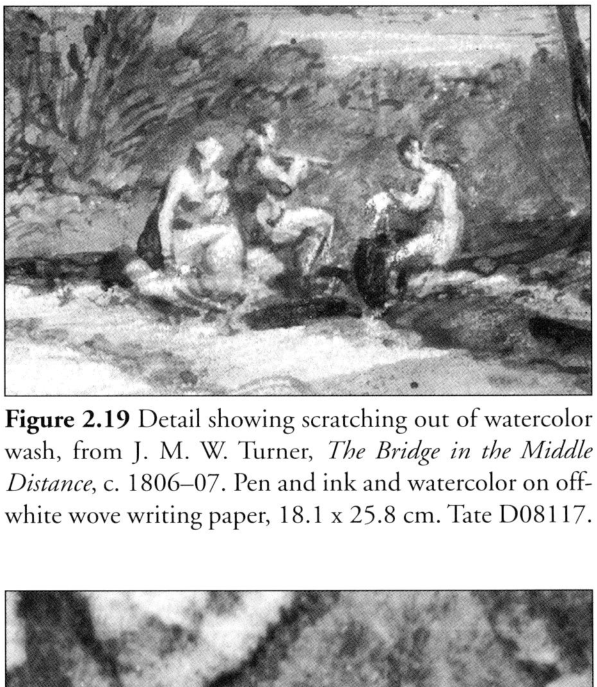

Figure 2.19 Detail showing scratching out of watercolor wash, from J. M. W. Turner, *The Bridge in the Middle Distance*, c. 1806–07. Pen and ink and watercolor on off-white wove writing paper, 18.1 x 25.8 cm. Tate D08117.

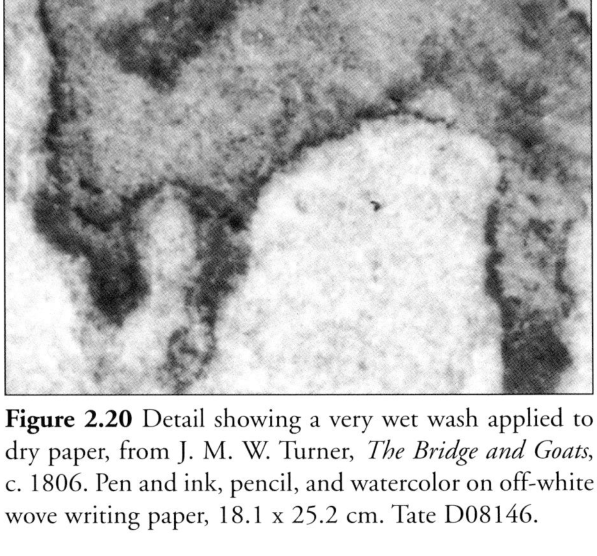

Figure 2.20 Detail showing a very wet wash applied to dry paper, from J. M. W. Turner, *The Bridge and Goats*, c. 1806. Pen and ink, pencil, and watercolor on off-white wove writing paper, 18.1 x 25.2 cm. Tate D08146.

Watercolor materials used by other artists

When I began to analyze Turner's materials in 1987, there was no practical method for the analysis of gums or proteins in the small sample that could be removed from a thinly painted watercolor. This lack led to a collaboration, still proceeding today, with a British university to improve such methods through an analysis of the materials and technique of William Blake (1757–1827) and his followers (Vallance 1997; Vallance et al. 1998). Gas chromatography with mass spectrometry (GC-MS) and high-performance liquid chromatography (HPLC) are being used to analyze gums and proteins, respectively, through the application of new derivitization methods to very small samples (Ormsby 2001a). Fortuitously, an artist and researcher in a neighboring city is working on the philosophy behind Turner's use of color in his watercolors, and has for many years been trying to emulate Turner's effects in watercolor through a simulation of his materials (Clark 2001). He has offered to try out Turner's pigments and media, as determined from analysis, on paper selected and glue-sized to resemble the types Turner used, in an attempt to reproduce Turner's subtle watercolor techniques. Currently he is experimenting with different mixtures of gums in order to match the gloss and hard or soft edges that Turner achieved in his watercolors, particularly in the late Swiss ones. He has found that gum arabic alone does not give the full range of effects which Turner could produce.

Near contemporaries of Turner's such as Blake used different combinations of gums, including gums arabic, tragacanth, cherry, and possibly sarcocolla, with plasticizers such as honey or unrefined sugar (Vallance et al. 1998). Analyses of a small number of samples of Turner's watercolor medium from his palettes and from watercolors painted in the 1840s revealed gum arabic, sugar or honey, and gum tragacanch or sarcocolla. The last two cannot be distinguished easily (Ormsby 2001a). There was also a drying oil used locally on thicker washes of color (Vallance 1996). These findings led to the hypothesis that gum formulations were bought as mixtures, rather than including gum arabic alone, as mid-twentieth-century sources have tended to suggest.

The surviving records of the English colorman Roberson (Ormsby 2001b) bear out the hypothesis. The recipes for their watercolor blocks include gums arabic, occasionally tragacanth, and senegal as mixtures, sometimes ox gall, honey, and/or candy for different colors, and occasionally isinglass, and are specific to a given color. That is to say, the gum formulation and the proportion of gum to pigment were different for each pigment listed, just as pastel recipes differed for each pigment (Burns 1994). The rationale in both cases would have been the production of compatible materials with similar handling properties. The market for watercolor materials, after all, included amateurs, ladies and gentlemen of leisure, as well as professional artists. The Roberson records refer to "soft" watercolors, which included honey or sugar, and "pan cakes" that had less sugar or honey in the formulation. The earliest mention of honey was in 1833 (Ormsby 2001c), though the records are largely lost for earlier decades.

Preliminary analysis of honey watercolors

The first step was clearly to obtain some genuine "honey" watercolors and to determine whether honey could be detected easily in the sample weight that would be obtainable from a real watercolor. This step has been taken, and the project is now actively seeking further watercolor boxes and palettes traceable to Turner, Blake, and their contemporaries. As yet no surviving material used by Blake has been located.

The box (239 mm wide, 200 mm deep, 78 mm high) that was analyzed is in a private collection. It includes a china palette and dishes, all cleaned of paint, and 22 full or fragmented watercolor blocks, some labeled VERY BEST HONEY COLOURS/COULEUR SANS DANGER (Fig. 2.21). These all have Ackermann labels. Some blocks have Winsor & Newton or Newman labels, but not all the Ackermann colors have the "honey" label. The box has a nineteenth-century style and appearance, and bears the original label R. ACKERMANN MANUFACTURER OF SUPERFINE WATER-COLOUR TO HIS MAJESTY. The actual date must then be pre-1837, since Queen Victoria reigned from that date to 1901, having been preceded by male monarchs from the later eighteenth century. The blocks were examined with EDX and optical and ultraviolet microscopy. They include a number of earth pigments; several madder lakes on a variety of substrates, including iron, as found in Turner's studio pigments; and a chrome yellow with the rounded particles seen in Turner's paint, in contrast to the more acicular particles seen today in the modern variety (Townsend 1993a). The Prussian blue pigment has the large flat particles seen in Prussian blue used by Turner for much of his life (Townsend 1993a) and by Sir Joshua Reynolds (1723–1792) for much of his (Jones 1998), though John Constable (1776–1837) used the fine-grained variety produced today (Cove 1991). Zinc white is present in two blocks, both of which are pale in color, so this may be the more transparent type of zinc white that preceded the invention of Chinese white in 1834 (Harley 1982). All the evidence from the pigments suggests that the box is contemporary with Turner's era.

GC-MS analysis for gums and sugars (Ormsby 2001a) showed that the "honey colors" include glucose as the main component and sometimes as the only one. This could derive from genuine sugar candy or natural honey, both of which have glucose as a major component. Both occur in the Roberson recipes (Ormsby 2001a). Nearly all the other blocks include

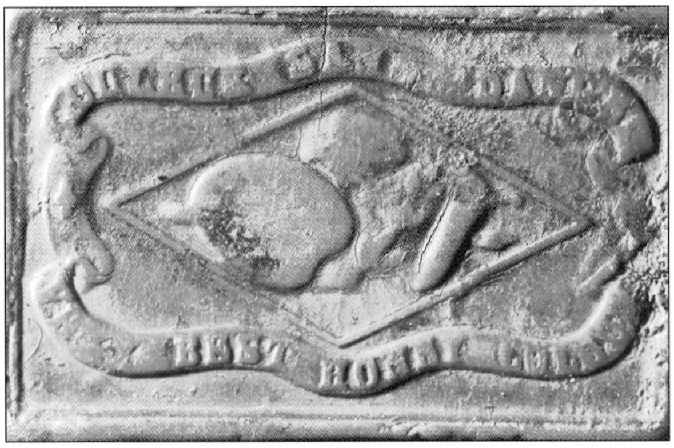

Figure 2.21 Honey label from a nineteenth-century Ackermann watercolor block.

detectable glucose as well, in quantities that vary from trace to moderate. Gum arabic is a major component in all cases.

Prospects for research

It is hoped that the combined efforts of analysts and a practicing artist, informed by a study of comparative materials, will lead to new insights into the methods and materials used by Turner. These findings will in turn provide information and enhanced understanding for those who admire Turner's watercolors, as well as for the conservators and curators who care for them.

Acknowledgments

The initial Turner research was funded by the Leverhulme Trust, U.K., and the Tate Gallery. Members of the Tate Gallery Conservation Department, and many others who have been acknowledged elsewhere (Townsend 1993b), gave support and advice. Bronwyn Ormsby's doctoral research is supported by Dr. J. Dean of the School of Applied and Molecular Sciences and Dr. B. Singer of the School of Humanities, both at the University of Northumbria at Newcastle, Newcastle upon Tyne, U.K., and the Tate Gallery Conservation Department. Dr. Fumio Yokoshima kindly made his watercolor palette available for sampling and provided illustrations of it. Paul Clark, Stephen Hackney, Bronwyn Ormsby, Elizabeth Sobczynski, and Dr. Fumio Yokoshima have reviewed this manuscript.

References

Bower, P. 1990. *Turner's Papers: A Study of the Manufacture, Selection and Use of His Drawing Papers 1787–1820.* London: Tate Gallery.

Bower, P. 1998. *Turner's Later Papers: A Study of the Manufacture, Selection and Use of His Drawing Papers 1820–1851.* London: Tate Gallery and Oak Knoll Press.

Burgio, L., and R. J. H. Clark 2001. "Library of FT-Raman spectra of pigments, minerals, pigment media and varnishes, and supplement to existing library of Raman spectra of pigments with visible excitation," *Spectrochimica Acta*, Part A, 57, 7, pp. 1491–1521.

Burns, T. 1994. "Chalk or pastel?: the use of friable media in early drawings," *The Paper Conservator* 18, pp. 49–56.

Casini, A., F. Lotti, M. Piccolo, L. Stefani, and E. Buzzegoli 1999. "Image spectroscopy mapping for non-invasive analysis of paintings," *Studies in Conservation* 44, pp. 39–48.

Clark, P. 2001. Doctoral research, Department of Philosophy, University of Durham; to be completed early 2002.

Cove, S. 1991. "Constable's oil painting materials and techniques," in L. Parris and I. Fleming-Williams (eds), *Constable.* London: Tate Gallery Publications, pp. 493–518.

Davey, R., D. J. Gardiner, B. W. Singer, and M. Spokes 1994. "Examples of analysis of pigments from fine art objects by Raman microscopy," *Journal of Raman Spectroscopy* 25, pp. 53–57.

Fletcher, S. and P. Desantis 1989. "Degas: The search for his technique continues," *Burlington Magazine* 131, pp. 256–65.

Harley, R. D. 1982. *Artists' Pigments c.1600–1835: A Study in English Documentary Sources*, 2nd edn. London: Butterworth Scientific, pp. 176–80.

Jones, R. 1998. Personal communication, Conservation Department, Tate Gallery.

Moon, T., M. R. Schilling, and S. Thirkettle 1992. "A note on the use of false-colour infrared photography in conservation," *Studies in Conservation* 37, pp. 42–52.

Ormsby, B. 2001a. Doctoral research, School of Applied and Molecular Sciences, University of Northumbria at Newcastle; to be completed early 2002.

Ormsby, B. 2001b. Personal communication re. Roberson Archive at the Hamilton Kerr Institute, University of Cambridge.

Ormsby, B. 2001c. Personal communication re. Roberson Archive reference 778/1993. Published with the permission of the Syndics of the Fitzwilliam Museum, University of Cambridge.

Shanes, E. 1997a. "Turner and the 'scale practice' in British watercolour art: Tint by tint by tint," *Apollo* 146, no. 429, pp. 45–51.

Shanes, E. 1997b. *Turner's Watercolour Explorations.* London, Tate Gallery, pp. 23–25.

Townsend, J. H. 1991. "The materials and techniques of J. M. W. Turner R. A. 1775–1851," Ph.D. diss., Courtauld Institute of Art, University of London.

Townsend, J. H. 1992. "Turner research project: information on materials *via* imaging techniques," *Journal of Photographic Science* 40, pp. 52–65.

Townsend, J. H. 1993a. "The materials of J. M. W. Turner: Pigments," *Studies in Conservation* 38, pp. 231–54.

Townsend, J. H. 1993b. *Turner's Painting Techniques.* London, Tate Gallery, pp. 28–38.

Townsend, J. H. 1994. "The materials of J. M. W. Turner: Primings and supports," *Studies in Conservation* 39, pp. 145–53.

Townsend, J. H. 1996. "Turner's 'drawing book,' the *Liber Studiorum*: Materials and techniques," in J. Bridgland (ed.), *ICOM-CC Preprints of the 11th Triennial Meeting Edinburgh.* London: James and James, pp. 376–80.

Townsend, J. H. 1998. "Analysis of pastel and chalk materials," *The Paper Conservator* 22, pp. 21–28.

Vallance, S. L. 1996. Personal communication, Royal Society of Chemistry, London.

Vallance, S. L. 1997. "Applications of chromatography in art conservation: techniques used for the analysis and identification of proteinaceous and gum binding media," *The Analyst* 122, pp. 75R–81R.

Vallance, S. L., B. W. Singer, S. M. Hitchin, and J. H. Townsend 1998. "Development and preliminary application of a gas chromatographic method for the characterisation of gum media," *Journal of the American Institute for Conservation* 37, pp. 294–311.

Winner, C. 1999. "Seeing blue: an investigation of the deterioration of J. M. W. Turner's blue drawing papers," *The Quarterly, the Journal of the British Association of Paper Historians* 3, pp. 11–17.

Zegers, P. K. 2002. "In the Kitchen with Paul Gauguin: Devising Recipes for a Symbolist Graphic Aesthetic." This volume, Chapter 3, pp. 138–44.

Paul Cézanne's Watercolors: His Choice of Pigments and Papers

Faith Zieske

A group of nine watercolor paintings by Paul Cézanne (1839–1906) in the collection of the Philadelphia Museum of Art was studied in order to identify the pigments in his watercolor palette and to learn more about his choice of papers. The green pigments were of particular interest because the emerald green paint that Cézanne used appeared to have darkened to a dull tan on several of his watercolors. More than 100 sites on nine watercolor paintings were analyzed by qualitative energy dispersive X-ray fluorescence (XRF)[1] to develop a profile of the artist's watercolor palette. Investigation of Cézanne's papers involved study of watermarks and sheet formation and analysis of fiber composition by polarized light microscopy. The paper analysis was relevant particularly because the loss of whiteness by discoloration of the paper had a significant impact on the appearance of the watercolors. A greater understanding of Cézanne's choice of watercolor pigments and papers will help us appreciate their original appearance and determine how they may have changed over the years.

The dates for the nine Philadelphia watercolors in this study were taken from Rewald (1983), unless another date was assigned subsequently.[2] Rewald cautions, however, that "many of the dates – albeit even the vaguest ones – mentioned in this catalogue are based on conjecture. There are only a few watercolors for which even an approximate date of execution is known." (Rewald 1983: 32). Many of the Philadelphia watercolors are landscape views, such as *House on the Hill Near Aix* (Fig. 2.22), that were painted in the environs of Aix-en-Provence, Cézanne's hometown. The beautiful sun-drenched landscape that Cézanne loved offered him the opportunity to paint daily from nature. The trees and undergrowth set within ochre-toned rocks depicted in the three watercolors of trees and rocks are favorite motifs that reflect scenes in cherished painting spots near Aix, such as the grounds of Château Noir and Bibémus

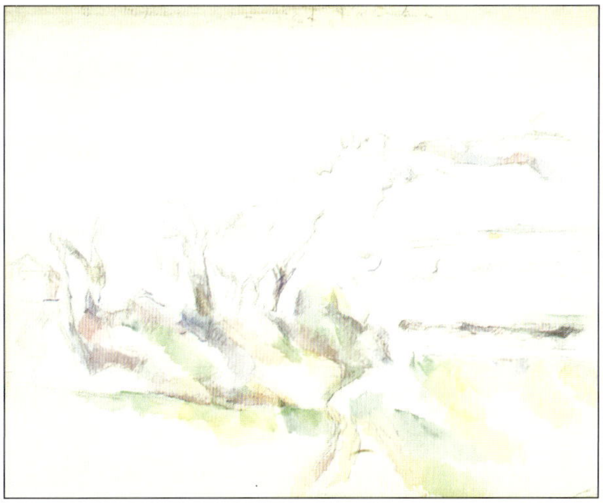

Figure 2.22 Paul Cézanne (French, 1839–1906), *House on the Hill Near Aix*, 1895–1900. Watercolor and graphite on wove paper, 47.7 x 57.2 cm. Philadelphia Museum of Art, The Louis E. Stern Collection, 1963-181-007. (Photograph credit: Photography Department of the Philadelphia Museum of Art.)

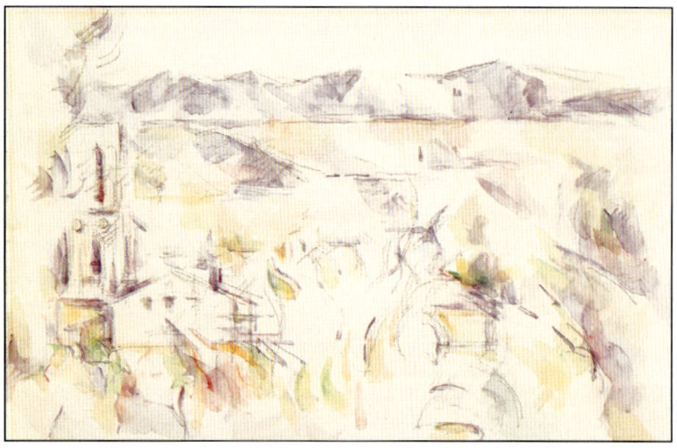

Figure 2.23 Paul Cézanne, *Aix Cathedral, View from the Studio at Les Lauves,* 1902–04. Watercolor and graphite on wove paper, 31.6 x 47.6 cm. Philadelphia Museum of Art, The Louise and Walter Arensberg Collection, 1950-134-035. (Photograph credit: Photography Department of the Philadelphia Museum of Art.)

quarry. With proceeds from the 1899 sale of Jas de Bouffan, his family home, the 63-year-old artist was able to build a studio at Les Lauves, in the hills north of Aix. There he worked from the fall of 1902 until his death in 1906. *Aix Cathedral, View from the Studio at Les Lauves* (Fig. 2.23) may have been painted from his studio window. Other late Philadelphia watercolors that were likely painted in Aix around this time include: *The Balcony* (Fig. 2.24), an image showing scrolled ironwork grills like those found in the windows of Aix's grand houses (Rishel 1983), and the engaging but discolored still life *Oranges on a Plate* (Fig. 2.25). Earlier work is represented by *Olympia,* c. 1875, one of Cézanne's variations after Manet's composition, and the sketchbook page *Carafe, Table Knife, and Unidentified Object,* c. 1882.

There is little evidence to indicate how Cézanne prepared his paper for painting other than small corner holes that suggest he simply tacked sheets to a board. Rewald (1983: 36–37) states that "Among the late works ... are many that show in their corners tiny tack holes (in some cases only two or three, in others up to a dozen). This may signify that the artist was in the habit of detaching them after every session and reattaching them to his drawing board whenever he wished to continue working on them. But it could also mean that he pinned them to the walls of his studio or his home so as better to contemplate them and think about what still remained to be done." Although adhesive residues around the edges of some watercolors suggest that they were mounted, this also could result from later picture framing. Cézanne may not have found it necessary to stretch his paper in the traditional manner, since his technique did not involve laying down broad areas of wash and thus may not have caused significant cockling of the paper support.

Cézanne's watercolor technique allowed the image to unfold through a subtle interplay of overlapping strokes of thin washes of color. Most of the sites analyzed by XRF indicated the use of pure pigments. Rarely were the

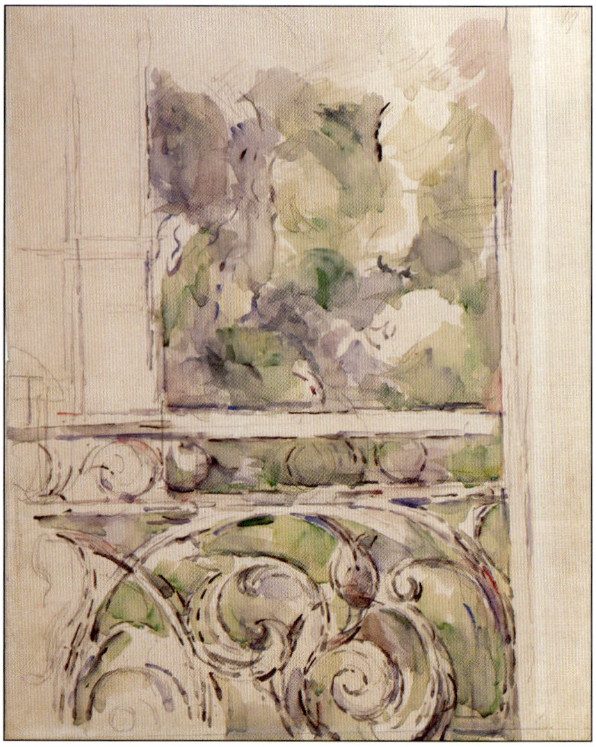

Figure 2.24 Paul Cézanne, *The Balcony,* c. 1900. Watercolor and graphite on wove paper, 61 x 45.1 cm. Philadelphia Museum of Art, The A. E. Gallatin Collection, 1943-075-001. (Photograph credit: Photography Department of the Philadelphia Museum of Art.)

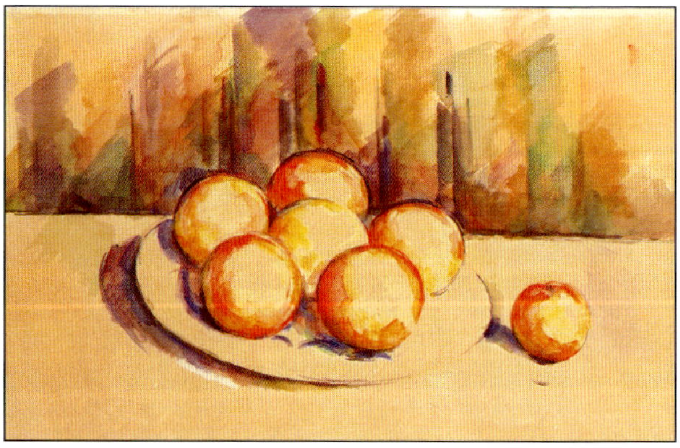

Figure 2.25 Paul Cézanne, *Oranges on a Plate,* c. 1900. Watercolor and graphite on wove paper, 31.5 x 47.7 cm. Philadelphia Museum of Art, Mr. and Mrs. Carroll S. Tyson, Jr., Collection, 1963-116-020. (Photograph credit: Photography Department of the Philadelphia Museum of Art.)

pigments mixed on the palette before application; instead, superimposed patches of translucent color create additional tints. This layering technique required restraint and practiced precision, as it did not allow for scraping out or erasing mistakes. Long drying times were necessary between overlapping strokes, for if the underlying colors were still moist, they would merge and run together. A fresh stroke of wash applied to dry paper or to a dried patch of color usually develops a ridge of

	Lead white	Vermilion	Organic red	Red lead	Iron oxide	Chrome yellow	Organic yellow or orange	Emerald green	Viridian	Mixed green (chrome yellow and cobalt blue)	Ultramarine	Cobalt blue	Organic-based purple (mixed)	Cobalt-based purple (mixed)
Aix Cathedral, View from the Studio at Les Lauves, 1902–04			X		X			X	X			X		
The Balcony, c. 1900			X					X	X			X		X
Oranges on a Plate, c. 1900	X	X			X	X		X				X		X
House on the Hill Near Aix, 1895–1900			X				X	X	X		X		X	
Trees and Rocks No. 1 c. 1895	X		X					X	X			X		
Trees and Rocks No. 2, c. 1890			X				X	X	X			X		
Rocks and Trees, 1880–85		X		X	X			X		X		X		X
Carafe, Table Knife, and Unidentified Object, c. 1882					X			X				X	X	
Olympia, c. 1875		X	X		X			X				X		

Table 2.4 Pigments indicated on nine Cézanne watercolors at the Philadelphia Museum of Art.

pigment particles around the edge as it dries – a phenomenon seen in many Cézanne watercolors. The artist also used soft pencil lines to structure the picture before painting and as loose hatch marks to create image tone and texture. He often incorporated unpainted areas of paper into the image, where they provide form or create space within the composition and lend emphasis to strokes of color or graphite lines.

Cézanne's choice of pigments

Cézanne often exchanged his paintings for paints and canvases with *père* Tanguy, a picture dealer and paint merchant who displayed Cézanne's paintings in his small shop on rue Clauzel, Paris (Rewald 1948). It is not known if Cézanne obtained his paint exclusively from Tanguy, or if he traded mainly for the more expensive oil paint and obtained watercolor paint elsewhere. Nor is it known who supplied him with materials after Tanguy's death, in around 1893. Dominique Sennelier, of Sennelier Beaux-Arts Distribution, has seen invoices for oil paints sold to Cézanne by his grandfather, who became a color merchant in 1887. According to Sennelier (1994), *les couleurs moites* available in France were manufactured by a relatively small number of companies: Sennelier, Lefranc & Cie, Bourgeois Ainé, P. C. Lambertye – J. M. Paillard (student grade), and English companies such as Winsor & Newton and Rowney & Co. English watercolor paints were readily available in France during this period: Sennelier catalogues advertise the sale of watercolors by English manufacturers, and Winsor & Newton sought out the French market with catalogues printed in French.

Lists of watercolor paints in French and English trade catalogues from the late nineteenth and early twentieth centuries helped confirm the commercial availability of pigments suggested by elemental analysis. Supplementary information was gathered from nineteenth-century manufacturers that remain in operation, such as Lefranc & Bourgeois, Sennelier, and Winsor & Newton. The pigments suggested by this analysis were further substantiated by lists of colors on the front and back covers and neighboring pages of Cézanne's sketchbooks (Chappuis 1973; Reff and Shoemaker 1989) and by colors mentioned in his published letters (Rewald 1976). Additional comparison was made to a pigment study of ten Cézanne oil paintings at the Philadelphia Museum of Art (Butler 1984).

Table 2.4 shows eleven individual pigments plus two or three mixtures that were suggested by elemental results of XRF on the nine Cézanne watercolors. The presence of organic pigments or dyes and ultramarine blue was extrapolated by the absence of detectable elements in analyzed areas.[3] Four of the watercolors studied are double-sided, with a second watercolor painting or drawing on the other side; care was taken to select sites for analysis that did not correspond to painted areas on the verso. Color swatches from a Winsor & Newton catalogue (Taylor 1887) were analyzed as reference standards. Supplemental analytical methods, including scanning electron microscopy (SEM), X-ray diffraction (XRD), Fourier transform infrared spectroscopy (FTIR), and high-pressure liquid chromatography (HPLC), were not applied for further identification. These methods require destructive sampling techniques that are not safely possible on thin watercolor washes. The following discussion of Cézanne's watercolor palette presents each of the pigments suggested by this study and includes brief remarks on their chemistry and historical background.

Green pigments

Green was of special interest, not only because often it had darkened, but also because Cézanne employed it liberally, particularly in his many landscapes. The three green pigments established by the elemental results of XRF were emerald green, viridian, and mixed green (Zieske 1995).

A representative XRF spectrum for emerald green (*vert Véronèse*, Schweinfurt green, Paris green), copper aceto-arsenite ($3Cu(AsO_2)_2.Cu(CH_3OO)_2$), is presented in Figure 2.26. Analysis identified generous amounts of copper and arsenic, indicating the presence of emerald green on all nine Cézanne watercolors in this study. A color valued by the artist, *vert Véronèse* is present among several of the lists of colors in his sketchbooks and is one of the most commonly found pigments on his oil

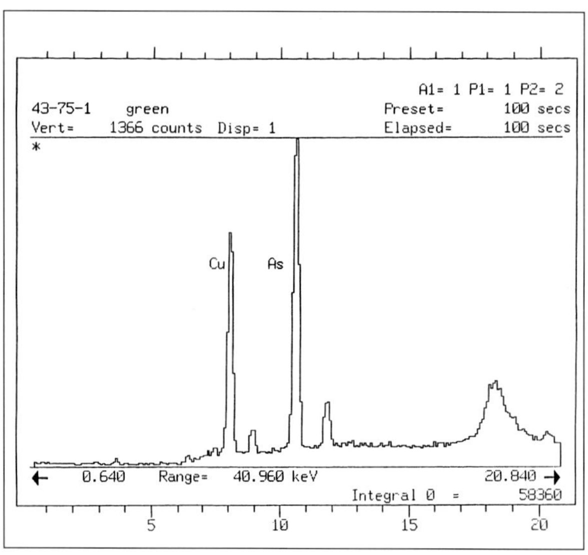

Figure 2.26 Spectrum of qualitative energy X-ray fluorescence detection of copper and arsenic, indicating emerald green pigment in Paul Cézanne's *The Balcony*, c. 1900 (Fig. 2.24).

paintings (Butler 1984). Elements for emerald green were detected both in bright green strokes and in drab green areas, suggesting that the pigment had darkened, to varying degrees, on many but not all of the Philadelphia watercolors.

According to Fiedler and Bayard (Fitzhugh 1997: 222), two producers developed emerald green in the first decades of the nineteenth century: Wilhelm Sattler and Friedrich Russ of Schweinfurt, Germany, and Ignaz von Mitis of Vienna. Sattler's firm produced the pigment as Schweinfurt green, keeping the formulation a trade secret. Once the manufacturing process was published in 1822, by Justus von Liebig and André Braconnot, the production of emerald green spread widely. Under the name "Paris green," it was introduced as the first chemical insecticide around 1867 (Haynes 1954). Its highly toxic nature led Winsor & Newton and other manufacturers to withdraw emerald green from production in the 1960s (Tate Gallery 1982). Emerald green was one of the watercolor pigments tested for permanence during the late nineteenth century. The Russell and Abney report of 1888 (Brommelle 1964) and Winsor & Newton Limited trade catalogues, beginning in 1892, rate it as moderately permanent, showing slight darkening. George Field stated that all copper greens have a tendency to blacken in impure air, although he considered emerald green to be the most durable of this class of pigments (Field and Davidson 1903). An 1889 report for George Rowney states that emerald green turns brown when exposed to hydrogen sulfide (Seward 1889). The trade catalogues of the period also warn that mixing copper greens with pigments containing sulfides would risk blackening the mixture. The validity of references to the darkening of emerald green was confirmed by recent experiments, which indicate that the pigment can be discolored by atmospheric sulfides,[4] although the chemical pathways are as yet unknown.

The watercolors in this study display emerald green in a fresh, mint-colored state and in various degrees of deterioration. The emerald green that can still be observed in early trade catalogues is similar to the brilliant color found in Cezanne's *House on the Hill Near Aix*, the best-preserved Philadelphia watercolor. Microscopic examination of emerald green washes often revealed a characteristic granular quality due to the paint's near insolubility in water. Deteriorated emerald green may appear as a tan wash, often with bright green particles that have accumulated and formed a ridge along the edge of the wash. In *The Balcony* and *Aix Cathedral, View from the Studio at Les Lauves*, for example, only the thinner washes of emerald green have discolored to a dull tan, while the paint has remained bright in the areas of thicker application. In several watercolors with darkened emerald green, there are other signs of deterioration, such as faded organic pigments or discolored paper, a condition evident in *Oranges on a Plate* and also in the three landscapes *Trees and Rocks No. 1*, *Trees and Rocks No. 2*, and *Rocks and Trees*. By contrast emerald green paint remains brilliant on the sketchbook drawing *Carafe, Table Knife, and Unidentified Object*, although the paper support has discolored severely, suggesting that it was sheltered from sulfide pollution but overexposed to light.

In some cases emerald green has affected the underlying paper, either protecting it from discoloration or causing discoloration on the verso. On three watercolors with highly discolored paper, *Oranges on a Plate, Trees and Rocks No. 2*, and the sketchbook page *Carafe, Table Knife, and Unidentified Object*, emerald green has protected the paper from discoloration, so that on the verso the paper remains lighter behind the painted area than in the darkened surrounding area. In two watercolors on thinner, less discolored paper, *Rocks and Trees* and *Trees and Rocks No. 1*, deteriorated emerald green has penetrated through the paper, staining the verso light brown. Of the remaining watercolors in this study, most of them on thicker or whiter paper, emerald green does not appear to have affected the paper support. These visual changes to the watercolor papers seem to be related to the degree of paper discoloration, the paper thickness, and the extent of emerald green deterioration. It is noteworthy that deteriorated emerald green has penetrated the paper and stained the verso of thinner, whiter papers of all rag content, whereas this pigment has protected the paper from darkening on the verso of more discolored papers containing greater amounts of non-rag fiber. The interaction of emerald green with the paper support is also among the topics requiring further investigation (Banik and Ponahlo 1982–83).

A useful technique for distinguishing emerald green is ultraviolet (UV) light examination.[5] Pale washes of the paint, whether bright green or discolored tan, absorb UV light in the same manner as thicker applications: both appear very dark. By contrast the other green pigments found in this study (viridian and mixed green) do not absorb UV light in the same manner. The dark appearance of thin washes of emerald green can be observed in the UV-light photograph of *Oranges on a Plate*, particularly in the pale green washes located in the center left of the plate and in the upper right background (see Fig. 2.27; compare to Fig. 2.25).

It is noteworthy that French and English terminology for emerald green led to its confusion with viridian. In French, *vert Véronèse* refers to emerald green, whereas *vert émeraude* refers to the English pigment viridian. Nineteenth-century trade catalogues (Taylor 1887; Bourgeois Ainé 1896) transpose the two names in their translations of the terminology. In Vibert (1892), a text translated from French to English, emerald green is incorrectly described as oxide of chrome (the chemical composition of viridian). However, by the early twentieth century, the translations of these terms as found in trade catalogues and artists' manuals appear generally accurate (Sennelier 1904; Robert 1911).

Viridian (*vert émeraude*, Guignet's green), hydrous oxide of chromium ($Cr_2O_3.2H_2O$), is typically a transparent bluish-green color and is considered

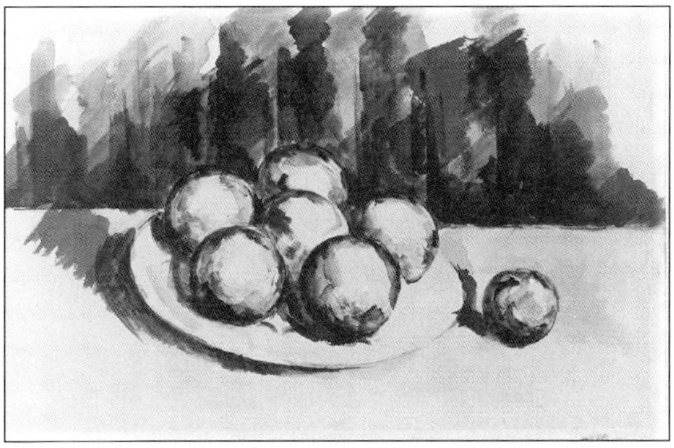

Figure 2.27 Ultraviolet light photograph of *Oranges on a Plate* (Fig. 2.25). (Photograph credit: Joe Mikuliak, Conservation Department, Philadelphia Museum of Art.)

permanent. It was first manufactured in 1838 in Paris by Pannetier and patented by M. Guignet in 1859. The determination of chromium by XRF indicated the presence of viridian on five of the watercolors analyzed: *Aix Cathedral, View from the Studio at Les Lauves, The Balcony, House on the Hill Near Aix, Trees and Rocks No. 1*, and *Trees and Rocks No. 2*. The pigment was not detected on watercolors in the study that date before 1890. In addition to chromium, significant peaks for copper and arsenic were found in analyzed areas on two watercolors, suggesting that emerald green may have been added to viridian on the palette. Cézanne employed viridian sparingly, to provide complementary accents to the more lavishly used emerald green. This may be due in part to the fact that viridian was almost twice as expensive as emerald green.

XRF detected cobalt, chromium, and lead in three isolated strokes of paint in one watercolor, *Rocks and Trees*, suggesting a mixed green paint composed of cobalt blue and chrome yellow. This is one of the few paints that Cézanne may have intentionally mixed on the palette; alternatively, he could have purchased it premixed under the name cobalt green. George Field described two types of cobalt green, a mixed compound of cobalt blue and chrome yellow, and a pigment prepared from cobalt and zinc called Rinmann's green. The former was likely a cheaper substitute for the latter (Field and Davidson 1903).

Blue pigments

Cobalt blue was found in all of the watercolors in this study except *House on the Hill Near Aix*. The element cobalt was detected by XRF in blue color areas, indicating the presence of cobalt blue pigment (*bleu de cobalt*), and cobalt aluminate ($CoO.Al_2O_3$). The discovery of cobalt blue was first published in France by Thenard in 1803; it was generally thought to be stable, although Field remained doubtful of its durability (Harley 1982). The spectra indicate that Cézanne may have mixed cobalt blue with other colors on the palette, possibly as a means of extending the costly pigment. Strong peaks for copper and arsenic, found in addition to cobalt, indicate emerald green additions on four pictures: *The Balcony*, *Trees and Rocks No. 1*, *Olympia*, and *Carafe, Table Knife, and Unidentified Object.* In two watercolors, *Oranges on a Plate* and *Trees and Rocks No. 1*, peaks for lead are present, suggesting an admixture of white lead. This was further substantiated by white particles visible in the blue paint layer and the detection of lead in an area of white paint in *Oranges on a Plate.*

In a number of blue sites on *House on the Hill Near Aix*, analysis did not yield elemental information, suggesting the presence of ultramarine blue (*bleu outremer*) or indigo. Ultramarine is the more likely candidate, because it was found in many of Cézanne's oil paintings (Butler 1984), and in his sketchbook color lists, whereas indigo does not occur in these sources. When compared to the hand-painted color swatches from an early Winsor & Newton catalogue (Taylor 1887), the blue wash of *House on the Hill Near Aix* appears closer in color to the swatch of ultramarine than to indigo. Factories for the production of synthetic ultramarine, sodium aluminum silicate with sulfur ($Na_{6\text{-}10}Al_6Si_6O_{24}S_{2\text{-}4}$), appeared in France shortly after 1828, the year that Jean Baptiste Guimet received an award for the manufacture of synthetic ultramarine from the Société d'Encouragement pour l'Industrie Nationale.

Purple pigments

Two different compositions of purple were inferred by XRF: cobalt-based and pigments that had no detectable elements revealed by XRF, such as organic-based or ultramarine blue containing pigments. Cobalt-based purple was found in three watercolors that also show areas of cobalt blue: *The Balcony*, *Oranges on a Plate*, and *Rocks and Trees.* These purple areas likely comprise a mixture of cobalt blue and organic red; however, they could also be cobalt violet, a pigment found on Cézanne's last great oil painting, *Large Bathers* (Butler 1984). The fading pattern on *Oranges on a Plate*, a watercolor that exhibits peaks for cobalt in both blue and purple washes, invites speculation about a mixed purple composed of cobalt and organic red: an area of purple that was partially protected by a mat has faded to blue in the exposed portion, as if the fugitive organic red component had been lost by overexposure to light.

Purples on *House on the Hill Near Aix* and *Carafe, Table Knife, and Unidentified Object* were analyzed, but they do not contain detectable elements, and are thought to reflect the presence of organic purple or a mixture of ultramarine blue and organic red. *House on the Hill Near Aix* also contains blue areas thought to be ultramarine. Harley (1987) cites mauve watercolor formulations by Winsor & Newton in 1866–68 consisting of mixtures of carmine and ultramarine. Neither mauve, cobalt violet, nor organic purples, such as purple lake, were found in Cézanne's sketchbook notes or letters. However, burnt lake (*laque brûlée*) was found among the lists in his sketchbooks, and this color might be similar to the dark violet hue of burnt carmine, as mentioned by Schweppe and Roosen-Runge (Feller 1986: 271) and verified by color swatches in a Winsor & Newton trade catalogue (Taylor 1887).

Red pigments

Three red pigments were suggested by the analysis: organic red, vermilion, and red lead. XRF analysis of red wash on six watercolors – *Aix Cathedral, View from the Studio at Les Lauves, The Balcony, House on the Hill Near Aix, Trees and Rocks No. 1, Trees and Rocks No. 2*, and *Olympia* – implied the presence of organic red by the absence of detectable elements. Cézanne mentioned various madders and lakes in his letters and sketchbooks: *laque garance foncée, garance rose, laque fine, laque brûlée.* While it is beyond the scope of this paper to determine which organic red pigment was used in a given painting, the following discussion includes three types of organic red pigments commercially available to Cézanne: carmine, natural red madder, and synthetic alizarin.

Carmine is a colorant extracted from the dried bodies of female scale insects (cochineal or kermes) that is used to make lake pigments, such as crimson lake. It was shown by the Russell and Abney report of 1888 (Brommelle 1964) to be one of the least permanent watercolors. Red madder lakes are made from the coloring material of the roots of herbaceous perennials (*Rubia tinctorum L.*, among others). The resulting shades of both madder and carmine-based lake pigments vary according to the constituents and methods of manufacture. Alizarin (1,2 dihydroxyanthraquinone), the main coloring material of natural madder, was first reported as synthesized by C. Graebe and C. Lieberman, around 1869 or 1870, in Germany. Rowney & Co. and Winsor & Newton first manufactured alizarin crimson as a watercolor in 1891 (Harley 1987). Although it was considerably more light-fast and less costly than natural madder or carmine-based pigments, many artists preferred the color of the natural pigments to that of synthetic alizarin. Alizarin is commonly present in watercolor lists in English trade catalogues of the period, but it is found in few French catalogue color lists, which remained dominated by madders and carmine lakes. Notably, visual examination by UV light revealed that the organic red washes did not fluoresce bright orange or pink but instead absorbed UV light and appeared dark, suggesting that natural madder is not present in the watercolors in this study.

Vermilion, red mercuric sulfide (HgS), displays peaks for mercury in XRF spectra. Mercury was detected in areas of red wash, suggesting the presence of vermilion on three watercolors: *Oranges on a Plate, Rocks and Trees*, and *Olympia.* In the first two examples, peaks for lead accompanied mercury. White lead and red lead were found on other areas of *Oranges on a Plate* and *Rocks and Trees*, respectively, and could have been mixed with vermilion by the artist. Alternatively, red lead could have been added by the manufacturer. Harley mentions that manufacturers sometimes adulterated vermilion with organics or red lead, and that vermilion was also added to carmine to improve the latter's permanence. A combination of vermilion and carmine was marketed as scarlet lake (Harley 1982).

XRF analysis identified lead in an area of red paint on *Rocks and Trees*, indicating the pigment red lead, red tetroxide of lead (Pb_3O_4). Red lead usually appears as *rouge de saturne* in French catalogues, but is not commonly found in Winsor & Newton trade catalogues. It is not a durable pigment; Church (1915: 210) deemed it "inadmissible" in watercolor. Red lead is found only once in Cézanne's oil paintings (Butler 1984) and just a single written notation occurs in the artist's color lists.

Yellow pigments

Three yellow pigments inferred by XRF analysis were chrome yellow, iron oxide yellow, and organic yellow. Chrome yellow (*jaune de chrôme*), lead chromate ($PbCrO_4$), shows peaks for chromium and lead in XRF spectra. Elements for chrome yellow were detected in a yellow highlight in *Oranges on a Plate* and in the mixed green in *Rocks and Trees*. According to Kühn and Curran (Feller 1986: 188–89), chrome yellow was introduced as a pigment by Vauquelin around 1804–09 and was used widely by the second quarter of the nineteenth century. It was not light-fast and was known to darken and turn greenish.

The presence of iron, indicating an iron oxide pigment, was confirmed by XRF in yellow and orange areas on five of the watercolors analyzed: *Aix Cathedral, View from the Studio at Les Lauves*, *Oranges on a Plate*, *Rocks and Trees*, *Olympia*, and *Carafe, Table Knife, and Unidentified Object*. Since peaks for iron were detected in all of the watercolor papers, the spectrum peaks indicating iron oxide pigments were necessarily stronger than the iron peaks shown by the paper alone. The iron oxide yellow is likely yellow ochre (*ocre jaune*), $Fe_2O_3.H_2O$, or if orange, burnt sienna (*terre de sienne brûlée*); both are found numerous times in the lists in Cézanne's sketchbooks.

The absence of elements detectable by XRF in yellow areas suggested an organic yellow. Gamboge (*gomme-gutte*) is the only organic yellow included in Cézanne's lists of colors. A gum resin produced from varieties of evergreen trees (genus *Garcinia*) found in southeast Asia, it is relatively fugitive to light. The organic yellow could also be a yellow lake, an occurrence mentioned in combination with iron oxide yellow (Butler 1984). Organic yellow was found on two watercolors: *House on the Hill Near Aix*, and *Trees and Rocks No. 2*. Orange areas that did not show detectable elements on *House on the Hill Near Aix*, by supposition, could be a mixture of organic yellow and organic red.

Cézanne's choice of papers

The paper surface – its tone and texture – contributes essential qualities to the appearance of Cézanne's watercolors. The characteristic luminosity created by his use of translucent washes depends on the white of the paper reflecting light back through the thin veils of color. The paper, visible through the translucent washes and in unpainted areas of the sheet, helps to articulate form or produce spatial effects. Cézanne used both laid and wove papers for his watercolors, with many of his later works on machine-made wove papers. The even surfaces of these machine-made papers allowed more fluid washes of color and smoother pencil marks, undisturbed by the texture of the laid lines. The poorer quality artists' papers introduced during the late nineteenth century, a consequence of the decreasing availability of rags and the increasing prevalence of machine-made paper, have a greater tendency to discolor over time. For example, the severe yellowing in *Oranges on a Plate* has caused a dramatic tonal shift in the large areas of unpainted paper within the composition and has masked the subtle nuances of transparent color washes.

Characteristics of the paper surface and color, the watermark, and fiber content for all nine of the Cézanne watercolors at the Philadelphia Museum of Art are summarized in Table 2.5 (p. 96). Watermarks found on laid papers used by Cézanne for the earlier watercolors include VIDALON, PL BAS, and the letter "S." The two watermarks found on wove papers used by Cézanne have been identified as CANSON & MONTGOLFIER VIDALON-LES-ANNONAY ANC<u>NE</u> MANU<u>RE</u> and MONTGOLFIER SAINT MARCEL LES ANNONAY. This study focussed on the wove papers that Cézanne often chose for his late watercolors.

These wove papers were produced by Montgolfier of Saint Marcel, and Canson and Montgolfier of Vidalon, papermills located in northern Ardèche, near Annonay, France, since the late seventeenth century.[6] The affordability and ready availability of relatively large sheets of smooth white machine-made paper must have made it an attractive watercolor support for Cézanne.

Characterization of the wove paper and impressed watermarks

The paper and watermarks were observed and systematically recorded on over 40 Cézanne watercolors in several collections in the United States and Europe.[7] This study confirmed that the two watermarks, CANSON & MONTGOLFIER VIDALON-LES-ANNONAY ANC<u>NE</u> MANU<u>RE</u> and MONTGOLFIER SAINT MARCEL LES ANNONAY, predominate in the wove papers chosen by Cézanne. The wove papers and watermarks repeatedly encountered in Cézanne's watercolors in other collections exhibit characteristics similar to those in the four late watercolors by Cézanne at the Philadelphia Museum of Art. The artist seems to

Title and Date	Dimensions	Thickness	Paper type	Watermark		Fiber content
Aix Cathedral, View from the Studio at Les Lauves, 1902–04	H 31.6 cm W 47.6 cm	0.28 mm	wove	MONTGOLFIER SAINT MARCEL LES ANNONAY		Rag fiber
The Balcony, c. 1900	H 61.0 cm W 45.1 cm	0.21 mm	wove	MANU^{RE} CANSON & MONTGOLFIER VIDALON		85–95% rag fiber 5–15% non-rag fiber
Oranges on a Plate, c. 1900	H 31.5 cm W 47.7 cm	0.20 mm	wove	FIER VIDALON-LES-ANNONAY ANC^{NE} MANU^{RE} CANSON & MONTGOLFIER		70–80% softwood (bleached sulfite) 20–30% rag fiber
House on the Hill Near Aix, 1895–1900	H 47.7 cm W 57.2 cm	0.28 mm	wove	FIER SAINT MARCEL LES ANNONAY MONTGOLFIER SAI		Rag fiber
Trees and Rocks No. 1, c. 1895	H 47.6 cm W 32.0 cm	0.18 mm	laid	S	Chain lines interval: 27 mm 8 laid lines per cm	Rag fiber
Trees and Rocks No. 2, c. 1890	H 44.9 cm W 28.8 cm	0.16 mm	wove	none		Rag fiber with approx. 20% non-rag fiber
Rocks and Trees, 1880–85	H 31.3 cm W 47.9 cm	0.18 mm	laid	PL BAS	Chain lines interval: 28 mm 8–9 laid lines per cm	Rag fiber
Carafe, Table Knife, and Unidentified Object, c. 1882	H 21.0 cm W 27.3 cm	N/A	wove	none		Rag fiber with moderate amount non-rag fiber
Olympia, c. 1875	H 24.7 cm W 28.2 cm	0.17 mm	laid	VIDALON	Chain lines interval: 26 mm 7–8 laid lines per cm	Rag fiber

Table 2.5 Papers chosen by Cézanne for nine watercolors at the Philadelphia Museum of Art.

have chosen these two papers with equal frequency. Both watermarks are found regularly on Cézanne's watercolors that date after 1890, with occasional occurrences in watercolors from the late 1880s. Other watermarks found on wove paper in this study include: J WHATMAN TURKEY MILL 1855, occurring on *The Curtain,* c. 1885 (R 193), and LEPAGE AINÉ TOCHON-LEPAGE SUCC^{P}, on *Standing Female Nude,* 1898–99 (R 387), both in the collection of the Musée du Louvre, Département des Arts Graphiques.

Unlike traditional wire-formed watermarks, both Montgolfier watermarks were cast in relief on the outside rims of small narrow wheels called *marques molettes* or *molettes gravées* (illustrated in Micoud 1933: 396, fig. 176). As the *molette,* or wheel, turned along the moving web of moist paper, the watermark was pressed into the paper, leaving embossed contours and compressed translucent areas; hence, they are called impressed watermarks (Labarre 1952). The watermark wheel was used on the continuous web of the Fourdrinier-type paper machine (Reynaud 1999) as the paper passed from the wet bed through the last press rolls or first drying cylinders. In many sheets the watermark does not read in its entirety, or may start in the middle of a word, suggesting that the spacing of the watermark impression was not adjusted consistently to the sheet size. These watermarks are clearly visible in transmitted light (see Fig. 2.28) and also in sharp raking light as embossed marks along the edge of the sheet (see Fig. 2.29).

Papers with the two impressed Montgolfier watermarks show many similar qualities and also marked differences. In both papers, the felt side is slightly textured, with smooth raised nubs, and the wire side of the sheet displays a fine screen impression. Cézanne painted on either the wire or the felt side of the sheet, the texture of the screen impression sometimes creating tiny unpainted gaps, called holidays, in the washes of color. In general, the sheet sizes of the two papers are similar, and it appears that larger sheets (approx. 63 x 47 cm) were halved to create the smaller sheets (approx. 31 x 47 cm). The watermarks are found parallel to one edge of the sheet, occurring along either the short or long side. All four edges of both watermarked sheets were cut or torn; no deckled edges were found. When tracings of the two Montgolfier watermarks were superimposed, the letter height (7 mm) and word spacing were the same, suggesting that the two marking wheels were of similar fabrication. However, the wheels appear to have been

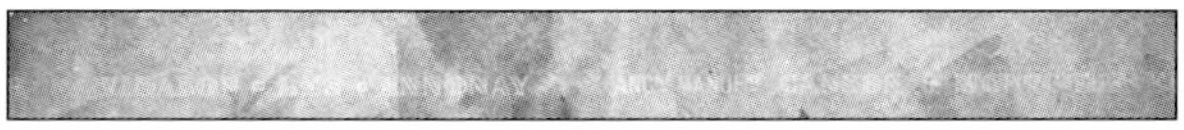

Figure 2.28 Paul Cézanne. *Oranges on a Plate*, c. 1900. Transmitted-light photograph showing watermark. (Photograph credit: Joe Mikuliak, Conservation Department, Philadelphia Museum of Art.)

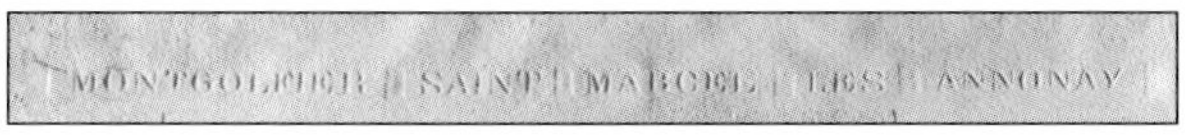

Figure 2.29 Paul Cézanne. *Aix Cathedral, View from the Studio at Les Lauves*. Raking-light photograph showing watermark. (Photograph credit: Joe Mikuliak, Conservation Department, Philadelphia Museum of Art.)

positioned somewhat differently on the papermaking machines, because although both watermarks read from the felt side of the paper, one is embossed into the felt side and the other into the wire side. They also differ according to whether the top or bottom edges of the words are oriented closer to the edge of the paper. Characteristics of the watermarks on the four late Philadelphia watercolors are as follows:

MONTGOLFIER SAINT MARCEL LES ANNONAY

This watermark is found on the papers used for *Aix Cathedral, View from the Studio at Les Lauves* and *House on the Hill Near Aix*. Executed in solid block letters with square shapes between the words, it is pressed into the felt side of the paper and reads from the felt side. The watermark is located close to the edge (approximately 8–15 mm from the edge) and oriented so that the bottoms of the letters are nearer to the edge of the sheet. On the wire side of the sheet, a prominent mesh impression occurs over the raised areas of the impressed watermark, visible in sharp raking light (see Fig. 2.30). *Aix Cathedral, View from the Studio at Les Lauves* (dimensions: 31.6 x 47.6 cm; thickness: 0.28 mm) was painted on the wire side of the paper. The watermark was sharply impressed along the longer edge, about 11 mm from the top of the sheet. Cézanne painted *House on the Hill Near Aix* (dimensions: 47.7 x 57.2 cm; thickness: 0.28 mm) on the felt side of the sheet. The watermark was sharply impressed along the shorter edge, about 11 mm from the left edge of the sheet. The strong relief of the watermark is visible along the left edge of the recto in raking light.

CANSON & MONTGOLFIER VIDALON-LES-ANNONAY ANC$^{\underline{NE}}$ MANU$^{\underline{RE}}$

The watermark is found on papers used for *Oranges on a Plate* and *The Balcony*. Executed in double-line typeface, it is pressed into the wire side of the paper and reads from the felt side; it is located approximately 15–30 mm from the edge, with the tops of the letters oriented nearer to the edge of the sheet. The watermark impression is usually visible on the wire side in raking light; however, sometimes the watermark embossing cannot be seen clearly on the felt side in raking light and requires transmitted light for reading. *Oranges on a Plate*

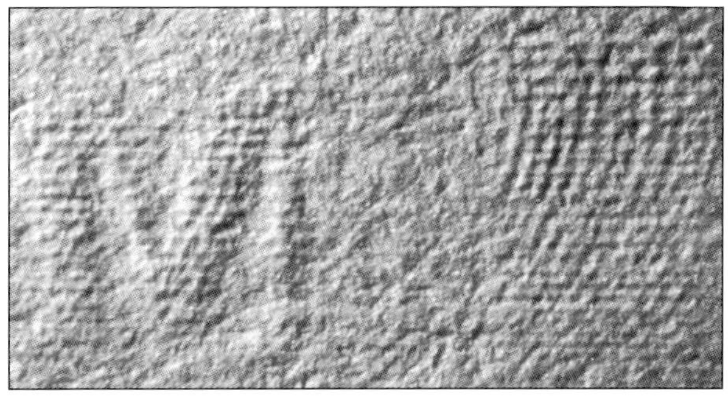

Figure 2.30 Paul Cézanne. *Aix Cathedral, View from the Studio at Les Lauves*. Photomicrograph at 6.7x magnification of mesh impression on wire side of sheet over watermark.

(dimensions: 31.5 x 47.7 cm; thickness: 0.20 mm) was painted on the felt side of the sheet. The softly embossed watermark runs along the long edge, about 3.2 cm from the top of the sheet. *The Balcony* (dimensions: 61.0 x 45.1 cm; thickness: 0.21 mm) was painted on the wire side of the sheet. The watermark was sharply impressed along the long left recto edge, which has been partially trimmed.

Characterization of the laid paper and wire-formed watermarks

Laid papers used by Cézanne for earlier watercolors at the Philadelphia Museum of Art display the following wire-formed watermarks: VIDALON and PL BAS. These watermarks were seen also on Cézanne's watercolors in other collections. The laid papers are thinner than the wove papers described above and show deckled edges. The watermarks are visible in both raking and transmitted light.

VIDALON

(Watermark dimensions: 1.0 x 10.3 cm; chain lines occur at 26 mm intervals; there are 7–8 laid lines per cm.)
Also made by Canson and Montgolfier, VIDALON watermarked paper may have been marketed as Raisen Ingres.[8] The wire and felt sides of the Vidalon paper used for Cézanne's *Olympia* (dimensions: 24.7 x 28.2 cm; thickness: 0.17 mm) are difficult to differentiate. The watermark reads from the recto and runs parallel to the right edge, measuring 7 cm from that edge. *Olympia* appears to be on a partial sheet with a deckled top edge. This watermark also appears in the center of two Cézanne watercolors on larger sheets (measuring approximately 31.5 x 47.5 cm) in other collections.[9]

PL BAS

(Watermark dimensions: 1.3 x 8.5 cm; chain lines occur at 28 mm intervals; there are 8–9 laid lines per cm.)
The paper with the PL BAS watermark is very similar to the VIDALON paper, having slightly wider chain- and laid-line intervals. *Rocks and Trees* (dimensions: 31.3 x 47.9 cm; thickness: 0.18 mm) was painted on the wire

side of a PL BAS watermarked sheet. The watermark reads from the wire side and runs parallel to the right edge, measuring 3.7 cm from that edge. All four edges are deckled, indicating that this is a full sheet.

Fiber analysis

Throughout the nineteenth century, papermakers searched for alternative fibers, new sizing agents, and processing methods that met the requirements of the papermaking machines. Machine-made papers using grass and wood fibers were developed in France around 1865–72, but they could not yet be produced on a large scale (André 1996). A decline in the quality of paper used in French books was documented as beginning around 1885 (Rutledge and Owen 1983). In 1831 Louis Piette became the first Frenchman to publish Illig's formulation of alum rosin engine size in his *Traité de la fabrication du papier.* Prior to this, trial-and-error experimentation for engine sizing was commonplace, and patents were jealously guarded. For example, Canson patented a method using wax and alum in 1827, and other methods combining starch and rosin were tried (Voorn 1961). New papermaking fibers and sizing agents were geared primarily toward large-scale production on the papermaking machine and did not address the long-term durability of the paper.

Fiber analysis using a polarized light microscope was performed on all nine Cézanne watercolors, and is summarized in Table 2.5 (p. 96).[10] Rag fibers were found in all the watercolor papers, with varying amounts of wood and/or grass in several of the sheets. The rag fibers were mixtures of linen and cotton, heavily beaten and short in length. While one paper contained a large amount of softwood, a number of other papers contained small amounts of hardwood and/or grass. Differentiating between the latter two fibers was difficult, due to the absence of sufficient diagnostic elements (such as vessels to distinguish hardwood, or epidermal and parenchyma cells to identify grass), thus it was not possible to determine if the papers contained hardwood, grass, or fibers of both. Trace amounts of non-rag fibers found in some of the samples were not numerous enough to be identified and perhaps were not conscious additions, but rather fibers left over from the manufacture of other papers in the mill. Some of the fiber samples contained small particles (stained yellow in Graff's "C" stain) that were identified as a proteinaceous material by FTIR.[11] This is an indication of gelatin sizing, which was most likely used in addition to alum rosin size, the standard sizing agent for early machine-made papers.

Watercolor papers in this study containing significant percentages of grass and/or wood pulp fiber also display greater degrees of discoloration, whereas the papers made of all rag fiber appear much whiter. For example, an estimated 70–80% softwood fiber (bleached sulfite) was detected in *Oranges on a Plate,* the most discolored paper in the study, whereas only rag fibers were found on *House on the Hill Near Aix,* the whitest of the group. Rag fibers were also found in other relatively light or slightly discolored watercolor papers, including *Aix Cathedral, View from the Studio at Les Lauves, Trees and Rocks No. 1, Rocks and Trees,* and *Olympia.* Analysis of *The Balcony* detected approximately 85–95% rag fiber and 5–15% hardwood and/or grass fiber. In this case the non-rag fiber was particularly difficult to identify with certainty, possibly indicating that grass and wood fibers are both present but not homogeneously distributed in the sheet. Two noticeably discolored watercolors, *Trees and Rocks No. 2* and the sketchbook page *Carafe, Table Knife, and Unidentified Object,* were found to contain mostly rag with moderate amounts of hardwood and/or grass fiber. In this study, the two papers watermarked CANSON & MONTGOLFIER VIDALON-LES-ANNONAY ANC$^{\underline{NE}}$ MANU$^{\underline{RE}}$ appeared discolored and contained non-rag fibers, whereas the two sheets with MONTGOLFIER SAINT MARCEL LES ANNONAY watermarks looked fairly white and were found to be all rag. Further investigation of Montgolfier papers is necessary to identify trends in fiber content and to characterize any associated tendencies to discolor over time.

Conclusions

The analysis of Cézanne's pigments and papers allows a greater understanding of his watercolors, and suggests how time and exposure may have altered the materials. XRF analyses of nine watercolors in the Philadelphia Museum of Art led to the conclusion that Cézanne's watercolor palette comprised about a dozen pigments, including a mixed green and possibly mixed purples (see Table 2.4, p. 91). The artist seemed to favor a restrained palette for his watercolors, averaging only five pigments on the individual paintings in this study. Pigment analysis detected elements for emerald green in many of the strokes of tan wash found on the watercolors, indicating that this wash was previously bright emerald green that has deteriorated and darkened. Other pigments suggested by this study that may have changed over the years include the organic yellows and reds, such as carmine lake, a pigment particularly fugitive to light exposure. Watercolor pigments found in this study were consistent with those indicated in other sources, including the analysis of Cézanne's oil paintings and notations in the artist's letters and sketchbooks. These other sources also list certain pigments not detected in the present study, such as orpiment, Indian yellow, *terre verte,* and Prussian blue.

Close examination of Cézanne's choice of watercolor papers produced as many questions as it did answers about late nineteenth-century papers and watermarks. Areas requiring further research include identification of the characteristics and manufacturing techniques of early machine-made paper, and additional diagnostic fiber and sizing analysis of papers of this period. Papers produced by Montgolfier are of special interest in the context of

Cézanne's oeuvre. Ongoing research includes documentation of impressed watermarks and the characteristic properties of the papers chosen by Cézanne for his watercolors.

The fugitive nature of watercolor paint has long been a concern to manufacturers, artists, conservators, and curators. Faded or darkened colors can have a profound visual impact on a picture, particularly if the paper has also discolored. Loss of paper whiteness can vastly change the appearance of the work of art, affecting not only the tone and hue of the translucent watercolor washes, but also causing a color shift in large areas of blank paper. It is a testimony to Cézanne's genius that even watercolors in poor condition, with darkened and faded colors or severely discolored paper, remain beautiful in their own muted way. However, to see many of them through the eye of the artist, one must envision fresh colors, particularly a brighter emerald green, sparkling on a sheet of creamy white paper.

Acknowledgments

This project would not have been possible without the analytical expertise of Andrew Lins, Beth Price, Janice Carlson, and Debora Dyer Mayer. I would also like to extend appreciation for the continued support of the Conservation Department of the Philadelphia Museum of Art, especially Nancy Ash, and Dana Mossman Tepper; and of Judith Walsh of the National Gallery of Art, Washington, D.C.

Endnotes

1. Energy dispersive x-ray fluorescence was executed by Janice Carlson and Beth Price using a Kevex Instrument 0750 at the Winterthur Museum Analytical Laboratory. The secondary target was carbon. This instrument detected elements with atomic weights including and greater than potassium. Pigments containing elements outside this detection range, such as organic pigments and ultramarine blue, were implied by the absence of detectable elements, the observed color, and other references. The number of sites tested on each picture varied from 6 to 19 areas, depending upon the complexity of the composition. The paper and mat board of each watercolor was also analyzed, and the discussion of the pigments is limited to the elements that are in addition to those found in these materials. Analytical reports on the nine watercolors, spectra of elemental results, and written summaries, are available through the Conservation Department of the Philadelphia Museum of Art.

2. Titles, dates, Reward catalogue numbers (R no.), and Philadelphia Museum of Art accession numbers (PMA no.) of the nine watercolors in this study are as follows: *Aix Cathedral, View from the Studio at Les Lauves*, 1902–04, R 581, PMA 1950-134-035; *The Balcony*, c. 1900, R 529, PMA 1943-075-001; *Oranges on a Plate*, c. 1900, R 553, PMA 1963-116-020; *House on the Hill Near Aix*, 1895–1900, R 464, PMA 1963-181-007; *Trees and Rocks No. 1*, c. 1895, R 418, PMA 1967-030-018; *Trees and Rocks No. 2*, c. 1890, R 328, PMA 1967-030-019; *Rocks and Trees*, 1880–85, R 233, PMA 1950-134-033; *Carafe, Table Knife and Unidentified Object*, c. 1882, R 202, PMA 1952-061-011; *Olympia*, c. 1875, R 135, PMA 1963-181-123.

3. See note 1.

4. Experiments to determine the cause of emerald green darkening were conducted by Andrew Lins and the author. Two sample sets of emerald green watercolor were prepared by grinding emerald green pigment (donated by D. Sennelier): the first set with gum arabic diluted in deionized water, the second set in deionized water alone. The gum arabic solution was painted in a graduated tone on swatches of Whatman filter paper; the water solution was dried and rubbed into swatches of Whatman filter paper. The sample sets were placed in closed chambers of 1000 cm^3; swatches from both sets were exposed over 0.2% w/v ammonium sulfide in deionized water, and another group was exposed to hydrogen sulfide gas (produced by dropping 2 ml of 10% w/v nitric acid into 0.2% w/v ammonium sulfide in deionized water). Exposure of both sample sets of emerald green watercolor to dilute ammonium sulfide fumes resulted in yellowish darkening after 30 minutes and exposure of both sets to hydrogen sulfide gas resulted in yellowish-brown darkening after 5–15 minutes.

5. UV examination was carried out using a hand-held lamp, model UVGL-25 MINERALIGHT® Lamp Multiband – 254/366 nm; excitation: 254 nm (short wave) and 366 nm (long wave).

6. In 1799 Baron Barthélémy Canson succeeded his father-in-law, Etienne Montgolfier, at the Vidalon mill; shortly thereafter, Jean-Baptiste Montgolfier established the neighboring papermills of Saint Marcel, 1807, and Grosberty, 1817. The Canson and Montgolfier families were forerunners in French papermaking technology; included among their accomplishments are the manufacture of the first French wove papers in 1777 and the installation at Vidalon of the first Fourdrinier-type paper machine (*machine à table plate*) in France, 1822. Canson also secured some interest in patents from Leger Didot, assuring Vidalon exclusive regional use of the Fourdrinier machine. In 1824, Jean-Baptiste Montgolfier obtained a patent for the Cameron, a cylinder mold machine (*machine à forme ronde*), for the mill at Grosberty. The machine-made paper industry grew rapidly, so that by 1872 there were five active papermaking machines in three establishments at Vidalon (Canson and Montgolfier 1872), and by 1900 there were eleven machines including both types – Fourdrinier and cylinder mold (Reynaud 1989, 1994, 1999, n.d.).

7. The Metropolitan Museum of Art, New York, The Museum of Modern Art, New York, The Princeton Art Museum, The Henry and Rose Pearlman Foundation, Inc., The Art Institute of Chicago, The Cincinnati Art Museum, Courtauld Institute Galleries, London, Musée du Louvre, Département des Arts Graphiques, Paris, and the National Gallery of Ireland, Dublin. I would like to acknowledge the assistance of conservators Margo McFarland, Niamh McGuinne, and Cecile Mear.

8. No. 38, Raisen C *à grain* paper (49 x 64 cm) in a 1875 Canson and Montgolfier catalogue was kindly shown to me by Reynaud (1999), who suggested that this paper might have been watermarked VIDALON.

9. VIDALON watermarks were found on two watercolors in other collections, both titled *Mount Sainte-Victoire*. One of 1900–02 (R 502) is held by the Musée du Louvre, Département des Arts Graphiques, Paris, and the other of 1885–87 (R 279) is in the collection of the Courtauld Institute Galleries, London.

10. Fiber analysis using a polarized microscope and Graff's "C" stain was carried out on all nine watercolors by the author at the Philadelphia Museum of Art. Additional diagnostic microscopic analysis was performed by Debora Dyer Mayer on *Aix Cathedral, View from the Studio at Les Lauves, The Balcony, Oranges on a Plate, House on the Hill Near Aix*, and *Trees and Rocks No. 2*. Analytical reports of the fiber analysis by Mayer are on file in the Conservation Department of the Philadelphia Museum of Art.

11. Samples from *Oranges on a Plate* and *Aix Cathedral, View from the Studio at Les Lauves* were analyzed by Beth Price of the Philadelphia Museum of Art using a Niclolet 510P FT-IR spectrometer with a Nic-Plan IR microscope equipped with a Spectra-Tech Micro sample diamond plan. IR data and discussion of band attribution are on file in the Conservation Department of the Philadelphia Museum of Art.

References

André, L. 1996. *Machines à papier en France 1798–1860.* Paris: École des Hautes Études en Sciences Sociales.

Banik, G., and J. Ponahlo 1982–83. "Some aspects of degradative paper caused by green copper-containing pigments," *The Paper Conservator* 7, pp. 3–7.

Brommelle, N. S. 1964. "The Russell and Abney Report on the action of light on water colours," *Studies in Conservation* 9, no. 4, p. 144.

Butler, M. 1984. "An investigation of the materials and techniques used by Paul Cézanne," *AIC Preprints, The American Institute for Conservation 12th Annual Meeting, Los Angeles*, pp. 20–33.

Canson et Montgolfier 1872. Canson et Montgolfier: papeteries de Vidalon-les-Annonay, *Notice historic*. Ardèche: Canson et Montgolfier.

Chappuis, A. H. 1973. *The drawings of Paul Cézanne: a catalogue raisonné*, 2 vols. Greenwich, Conn.: New York Graphic Society; London: Thames and Hudson.

Church, A. 1915. *The Chemistry of Paints and Painting*. London: Seely Service & Co. Ltd.

Feller, R. (ed.) 1986. *Artists' Pigments: A Handbook of Their History and Characteristics*, Vol. 1. Washington, D.C.: National Gallery of Art.

Field, G., and E. A. Davidson 1903. *A Grammar of Colouring*. London: Crosby Lockwood and Son.

Fitzhugh, E. W. (ed.) 1997. *Artists' Pigments: A Handbook of Their History and Characteristics*, Vol. 3. Washington, D.C.: National Gallery of Art.

Harley, R. D. 1982. *Artist's Pigments c.1600–1835: A Study in English Documentary Sources*, 2nd edn. London: Butterworth Scientific.

Harley, R. D. 1987. "Some new water colours in the nineteenth century," *The Conservator* 11, pp. 46–50.

Haynes, W. 1954. *American Chemical Industry*. New York: D. Van Nostrand Co., Inc., p. 363.

Labarre, E. J. 1952. *Dictionary and encyclopedia of paper and paper-making*, 2nd edn. Amsterdam: Swets and Zeittinger.

Micoud, H. 1933. *Propriétés Physiques des Papiers*. Grenoble: Imprimerie Saint-Bruno.

Reff, T., and I. Shoemaker 1989. *Paul Cézanne, two sketchbooks*. Philadelphia: Philadelphia Museum of Art.

Rewald, J. 1948. *Paul Cézanne, A Biography*. New York: Simon & Schuster.

Rewald, J. (ed.) 1976. *Paul Cézanne Letters*. Oxford: Bruno Cassirer.

Rewald, J. 1983. *Paul Cézanne: The Watercolors, A Catalogue Raisonné*. Boston: Little, Brown and Company.

Reynaud, M. 1989. *Une histoire de papier les papeteries Canson et Montgolfier*. Annonay: Canson.

Reynaud, M. 1994. "Le groupe Montgolfier, des papeteries a travers le monde," *IPH Congress Book, Papers of the 22nd International Congress of Paper Historians* 10, pp. 82–87.

Reynaud, M. 1999. Personal communication. Archives du Musée, Papeteries Canson et Montgolfier, Annonay, France.

Reynaud, M. n.d. *Au temps de las fabricaion du papier à la main*. Annonay: Musée des Papeteries Canson et Montgolfier, p. 20.

Rishel, J. 1983. *Cézanne in Philadelphia Collections*. Philadelphia: University of Pennsylvania Press.

Robert, K. 1911. *L'aquarelle*. Paris: Henri Laurens.

Rutledge, J., and W. Owen 1983. "Changes in the quality of paper in French books, 1860–1914: A study of selected holdings of the Wilson Library, University of North Carolina," *Library Resource & Technical Services* 27, no. 2, pp. 177–87.

Sennelier, D. 1994. Personal communication. Sennelier Beaux-Arts Distribution, Paris.

Seward, H. 1889. *Manual of colours showing the composition and properties of artists' colours with experiments on their permanence*. London: Geo. Rowney & Co., pp. 68–70.

Tate Gallery. 1982. *Paint and Painting*. London: Tate Gallery.

Taylor, J. S. 1887. *A descriptive handbook of modern water-colour pigments illustrated with seventy-two colour washes skilfully (sic) gradated by hand on Whatman's drawing paper with an introductory essay on the recent water-colour controversy*. London: Winsor & Newton Limited.

Vibert, J. G. 1892. *The Science of Painting*, transl. by Percy Young. London: Percy Young.

Voorn, H. 1961. "A brief history of the sizing of paper," *The Papermaker* 30, no. 1, pp. 47–52.

Zieske, F. 1995. "An investigation of Paul Cézanne's watercolors with emphasis on emerald green," *The Book and Paper Group Annual of The American Institute of Conservation* 14, pp. 105–15.

Trade catalogues

Bourgeois Ainé 1896. *Catalogue illustré couleurs fines et matériel pour l'aquarelle, la gouache, le pastel, le dessin, le modelage, la peinture à l'huile* ... Paris.

Lefranc & Cie 1898. *Fabrique de couleurs & vernis couleurs fine et matériel pour la peinture à l'huile, l'aquarelle, la gouache, le pastel* ... Paris.

Sennelier, G. 1904. *Catalogue général illustré de G. Sennelier fabricant de couleurs fines et matériel d'artistes*. Paris.

Winsor & Newton [188-]. *Couleurs à l'aquarelle modernes des Winsor & Newton*. London.

Winsor & Newton Limited. [188-]. *Winsor & Newton's Water colours*. London.

Winsor & Newton Limited. 1883. *Manufacturing artists' colourmen, drawing paper stationers, artists' pencil and brush makers, publishers of works on the fine arts. (For the Trade Only)*. London.

Winsor & Newton Limited. 1892. *Composition of Pigments*. London.

Winsor & Newton Limited. 1901. *Manufacturing artists' colourmen*. London.

Winslow Homer's Watercolor Pigments

Barbara Berrie, Yoonjoo Strumfels, and Carol Tolocka

Winslow Homer (1836–1910) chose the watercolor medium to create hundreds of much-loved images. In the few statements he made about his art, he spoke primarily of the need to be truthful to what he saw in nature. After Homer's death, his studio in Prout's Neck, Maine, was retained by his family, preserving the studio materials that today offer us a glimpse of Homer's working methods. Some of these materials remain in his studio, while others have been sold to various museums or otherwise dispersed. Among these materials are three watercolor paint boxes. The pigments in the watercolor box belonging to Bowdoin College Museum of Art were analyzed by Richard Newman and Craigen Weston (Newman and Weston 1980). The Portland Museum of Art box was analyzed at the National Gallery of Art, Washington, D.C.; the box displayed in Homer's Prout's Neck studio has yet to be analyzed.[1] This essay treats the Portland Museum example, the analysis of which was augmented by examination (accomplished without taking samples) of Homer's watercolor paintings in the National Gallery's collection to investigate how he chose and used pigments.

Pigments in the Portland Museum watercolor box

The Portland Museum paint box (Fig. 2.31) is a "hand-japanned tin box" manufactured by Winsor & Newton in England in the nineteenth century. In his biography of Homer, Beam (1966) writes that Winsor & Newton brand art materials were Homer's favorite type. The exterior of the box is finished with black enamel and the interior with white enamel. The box opens to reveal three parts: a middle tray and two palette wings. Half of the middle tray is fitted with a partition designed to secure 10 whole-size watercolor pans; the other side is a non-partitioned trough. Currently, the box contains 20 "moist watercolor" cakes, 10 in the individual partitions and 10 in the open trough.

Homer signed his name on this box and made nine distinct notations of pigment names in what appears to be black wax crayon. Two notations are complete and seven are abbreviations; parentheses enclose possible pigment names: "Indian purple," "Scal lake" (scarlet lake), "B sien" (burnt sienna), "P gr" (Prussian green or Payne's gray), "Hook green" (Hooker's green), "— Gr y" (Payne's gray or charcoal gray), "C orange" (chrome orange or cadmium orange), "Aureolin," and "Scarle" (Scarlet). Three of these colors – Prussian green, Hooker's green, and aureolin – were shown by analysis to be present in the box. The pigments designated by the other notations are not present in the paint box.

The results of pigment analysis showed that the Portland Museum paint box contains three different red pigments, including three separate cakes of vermilion; an iron earth of a purple-brown hue that can be identified as Indian red; and carmine, a rich, intense, and transparent lake.

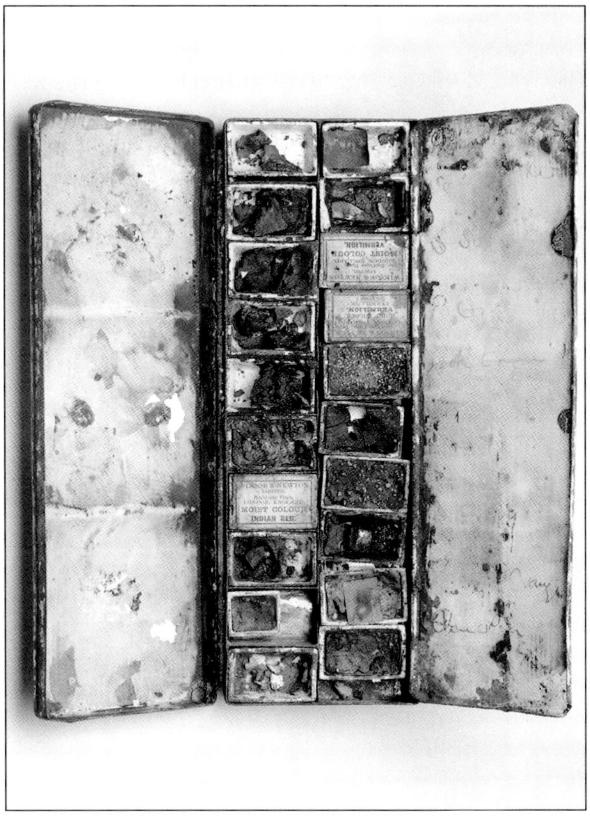

Figure 2.31 Winsor & Newton (maker), Winslow Homer's watercolor box, 1910 (latest). 7 x 19 x 21.4 cm (irregular case). Portland Museum of Art, Maine, gift from Mr. and Mrs. L. Robert Porteous, Jr., in honor of Bennett C. Porteous, 1979.15.1. (Photograph courtesy of the National Gallery of Art, Washington, D.C.)

The only yellow pigment present is aureolin. This pigment, which was discovered in 1860, is a semi-transparent color, making it useful for glazes and well suited to the watercolor technique.

There are five green cakes, each a combination of Prussian blue and gamboge. Two cakes may be characterized as Hooker's greens, since they contain a larger percentage of gamboge than the others, which gives them a grassy hue. Hooker's green was offered in different tones: Hooker's green No. 1, a lighter green; and Hooker's green No. 2, a darker one. The other three cakes, which may be termed Prussian greens, contain a smaller proportion of gamboge, resulting in a deep blue-green hue.

Cobalt blue and indigo are the only blue pigments present in this box, despite the variety of blue colorants seen in Homer's late watercolors. Cobalt blue is a pure blue color that contemporaries of Homer regarded as inferior to ultramarine in both depth and transparency, although it was thought to be a better pigment for use in washes. Indigo, a much darker blue pigment, possesses a blue-black tone similar in hue but duller than Prussian blue.

One purple pigment found in the Portland Museum paint box has been characterized as purple lake. This cake is set in a china pan labeled Indian purple and is placed next to Homer's notation "Indian purple." Analysis indicated carmine (Halpine 1996), but no copper, which Winsor & Newton claims was in the formulation of Indian purple.[2] This cake instead contains a large amount of calcium, which suggests that it may be the pigment sold by Winsor & Newton under the name purple lake.

Four cakes can be identified as organic brown earths, either Vandyke brown or Cologne earth, both of which are deep, semi-transparent colors. Vandyke brown is sometimes considered slightly lighter in tone than Cologne earth, but the distinction between the two pigments is not well defined. Taylor (1887: 61–62) stated, "Vandyke brown has an awkward habit of 'washing up'; and when it is necessary to lay on a large body of colour, the moist tube colour should be used in preference to the cake." Muckley (1893: 23–24), on the other hand, encouraged the use of this brown pigment: "It [Vandyke brown] is permanent, and excellent glazing colour, for which purpose it is mostly used, and often in combination with other transparent colours. The palette should never be set without it."

There are two cakes of black pigments, which could be either lamp black or blue black. Lamp black is a very finely divided pigment considered inferior to other blacks in transparency and intensity. The less dense blue-black pigment provides a contrasting working quality, recommended when the artist wishes to avoid a sooty effect (Taylor 1887: 65).

It is important to keep in mind that the pigments found in the Portland Museum paint box present a catalogue of the contents as they existed at a particular moment. Homer's handwritten labels for colorants no longer present in the box lead us to assume that at one time these colorants, or at least pigments sold under those names, were contained in the watercolor box, and were used by Homer. A wide array of pigments was available during Homer's time: over 100 moist watercolor cakes are listed in Winsor & Newton artists' materials catalogues of the nineteenth century (Taylor 1887). Indeed, a survey of the work of 46 artists conducted by Russell and Abney in 1888 reported that as many as 101 different colors, or colors sold under different names, were in use by watercolorists of the day (Russell and Abney 1888).

Several pigments in the Portland Museum box were known to be light fugitive. Fading of the watercolor medium was a well-known phenomenon both in Europe and the United States. Despite their impermanence, Homer used these pigments throughout his career. Carmine, for example, notorious for its tendency to fade, was described as "the brightest and most attractive of the beautiful cochineal colors. Not a favorite with professional artists on account of its want of permanency, yet it is quite popular with amateurs in flower painting" (Herrick 1882: 46). Church (1890: 166) denounced cochineal colorants thus: "Beautiful and rich as are the colours prepared from cochineal, not one of them should ever find a place upon the palette of the artist. They all became brownish, and ultimately almost disappear after short exposure to sunlight or the more prolonged attack of strong diffused daylight."

Vandyke brown, indigo, and Indian red were also thought to be fugitive pigments. Hooker's green and Prussian greens become more blue as their gamboge component fades. The probability that Homer was knowingly using unstable pigments suggests that the immediate chromatic effect and particular working properties were of paramount importance to him. Once asked if he modified the colors in nature when he painted, Homer replied, " Never! Never! ... When I have selected the thing carefully, I paint it exactly as it appears. Of course you must not paint everything you see ... You must wait, and wait patiently, until exceptional, the wonderful effect, or aspect comes. Then, if you have sense enough to see it – well, that's all there is to that" (Gould 1962: 283). These words reveal an artist determined to express a carefully selected "wonderful effect" with his colors, his primary tools.

Examination and analysis of Homer's watercolor paintings

Homer's 30 watercolor paintings in the National Gallery of Art's collection were analyzed with X-ray fluorescence spectroscopy (XRF), infrared reflectography (IR) and photography, and microscopic examination.[3] While XRF and IR helped to identify inorganic colorants, microscopic examination of the paintings was effective in characterizing some organic colorants and, more importantly, in understanding how Homer used his pigments.

The watercolors were examined systematically by subject matter and by date to determine trends in the changes of Homer's palette. It became clear that his choice of specific pigments was a response to his evolving painting style. In his early gouache paintings from the 1870s, Homer mixed zinc white into almost all his colors. As he moved on to paint in a broader and more transparent wash manner, he began to choose colorants for a wider variety of properties, such as transparency or lack of it, and flow of the wash.

The Sick Chicken and *Fresh Eggs*, dating to 1874, and painted at Walden, New York, are the earliest watercolors in the National Gallery's collection. Homer painted them during the short period when he worked in a gouache manner, utilizing zinc white for a wide variety of effects. In *Fresh Eggs*, Homer mixed zinc white into most colors and covered almost the entire sheet. He used the white pigment by itself to stipple in extensive heightening and to break up solid colored areas over the haystack and barn floor. He also mixed in thick white paint and used the stiff mixture for textural descriptions of his objects. The artist used the stiff mixture of zinc white, vermilion, and red lake to portray the dotted pattern on the woman's dress, and the strutting rooster in the foreground was painted with pasty overlapping strokes to suggest the feathers. Closer examination of these areas reveals that uneven zinc white agglomerates were mixed in with resin-like amber color drops (perhaps gum binder), suggesting that Homer made his own paint by crudely mixing ground zinc white with gum (see Fig. 2.32).[4] Homer, who is known to have ground his own pigments (Beam 1966), painted this watercolor at the farm of his major patron, Lawson Valentine, a varnish manufacturer. Valentine also maintained experimental farms staffed with chemists, and it is possible that Homer mixed his paint with what was available to him in this facility (Cikovsky 1995). However, Homer would soon cease to work in the gouache manner, as he began to use a more fluid painterly style.

In the same year, Homer used gum in *The Sick Chicken* to modify the appearance of his pigments. He was precise in placing this medium to achieve subtle yet profound visual effects. He applied a minute stroke of gum coating over the girl's lower lip to saturate the color, and to create different tones with one pigment (see Fig. 2.33). He used this medium for similar effects on the cock's comb and on the girl's black ribbon. Homer glazed these passages with gum to saturate the color, and to allow the painted surface to catch light differently. The fact that these touches are scarcely visible to the unaided eye attests to Homer's attention to details.

Homer painted the watercolor *On the Sands* (Fig. 2.34) in 1881, while he lived in the fishing village of Cullercoats, near Tynemouth, England. It is a good example of his transparent watercolor style from this period (1881–82). Homer adjusted his palette, using less opaque pigments: chrome green was now used sparingly, replaced by more transparent greens, such as Hooker's green and Prussian green. From this period on, Homer would reserve chrome green for opaque passages.

Homer's remarkable skills in maneuvering pigment are seen with his use of a pigment called charcoal gray in *On the Sands*. Charcoal gray was identified in Homer's watercolors through microscopic examination, and through comparison with the sample color washes of

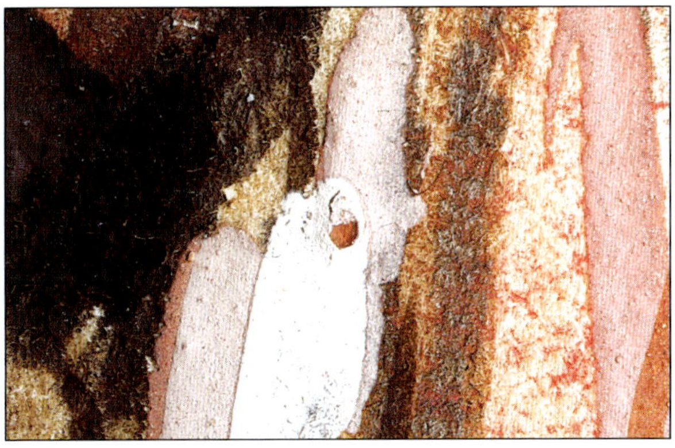

Figure 2.32 Winslow Homer (American, 1836–1910), *Fresh Eggs*, 1874 (detail). Watercolor, gouache, and pencil on paper, 23.7 x 19.3 cm. National Gallery of Art, Washington, D.C., Collection of Mr. and Mrs. Paul Mellon, 1994.59.26. Detail shows crude paint mixture of zinc white and amber color gum residue.

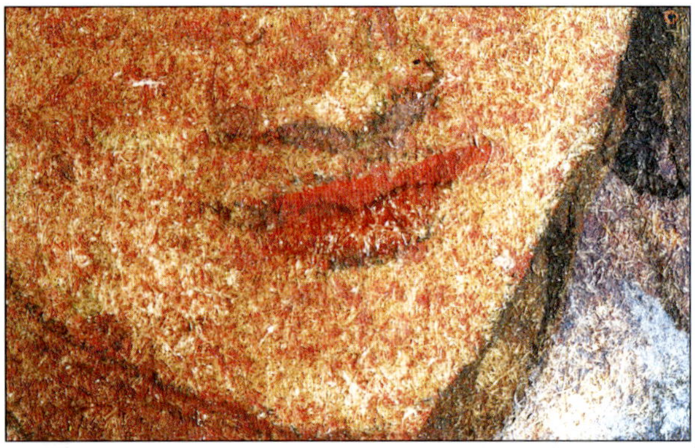

Figure 2.33 Winslow Homer, *The Sick Chicken*, 1874 (detail). Watercolor, gouache, and graphite on paper, 24.7 x 19.7 cm. National Gallery of Art, Washington, D.C., Collection of Mr. and Mrs. Paul Mellon, 1994.59.21. Note the gum glaze on the girl's lower lip that saturates the vermilion red.

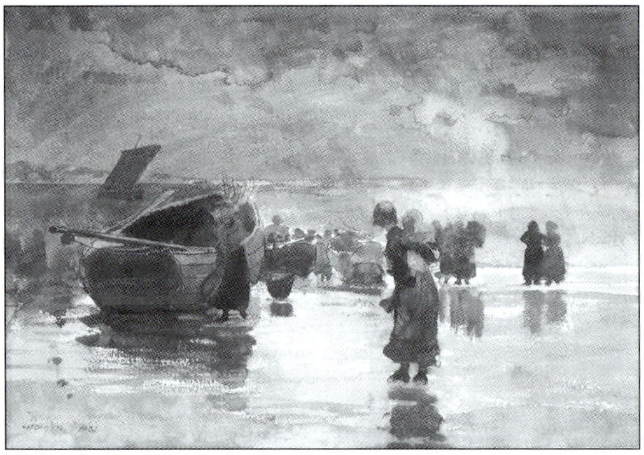

Figure 2.34 Winslow Homer, *On the Sands*, 1881. Watercolor and pencil on paper, 33.5 x 47.5 cm. National Gallery of Art, Washington, D.C., Bequest of Julia B. Engel, 1984.58.1. (© 2000 Board of Trustees, National Gallery of Art, Washington, D.C.)

charcoal gray in the nineteenth-century catalogues of artists' materials. In colormen's manuals, charcoal gray is described as a "prepared charcoal" watercolor.[5] The particles are large, angular, flat, and sparkly under the microscope. They have a tendency to stay suspended in water, giving the painter more time to manipulate a wash. Homer exploited this property of charcoal gray throughout *On the Sands*. While the particles were floating in water, he manipulated the wash, allowing some particles to settle in the paper grain and others to flow along the contours of his underdrawing, so that they would dry to "hard edges" that enlivened his drawing lines (Cohn 1977; Walsh 1987). Such "hard edges" appear to dramatic effect in the sky, where Homer orchestrated complex linear configurations to manifest the rolling clouds, and on the skirt of a fisher-wife standing on the shore with a baby on her back.

On his return from England, Homer settled in Prout's Neck, Maine, where he lived until his death in 1910. In 1884 the artist began to make short trips to tropical regions to escape the winter months of Maine. Homer's tropical watercolors in the National Gallery's collection span the two decades between 1884 and 1904. In these paintings he occasionally used bright blues such as cobalt, ultramarine, and cerulean, in addition to his familiar Prussian blue and indigo, to depict brilliant seascapes. A copper-based green pigment was found in only one of the works in the National Gallery's collection – *Salt Kettle, Bermuda* (1899) – where Homer used it for the window shutters, and their reflections on the sea.

In *Native Huts, Nassau*, of 1884–85, Homer mixed charcoal gray with carmine and vermilion to create a purple hue found in the thatch roof, foreground, and clouds. Analysis of paintings in the National Gallery's collection indicates that Homer preferred to make his own purple mixture for broad areas, while using manufactured purple pigments, such as Indian purple or purple lake, for smaller touches. In *Native Huts, Nassau*, he mixed charcoal gray with greens to create a cooler green for the foliage. The work's overall cool appearance was achieved by using charcoal gray throughout the watercolor. Homer used bone black, the warmer and more opaque black pigment, to paint the shadow beneath the hut. He also used this warm black, mixed with opaque red pigments, to paint the children's dark skin. Homer chose each black colorant for its visual effects and working properties, employing them for specific uses to obtain the exact effect he wanted.

In *Santiago de Cuba* (1884–85), Homer utilized colors to describe the stifling heat he encountered on this trip. He mixed carmine and vermilion into blue and green. In a nod to his earlier work in gouache, Homer returned to zinc white, scumbling a layer of white over red paint at the horizon to give the impression of haze hanging over the town. Unfortunately, this painting has suffered the consequences of Homer's choice of the light-fugitive pigment carmine, which he used to tint the sky in a pink hue and to suffuse the mountains with a warm sunlight. Microscopic examination was needed to detect the presence of a few remaining particles of carmine, hinting that passages surviving as dilute blue would have been purplish red. Between these faded passages, the center is profuse with a medley of colors that now make the composition appear compressed and unbalanced. XRF analysis confirmed that Homer used stable pigments for the town, including Indian red, Prussian blue, ultramarine, chrome yellow/orange, and vermilion. Due to his extensive use of carmine, it is not uncommon to find fading in Homer's watercolors, particularly in broadly washed areas such as sky and water. In his renderings of English seashores, he mixed carmine into indigo and Prussian blue to depict purple skies. *On the Sands*, for example, now has a drab gray sky, but the edges that were concealed behind a picture mat retain a purple hue. Now faded, they give the watercolor its present forlorn atmosphere.

Homer's interest in Chevreul's color contrast theory is well known (Hoermann 1988). His copy of Chevreul's *The Laws of Contrast of Colour*, found in his studio after his death, had been "read to pieces, so worn was it with use" (Beam 1966: 226). Passages in the watercolors appear to show how Homer put Chevreul's theories into practice. In *Two Scouts* (1887), a close look at the soldier's uniform buttons discloses Homer's precise placement of yellow, juxtaposed with contrasting blue violet, so that each color appears more intense. White paper surrounding the strokes enhances this vibrancy. A different application of Chevreul's theories can be observed in the foliage in *Salt Kettle, Bermuda.* In this painting, rather than placing complementary colors adjacent to each other, Homer laid pure red over contrasting bright green to achieve an optical vibrancy (see Fig. 2.35).

Homer's watercolors of the Adirondacks in the National Gallery's collection include the most extensive palette of his career. Although chrome yellow is a dominant yellow in his paintings of this period, a small touch of cadmium yellow was identified for the first time in *Lone Fisherman* (1889). Indian purple was detected, through the presence of copper, in *Sketch for Hound and Hunter* of 1892. This is the earliest watercolor in the National Gallery's collection that contains evidence for Homer's use of Indian purple.

One of Homer's Adirondacks watercolors – *A Good Shot, Adirondacks* (1892) – is an emotionally engaging work in which a serene, foliated backdrop contrasts with the image of a struggling wounded deer (Fig. 2.36). This painting shows the artist's ability to use his pigments in precise combinations, yet it appears effortless. Although Homer had at his disposal a great number of colorants, he chose certain ones for specific effects. For example, he relied on the two blue colorants, Prussian blue and indigo, for broad areas in the forest backdrop. These two chemically different blue pigments actually have similar working properties. Indigo is "physically" likened to Prussian blue, and Prussian blue has been described as inky and dye-like. Both colorants are composed of finely divided particles, and the paints have excellent flowing properties. In the background, Homer exploited the opacity of cadmium yellow by scumbling the pigment profusely over the deep blue foliage to create the impression of green mist rising off the water. A delicate, subtle effect is seen in the dark foliage, where Homer scumbled a thin layer of bright chrome orange over a dark blue-green area. By painting these opaque pigments in scumbles, Homer retained the purity of each hue and achieved a vibrant result. His ability to create an abundant chromatic effect with only a few colorants is astonishingly evident.

Compared to the limited colors with which he painted the background, Homer introduced a greater range of colors for the foreground, using an assorted palette of indigo, Prussian blue, cobalt blue, Prussian green, purple lake, and black. Over strokes of these colors, Homer scumbled yellows and reds for a subtle vibrancy. Infrared photography dramatically shows how Homer placed strokes of colors on paper like mosaic

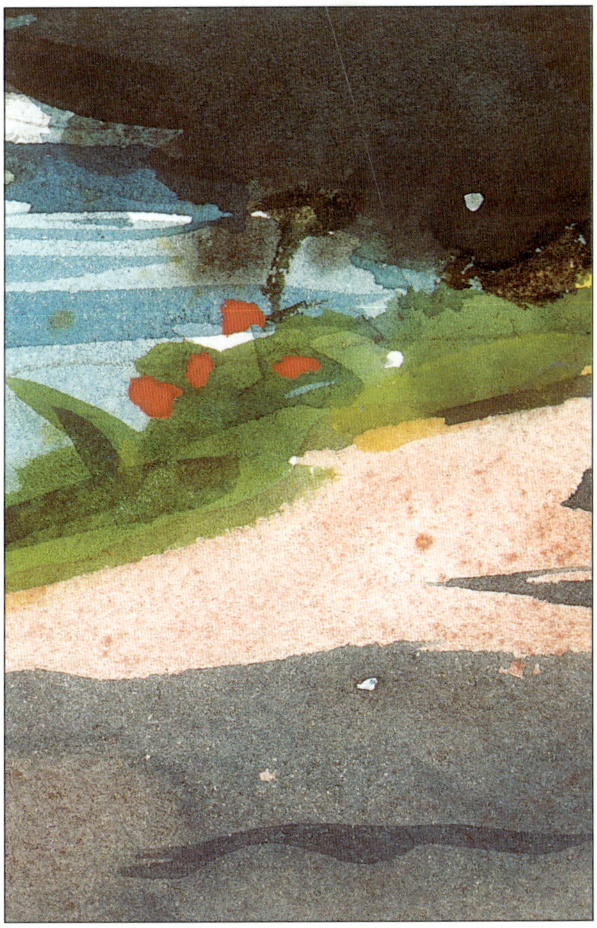

Figure 2.35 Winslow Homer, *Salt Kettle, Bermuda*, 1899 (detail). Watercolor and pencil on paper, 35.5 x 53.3 cm. National Gallery of Art, Washington, D.C., Gift of Ruth K. Henschel in memory of her husband, Charles R. Henschel, 1975.92.15. Homer's application of Chevreul's contrasting color theory was applied here by placing spots of red on a field of its complementary green, the choice of opaque pigments maximizes the effect of the contrast since there is no optical mixing of the colors.

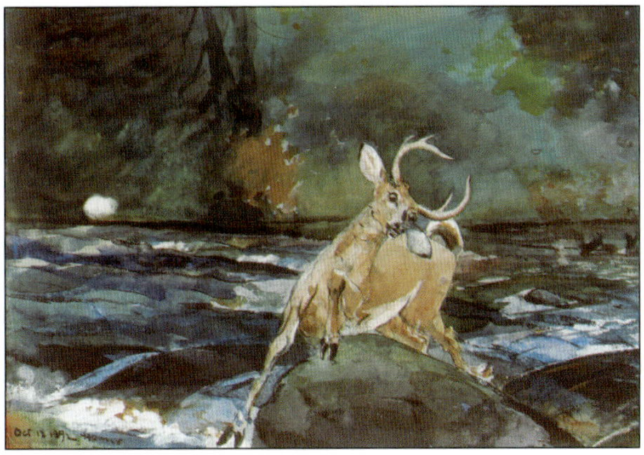

Figure 2.36 Winslow Homer, *A Good Shot, Adirondacks*, 1892. Watercolor and graphite on paper, 38.2 x 54.5 cm. National Gallery of Art, Washington, D.C., Gift of Ruth K. Henschel in memory of her husband, Charles R. Henschel, 1975.92.5. (© 2000 Board of Trustees, National Gallery of Art, Washington, D.C.)

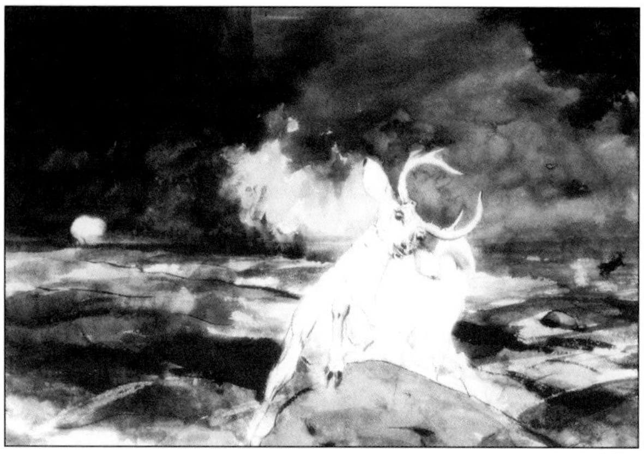

Figure 2.37 Infrared photograph of *A Good Shot, Adirondacks* (Fig. 2.36). (Photograph courtesy of the National Gallery of Art, Washington, D.C.)

tiles, sometimes in overlapping layers (see Fig. 2.37). It is eye-opening to discover that he used many distinct colors and tones for this passage, since their different appearances are not readily apparent. We came to the conclusion that Homer was painting exactly what he was seeing. He perceived all the colors of nature and used just as many (or as few) different pigments as he needed to capture the moment; in this case, the multifaceted appearance of the broken water's surface, and the subtle effects of vegetation flowing beneath.

What appears to be simple, quick, and effortless was in truth well thought out, with precise choice and placement of colors. Perhaps Homer wanted us to look very closely at his watercolors and discover his observations of nature. His intention notwithstanding, that is what he achieved. We have seen Homer's work, almost as it flowed from his hand with poetic and technical mastery. Through his watercolors, we are able to see glimpses of nature that he so keenly perceived.

Acknowledgments

We would like to thank the Portland Museum of Art for granting permission to analyze the paint box, the National Gallery of Art for its support, and The Art Institute of Chicago for organizing the conference. Many people helped us and we gratefully acknowledge the contribution and support of Shelley Fletcher, Judith Walsh, Connie McCabe (photography), Glenn Gates (analysis of pigments in the watercolor box), Lisha Glinsman (XRF), Lehua Fisher, Division of Imaging Visual Services, and George Strumfels.

Endnotes

1. To determine the inorganic elements present in the cakes in the watercolor box without taking samples, energy dispersive X-ray fluorescence (XRF) was used. XRF analysis was done by Glenn Gates, using a Kevex 0750A spectrometer equipped with 6 mm collimators and $BaCl_2$ secondary target. Data were collected for 100 live time seconds at a 60 kV anode voltage and 0.4 mA anode current. For analyses that required sampling, small samples were taken with a tungsten needle and/or fine-tipped tweezers from the underside of the cakes. The choice of this sampling site not only reduced the possibility of contamination from other paints spattered over the surface of the box and its contents, but was also inconspicuous. Of the 17 cakes that underwent analysis, each had a sample taken for high-pressure liquid chromatography (HPLC), and 12 had samples removed for polarized light microscopy (PLM) and X-ray powder diffraction (XRD) analysis. PLM and XRD provided information on specific crystalline compounds present. PLM was conducted using a Leitz Orthoplan microscope. The samples were mounted in Arcolor 5442 and observed at a magnification of 160x to 320x. XRD analysis was performed by Glenn Gates, on a Philips XRG 3100 X-ray generator with a tube equipped with a copper anode and nickel filter and 57.3 mm radius Gandolfi cameras. Exposures were made at 45 kV anode voltage and 25 mA anode current. HPLC was used to identify organic dyes. HPLC analysis was completed by Susana Halpine following extraction of the organic pigment with a 1:1 solution of methanol and 3N hydrochloric acid. The solid extract was redissolved in methanol and commercial grade HPLC water. Analysis was done on a Hewlett Packard 1090 chromatograph fitted with a Brownlee C-18 column and a RP-18 Newguard pre-column. The solvent system consisted of commercial HPC-grade water and acetonitrile with 0.1% trifluoroacetic acid added to each. The 40-minute gradient was programmed to start at 5% acetonitrile, changing to 30% in 24 minutes and 60% in another 15 minutes. The column temperature was set at 38°C and the solvent flow at 0.6 ml/min. The analytical runs were monitored at 275 nm using a diode array detector installed in the instrument. All spectral data were collected from 200–600 nm providing spectra for all peaks in the chromatogram.

2. Apparently, Indian purple has been prepared differently at times by colormen for different reasons. Taylor described Indian purple as "prepared by precipitating the colouring matter of a decoction of cochineal on a base of copper. A very deep-toned but rather cold and subdued purple, apt to blacken under ordinary conditions of exposure" (Taylor 1887: 59–60). Later on, this colorant was prepared differently in order to improve its permanence. In 1898 Allen listed a number of pigments for which the original recipe was different from that used in 1898, although the 1898 pigments had acquired the name given to the earlier pigments that had a similar hue. Here, Indian purple was described as "Originally a fugitive color made of Cochineal Lake with a base of Copper. A permanent substitute is now manufactured from Madder Lake and French Ultramarine" (Allen 1898: 56). Upchurch described another purple colorant named purple lake as a mixture of carminic acid and calcium oxide (Manager of Educational and Technical Services, Winsor & Newton, New Jersey, personal communication, 1993). Although we cannot know when Winsor & Newton first prepared purple lake in this manner, we can identify the purple pigment in the Portland Museum box as such.

3. For infrared photography, watercolor paintings were photographed with Kodak Wratten Filter 87 and Kodak infrared high speed 4143 film.

4. Winsor & Newton introduced "Chinese white," a dense zinc oxide paint, in 1834 to replace lead white, which darkens in polluted environments. The manufacture of zinc oxide depends on a vapor process in which fumes of burnt zinc ore are gathered to make the pigment. This process produces very fine paint particles that can be easily mixed to give a smooth paint. Therefore agglomerates of white particles in these watercolors may suggest that Homer himself prepared this paint with raw materials. Mixed with the white paint are unevenly sized amber colored droplets that might be the residue of gum that Homer incorporated into this rather crudely prepared paint.

5. George Rowney & Co., Ltd., *Geo. Rowney & Co.'s artists' water colours.* London, n.d.

References

Allen, G. B. 1898. *Watercolor painting.* Boston: Lothrop, Lee & Shepard Co.

Beam, P. 1966. *Winslow Homer at Prout's Neck.* Boston: Little, Brown and Company.

Berrie, B. H., and Y. Strumfels, forthcoming. "Winslow Homer's watercolor palette," in *Conservation Research Studies in the History of Art 2002/2003.* Washington, D.C.: National Gallery of Art.

Church, A. H. 1890. *The Chemistry of Paints and Painting.* London: Seeley and Co. Limited.

Cikovsky, N. 1995. "Reconstruction," in *Winslow Homer.* Washington, D.C.: National Gallery of Art; New Haven and London: Yale University Press, pp. 95–107.

Cohn, M. B. 1977. *Wash and Gouache: A Study of the Development of the Materials of Watercolor.* Cambridge, Mass.: The Center for Conservation and Technical Studies, Fogg Art Museum.

Gould, J. 1962. *Winslow Homer: A Portrait.* New York: Dodd, Mead and Co.

Halpine, S. M. 1996. "An improved dye and lake pigment analysis method for high-performance liquid chromatography and diode-array detector," *Studies in Conservation* 41, pp. 76–94.

Herrick, H. W. 1882. *Water color painting: Description of materials with directions for their use in elementary practice.* New York: F. W. Devoe & Co.

Hoermann, K. 1988. "A hand formed to use the brush," in M. Simpson (ed.), *Winslow Homer Paintings of the Civil War.* San Francisco: Bedford Arts, Publishers; The Fine Arts Museum of San Francisco, pp. 103–19.

Muckley, W. J. 1893. *A handbook for painters and art students on the character, nature, and use of colours.* London: Baillière, Tindall and Cox.

Newman, R., and C. Weston 1980. "Analysis of watercolor pigments in a box owned by Winslow Homer," *Journal of the American Institute for Conservation* 19, pp. 103–04.

Russell, W., and W. Abney 1888. *Report to the Science and Art Department of the Committee of Council on Education on the Action of Light on Water Colours.* London: Eyre and Spottiswoode.

Taylor, J. S. 1887. *A descriptive handbook of modern water-colour pigments illustrated with seventy-two colour washes skilfully (sic) gradated by hand on Whatman's drawing paper with an introductory essay on the recent water-colour controversy.* London: Winsor & Newton, Limited.

Walsh, J. C. 1987. "Observations on the watercolor techniques of Homer and Sargent," in S. E. Strickler (ed.), *American Traditions in Watercolor, The Worcester Art Museum Collection.* New York: Worcester Art Museum, Abbeville Press Publishers, pp. 44–65.

The History and Use of Colored Inks

Carlo James

The term colored ink is not a very happy one – except for invisible inks, all inks are colored. When we talk of ink, we understand pen and ink. I have seen only one or two Old Master pen drawings where colored inks other than black and brown have been used; I know that other such drawings exist, but they are extremely rare. It is impossible to talk about colored inks without talking about ink washes as well, because many black and brown pen and brush drawings include washes using the same or other colors. Because different materials in a drawing are almost always imbricated and interdependent, we cannot talk of one without talking of the others. In the simplest terms, one can say that an ink is a colored solution or dye, and watercolor is a suspension of extremely fine pigment particles in gummed water. Things become more complicated with the use of some intensely colored and opaque washes, called body colors. These are found most commonly in the drawings of German painters such as Albrecht Altdorfer and Albrecht Dürer. The combination of transparent and opaque media often results in a certain imbalance in their drawings due to the different covering capabilities of the two mediums.

Thus it is essential, before proceeding further, to define historically the terms ink and watercolor. It is generally believed that "ink" derives from the Italian *inchiostro*. According to Winslow Ames, *inchiostro* and *ink* derive from the Greek *enkaiein*, which means "to burn in."[1] This suggests that the ancient Greeks added a mordant to inks to make them more permanent. *Watercolor*, the English term for *acquarello* used by Cennini, may be interpreted as "lightly colored water" or "wash." Cennini explained that you may "aombrare le pieghe d'acquerelle d'inchiostro,"[2] by which he meant that you can dilute your ink with water to shade the folds of drapery. It is clear that the term as used by Cennini did not have the same meaning as it has today.

Colored inks and watercolors have different fluidities. An ink may be used as a wash, but a wash cannot be used with a pen as an ink. However, these definitions are much too restrictive. Watrous characterizes ink much better: "Regardless of their compositions or colors, all inks of good quality had to possess certain physical characteristics in order to be useful for pen or brush drawing, or if diluted with water, for wash drawing. Above all, an ink had to flow evenly in order to assure the artist of uninterrupted lines or strokes."[3] Thus inks may be prepared from dyes, as well as from finely grained pigments with strong coloring power.

A dye colors the paper fibers; the colored grains of a watercolor are dispersed on and between the fibers. Watercolor, according to our modern definition, is a technique of transparency. The light falling on the surface of the artifact is diffused by the colored grains, and reflected by the paper surface. If the pigment particles are too coarse, the wash gets strongly tinted and loses its transparency. This kind of wash is called body color. If the particles are too fine, their diffused surface may be insufficient to produce a well-defined color. In this case, if the suspension is sufficiently concentrated and gummed, the paint may have a fluidity that makes it suitable for pen drawing.

Tempera derives from the Italian term *stemperare,* which means mixing a pigment with a binding agent in water. Cennini advises the use of "colors tempered with gum, or with clear white of egg well beaten and liquified."[4] Our modern definition of tempera exclusively mentions the use of egg white or the yolk of the egg as a binder for pigments that produce semi-transparent, fast-drying colors. Cennini recommends lead white to lighten all colors,[5] but not to give them more covering power. However, he painstakingly describes how to mix pigments with a carbonate and a binder for preparing paper and parchment. Here he seems to imply also that these colors were used as gouache. Tempera and gouache cannot be used as inks. However, they were masterfully used in the Early and High Renaissance with the point of the brush on prepared paper, and sometimes careful observation is needed to distinguish them from pen drawing.

I fear that this introduction to colored inks, for those who are not very familiar with these matters, has created more confusion than clarification of the problems. I regret this, but want to emphasize the danger of categorizing the different techniques according to our modern definitions, particularly regarding drawings from the Late Middle Ages and Early Renaissance, a period that saw the birth of creative drawing, the introduction of paper, and the adaptation of existing techniques to this new substrate.

A very short history of the evolution of drawing and drawing media

Most art historians agree that modern drawing may be defined by its great spontaneity in the search for form and tension in terms of line. Meder states that the basic characteristic of drawing is its sketchy character.[6] Because few late medieval and Early Renaissance drawings survive – essentially in the form of book illustrations and modelbooks – Meder, like Berenson, concludes that drawing did not yet exist at that time, and that it was only around the middle of the fifteenth century that drawing acquired its status as an autonomous activity. All earlier graphic expressions originating in forms of medieval art had to be repudiated to make possible a rupture with trecento traditions, thus giving birth to modern drawing. As a consequence these authors exclude practically the whole graphic production of the trecento, a great part of the quattrocento, and completely omit the illustrated book.[7]

More recently Degenhart and Schmitt and others[8] have advocated a more evolutionary vision that goes back into the late Middle Ages. They contend that form was not developed solely through the execution of the artwork itself, but was the result of a progressive approach to form by drawing. So few trecento and quattrocento drawings survive because they were not yet considered important pieces, and were lost in the creative process. The extant material is not sufficient to provide a clear understanding of early drawing and drawing methods. Fortunately, we can see them reflected through the evolution of modelbooks. Indeed, Jenni notes that "the medium of drawing, i.e. the various modelbooks and individual pages from modelbooks, is especially sensitive in registering change in artistic practice since it is so closely allied to the very process and method of artistic production."[9] Scheller explains more precisely that:

> In the first stage the model function was overshadowed by the nature study. Then the programmes and formulae became restricted to individual workshops, where they were regarded as the master's personal property. This restriction became even tighter when the artist used the formulae solely for his own work, although he continued to utilise them throughout his career ... The final stage is reached when the drawing was associated with just a single commission. It was at that point that the modern sketch or study was born, the model being eliminated from the creative process.[10]

However, it is necessary to emphasize the fact that the evolution of the modelbook is a reflection and codification of what was happening under the surface long before.

As I noted in the introduction, our modern definitions of different techniques do not correspond at all with those of these early times. It is generally presumed today that modern graphic media derive from *sinopia* and from manuscript illumination. *Sinopia* and underdrawings for panel paintings may account for the development of chalk drawing and cartoons. Pen and colored wash drawing derive from the techniques used in illumination and from the illustrated book.

Cennini was fairly conservative in his artistic and technical conceptions. This is why we may consider his *Il libro dell'arte* of 1435 as a contemporary snapshot of the technical evolution of drawing in the early fifteenth century. Copying was, for Cennini, a means of ensuring the continuation of the trecento traditions in which he was trained. Inventive drawing on paper was not mentioned at that time, probably because it was self-evident. A drawing by the Florentine artist Giovanni del Ponte (1385–after 1437) (see Fig. 2.38), who died the year Cennini published his book, illustrates the moment of technical evolution that Cennini described. It is executed using the point of the brush to apply wash and bistre ink and lead white over a preparatory drawing of silverpoint on blue prepared paper. Cennini painstakingly described how to draw copies on prepared paper, wood, or parchment tablets, which could be corrected or erased with saliva. He also mentioned that

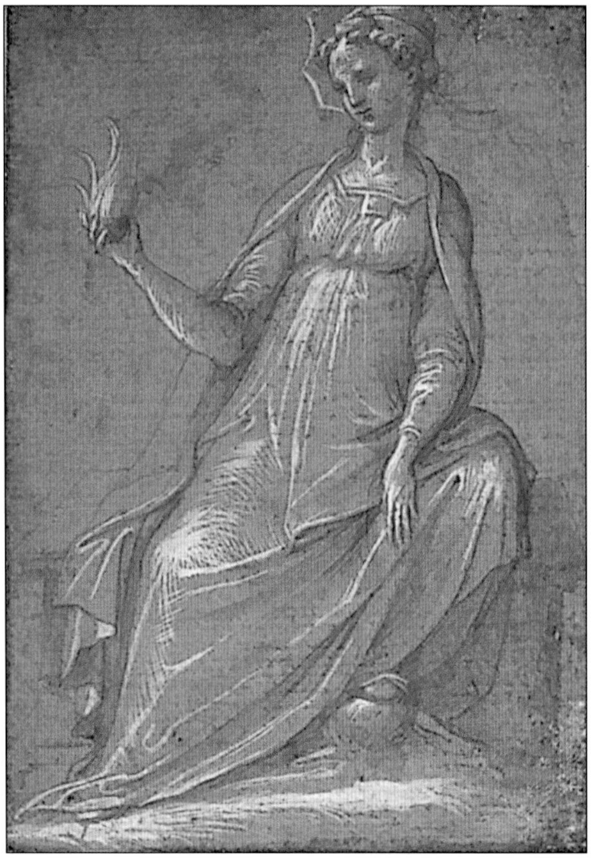

Figure 2.38 Giovanni del Ponte (Italian, 1385–after 1437), *Allegorical Figure of Charity.* 13.8 x 9.6 cm. Fondation Custodia, Collection F. Lugt, inv. no. 1170. (Photograph credit: Marie Dominique Jarry James.)

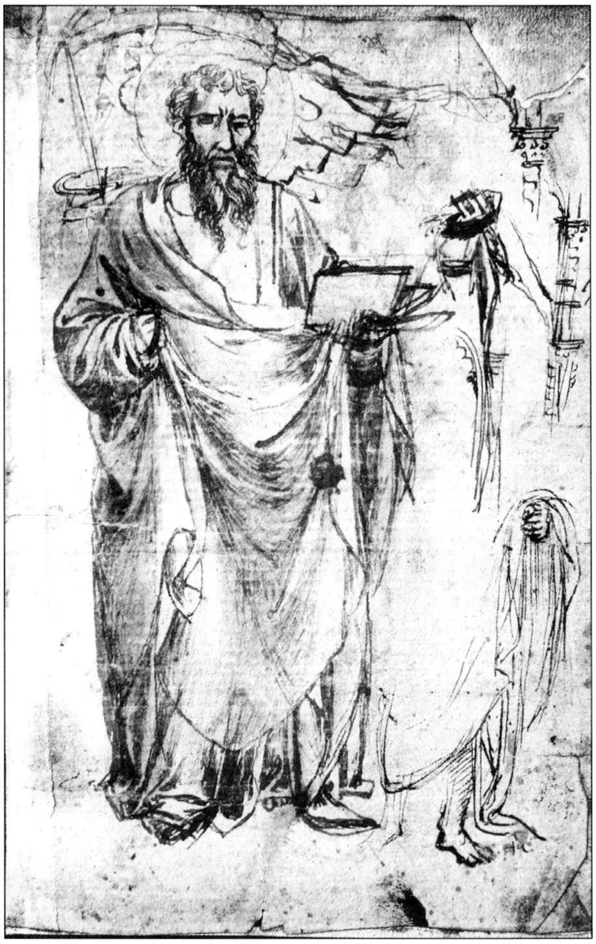

Figure 2.39 Lorenzo di Bicci (Italian, c. 1340–1427), *Saint Paul (?), Full-length, Standing,* c. 1380 (verso). 26.5 x 16.5 cm. Fondation Custodia, Collection F. Lugt, inv. no. 2072. (Photograph credit: Marie Dominique Jarry James.)

you can draw on paper with a lead point and correct errors by rubbing the lines away with breadcrumbs.

We may wonder why almost all surviving Early Renaissance drawings are done on parchment and prepared paper, rather than white or colored paper. I believe that herein lies the fundamental question to understanding the evolution of drawing techniques. Although I will not address art-historical considerations, it is necessary to keep in mind what Degenhart and Schmitt point out: "One cannot really talk of a history of drawing techniques between 1300 and 1450; indeed, the change of their use reflects the evolution of style, because the real historical evolution of drawing in this period is that of style."[11]

Late medieval illumination was exclusively done on very expensive parchment. The paper manufacturer's aim was to produce a substrate that could compete, in surface structure and quality, if not in strength, with that of parchment. An equilibrium between the use of paper and parchment was attained in the second half of the fourteenth century and by 1500 "paper had won the day."[12] Francis Ames-Lewis maintains that the rarity of surviving drawings on paper is partly due to its high cost, which made it less available for trecento artists, who instead preferred a reusable surface.[13] I am doubtful that this was the real reason.

We have no positive information on what the first European paper molds looked like, but Loeber[14] conjectures that late medieval molds consisted of rigid frames and metal wire. Briquet notes that "the laid wires are fine and closely spaced by the thirteenth century; they become thick and broadly spaced toward the middle of the fourteenth, and then become fine and close again."[15] *Medieval* laid paper counts some 20 laid lines per 62 mm, and half a century later, *antique* laid paper counts up to 20 laid lines per 20 mm.

Medieval and early fifteenth-century laid papers were certainly not adapted for traditional illumination techniques. The surface structure was rather rough, absorbent, and unsuitable for pen and ink, even less so for tempera. It is tempting to conjecture that, on the one hand, by preparing the laid paper with a ground, one obtained a surface much nearer to that of parchment and suitable for miniature techniques. On the other hand, practitioners of the transparent technique of colored inks, brush and wash (derived from illumination and developed for these prepared paper surfaces in response to stylistic evolution) gradually chose unprepared paper, which produced a more vigorous and expressive result. As paper became smoother and better sized, the brush could be substituted by pen, ink, and wash. A pen and brush and carbon ink drawing by Lorenzo di Bicci (c. 1340–1427) (Fig. 2.39), dated around 1380, is a pure example of early creative drawing with studies of details and a liberty of brush strokes that we find only a century later in modelbook drawing. It is done on medieval laid paper. One realizes the evolution of drawing techniques when one compares this sheet to another of the Veronese school of the first half of the fifteenth century, drawn in pen and carbon ink on antique laid paper (Fig. 2.40).

The brown and black inks used for drawing from the late Middle Ages, to the Early Renaissance, and onward into the eighteenth and nineteenth centuries are: the brown inks bistre and sepia, and the black inks iron gall and carbon black. Their physical characteristics are as varied as the papers on which they were applied.

Brown inks

Bistre and sepia were the most commonly available brown inks. Even though it is rarely described as an ink, the term bistre appears in artists' manuals of the seventeenth century. Yet bistre had already been mentioned in 1431, under the name *caligo* or *fuligo* in Jean le Begue's *Table of Synonyms*. It is not clear if *fuligine* refers to brown or black soot.[16] The term *saepia* is found

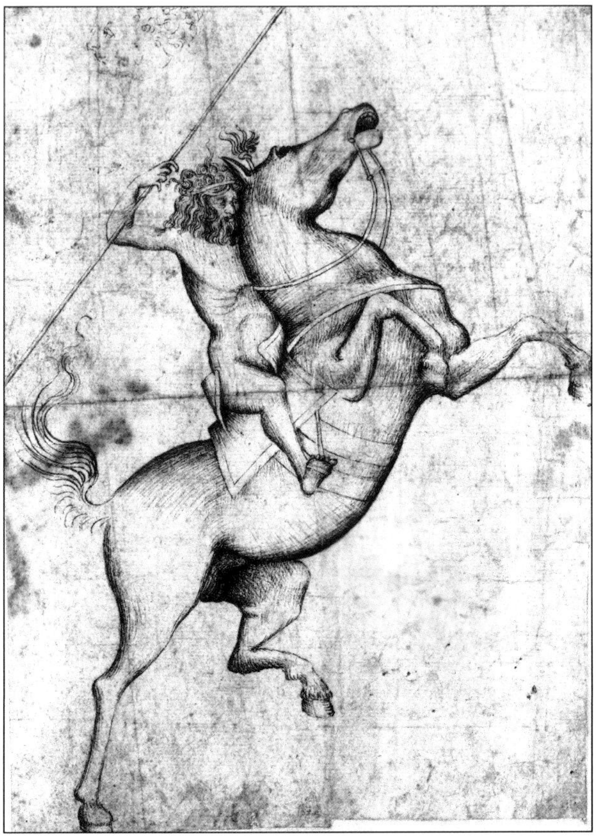

Figure 2.40 School of Verona (first half of the fifteenth century), *Rider on a Rearing Horse.* 25.2 x 17.7 cm. Fondation Custodia, Collection F. Lugt, inv. no. 5082. (Photograph credit: Marie Dominique Jarry James.)

in the writings of Pliny the Elder, without specifications for its use.

Bistre

The various recipes for bistre call for scraping the soot formed by burning wood from chimney flues and boiling or steeping it in hot water. The resulting liquor is filtered to remove any insoluble residue. Depending upon the wood that is burned, the color of bistre can vary from light yellow to dark golden brown. The main characteristics of bistre are transparency and luminosity. Another is the presence of some particulate residue, a consequence of imperfect filtering, which deposits as grayish-black grains on the paper. As noted by Cohn, variations in the sizing of the paper support can produce enormous variation in the color of a bistre ink: "Hard-surfaced paper will tint golden while a relatively more absorbent sheet will show gray. It is conjectured that on the well-sized paper all the wash remains on the surface, while on the more absorbent sheet the yellow liquid soaks in, leaving only black particles on the surface."[17]

Occasionally other decoctions were used, some made from walnut shells, catechu (a brown dye), or other substances that are extremely difficult to analyze. Even walnut cordial, which has a greasy appearance, can be confused with the diluted brown printing ink that was sometimes used in the nineteenth century to copy or imitate Old Master drawings done in iron-gall ink. Sometimes artists modified inks to obtain special effects. They added red chalk to their bistre ink to give it a reddish tinge, and as a result their color can be confused with iron-gall ink. Bistre is a dye, and the addition of a binder is rarely mentioned in recipes. The addition of gum arabic is recommended to add more brilliance to the color.

Sepia

Sepia is extracted from the dried ink bladders of cuttlefish by boiling them in water. The liquor is decanted several times, the resultant sediment is finely ground, and gum arabic is added. Sepia is a dull, cool, violet-brown color. Unlike bistre, when sepia is used as a wash it lacks transparency.

According to Meder, sepia appears in artistic drawing only at the end of the eighteenth century.[18] In the nineteenth century, the use of sepia ink became widespread – so much so, in fact, that it became the generic term for most brown inks, thus creating the probable source of present-day confusion in the identification of brown drawing inks. However, Meder is not entirely correct in his suggested dating. I have found some early Italian, Dutch, and Flemish works that do not contain traces of free bivalent iron, which behaves as a dye, and whose colors do not correspond with any ink other than sepia.

Black inks

Carbon black ink is probably the first known ink ever used. Its appearance in China dates to approximately 2500 B.C. Because of its easy preparation and its permanence, it has been used through the centuries and up to the present day for writing and drawing. In the Western world, carbon black ink consists, more or less, of pure carbon and an aqueous binder. The Eastern equivalent is *sumi*, an ink made of soot, animal glue, and perfumes.

In early times, a mordant was added to carbon black inks, however, it is only at the beginning of the twelfth century that one finds black inks other than those derived from carbon mentioned for writing purposes. These include iron-gall ink, or inks referred to as common inks. These inks were certainly also used by artists, although the first mention of drawings in iron-gall ink did not occur until the end of the seventeenth century. The first specific recipes for this sort of ink appear at the beginning of the thirteenth century, but presumably their preparation and refinement evolved slowly from the tenth century onward. This produced an assortment of mixtures with a lamp black base supplemented by metallic salts or vegetable decoctions of husks or galls. In due course, preparation became less empirical, with the goal of creating easily available inks containing a mordant.

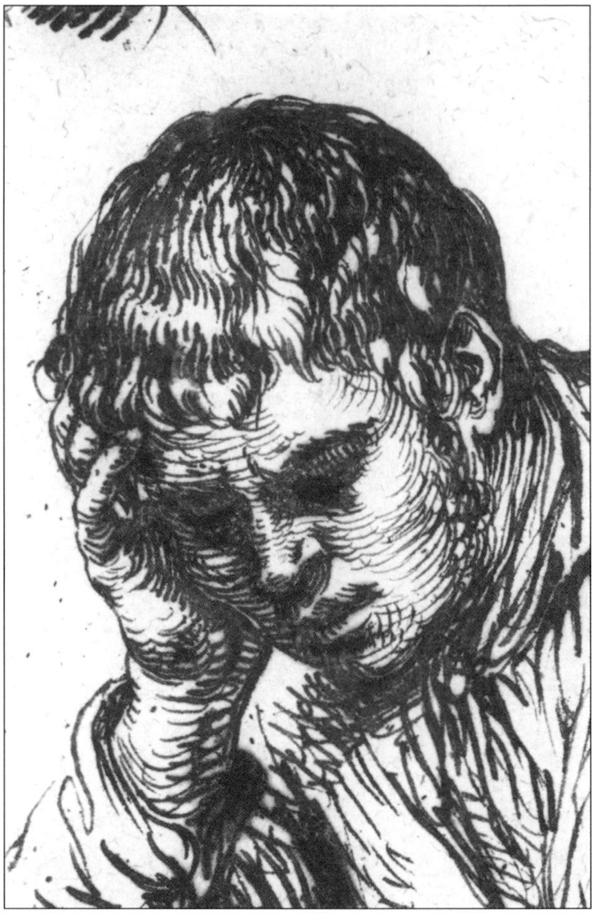

Figure 2.41 Jacob de Gheyn (Dutch, 1565–1629), *Study of a Young Man Resting his Head on his Hand* (detail). 29.8 x 19.6 cm. Fondation Custodia, Collection Frits Lugt, inv. no. 1979. (Photograph credit: Marie Dominique Jarry James.)

Carbon black ink

Carbon is obtained by burning oils, fats, resins, bone, ivory, and fruit pits. The resulting soot is ground into a fine powder and mixed with water, supplemented with various gums or glues. The latter improved the binding properties and stability of the mixture and held the pigment in suspension. Other components such as sugar candy, honey, and ox gall were sometimes added to modify texture, color, or luster, or to increase its fluidity. Recipes for carbon black inks vary tremendously, and the quality of the product is related to the care taken in its preparation. The difference between Western carbon inks and Chinese and Japanese inks seems slight in pen drawings, but can be seen clearly in most ink washes. Chinese inks, the finest known, have a bluish tinge. The finest Japanese inks are brownish. To imitate these renowned Eastern inks, Western artists added indigo blue, bistre, or logwood extract to coarser local carbon black inks.

Iron-gall ink

The principal raw materials of iron-gall ink are the gall nut and iron or copper sulfate. Unlike carbon black ink, its color and intensity derive from a chemical reaction, which produces a violet-black liquid. The corrosive action of iron-gall ink varies according to the proportions of gallo-tannic acids and iron sulfate. The resulting gallo-tannate of iron is only faintly colored at first but oxidizes quickly in the open air, changing from the ferrous to the ferric state.

There are many recipes for the manufacture of iron-gall ink; the recommended materials are not always of the highest quality, and the proportions of gallic acid, iron, or copper sulfate vary from recipe to recipe. These factors contribute to and explain the great variation in the color and condition of the inks on documents, drawings, and manuscripts. Ironically, medieval scribes valued iron-gall ink for its corrosive character, which rendered their writings less susceptible to tampering and, they thought, more durable. In reality, iron-gall inks may change quickly from black to rusty brown, especially when they are exposed to light and humidity. They also destroy the fibers of the paper to the extent that in some cases there is a total destruction of paper by the ink lines, leaving a corroded lacework behind. As a consequence some drawings in iron-gall ink are extremely fragile.

Even if iron-gall inks are always described in catalogues as *brown* inks (with the sole justification that they present that appearance today), it is still necessary to stipulate that originally they were *black*. Surprisingly, no study has been done about the highly significant aesthetic consequences of this color change. Iron-gall inks were appreciated by artists for their fluidity and ease of use with pen and brush. Sometimes artists used a mix

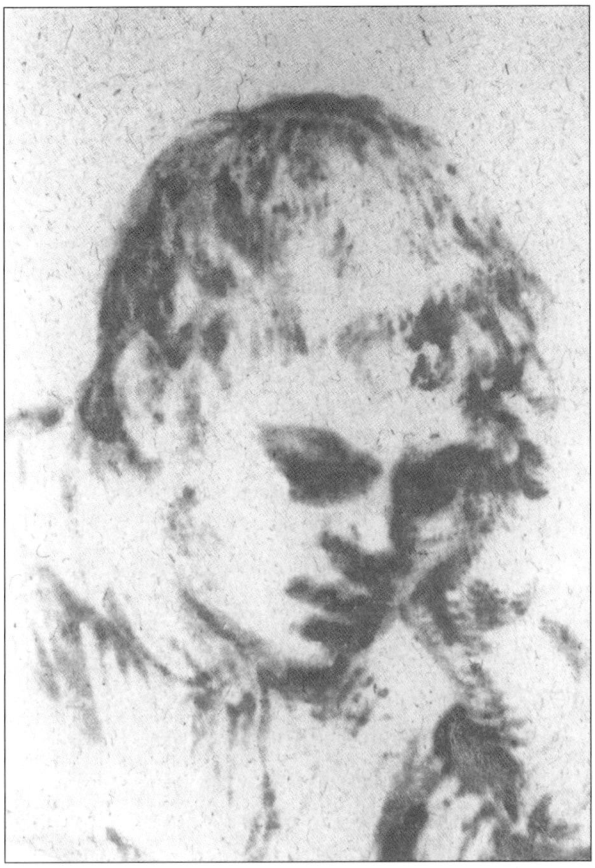

Figure 2.42 Jacob de Gheyn, *Study of a Young Man Resting his Head on his Hand* (detail of verso). (Photograph credit: Marie Dominique Jarry James.)

of techniques, for example drawing pen lines in iron-gall ink and applying washes with carbon black ink: the pen lines have become brown, but the washes have not changed color. This gives some Old Master drawings their peculiar bicolored appearance.

The aging process of iron-gall inks is very complex: the role of one or another component of the mixture is as difficult to identify as the nature of the various other substances in the recipes. According to Neevel the proportions of a balanced ink should be 1 part iron to 3.6 parts tannin.[19] According to Banik an excess of iron sulfate causes ink decay and oxidation of the paper fibers. When an excess of tannin is present, the ink will change to brownish-yellow with hydrolysis of the cellulose chain by acids.[20] These details of a drawing by Jacob de Gheyn (1565–1629) (Figs. 2.41 and 2.42) show a very dark and corrosive iron-gall ink. It has diffused through the paper and beyond the drawn lines to such an extent as to confound them.

Colored inks other than brown and black

Cennini made a clear distinction between "certain colors that have no body" and those that "one must grind finely."[21] Two and a half centuries later, Gautier de Nîmes – who had a clear perception of the difference between a dye and a pigment – denied flatly that such a difference exists: "Most engineers who wash their plans make rather important distinctions between the colors they use for them, that is, those who have a body and those who have not. But they may say what they want, body they have all."[22]

Watrous shows that from the seventeenth century onward, the number of organic dyes derived from colored plant extracts that could be used for drawing increased significantly. As a result it becomes more and more difficult to differentiate among them. One would expect that berries, leaves, flowers, and insects of a certain color would give a colored substance of a similar tint. This is approximately so for saffron, red woods, and carmine. In other cases, one obtains the colored substance by fermentation and/or oxidation of the extracted colored liquid, such as with the leaves of the indigo plant. Plant extracts may be made into a lake by fixing a dye onto a substrate like calcium carbonate or alum to produce a pigment. Some dyes are light sensitive while others are sensitive to pH. Dyes tend to seep through paper, and pen strokes and washes can have sharp borders. A drawing by the Flemish artist Pieter Stevens (c. 1567–after 1624) (Figs. 2.43 and 2.44), executed in pen and brown ink, with red, blue, gray, and gray-brown washes, illustrates the difference. On the verso the red, blue, and brown washes have seeped through the paper while the gray wash has not. This indicates that the former are dyes and the latter is a diluted carbon black ink.

In the Middle Ages and Early Renaissance, ground minerals, which produce grainy pigments, were used in oil paint, in tempera for illuminations, and to color grounds for metalpoint drawings. Through decantation, ground pigments were homogenized. Decantation is a simple process whereby ground pigment is put in suspension in water, vigorously stirred, and left to settle for a while to allow the coarsest particles to sink to the bottom. The colored water is drawn off, left to settle, and drawn off again. The different sediments that remain produce different grades of pigment, colored from deep and bright to pale.[23]

The finer pigment particles were used in suspension in gummed water or in the form of egg tempera using egg white or yolk. They produced colors with very different appearances, ranging from delicately transparent washes (which we would call watercolor today), to opaque, strongly tinted, and grainy strokes (which we would call body color today) that tend to rub off if they do not contain enough binder.

Cennini frequently specified that several pigments were "artificiati e fatti d'archimia" (artificial and made by alchemy). It is clear that by "archimia" he meant "chemically"; the term had not yet acquired the dubious distinction it would inherit in later centuries. He also maintained that "all colors you use for panel painting may be used on paper."[24] This means that we cannot always be sure if a color is a dye or a pigment and cannot make the determination without the aid of analytical equipment. In

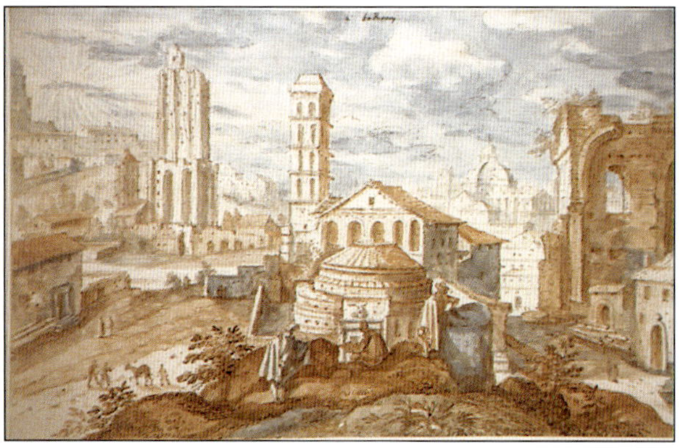

Figure 2.43 Pieter Stevens (Flemish, c. 1567–after 1624). *The North Side of the Foro Romano, with a view of Saints Cosma e Damiano.* 19.4 x 31.2 cm. Fondation Custodia, Collection F. Lugt, inv. no. 8149. (Photograph credit: Marie Dominique Jarry James.)

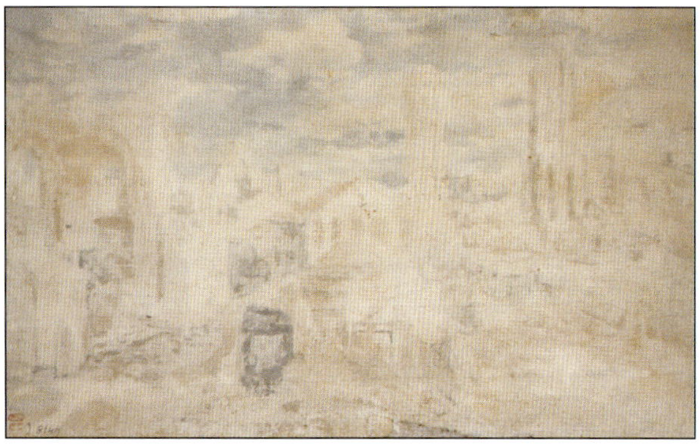

Figure 2.44 Pieter Stevens, *The North Side of the Foro Romano, with a view of Saints Cosma e Damiano* (verso). (Photograph credit: Marie Dominique Jarry James.)

their earliest applications, most of the following pigments were used for illumination and to color grounds for metalpoint drawings. The dyes were used interchangeably in painting or as washes in drawing.

Organic dyes

According to Brunello, Cennini's *verzino*, or red wood, may have been extracted from *Caesalpina sappan*;[25] all red woods, like Brazil wood, Campeachy (logwood), etc., contain the coloring substance *brazilin.*[26] Carmine is a carminic acid, extracted from *Dactilopius coccus cacti* and the female cochineal insects. It was imported to Europe from Mexico after 1512. Turnsole is a dye of an intense violet color, extracted from *Croton* or *Crozophora tinctorum.* It turns red under the action of acids and blue with alkali, and it is still used as a chemical reagent and as a pH indicator. Saffron yellow is extracted in water from the orange-colored pistils of the *Crocus sativus.* In high concentration it takes on a reddish tint. When used for tempera painting, it was extracted in egg white, sometimes with the addition of some wine or vinegar. Fixing saffron on lead white or alum gives it greater covering power, yet it was not used in panel painting because of its poor light stability. Weld, or "dyers weed," called *arzica* by Cennini, was extracted from *Reseda luteola* by boiling the herb in water to extract the yellow dye it contains. It was used only for illumination because of its sensitivity to light. Indigo is an organic dyestuff extracted in the Far East from the *Indigofera tinctoria.* It was imported in Italy through Venice. In Europe it was extracted from the *Isatis tinctoria,* a plant with a lower concentration of indigo, called pastel. This name originated in Provence in the fifteenth century. With this plant extract, fixed on calcium carbonate, sticks were fabricated for drawing which took the name of pastels by extension.[27] Indigo is dark blue and color-fast, and Cennini recommended grinding it to perfection and mixing it with lead white to obtain a color that looks like azurite.[28]

Mineral pigments

Cinnabar is a natural red mineral, mercury sulfide (HgS). Sinoper, a red chalk, *sinopia* (red ochre), or hematite (Fe_2O_3), is similar to sanguine and was used in fresco painting for preparatory drawing. The earth mineral sinoper, as well as yellow ochre, was essentially used for preparing colored grounds on parchment or paper. These were ground finely, mixed with lead white, and tempered with saliva, gum, or egg white. Cennini does not mention sanguine, or red chalk, a clay containing hematite (like sinoper) but with a more orange tinge. The yellow mineral pigments include yellow ochre or lemonite, hydrated iron ($FeO.OH.nH_2O$), and orpiment. Orpiment, arsenic trisulfide (As_2S_3), is very poisonous. It is not mentioned as an ink in Cennini, but it was mixed with indigo or ultramarine to make a green ink. Azurite, basic copper carbonate ($2CuCO_3.Cu(OH)_2$), is a blue mineral found in copper-containing rocks. During the decantation process, ashes were sometimes added to azurite, producing the pigment called "blue ashes." Ultramarine or lapis lazuli, a complex sulfur containing sodium aluminum silicate, was considered the most precious and most expensive pigment. It was used primarily for illumination. Malachite, basic copper carbonate ($CuCO_3.Cu(OH)_2$), is a green mineral found in copper-containing rocks. Green earths (*terre verte*), are natural silicates containing glauconite and celadonite.

Artificial pigments

Lead white, the Italian *biacca,* basic lead carbonate ($2PbCO_3.Pb(OH)_2$), was used for white heightening, to prepare grounds, and to lighten colors and give them some covering power. It was already known to Cennini that lead white can turn black. Vermilion, synthetic mercuric sulfide (HgS), is the artificial equivalent of cinnabar. Another red pigment, minium, lead orthoplombate (Pb_3O_4), was known to blacken easily and therefore was used mostly for illumination. Naples yellow, lead antimonate ($Pb_3(SbO_4)_2$), was not easily available. It was already being artifically produced by the Middle Ages. Copper green (verdigris), basic copper acetate ($Cu(C_2H_3O_2)_2.[Cu(OH)_2]_3.2H_2O$), is formed through the reaction of copper with an acid. It is extremely aggressive and can easily damage paper substrates. A diffuse bright green color or shine on the back of a drawing usually indicates a copper content in the ink used on the recto. Watrous mentions numerous mixtures of copper-containing solutions with organic colors.

Sometimes it is enough to look at the color of the media or at the back of a drawing to know which materials and techniques were used; often it is not. To learn more about drawing techniques, we must train our eyes and refine our judgment. I believe there is no better way to exercise the eye than to look closely at as many drawings as possible, and to compare them. One will realize that inks or washes made with a given material will have characteristic colors and transparencies. Others have specific physical behaviors. Certain inks are acid and others are not; pen lines and washes may be diffuse or sharp. Historical information may help to exclude a certain hypothesis or may confirm it. However, at a certain point visual identification alone is not adequate, and other analytical methods are needed. Visual identification of graphic materials is an essential aspect of what the conservator does, and not only for the sake of accuracy. The differences in color, texture, and reflectance of colored media on paper before and after treatment are extremely subtle. The balance between an aged substrate, and the various aged colored media applied to it may be easily disrupted. Only an experienced eye may be able to interpret the uncomfortable feeling that something is just not right in a drawing. Whether it be a paper that is just too white, a graphic material that has lost some of its freshness or luster, a drawing with severely aged colored media on a sheet of paper that is just too clean and healthy – observations like these will be informed by years of looking at drawings and will inform future conservation decisions about any drawing taken to hand by the responsible conservator.[29]

Acknowledgments

I gratefully acknowledge the editorial assistance of N. Kersten-Pampiglione and the photography of M. D. Jarry James.

Endnotes

1. J. Meder, *The Mastery of Drawing,* 2 vols, transl. by W. Ames (New York, 1978), Vol. 1, p. 51.

2. C. Cennini, *Il libro dell'arte,* ed. by F. Brunello (Vicenza: Neri Pozza Editore, 1971), p. 12.

3. J. Watrous, *The Craft of Old-Master Drawings* (Madison, Wis., 1957), p. 66.

4. C. Cennini, *Il libro dell'arte*, transl. by D. V. Thompson, Jr., as *The Craftsman's Handbook* (New Haven, Conn., 1933; reprint, New York: Dover Publications, Inc.), p. 7.

5. Cennini 1971 (as in note 2), p. 61.

6. Meder 1978 (as in note 1), Vol. 1, p. 10.

7. B. Degenhart and A. Schmitt, *Corpus der Italienischen Zeichnungen 1300–1450* (Berlin: Gebr. Mann Verlag, 1968), I-1: L.

8. See ibid.; F. Ames-Lewis, *Drawing in Early Renaissance Italy* (New Haven and London: Yale University Press, 1981); R. W. Scheller, *Exemplum* (Amsterdam: Amsterdam University Press, 1995); S. Currie, *Drawing 1400–1600. Invention and Innovation* (Brookfield, Vt.: Ashgate Publishing Company, 1998).

9. U. Jenni, "The phenomena of change in the modelbook tradition around 1400," in *Drawings Defined* (New York: Abaris Books, Inc., 1987), pp. 35–47.

10. Scheller 1995 (as in note 8), p. 81.

11. Degenhart and Schmitt 1968 (as in note 7), Vol. 1, XL.

12. V. W. Clapp, "The story of permanent/durable book-paper 1115–1970," *Restaurator*, supplement 3 (1972), pp. 1–50.

13. Ames-Lewis 1981 (as in note 8), p. 14.

14. E. G. Loeber, *Paper Mould and Mould-Makers* (Amsterdam: The Paper Publication Society, 1982), p. 20.

15. C. M. Briquet, *Les filigranes. Dictionnaire historique des marques de papier* (New York: Hacker Art Books, reprint of the 2nd Leipzig edition, 1996), Vol. 1, p. 7.

16. Watrous 1957 (as in note 3), p. 74.

17. M. B. Cohn, "A note on media and methods," in *Tiepolo, a Bicentenary Exhibition 1770–1970* (Cambridge, Mass.: Fogg Art Museum, Harvard University, 1970), pp. 211–21.

18. Meder 1978 (as in note 1), Vol. 1, p. 47.

19. J. G. Neevel, "The development of a new conservation treatment for ink corrosion, based on the natural anti-oxidant phytate," *IADA preprints, 8th International Congress of IADA* (Tübingen: IADA), pp. 93–100.

20. G. Banik, H. Stachelberger, and K. Messner, "Untersuchung der destruktiven Wirkung von Tinten auf Schriftträgermaterialien," *Restauro* 100, no. 4 (Oct. 1994), pp. 302–08.

21. Cennini 1971 (as in note 2), p. 168.

22. H. Gautier de Nîmes, *L'Art de Laver ou nouvelle manière de peindre sur papier* (Lyon: Thomas Amaulry, 1687; reprint Portland, Oregon: Collegium Graphicum), p. 32.

23. Cennini 1933 (as in note 4), p. 35, note 1.

24. Cennini 1971 (as in note 2), p. 168.

25. Ibid., p. 67, note 9.

26. Cennini 1933 (as in note 4), p. 39, note 5.

27. F. Viatte, "Lavis d'indigo et papiers bleus," in *Sublime Indigo* (Marseille: Musée de Marseille/Office du Livre, 1987), pp. 75–78.

28. Cennini 1993 (as in note 4), p. 36.

29. See also C. James, C. Corrigan, M. C. Enshaian, and M. R. Greca, *Old Master Prints and Drawings. A Guide to Preservation and Conservation*, transl. and ed. by M. B. Cohn (Amsterdam: Amsterdam University Press, 1997), pp. 74–84, 120–23.

Aging of Paper and Pigments Containing Iron and Copper: A Review

Vincent Daniels

Conservators and conservation scientists see many deteriorated, aged paper objects. The observed degradation phenomena can usually be explained using the known degradation mechanisms once the object's composition and storage history have been ascertained. Biodeterioration and photochemical deterioration are important for objects stored in bad conditions, i.e. high relative humidity (RH) and high illumination. However, the two main chemical mechanisms for paper-based objects in museum and gallery collections, which are stored in low illumination or darkness are: (a) acid-catalyzed hydrolysis; and (b) oxidation, which may be metal-catalyzed. Although other degradation mechanisms exist, most phenomena can be explained by one of these two pathways.

Not all metals will catalyze degradation of cellulosic substrates; indeed, magnesium and calcium compounds have a well-established reputation for stabilization. The two metals most frequently cited as causing degradation are iron and copper. These metals, in common with others known to cause metal-catalyzed oxidation, have a variable oxidation state. Apart from their metallic state, Cu(0) and Fe(0), copper and iron can occur as Cu(I), Cu(II), Fe(II), and Fe(III), respectively (alternatively written Cu^{+}, Cu^{++}, Fe^{++}, and Fe^{+++}).

The ability of these metals to transfer reasonably easily from one oxidation state to another gives them the capacity to initiate oxidation by the acceptance or donation of an electron (e) with a single negative charge, e.g.

$$Cu^{+} \rightleftarrows Cu^{++} + e$$
$$Fe^{+++} + e \rightleftarrows Fe^{++}$$

This property enables these metals to initiate free radical chain reactions, of which oxidation at room temperature (autoxidation) is an example. The reactions occur with all manner of polymers and other organic compounds. In this paper, cellulose and pigment binders are of principal relevance. However, for a detailed report of how metal ions can catalyze oxidation of polymers generally, readers are referred to Ranby and Rabek (1975). Three of the more important routes are:

(i) production of a free radical of the polymer (Cell OH = cellulose),
e.g. $Cu^{++} + \text{Cell OH} \rightarrow Cu^{+} + \text{Cell O}^{\bullet} + H^{+}$

(ii) decomposition of hydroperoxides formed by the autoxidation process,
e.g. $Cu^{+} + \text{Cell OOH} \rightarrow Cu^{++} + \text{Cell O}^{\bullet} + HO^{-}$

(iii) direct interaction with oxygen, in acid conditions,
e.g. $Cu^{+} + O_2 + H^{+} \rightarrow Cu^{++} + HO_2^{\bullet}$

The superscript dot denotes an unpaired electron, and the group of atoms it is attached to is a free radical. Reaction (iii) above produces the highly reactive hydroperoxyl radical $HO_2^{\bullet}$, which can abstract a hydrogen atom from a molecule to form hydrogen peroxide and another free radical. The study of how free radical reactions initiated by copper and iron may influence the development of inflammatory disease, aging, cancer, rancidity of edible fats, and ripening of fruit is a fascinating one, and readers are referred elsewhere (Halliwell and Gutteridge 1985).

Metal-catalyzed oxidation is sometimes attributed to having "Fenton chemistry," i.e. relating the chemistry to the reaction first studied by Fenton in 1894. Iron(II) reacts with hydrogen peroxide to give a reactive hydroxyl radical:

$$Fe^{++} + H_2O_2 \rightarrow Fe^{+++} + OH^{\bullet} + OH^-$$

The iron can also be reduced back to Fe(II) by the hydrogen peroxide:

$$Fe^{+++} + H_2O_2 \rightarrow Fe^{++} + O_2^- + H^+$$

Copper salts react in a similar way.

Compounds with primary hydroxyl groups, e.g. cellulose, generally oxidize to form products that are aldehydes and ketones, and can further oxidize to form carboxylic acids. Aldehydes are often also precursors of discolored compounds (Lewin 1965). However, acids catalyze hydrolysis of cellulose, and thus in some cases oxidation and acid-catalyzed hydrolysis of paper may occur simultaneously. Metals can exchange with hydrogen ions on carboxylic acid groups in oxidized cellulose.

The paper industry has been interested in the metal-catalyzed oxidation of paper for a long time. When alkaline peroxide or ozone bleaching of paper pulps is practiced, the presence of salts of cobalt, iron, manganese, and copper as impurities accelerate unwanted oxidative depolymerization of cellulose (Ericsson et al. 1971). The addition of magnesium ions in such processes (e.g. as magnesium sulfate), greatly reduces the effect of the impurity metals. Several mechanisms have been proposed for the stabilization of cellulose pulps by magnesium ions (Liden and Öhman 1997).

Although the paper industry has produced a large body of research into the deterioration and stabilization of paper fibers by metal ions, these studies have been performed with water-saturated pulp. Studies on dry paper are much more relevant to aging of papers in ambient conditions, and these will now be described.

Williams (Williams et al. 1977) treated paper with copper acetate and showed that it accelerated the degradation of paper. Humid aging (90°C, 50% RH) produced faster aging than at 100°C in a convection oven. Residues of magnesium carbonate, applied initially as the aqueous bicarbonate, produced the best stabilization, followed by calcium carbonate applied by the Barrow two-step treatment (Daniels 1987).

Shahani and Hengemihle (1986) studied the effects of iron and copper (as sulfates) on paper, initially in the pH range 5.2–6.0. It was found that humid aging (90°C, 50% RH) produced degradation more quickly than dry aging at 100°C; however, comparing the experiments performed at different RHs, there were significant differences in the extent to which the metals accelerated the degradation. The type of compound in which the metal was applied to the paper made a large difference. For example, the effects of iron and copper acetyl acetonate complexes were studied, showing that while the copper compound was an efficient catalyst for degradation, the iron complex was not. In an attempt to isolate the degradative effect of the metal ions from that of acidity, sodium bicarbonate was used to neutralize papers, followed by washing in water. It was concluded that the metals catalyzed the degradation, and that neutralization slowed down the rate of degradation.

However, it was subsequently found that bicarbonate solutions were efficient at dissolving copper from paper, and that immersion in calcium and magnesium bicarbonates stabilized paper by this mechanism (Shahani and Hengemihle 1994). Spray-deacidified paper was not stabilized. It is not clear from the literature whether this finding alters the conclusions from their earlier work, but these findings do cast some doubt on the generally accepted ability of magnesium to stabilize iron and copper in paper.

Bicchieri and Pepa (1996) have studied the mechanism of cellulose deterioration by Fe(III) and Cu(II) chlorides. They concluded that there was a fundamental difference between the degradation mechanisms. Iron ions tend to cleave the 1–4-β glucosidic bond, while copper ions catalyze oxidation of the anhydroglucose ring by accepting electrons from the lone pair on the anhydroglucose unit or the β-glucosidic oxygen. In both cases bond breakage results. Mechanisms for the oxidation of cellulose are also applicable to the oxidation of other polysaccharides in paper, e.g. hemicelluloses.

Metal content of paper

All plants contain trace metals, and the trees from which most paper cellulose is made are no exception. Around 50–100 ppm is normal (Wise 1944). During processing, paper pulp and wet paper have further opportunities to absorb metal ions from solution and to pick up metal particulates from the wear of manufacturing equipment. Most commercial paper contains a significant level of contaminants. The level of contamination can increase twelve-fold if no effective treatment is applied (Shahani and Hengemihle 1986). Tang (1978) has measured the iron and copper levels of paper from several eighteenth- and nineteenth-century books; typical results range from 39–300 ppm iron and 5–40 ppm copper. Grant (1960) reports iron contents of 4–60 ppm for newsprint, 3–30 ppm for periodical papers, and 8–25 ppm for book papers.

Manganese may be unintentionally introduced into paper by conservators. In the past, bleaching with potassium permanganate was practiced. At the end of the bleaching process, all the residual deep purple permanganate should be washed out, but oxalic acid was often used to aid this process by formation of Mn(II) salts, which are very pale pink and thus hardly visible. If washing is insufficient, the paper can contain residual Mn(II). In one example paper that had been treated in this way was treated with an oxidizing bleach. The Mn(II) salt was oxidized to black MnO_2, producing an unexpected and dramatic effect. The manganese dioxide was removed with hydroxylamine hydrochloride solution.

Copper-containing pigments

Copper usually forms green or blue compounds, and several of these, which are found naturally as minerals, may be ground and used as pigments. Others containing acetate groups can be manufactured (Mairinger et al. 1980) but may be of variable composition and may change their composition on aging. The name "verdigris" is often attached to these copper-containing pigments.

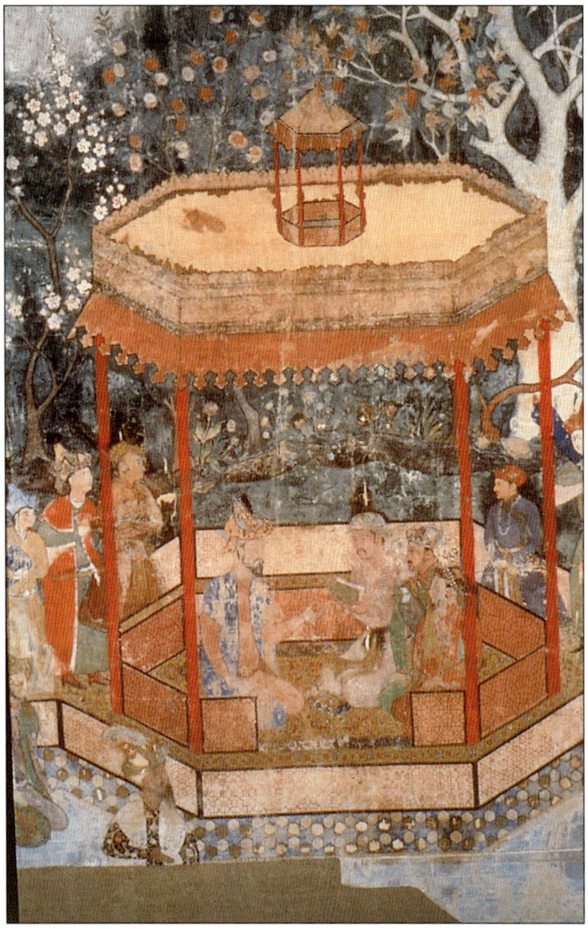

Figure 2.45 A detail from the Indian miniature painting *Princes at the House of Timur*. The pavilion roof was painted in verdigris and is now missing. (Photography by T. Springett. © Trustees, British Museum, London.)

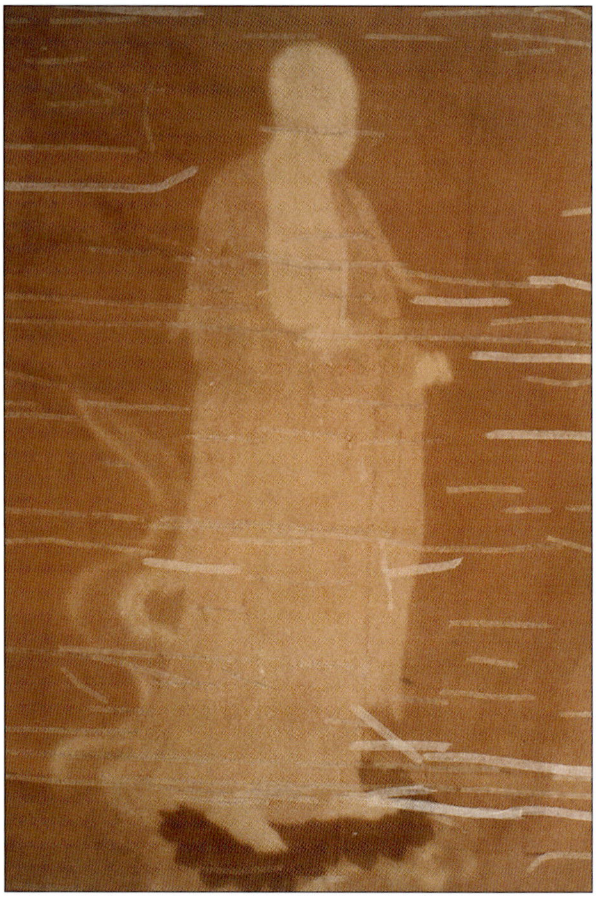

Figure 2.46 The verso of a painting on paper showing simultaneous stabilization of the paper by calcium in a white paint and destabilization of the paper by copper in a degraded brown paint. (© Trustees, British Museum, London.)

However, modern usage of the word is limited to normal or basic copper acetates (Kühn 1992).

The type of damage produced by copper-containing pigments is seen in Figure 2.45, where the paper has been weakened enough to fall out of the page, leaving a hole. At a lesser extent of damage, browning of the paper can be seen (Abe and Thomson, this volume, pp. 207–13). The backs of paintings can show simultaneous stabilization by calcium carbonate and destabilization by copper-containing pigments (see Fig. 2.46). Scientific study of the types of pigment that cause deterioration of paper is hindered by the difficulty of identifying them, as they lose their initial properties on aging and, even when new, are often poorly crystalline with no X-ray diffraction pattern.

Banik (Banik and Ponahlo 1982–83) has analyzed the products of the degradation of paper and found a variety of products consistent with a Fehling reaction, i.e. the reduction of Cu(II) by reducing agents. However, this type of reaction is more common in alkaline environments. Banik suggests that although free radical acidic conditions may prevail at the start of the degradation, alkali hydroxides are formed by the aging and the oxidation of the cellulose proceeds in alkaline conditions. Magnesium bicarbonate is suggested for inhibiting the deterioration, and low RH storage is recommended.

Some of the copper-containing acetates have an appreciable solubility in water, and damage to the cellulose fibers in paper will be enhanced by easy access. Normal copper acetate is particularly water-soluble. Banik (1989) emphasizes that solubility is the key word for chemical reactivity of copper compounds with organic substrates. He reports that copper seems to accumulate in the interior of fibers, where the disintegration starts. Azurite converts to malachite in humid conditions; the latter is more soluble and thus more reactive. Some malachite may react with sulfur dioxide to form the even more soluble basic copper sulfate. Paint recipes in ancient manuscripts suggest that saffron was used to stabilize copper acetate in Persian miniature paintings (Barkeshli 1999).

Some of the earliest examples of copper-pigment-catalyzed degradation can be seen on ancient Egyptian papyri and cotton shrouds. I have seen several examples, but few of the pigments have been identifiable by X-ray diffraction. However, degradation is sometimes seen where the easily identifiable Egyptian blue is present. Egyptian blue, the earliest synthetic pigment, contains silica, a blue copper-containing glass, and cuprorivaite, a blue copper-containing mineral. Deterioration of the cellulosic papyrus is sometimes associated with Egyptian blue (see Fig. 2.47). However, at present it is not known whether the deterioration of the papyrus is caused by copper released from deteriorated Egyptian blue or by another copper-containing pigment mixed in to modify the hue.

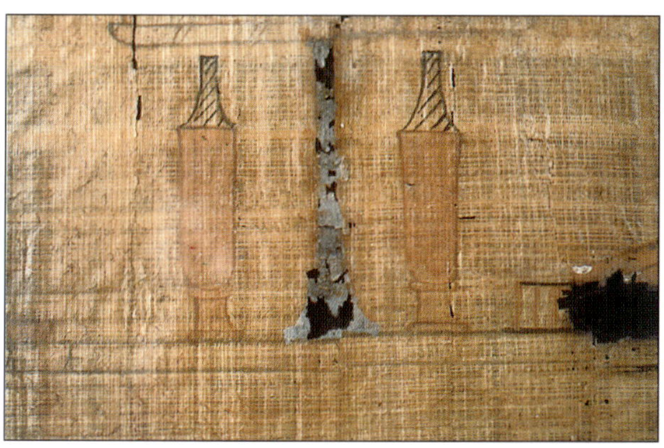

Figure 2.47 The verso of an Egyptian papyrus showing degradation of the cellulosic substrate caused by copper-containing pigments. (© Trustees, British Museum, London.)

There is little information on the degradation products of copper-containing pigments. Banik believes black copper oxide is formed from many verdigris-type pigments, while others have reported basic carbonates. Basic copper chloride and a material similar to malachite have been reported as degradation products of Egyptian blue (Lee 1995). Copper from pigments may also cause degradation of paint binders, especially polysaccharides. For example, accelerated aging of gum arabic with a trace of copper(II) sulfate produces a dark brown product (70°C, 28 days).

Foxing

The occurrence of brown spots on paper is referred to as "foxing." At one time a controversy raged between those who thought foxing was due to microbiological factors or those who attributed it to iron particles. Nowadays, the general opinion is that either mechanism can be involved. Certain categories of foxing are unquestionably due to metal particles in the paper, and in some cases the elemental concentration in the paper can be mapped using a scanning electron microscope (SEM). Work by the author (Daniels and Meeks 1992) has shown that in one contaminated sample, the nucleus of the iron-containing stain contained 4.2% iron and 0.7% copper, compared with a background concentration of 0.3% iron and 0.3% copper. In another sample, the metallic nucleus could be seen as a lump with 58% tin, 6% iron, and 3% aluminum, with no detectable chlorine in the paper. Deterioration of the paper was evident.

Copper-alloy particles can also cause a type of foxing. A particularly interesting example (Daniels and Meeks 1994) produced a spreading dendritic pattern of black corrosion products. All parts of the foxing contained various levels of copper, zinc, sulfur, and chlorine, but none of these elements was detectable in the surrounding unaffected paper. It is hypothesized that water-soluble copper- and chlorine-containing corrosion products migrated along the fibers and then formed black copper sulfide in the fibers, producing the highly characteristic dendritic appearance (Fig. 2.48). There was little evidence of deterioration of the surrounding paper, except for a slight discoloration.

Tang (1978) has demonstrated a positive correlation between brown stains on paper and the local presence of iron or copper in the paper. However, the concentrations detected covered a wide range; for iron from a few hundred ppm to 2.95%, the latter due to contamination by a rusty staple. However, for most foxing, values in the range 0.1–1.0% were most common.

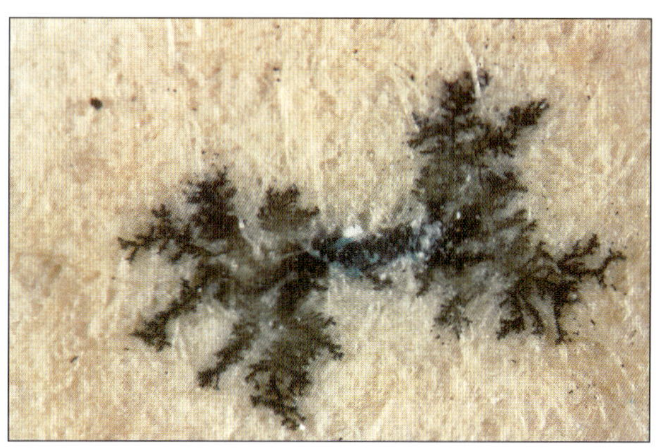

Figure 2.48 Optical micrograph of a foxing spot caused by corrosion of a metal particle. Copper, zinc, chlorine, and sulfur were present. (© Trustees British Museum, London.)

In many cases the brown staining can be reduced significantly by the use of iron-solubilizing reagents. However, any degradation of the paper that has produced brown organic chromophores in the cellulose will probably not be removed, and conventional oxidizing or reducing bleaches will need to be employed.

Iron-gall ink

When iron salts react with plant polyphenols (tannins), a black dye can be produced. If plant galls are used as the polyphenol source (as they often are), the ink is called iron-gall ink. Many drawings and manuscripts containing these inks show signs of severe degradation. In the worst cases, the paper has cracked in the inked area and may even have fallen away. At the same time, the ink has often changed color to brown. Adjacent areas can also be discolored. Iron(II) sulfate is frequently used as the iron source, and when the polyphenol has reacted with the iron, sulfuric acid is produced. Indeed, some iron-gall inks have sulfuric acid as an ingredient added to stabilize the liquid ink (Mitchell 1937). Gum arabic is often added to inks as a protective celloid for the iron/polyphenol complex. The ink sometimes becomes brown or black as the gum degrades (Sistach et al. 1999). Inks that fade to pale brown have a low polyphenol content (Sistach 1993). Analysis by gas chromatography/mass spectroscopy has shown that the gallic acid content of inks decreases on aging.

Iron-containing pigments, such as iron oxides, have no apparent effect on the deterioration of paper. A reasonable explanation for this is that if a metal is to have an effect on the stability of cellulose, it must have a means of gaining access to all parts of the fiber; water-insoluble pigments cannot do this. In contrast, deteriorated iron-gall ink might form ions that become soluble, particularly if an additional anion can be found, e.g. provided by an acid. Water solubility provides the mobility the ions need to migrate through the hydrophilic cellulose fiber, especially if the RH is high.

In practice, it is difficult to distinguish between the effects of the iron-gall dye and that of intentionally added acid in an iron-gall ink drawing. However, black dyed textiles provide the opportunity to study separately the effect of the neutral dye. In New Zealand, Maori weavers use cellulosic yarn derived from the leaves of the *Phormium tenax* plant; it is dyed black by iron/plant polyphenol complexes similar to those found in iron-gall ink. The dye obtains its iron from a black iron-rich mud that contains no or very little water-soluble iron, and the transfer of iron probably takes place by reduction of Fe(III) at the contact point between the polyphenol and the iron in anaerobic conditions (Daniels 1996). In any case the freshly dyed fiber contains little or no soluble iron or acid, particularly as it is washed well after dyeing, unlike ink drawings.

On aging, degradation of fiber and dye similar to that seen on iron-gall ink drawings takes place, demonstrating that the initial presence of free acid and iron ions is not necessary for degradation to take place. Release of hydrogen and iron ions into the fiber probably takes place after aging, when the cellulose substrate and the iron/polyphenol complex are degraded. Migration of acids in iron-gall ink has been demonstrated by mapping the migration of sulfur (from sulfuric acid) away from the ink line, although the iron in the ink did not move. Fe(II) ions were a major factor in the degradation (Neevel 1999). When chemically active iron is released from an iron/polyphenol complex, the remaining degraded polyphenol may still participate in accelerating the degradation by acting as a reducing agent for Fe(III). The fiber becomes brittle and acidic, while the black dye loses its color and becomes brown.

Wunderlich (1994) has studied iron-gall ink and finds that pH 5 is the lower limit at which iron-gall inks are stable; below that Fe(III) salts become mobile and function as a redox catalyst. Irreversible electron transfer, forming a gallate radical, can take place with subsequent oxidation, forming a complex brown polymer.

Neevel (1995) has demonstrated that a combined deacidification and complexation process can be used to stabilize iron-gall ink drawings. Magnesium bicarbonate is used for deacidification, while sodium phytate is used to complex the iron and thereby deactivate it.

Offset images from metals

Occasionally a manuscript may produce an offset image due to paper degradation initiated by metal ions from inks or metal foils. Even gold paint can be seen to do this when copper is alloyed in the gold. Oriental papers often contain intentional inclusions of gold, silver, and even iron metal alloys, and can cause offset spots on book pages.

Of all the metals that catalyze redox reactions, the most effective catalyst is platinum. This metal occurs rarely in museum collections, with the exception of one group of objects, platinotype photographic prints. The image is made on a paper support sensitized with potassium chloroplatinate and iron(III) oxalate. After exposure to light, the Fe(III) is reduced to Fe(II), which is developed with potassium oxalate to produce an image composed of platinum. A rinse in citric or hydrochloric acid removes residual excess iron salts. Platinotypes often produce offset images in albums, an effect that Reilly (1986) attributes to platinum catalyzation of cellulose degradation. The effect is also seen in silver prints with platinum toning.

Conclusions

The chemistry of the metal-catalyzed degradation of paper is complex. Iron and copper evidently degrade cellulose by different mechanisms, or at least use the available mechanisms to different extents. Iron and copper as trace elements ordinarily occur in paper and may not be as reactive as impurities added in experiments as readily soluble chloride, acetates, or sulfates. At present it is necessary to extrapolate the results obtained from model systems to real paper in ambient conditions. There is a need for further work in this area, particularly in the differences between degradation caused by iron and copper.

The mere presence of iron or copper in paper or a pigment is not sufficient for destabilization; access to the fiber is important, as is the activity of the metal (i. e. the ease with which it can participate in the deterioration processes). The chemical environment – e.g. pH, presence of other metal ions, and water concentration – greatly affects reactivity.

Examination of degradation phenoma caused by ink and pigments often reinforces the hypotheses produced by the study of paper containing small amounts of metals present as impurities.

References

Abe, M., and S. Thomson 2002. "The Conservation of Japanese *Fusuma*." This volume, Chapter 4, pp. 207–13.

Banik, G. 1989. "Discoloration of green copper pigments in manuscripts and works of graphic art,' *Restaurator* 19, pp. 61–73.

Banik, G., and J. Ponahlo 1982–83. "Some aspects of degradative paper caused by green copper-containing pigments," *The Paper Conservator* 7, pp. 3–7.

Barkeshli, M. 1999. "The presence of saffron in Persian miniature paintings and its use as an inhibitant for the destructive effect of verdigris," in J. Bridgland (ed.), *12th Triennial Meeting Lyon, ICOM Committee for Conservation.* London: James and James, pp. 489–94.

Bicchieri, M., and S. Pepa 1996. "The degradation of cellulose with ferric and cupric ions in a low-acid medium," *Restaurator* 17, pp. 165–83.

Daniels, V. 1987. "Alkaline buffering," in G. Petherbridge (ed.), *Conservation of Library and Archive Materials and the Graphic Arts.* London: Butterworths, pp. 109–15.

Daniels, V. 1996. "Maori black dyeing of Phormium tenax fibre," *Dyes in History and Archaeology* 14, pp. 9–13.

Daniels, V., and N. Meeks 1992. "An investigation into foxing phenomena with particular reference to the inorganic component," in K. Toishi, H. Arai, T. Kenjo, and K. Yamano (eds), *Biodeterioration of Cultural Property* 2. Tokyo: International Communications Specialist Inc, pp. 292–304.

Daniels, V. D., and N. Meeks 1994. "Foxing caused by copper alloy inclusions in paper," in H. Burgess (ed.), *Proceedings of Symposium 88.* Ottawa: Canadian Conservation Institute, pp. 168–69, 229–33.

Ericsson, B., B. O. Lindgren, and O. Theander 1971. "Factors influencing the carbohydrate degradation under oxygen-alkali bleaching," *Svensk Papperstidning* 54, pp. 757–65.

Grant, J. 1960. *A laboratory handbook of pulp and paper manufacture.* London: Edward Arnold, pp. 354–56.

Halliwell, B., and J. M. C. Gutteridge 1985. *Free radicals in biology and medicine.* Oxford: Clarendon Press.

Kühn, H. 1992. "Verdigris and copper resinate," in A. Roy (ed.), *Artists' Pigments: A Handbook of Their History and Characteristics,* Vol. 2. Oxford: Oxford University Press, pp. 131–58.

Lee, L. 1995. "Recent analysis of pigments from Ancient Egyptian artefacts," in C. E. Brown, F. Macalister, and M. Wright (eds), *Conservation of Ancient Egyptian Collections.* London: Archetype Publications, pp. 85–91.

Lewin, M. 1965. "The yellowing of cotton cellulose Parts II and III," *Textile Research Journal* 10, pp. 935–42, 979–86.

Liden, J., and L. O. Öhman 1997. "Redox stabilization of iron and manganese in the II oxidation state by magnesium precipitates and some amionic polymers," *Journal of Pulp and Paper Science* 23, pp. 193–99.

Mairinger, F., G. Banik, H. Stachelberger, A. Vendle, and J. Ponahlo 1980. "The destruction of paper by green copper pigments demonstrated by a sample of Chinese wallpaper," in N. S. Bromelle, G. Thomson, and P. Smith (eds), *Conservation Within Historic Buildings.* London: International Institute for Conservation of Historic and Artistic Works, pp. 180–85.

Mitchell, C. A. 1937. *Inks: Their Composition and Manufacture.* London: Charles Griffin, p. 179.

Neevel, J. G. 1995. "Phytates a potential conservation agent for the treatment of ink corrosion caused by iron-gall inks," *Restaurator* 16, pp. 143–60.

Neevel, J. G. 1999. "The behaviour of iron and sulphuric acid during iron-gall ink corrosion," in J. Bridgland (ed.), *12th Triennial Meeting Lyon, ICOM Committee for Conservation.* London: James and James, pp. 528–33.

Ranby, R., and J. F. Rabek 1975. *Photodegradation, photo-oxidation and photo-stabilization of polymers.* London: John Wiley.

Reilly, J. 1986. *The care and identification of 19th century photographic prints.* Rochester, N.Y.: Kodak.

Shahani, C. J., and F. H. Hengemihle 1986. "The influence of copper and iron, on the permanence of paper," in H. L. Needles and S. H. Zeronian (eds), *Historic textile and paper materials.* Washington, D.C.: American Chemical Society, pp. 387–410.

Shahani, C. J., and F. H. Hengemihle 1994. "Effect of some deacidification agents on copper-catalyzed degradation of paper," in H. Burgess (ed.), *Proceedings of Symposium* 88. Ottawa: Canadian Conservation Institute, pp. 263–68.

Sistach, M. C. 1993. "Organic and inorganic components of iron gall inks," in J. Bridgland (ed.), *10th Triennial Meeting. Washington, D.C., 22–27 August 1993 preprints.* Paris: ICOM Committee for Conservation, pp. 485–90.

Sistach, M. C., J. M. Gibert, and R. Areal 1999. "Ageing of laboratory iron-gall inks studied by reflectance spectroscopy," *Restaurator* 20, pp. 151–66.

Tang, L. 1978. "Determination of iron and copper in 18th and 19th century books by flameless atomic absorption spectroscopy," *Journal of the American Institute of Conservation* 17, pp. 19–32.

Williams, J. C., C. S. Fowler, C. S. Lyon, and T. C. Merrill 1977. "Metallic catalysts in the oxidative degradation of paper," in J. C. Williams (ed.), *Preservation of paper and textiles of historic and artistic value.* Washington, D.C.: American Chemical Society, pp. 37–61.

Wise, L. E. 1944. *Wood Chemistry.* New York: Reinhold, pp. 434–37.

Wunderlich, C. H. 1994. "Geschichte und chemie der eisengallustinite," *Restauro* 100, pp. 414–21.

Effects of Aqueous Treatment on Iron-Gall Ink – Monitoring Iron Migration with the Iron(II) Indicator Test

Elmer Eusman

There is little consensus in the international field of paper conservation concerning the treatment of objects suffering from iron-gall ink corrosion. This was well illustrated in a survey by R. van Gulik and N. Kersten-Pampiglione published in 1994.[1] Not only did the survey show that conservators use different methods to treat similar problems, but it also exposed the lack of research on which these decisions are based.

One of the main concerns when deciding on a treatment is the potential migration of iron in the form of Fe(II) ions originating from the ink into the surrounding paper during treatment. Recent research has clearly demonstrated the destructive role of Fe(II) in the iron-gall ink corrosion process.[2] Its capability to catalyze the oxidative breakdown of cellulose through the "Fenton Reaction" is possibly more important than the inherent high acidity of iron-gall ink. The correct proportions of iron sulfate and tannins to produce a stable iron-gall ink were unknown for centuries, and iron sulfate was often added in excess to the ink mixture. The excess portion of iron sulfate did not react with the tannins to form the ink complex, and this typically resulted in a high concentration of residual Fe(II) in the applied ink. The portion of iron sulfate that did react with the tannins was oxidized by atmospheric oxygen to Fe(III) to form the colored iron-gall ink complex. The production of this pyragallol complex and its chemical structure have been described by Krekel.[3] Fe(III) is not water-soluble and, more importantly, does not have the same destructive properties as Fe(II). Since Fe(II) is water-soluble, the fear for the migration of Fe(II) seems legitimate. However, it is unclear to what degree this process takes place using different treatment techniques. It is also uncertain if, and to what degree, the iron will contribute to the accelerated degradation in the areas into which it migrated. In other words: what is the significance of iron migration?

History

Research comparing different treatment techniques was conducted in 1992 by Schreiner and Wächter.[4] Using X-ray fluorescence (XRF) in combination with energy dispersive X-ray spectroscopy (EDX), they determined the amount of iron in the exact same area before and after various treatment procedures. The amount of iron (Fe(II) and Fe(III)) in the paper was measured both in areas within the ink line and directly outside. The emphasis of this research was to evaluate the results of treating iron-gall ink on paper using electrolysis versus washing in hot (boiling) water. The results using the latter type of treatment showed spectacular drops in the iron count within the ink line, while iron in the surrounding paper did not change significantly. Other research has shown that some inks can withstand this type of treatment very well.[5] However, a follow-up project by Heller et al. warns of the risks of bleeding of the ink line, especially when the ink's acidity is high.[6] A study by Reissland found the paper underwent severe changes after washing in hot water.[7] Heller et al. also found a decrease of iron in the ink line on an object subjected to electrolysis treatment, but to a much lesser degree. Even less iron was removed from the ink line with a conventional immersion treatment in various aqueous solutions.

The research project in the Museum Boijmans Van Beuningen in Rotterdam is concerned with the immediate and long-term effects of treating objects containing iron-gall ink using aqueous solutions. Treated samples were examined using a scanning electron microscope with an energy dispersive spectrometer (SEM-EDX) and the newly developed Fe(II) indicator test. They were also aged artificially.[8] The results follow.

The iron(II) indicator test

The non-bleeding Fe(II) indicator test was developed by Dr. Han Neevel at the Netherlands Institute for Cultural Heritage (ICN) in Amsterdam.[9] It employs a sheet of filter paper impregnated with bathophenanthroline which forms a pink-colored complex with Fe(II) ions. To perform the test, a small piece of the impregnated filter paper is cut off from the larger sheet and wetted with deionized water. Excess water is removed by holding the strip against a dry blotter for a few seconds. The now-damp indicator paper is brought into direct contact with ink on a paper or parchment substrate. The Fe(II) ions present in the object are soluble in water and will move from the object into the test strip, react with the indicator, and create a bright pink-colored complex. The intensity of the pink color depends on several factors: the amount of Fe(II) ions in the ink, the amount of ink on the paper or parchment, the surface sizing, and the amount of time the indicator paper is brought into contact with the sample (see Fig. 2.49). Since neither the compound nor its Fe(II) complex are water-soluble, there is no risk of the colored complex bleeding to the object during contact.

This test was originally developed to evaluate the effectiveness of treating iron-gall ink with a phytate solution.[10] It is hoped that the test can now be used as part of a so-called early warning system. In combination with an ink corrosion rating system, it could be used to determine the condition of individual objects and collections affected by iron-gall ink corrosion.[11] A high concentration of free iron in an ink could give an indication of the long-term stability of the iron-gall ink.

The indicator and its pink Fe(II) complex are fluorescent under ultraviolet (UV) light. This characteristic makes it easy to determine whether any indicator products are left behind on the paper substrate after testing. During the course of this project, no fluorescence corresponding to the test strip was found on any of the test samples. Using the recommended contact time of 30 seconds, a small weight or the index finger, and a piece of Mylar® to keep the indicator paper in place, usually was enough to determine the presence of Fe(II). Generally, because of the short duration of the test, water-soluble ink or paper degradation products did not move. However, great care should be taken to avoid this from happening, especially when the indicator paper is left on the object for a longer period of time.

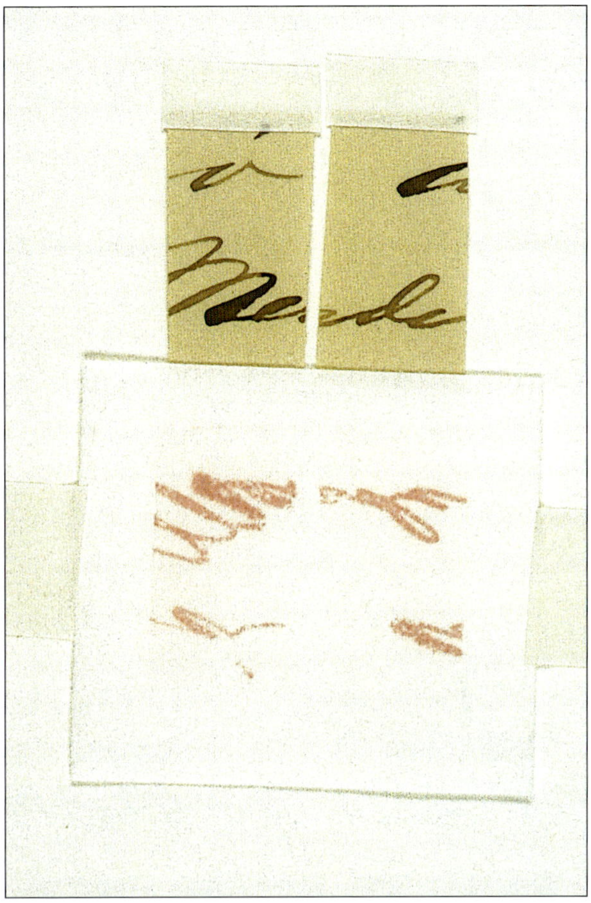

Figure 2.49 Example of the Fe(II) indicator test.

One should also keep in mind that, depending on the type of ink examined, particles of ink can become attached to the indicator strip. This depends mainly on the time period between the production of the ink and its use. When a historic iron-gall ink was used soon after it was produced, the ink would penetrate the paper deeply, and as a result, today the loss of particles during testing is negligible. However, ink that was stored for a longer period of time would contain many more solid ink particles. After application, the ink would largely remain on top of the surface of the paper. It is with this type of ink that particles can become detached during examination and testing. The amount of particle loss is usually small, and does not leave noticeable changes in the texture and color of the ink line.[12]

It is important to note that the indicator paper detects only the presence of Fe(II) and not Fe(III). The test results, therefore, give no indication of the total amount of iron in the ink line. It is also important to note that the absence of Fe(II) says little about the stability of the ink or about the total amount of iron present in the ink. As discussed below, washing a sample of iron-gall ink on paper will remove most excess iron from the sample. However, long-term effects are hard to predict because new Fe(II) can be formed over time. The indicator strips are currently being tested as a tool for the identification of brown inks.[13]

Experimental methods

The first set of treatments tested in this project included conventional immersion and humidification treatments. It was hoped that results of the Fe(II) test would give an indication of which treatments could be considered safe to use, and which treatments could potentially cause migration of Fe(II).

Finding original sample material for destructive testing is generally very difficult, and I was fortunate to obtain five paper samples to use for this project. These were documents from the end of the nineteenth century that had all been executed in pen and ink. All samples contained iron-gall ink that had partially transferred to the verso, but none showed any tears or losses in the inked areas. One paper sample was a "copy-paper," originally used to make copies of newly made documents by putting the sheet in direct contact with the original. The remaining four samples were all machine-made papers. Eight strips were cut from these papers to be treated individually. The strips were cut into smaller parts (see Fig. 2.50), treated, and examined using different techniques. The results obtained using the Fe(II) test are presented here.

The samples were subjected to seven different kinds of treatment:

1. Untreated (control sample)
2. Washed, 30 minutes in deionized water
3. Washed, 30 minutes in 50% deionized water and 50% ethanol
4. Washed, 30 minutes in 33% deionized water and 66% ethanol
5. Washed, on the suction table using 66% deionized water and 33% ethanol
6. Humidified, using Gore-Tex® and wet blotting papers, 60 minutes
7. Humidified, using a water spray
8. Humidified, using a cold humidification chamber, 120 minutes

The pieces were treated individually and later tested using the indicator test strips. The indicator test was used to determine the presence of iron not only in the ink, but also in the washing solution and in the blotting paper underneath the samples that were treated on the suction table.

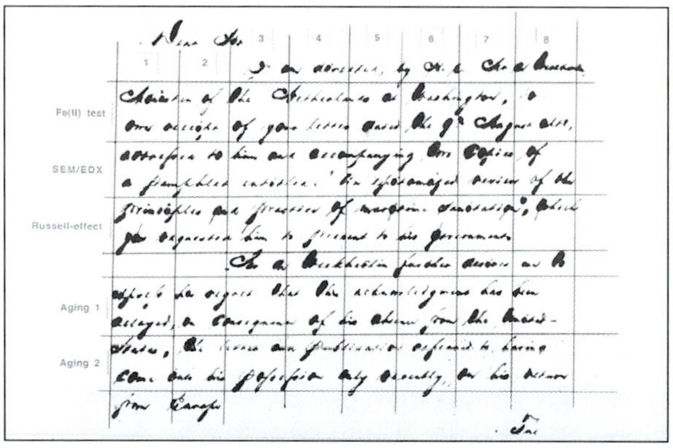

Figure 2.50 Schematic picture of cut up sample paper.

Results

Treatment by immersion

The results presented in Figure 2.51 were obtained by leaving indicator papers in the various solutions in which the samples were washed. The columns represent the five paper types, and the three rows correspond to the washing solution used. The top row shows the most intense pink color, indicating that washing of an object containing iron-gall ink in 100% deionized water resulted in the most migration of Fe(II) into the solution. Increasing the ethanol ratio in the washing solution decreased this amount. This is shown, in turn, by the results of using a solution of 66% ethanol and 33% deionized water (represented in the bottom row), which show a much lighter pink color. But even using a smaller percentage of water in an aqueous solution resulted in diffusion of Fe(II) from the ink into the washing solution.

These observations correspond with the test results obtained from examination of the ink surface. Figure 2.52 shows the indicator test results from an untreated sample and three washed samples. The untreated sample contains substantial amounts of Fe(II), shown by the bright pink color in the indicator strip. However, the test strips used on the washed samples show a substantially lighter color. In some samples washed in 100% deionized water, Fe(II) was not detected at all. When more ethanol was used in the washing solution, more iron remained in the samples. This outcome was representative of all the paper types tested. The diffusion of Fe(II) did not produce any fading or change in color.

The iron test did not show any reaction with regions outside of the ink lines, either in treated or untreated samples. Time constraints did not allow for determination of the sensitivity of the indicator test, and it remains unclear whether there were any iron deposits in the paper after washing. Another complicating factor is that if iron was deposited onto the remaining paper, it could have been in the form of Fe(III). Higher pH values

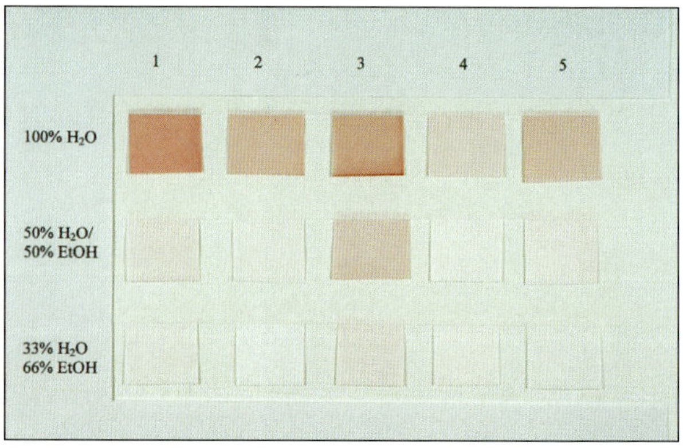

Figure 2.51 Overview of indicator papers used for the washing solutions.

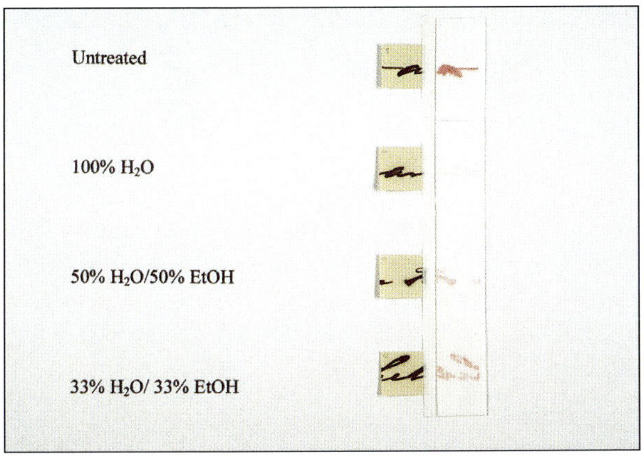

Figure 2.52 Sample and indicator paper after washing in various solutions.

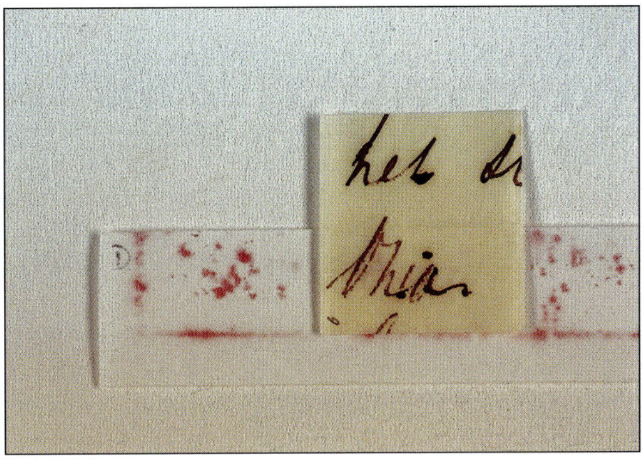

Figure 2.53 Sample and indicator paper after treatment on the suction table.

outside the inked areas promote the oxidation of Fe(II) ions into Fe(III) ions.

Treatment using a suction table

A Lascaux-Restauro suction table powered with a compressor manufactured by Gebr. Becker was used during this treatment. The pressure was kept at 100 mbar and the samples were treated three times with five spray applications of a 66% deionized water and 33% ethanol solution. Most papers were well sized, and ethanol needed to be added as a surfactant. The indicator paper was placed underneath the test sample to monitor any Fe(II) movement through the paper.

All samples showed partial transfer of the free iron through the paper during treatment. Examination of the treated samples at the ink surface showed that not all excess Fe(II) had been removed, but treating the sample on the suction table for a longer period of time might well have resulted in complete removal of excess Fe(II). Some samples showed partial overflow of the washing solution, transporting free iron ions over the surface of the paper and depositing them at the edges of the sample. Figure 2.53 shows the indicator paper for a paper sample that had been treated on the suction table. The washing solution only partially penetrated the samples, shown by the spotted pattern of the pink indicator complex in the test strip placed beneath the sample. This underscores the need to test the paper's behavior on the suction table before actual treatment. To ensure the transport of Fe(II) ions through the paper and not over its surface, both the washing solution and the amount of pressure introduced by the suction table need to be properly adjusted.

Humidification treatments

During this project it became clear that the samples chosen were fairly stable types of iron-gall ink. Except in the case of the copy-paper sample, no dilution of the ink, very little halo formation, and no change in color was observed after any of the humidification treatments. However, with both the Gore-Tex® and the spray-treated samples, a decrease of free iron content in the applied ink occurred. In the case of the Gore-Tex®, this might be explained by direct contact with the Teflon layer, which in some cases resulted in ink loss and therefore loss of Fe(II). This explanation did not apply to the spray-treated samples. Whether the small decrease of Fe(II) in these samples was due to migration or oxidation into Fe(III) is unclear. Here too the iron test did not seem to be sensitive enough to indicate small amounts of iron deposited outside the ink line.

Judging by the bright pink color of the indicator strip taken from the test sample, use of a cold humidification chamber for two hours is the safest procedure to use with this set of samples. No significant difference in indicator color between the humidified and the untreated sample was found. Yet it cannot be stated that this observation is significant for the treatment of iron-gall ink objects in general. Another sample of a completely different ink was examined at a later date using SEM-EDX and it did show a considerable amount of iron migration using cold humidification.

Using Gore-Tex® humidification on the copy-paper samples produced dramatic results. Fe(II) presence was not detected outside of the ink lines before humidification. This is an interesting observation, since originally the copy paper was humidified before it was pressed against an original document to make the copy. These humid conditions apparently did not result in the presence of Fe(II) over the entire surface of the paper. However, after humidification using Gore-Tex® (and to a lesser degree the spray), Fe(II) migration was significant. This was illustrated by large areas of pink on the test strip that corresponded to areas beyond those of the applied ink.

Additional research

Two sets of samples were aged artificially using two different methods. The first set was aged for 12 days at 90°C and a constant RH of 50%. The second set was also aged for 12 days at 90°C, but with RH cycling between 35% and 80%. The most important observation from these aged samples was the positive effect of washing the samples in 100% deionized water. Both the ink and paper of all five washed test samples were less degraded than their untreated counterparts. The fear that Fe(II), present in the washing solution, could be deposited over the surface of the paper was not confirmed by the results of age testing.

Subsequent examination of unaged samples using SEM-EDX reinforced the notion that iron does not necessarily get deposited on the paper during washing. Both treated and untreated samples were analyzed, and except in the case of the untreated copy paper, no significant iron deposit was found in any of the samples. The humidified samples did show migration of a compound that, while not visible directly after treatment, did create a halo after aging. Where the untreated and washed samples showed no or very little brown discoloration around the ink, most humifidied samples did show a brown halo directly outside the applied ink. Since none of this discoloration was visible before aging, it can be speculated that the cause for the discoloration was Fe(II) (see Fig. 2.54), which is colorless and can accelerate oxidative degradation of the cellulose. The results from the SEM-EDX measurements were presented at the First Triennial Northumbria Conservation Conference in Newcastle and published in the postprints.[14]

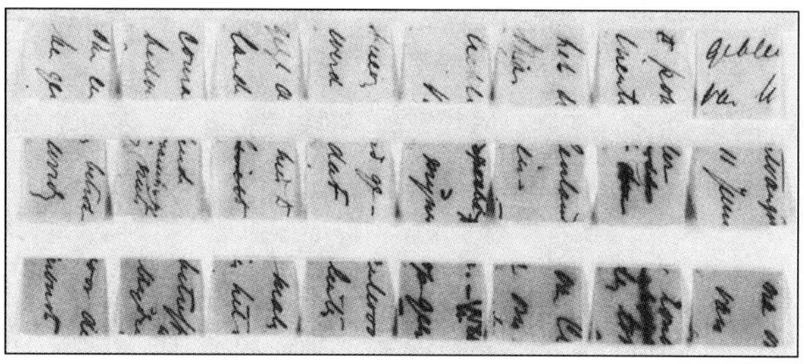

Figure 2.54 Comparison of the treated unaged samples with the treated artificially aged samples. The columns represent the various treatments as described. The first row of samples was not artificially aged. The second row of samples was aged for 12 days at 90°C and a constant RH of 50%. The third row of samples was aged for 12 days at 90°C and a cycling RH of 35 to 80%.

Conclusions

The number of samples tested does not provide us with enough data to make conclusive remarks about the behavior of iron-gall ink during treatment in general. The type of ink and support, as well as the conditions in which the objects were stored, are far too diverse to make any predictions on how all objects will behave during treatment. However, there are a number of observations that will likely apply to most objects containing iron-gall ink. First, for certain purposes the Fe(II) test has proven to be a great tool for the conservator. It can be used to monitor the presence of free iron on the paper and also in the washing solution. However, it cannot be used to make predictions about the long-term stability of the ink. Second, testing of the washing solution for the presence of Fe(II) clearly showed that large portions of Fe(II) migrate from the ink into the washing solution. This was confirmed by using the Fe(II) test on the paper surface. A decrease of Fe(II) in the ink line was detected with those samples that had been washed in an aqueous solution. Third, the indicator strip can also be used successfully in combination with a suction table or disc to monitor the transport of free iron through the paper and also over the paper surface. The latter is confirmed by the iron deposits detected at the outer edges of the object.

Acknowledgments

I would like to thank Han Neevel and Birgit Reissland of the Netherlands Institute for Cultural Heritage, Amsterdam (ICN), for their help with the Fe(II) test and the artificial aging of the test samples; Kees Mensch and Wim Genuit of the Shell Research and Technology Centre in Amsterdam (STRCA) for the SEM-EDX measurements and John Havermans of the Netherlands Organization for Applied Scientific Research, Delft (TNO), for the artificial aging of the test samples. Thanks to Ted Steemers and Gerrit de Bruin, the Netherlands National Archives, The Hague (ARA), for providing me with the always hard-to-find sample material, and Louis Damen, Museum Boijmans Van Beuningen, Rotterdam, for making this research project possible in the first place.

Endnotes

1. R. Van Gulik and N. Kersten-Pampiglione, "A closer look at iron gall ink burn," *Restaurator* (1994) 15, pp. 173–78.

2. J. G. Neevel, "Phytate: A potential conservation agent for the treament of ink corrosion caused by iron gall inks," *Restaurator* (1995) 16, pp. 143–60.

3. Chr. Krekel, "Chemische Struktur historischer Eisengallustinten, in: Tintenfraßschäden und ihre Behandlung," in G. Banik (ed.), *Tintenfraßschäden und ihre Behandlung*, Landesarchiv Baden-Württemberg, Werkhefte, Serie A, Heft 10 (Stuttgart: Kohlhammer, 1999), pp. 25–36.

4. M. Schreiner and O. Wächter, "Ein erweiteres System der Röntgenfluoreszenzanalyse zur zerstörungfreien Analyse von Künstwerken und seine Anwendung bei Tintenfrassgeschädigten Papierobjecten," *Restauratorenblätter zum Thema Papier und Graphik* 14, (1994), pp. 43–55.

5. J. Biggs, "Controversial treatment options applied to the conservation of a sketchbook in iron gall ink by George Romney," in J. Egan (ed.), *The Institute for Paper Conservation Conference Papers* (London: Institute of Paper Conservation, 1997), pp. 175–81.

6. F. Heller, F. Mairinger, M. Schreiner, and O. Wächter, "Die Migration der Eisenionen bei konservatorischen Nassverfahren," *Restauro* (1993) 99, pp. 115–21.

7. B. Reissland, "Ink corrosion: Comparison of currently used aqueous treatments for paper objects," in M. S. Koch (ed.), *Preprints of the 9th International Congress of IADA – Copenhagen* (Copenhagen: The Royal Danish Academy of Fine Art School of Conservation, 1999), pp. 121–29. It should be noted that in her experiments B. Reissland did not take similar precautionary measures when treating her samples as was done with the treatment of originals conducted by J. Biggs.

8. Results obtained from SEM-EDX and the artificial aging of the samples will be published at a later date.

9. Development of the test and procedures are still under way. All questions regarding the exact nature of the indicator should be addressed to: Dr. Han Neevel, Netherlands Institute for Cultural Heritage, P.O. Box 76709, NL-1070 KA Amsterdam; e-mail: han.neevel@icn.nl

10. Phytate is capable of complexing Fe(II), neutralizing its ability to accelerate oxydative breakdown of cellulose. The complexed Fe(II) will no longer react with the indicator.

11. B. Reissland, "Neue Restaurierungsmethoden fur Tintenfrass auf Papier mit wässrigen Phytatlösungen – Möchlichkeiten und Grenzen" (Diplomarbeit, Staatliche Akademie der Bildenden Künste, Stuttgart: Studiengang Restaurierung und Konservierung von Graphik, Archiv- und Bibliotheeksgut, 1997).

12. Sometimes transported brown discoloration onto the indicator strip will obscure a clear result of Fe(II) presence. By wetting the strip with water and removing excess water with blotting paper, the discoloration will dissolve but the colored complex will remain in the test strip.

13. I hope to present my findings identifying iron-gall ink using this indicator test in the near future. The difficulty with this test is that the presence of Fe(II) does not automatically mean that the examined ink is an iron-gall ink. A bistre or sepia ink, stored in an iron container might very well show Fe(II) when tested. However, absence of Fe(II) does not mean that the ink is not iron-gall ink. A stable iron-gall ink, one that does not contain excess Fe(II), will not test positive using this indicator strip.

14. E. Eusman and K. Mensch, "Ink on the run – measuring migration of iron in iron gall ink," in A. J. E. Brown (ed.), *Postprints of the First Triennial Northumbria Conservation Conference, University of Northumbria* (Newcastle, 2001), pp. 115–23.

All the Colors of White: The Changing Nature of White Papers

Peter Bower

Millions of different nominally "white" papers have been produced in Western Europe since the thirteenth century, and their rich variety has contributed much to our visual heritage. For most of papermaking history, none of these white papers were designed for watercolor or drawing: artists worked on whatever suited their particular needs. Many of the same factors affecting the final tone of colored papers also apply to white papers and, as with colored papers, determining the original color of some white papers can be problematic. Some of the factors involved, in particular the water used in the making, are more important in determining sheet color in white papers than in colored.

The preparations given to white papers at the papermill, by artists' colormen or stationers, or more interestingly by the artists themselves, are crucial to our understanding of how a drawing or painting actually works. Such interventions include the coloring and glazing of surfaces and the resizing or toning of sheets. Most artists seem to choose their papers by a mixture of conscious thought and past experience, understanding sometimes instinctively and sometimes through deliberate experience which paper will work in what situation. The reasoning behind an artist's decision to prepare a particular paper is very different from that of an artist's colorman when preparing a batch of paper, but regardless of that, the issues for us years later are whether we appreciate the nature of those choices, and, quite simply, what color was the paper originally? The colorman makes a conscious and commercial decision to market this or that paper in whatever color because he thinks the market will bear it; the artist prepares his paper simply because what he has in front of him is not what he wants for the work in hand.

The following factors influence sheet color: the choice of raw materials and their preparation; the effect of fermentation on soap and lye residues remaining in the rags; methods and degree of beating and formation; drying (i.e. airflow, gravity, and the use of heat); bleaching, whether by sun or by chemicals; sizing; loadings; and the effects of residues remaining in the sheet. Changes that can occur during conservation treatment can also alter this effect.

The wonderful variety of tones and surfaces in the range of nominally white papers used by artists was the result of the small scale and local nature of papermaking by hand throughout most of papermaking history. Most mills were small, often operating only one, or possibly two, vats and producing perhaps 500 to 1,000 sheets a day, depending on the size and weight of the sheet. Local and regional variations in papermaking practice – particularly regarding the collection, washing, and preparation of the rags, as well as variations in the cleanliness and quality of the river or spring water used by the mill – made an extraordinary range of subtly different papers available.

The production of papers specifically for drawing began some time in the 1770s but was not to become a major part of the hand-made paper industry until the early years of the nineteenth century. Until this time "drawing" papers, which included papers for watercolor as well as those for pencil, pastels, chalks, and inks, were quite simply those papers that artists found they could work on, regardless of the use for which papermills had made them. After a slow start, the English paper industry began a rapid period of expansion from the early 1700s onward. Working from the Excise returns for the period 1714–18, it has been calculated that England and Wales were producing 300,000 of the 418,000 reams consumed.[1] A contemporary report states that

> Before the Revolution[2] there was hardly any other paper made in England other than brown: the war economy, and duties being laid down from time to time on Foreign paper, it gave much encouragement to the paper-makers, that most of them began to make white paper fit for writing and printing, and they have brought it by degrees to so great perfection, both for quantity and goodness, that they make now two-third of what is consumed in Great Britain; and several of them make it as white and as well-being as any comes from abroad.[3]

This eulogy on both the quality and quantity of white paper being made at this period is perhaps a little exaggerated. Much of the recorded increase in production actually came from brown and blue paper production rather than from any massive increase in the production of white paper. However, by the middle of the century, the trade was well enough established to be included in published advice for young men wanting to enter various professions. One author described how "several of the principal Stationers of London have mills with which they make their own paper," and went on, echoing the remarks of 50 years earlier, "This art which has been lately so greatly improved, as almost entirely to exclude the importations of Foreign paper, requires both ingenuity and strength."[4] Yet examination of works on white paper produced by artists working in England during the middle years of the eighteenth century shows them just as likely to be working on imported Dutch, French, or Italian papers as the products of the new English mills.

Cost must have often played a significant part in the choice and purchase of particular papers for many artists. Much of the paper used by artists throughout history is not of the first quality, despite the often illustrious makers' names seen in the watermarks. Quality control varied greatly from mill to mill and even within a particular mill over time. Many of the sheets would have been sold as *retree* or *outsides*. These are sheets with something substandard about them; they were often used as the top and bottom quires of a packed ream in order to protect the good sheets that made up the bulk of the ream, and were generally sold at half price. But, despite their imperfections, close examination shows that they are generally sound, particularly as regards sizing. What deficiencies they have were obviously never serious enough to inhibit artists from working on them.

Papermakers explored new raw materials and production methods; developed new qualities and characteristics; and produced new surfaces, sizings, strengths, tones, and colors. Artists seized on these developments, not always using the new papers in the ways the makers expected, but grateful for the opportunities they provided. By the 1790s, the London colormen and stationers were handling a greater variety of sizes, weights, and textures of increasingly high-quality paper that was sometimes made specifically for drawing. Paper was usually described quite simply: for writing, printing, or wrapping. It was either thick or thin, large or small, sometimes stout, sometimes coarse. Colors were white, brown, blue, or drab – terms that covered a small range of browns, buffs, gray, blue-gray, and olive-green. Quality might be described as superfine, fine, *retree*, or bastard.[5] Until well into the nineteenth century, the industry was still somewhat haphazard in its description of many of its products.

Before 1800 English artists, whether working on English-made or Continental papers, generally used fine laid writing papers. These papers had the necessary surface strengths and light-reflecting capabilities, unlike printing and plate papers, which were too soft-sized. Many wrappings were also just not strong enough, although artists' colormen did stock the stronger grades of buff, blue, and gray wrapping papers for artists. Artists have always chosen what suits their particular needs; if they cannot find what they want, then they simply alter the sheet. As one might expect, examples of prepared grounds are found in the work of many artists. While it was possible to buy toned and colored grounds, such as bistre papers, from the colormen,[6] the highly individual nature of many of these prepared grounds suggests that artists often prepared the papers themselves. It was just this kind of individual usage that gradually filtered back to the papermakers, via the colormen and merchants, in some cases eventually leading to changes in papermaking practice.[7]

Although cost played a part in the choice of a paper, the real criteria for their choices probably derived more from a blend of availability (particularly for those artists traveling on the Continent or to out-of-the-way places), and differences in individual working methods. Some artists preferred hard-sized smoother papers; others preferred slightly coarser and softer-surfaced sheets. The media being used and the speed and pressure of the artist's hand all contribute to the success, or otherwise, of a particular sheet for an individual artist.

A major contributing factor to the finished color of papers was the source of the rags. An international industry – the import and export of rags – grew up alongside the European paper industry. In the sixteenth and seventeenth centuries, rags were classified quite simply, but by the end of the nineteenth century, a complex system of classification had arisen among merchants and papermills. The increasing use by British

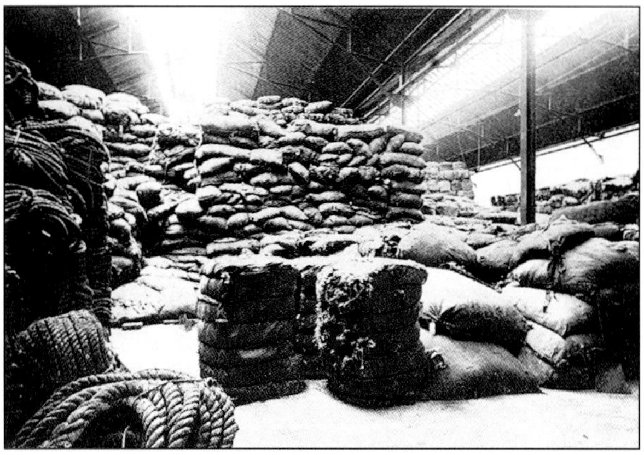

Figure 2.55 The rag store at Caldercruix Mill, Scotland, just after the First World War, showing the bales of rags and coils of rope waiting to be sorted.

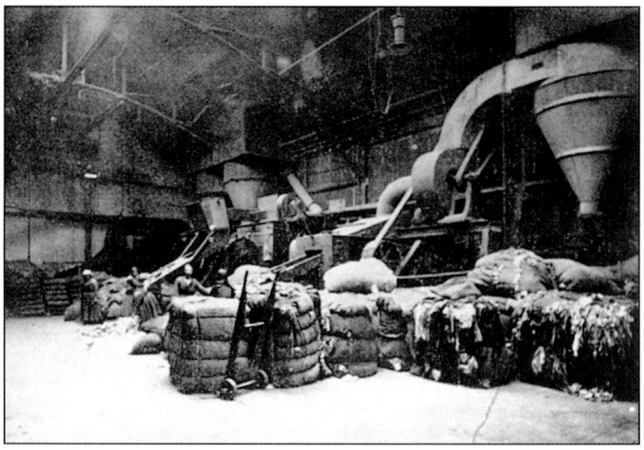

Figure 2.56 The rag-dusting house at Caldercruix Mill, Scotland, just after the First World War, showing the bales ready for their first cleaning in rag willows and dusters.

mills of imported rags provided a much wider choice: all these rags were of varying qualities, differences in methods of preparing the fiber for weaving, methods of washing, even the different climatic and soil conditions in which the original plants were grown, all contributed something subtle to the finished paper (see Figs. 2.55 and 2.56).

The profitability of the rag industry is attested by John Houghton, who recorded that rags are "gathered by the rag-folk; and after a great many of separations of these raggs [*sic*] they are sold to several sorts of tradesmen, and several there be that have got estates out of them."[8] Houghton, a fellow of the Royal Society, also published figures for the import into England of hemp and flax and other materials for the year 1694–95, recording: Holland, 34 casks; Danzig, 79; Russia, 1,123; Sweden, 70,538; and Germany, 35 tons.[9] The figures for the years 1725–1800 show a massive increase in imported rag throughout the century. Even when allowing for weight loss in sorting, dusting, washing, and later bleaching, much of this rag must have gone for other uses, since during the same period paper output in England and Wales went from 3,500 tons to just over 10,000 tons.[10] By the 1830s the sources of rags were: Germany, 4,436 tons per year; Italy, 2,306 tons; Russia, 1,104 tons; other countries, 667 tons.[11]

The literature regarding the color of paper, particularly those aspects other than the use of dyes and pigments that affect the final color of the sheet, is very sparse. But the classification and preparation of rags constantly exercised the minds of papermakers. One of the earliest published works on the use of rags is the Dutch scholar Gerard Meerman's pioneering study of the origins of rag-based paper, *De chartae vulgaris seu linea origine*,[12] which includes the correspondence between Meerman and other scholars who had sent him the results of their research together with specimens of paper from all over Europe, following his offer of a 25-ducat prize to the person who could establish the earliest making of rag papers. The Pennsylvania State Archives at Harrisburg houses the journals of the American papermaker Joshua Gilpin,[13] recording his travels and observations on papermaking and papermakers, as well as many other industries, during two trips to Europe (1795–1801 and 1811–14). Among the many fascinating insights into late eighteenth-century papermaking practice gathered by Gilpin is a letter from Anthony Cesari de Poggi to James Whatman the younger, discussing the various merits of English and continental printing papers and their production. The letter is not in Gilpin's hand and was probably copied into his journal for him, either when Gilpin met Whatman at Maidstone in March 1796 or perhaps sometime in the summer of 1797, prior to his visit to France. Poggi's original letter must date from some years earlier and could well have been given to Gilpin for its discussion of the differences in English and French manufacture. By 1794 people already believed that the Whatman printing papers were superior to the French:

> I came home with the Alderman [the printer Boydell] who told me Whatman now makes printing papers equal in quality to French paper, and has an advantage from being manufactured more neatly. He does not think we should again apply to France for the article of Paper.[14]

Poggi is an elusive character, variously described as a painter, engraver, publisher of prints, art dealer, and collector of Old Master drawings and engravings.[15] He seems to have been a man of many parts: he filed a patent for "Constructing and using Ordnance, both for Sea and Land" in 1800.[16] By 1808 he had left England under a cloud, having spent his wife's fortune.[17] Perhaps he was a scoundrel, but his letter – while concentrating on his prime interest, the perfecting of the manufacture of printmaking papers – raises some interesting and little-considered points regarding the "whiteness" of white paper, and the possible causes for the vast range of different tones seen in white papers across Europe. It is worth quoting at some length:

All linen which becomes rags in the Northern provinces of France, in England, Holland, Germany, having been washed by dint of soap alone does retain an inflexible quality and substance which renders it unfit for our purpose. The greatest part of the French, Spanish and Italian linen is consumed and brought to rags by a process totally different from the English, Dutch and German. They take in those countries a quantity of wood ashes in proportion to the linen they have to wash, which ashes they boil in water for some hours until it becomes a strong lees. Then they pour it hot over the linen which are first put in a tub and covered with a double coarse cloth that serves as a strainer to retain the ashes from the linen. Thus it remains for eight or ten hours, when this first lees is drained off, and other added in the same manner as before. By this process the linen is almost clean before any soap is used, which I should add, is only for the finest linen.

The whiteness of paper preoccupied many by the end of the eighteenth century. Hector Campbell, who had been heavily involved in the development of chlorine bleaching for the paper industry,[18] took advantage of the brief peace that followed the Treaty of Amiens to venture into France and compare the relative states of the French and English papermaking industries. Campbell considered it possible that "coarse cotton rags could be brought to an equality with fine linen rags" by the careful use of chlorine bleaching, but that it depended upon the care with which the rags were sorted and prepared. He made a very interesting point about the effect of chlorine on different sorts of rags, something that has an important visual effect on the finished paper:

... bleaching gas is much better adapted to coarse cotton rags than coarse linen or hempen rags; because the former is without ligneous particles, and the latter abounds with them, and these particles, called by the papermakers *sheaves*, are made more conspicuous by bleaching.[19]

Different rag-boiling liquors have varying effects on all characteristics of the finished sheet, some of which are visual, some of which continue to affect the sheet long after it is made. As with the fermentation of rags, the length of time the rags are processed has an important role. How long the rags are allowed to lie after boiling and before washing has a decided influence on the ultimate character of the half stuff and on the finished sheet.

Rag-boiling liquors were based on alkali chemistry: caustic soda, sodium carbonate, or caustic lime (calcium hydrate) were used separately and in various combinations depending on the preferences of particular mills. Each of them affects the actual color of the fiber, even if they are all nominally white. Caustic soda, often favored by mills with hard water, attacks natural waxes, oils, and tar, and will affect heavily colored rags. High concentrations in conjunction with high pressures and temperatures in the rag boiler will "burn" the cellulose and seriously degrade the ultimate fiber strength and alter the color of the paper. Sodium carbonate, less active than caustic soda, was more often used with the higher grades of rag for papers where good fiber strength was really important, such as for fine writing, printing, and drawing papers.

Barry Watson published information on the use of various rag liquors at Croxley Mill (Hertfordshire, U.K.) when he worked there, making comparisons with information he obtained from other mills. A very wide range of responses by different mills to the same raw material is evident.[20] Some mills only operated their rag boilers at a constant pressure, while other mills varied the boiler pressure depending on the fiber being processed. The importance of washing out bleach cannot be overstressed, but many mills, particularly in the early years of the use of bleach, were careless in this respect: chlorine residues in particular led to degradation of the cellulose. Sometimes an antichlor, such as sodium hyposulfite, was used to neutralize excess bleach. Color change could also occur after excess bleach was removed. The stuff was dumped into drainers to drain away the water and any remaining water-soluble impurities. In earlier years the rag was often left in the drainer for some weeks; the resulting rag was mellow and easily beaten, but somewhat lacking in strength, because the small amount of liquor left in the rag continued to bleach it. This long drainage time often also led to bacterial build-up and consequent slight color shifts in the rag. By the middle of the twentieth century, rags were retained in the drainer for no more than a week or so before going to be beaten, lessening such bacterial growth and preserving fiber strength.

By the end of the nineteenth century, the classification of rags had become very complex: hundreds of different types were now sold by the rag merchants. Consistency of supply in terms of both quantity and quality has always been problematic. Availability and price were the basic considerations that always guided papermakers when they bought their rags, whether in the sixteenth or the twentieth century. In addition to these two essential factors, other aspects were crucial when choosing rags, whatever the fiber they were made from, as they always played a part in the look, feel, handle, and behavior of the finished sheet. They included the length of the staple or fiber; coarseness or fineness of quality; and whether the rags were made from bleached, half-bleached, or unbleached fiber. In addition to any bleaching of the fiber when it was cloth, rag merchants also developed bleaching to a fine art during the nineteenth century. They could realize better prices for bleached rag – the general color of the consignment was more likely to appeal to the papermaker, and there would be less work for the mill in the early stages of the papermaking process. Still other aspects include whether the rags were clean, dirty, or very dirty; whether they were strong or tender; whether they were hard or soft; if colored, whether the dyes were light or dark; if printed, were they printed on one side or both and could this color be removed by bleaching? In some mills, colored rags would be beaten separately, without bleaching, and either made up into paper of that color or blended with other stock to produce a desired tone and color. The amount of preliminary work needed on the rags, i.e. splitting of seams, removal of buttons and clips, etc., varied; and the average size of rags in the consignment also made a difference, as large ones are quicker (and thus cheaper) to sort than small. The bulk of the bale, as opposed to its weight, was particularly important when considering how much extraneous matter the bale might contain. The rag willow and duster played an important part in the finished color of the paper as well. Unless all the detritus and non-cellulose particles are removed, they contribute both to the actual color in the sheet (whether white or colored), and, more importantly, they also contribute to, or indeed are the cause of, some of the changes that the sheet will undergo during its life (see Figs. 2.57 to 2.59).

For most of papermaking history, hemp has been an important raw material. Although difficult to work in its raw state, as waste cloth, netting, rope, and cordage, its versatility and strength has proved popular with papermakers all over the world. Fiber analysis of papers

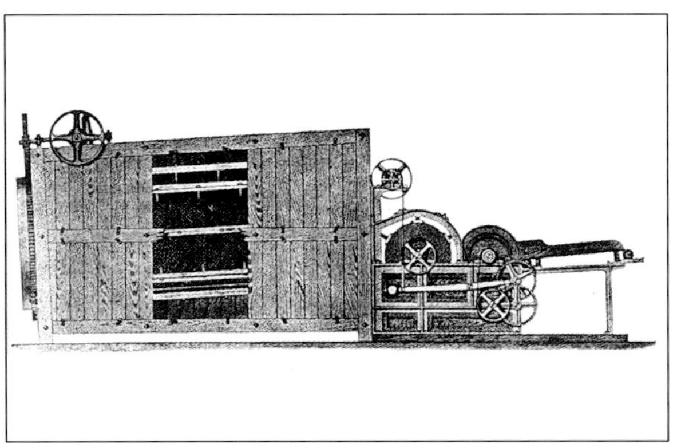

Figure 2.57 Rag willow and duster with part of the cover removed to show the wire-covered drum and spikes within the machine.

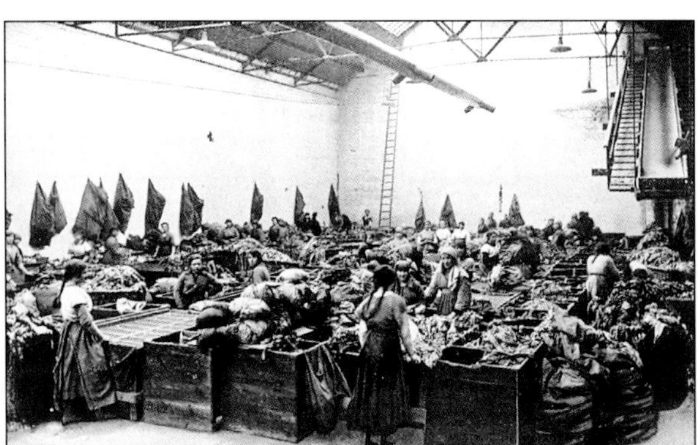

Figure 2.58 The rag-sorting house at Caldercruix Mill, Scotland, just after the First World War, gives a clear indication of the scale of the operation.

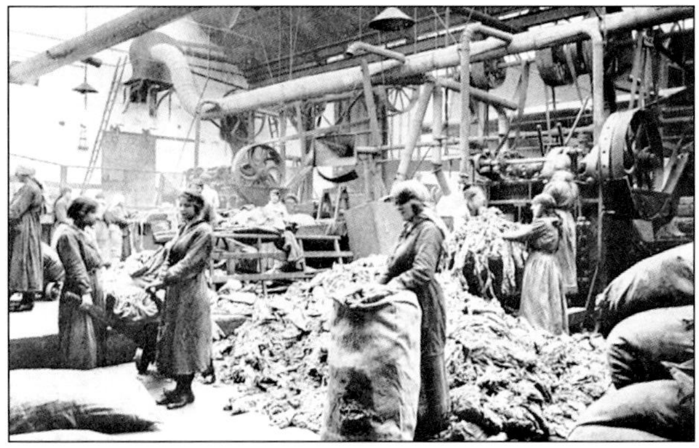

Figure 2.59 The rag-cutting room at Caldercruix Mill, Scotland, just after the First World War. In earlier centuries rags were cut by hand but by the end of the nineteenth century the process had been mechanized in many mills.

made in Britain and most of Western Europe during the sixteenth, seventeenth, and eighteenth centuries shows considerable amounts of hemp in use for papermaking, not just in the "rope browns" and wrappings one might expect, but also in smaller amounts in many fine white papers as well. Most of this fiber came to papermakers in the form of old rope or worn sailcloth. The rope would have needed considerable cleaning before use and was generally destined for brown papermaking, but much of the sailcloth, although often very dirty, would have been well bleached by sun and sea by the time it reached the papermaker.

Since the 1760s various makers, such as Schäffer in Germany[21] and Koops in London,[22] experimented with new materials to solve the problem that increasingly faced papermills all over Europe: because of demand, consistent supplies of linen rags, and the more recently introduced cotton ones, were increasingly difficult to obtain.[23] Several makers experimented with the use of other fibers, including straw, either on their own or mixed with rag, as the basis for a range of papers. The papers they produced were often used by artists, for both drawing and watercolor, as well as by printers for engravings (and later lithographs and etchings) and books. These white papers often display a very marked "color."

Recently the only surviving copy (as far as is known) of an extraordinary book by Ludwig Piette de Rivage, papermaker of Dillingen in the Saarland, Germany, has come to light. Published in 1838 and entitled *Die Fabrikation des Papiers aus Stroh und vielen andern Substanzen in Grossen*, it documents the research into the use of straw and other alternatives to rag in detail.[24] An earlier work had appeared in 1833: *Handbuch der Papierfabrikation* was published in German, translated from Piette's original French text.[25] The preface to the *Handbuch* describes Piette as a Frenchman who had begun his working life as a lawyer but was later forced to abandon his career to become head of the papermill at Dillingen. Trying to learn about the papermaking process, he had found very little suitable published material and consequently had decided to write such a manual himself. Piette went from Luxembourg to Dillingen in 1819 and operated the mill until 1854, when he moved to Belgium. The mill itself ceased to exist in 1860, although some ruins remain.[26]

Although the title page of the *Handbuch* records that Piette made "160 samples of various paper qualities," this particular copy actually contains 185 samples, made from a wide range of materials, either alone or in combination. These include straw from rye (Piette's preferred choice), wheat, barley, oats, and maize, as well as 14 other plants. He went into considerable detail about different types of straw, the manner of pre-treatment, methods of bleaching, and making paper from different mixtures of straw and rags. The samples themselves are remarkably well preserved and show the various cereal straws to advantage.

In his introduction, Piette described his research as "tiresome and time-consuming but finally successful trials," and stated that

> The progress of civilization forces us to do everything to avoid obstacles due to the imminent shortage of papermaking raw materials and, at a moderate price, to put into the hands of the poor also this indispensible material for all cultural developments in art and sciences.[27]

Straw was never the major alternative to rags: it was always difficult to obtain consistent quantities, as farmers often did not want to part with it, and the processing involved considerable pollution problems resulting from the high silica content of the stems. Only in times of real shortage, such as war, when the normal international trade in pulp was regularly interrupted, has straw been used in relatively large amounts. The various wood pulps developed over the past 150 years (and, in Britain at least, esparto – in use for just over a century) proved much more able to satisfy the increasing demand for paper.

Gelatin can play a very significant part in determining the color of the newly made sheet of paper, and its effects are particularly obvious in white paper. John Balston describes how gelatin's various animal sources, as well as its processing, using white vitriol (zinc sulfate), or worse, copperas (ferrous sulfate) to bleach the fiber, can produce very different tones in the sheet.[28] We have all seen papers from various centuries that have discolored in extraordinary ways, for instance books with some signatures where every sheet is brown while other signatures with sheets bearing the same watermark and made on the same molds are as white as when they were made.

In the eighteenth century, Du Monceau advised the use of glues from the hides of mature animals for strength – "the older and leaner the animal the stronger the glue" – but for painting he preferred younger animals because the size produced was both clearer and weaker.[29] This seems to have been echoed by James Basire in his specifications to James Whatman the younger in 1772 regarding the manufacture of antiquarian paper. He specified that

> it be not sized with parchment, nor any allum used for whitening it ... That the sizing be made of Kid-Leather; and the outward surface of the Paper to approach as near as may be, to that of the French.[30]

Whatman's preferred source of gelatin for most of his papers was *scrowls*, parchment, vellum, or very good quality leather scrapings and strips. He bought some 13 tons a year at a price of around £300 per annum, as well as 8 tons a year of the much cheaper *pieces*, from rabbits, calves, bullocks, and oxen, costing around £90 per annum.[31]

Another factor that is rarely considered when discussing gelatin sizing is the effect of temperature and humidity on the size. As early as the seventeenth century, Houghton described how papermakers

> in spring or fall take a fine dry temperate day, on which they put into a copper about two barrels of water, and when it is just warm, they put into it sixty pounds weight of clean parchment or vellum shavings, and it is boiled till it comes to a perfect size ... the direct rays of the sun must not come nigh until it is dry, for otherwise the size would be drawn out.[32]

One of the curiosities of some linen rag- and hemp-based papers is the effect of overexposure to light on some of them. Both linen and hemp are actually bleached by light rather than discolored, and yet we regularly find works of art on paper made from linen rag or linen and hemp blends where the paper has not lightened but darkened from too much exposure to light.

Many of the white and off-white papers made in earlier centuries do not simply rely on the color of the rags from which they were made. Rag pulps were often slightly tinted, usually during beating, either to make the sheet appear whiter or to warm the tone. The addition of inappropriate dyestuffs at this stage could have a long-term effect on the color of the sheet. I have mentioned the deleterious use of vitriol in sizing, but two other substances can be equally destructive: the fugitive indigo added in place of smalt as a brightening agent, or the highly acidic tea used to give a sheet a much creamier warm tone. Both these processes are very visually effective when first done, but their inherent instability regularly defeated the papermaker's best intentions.

A paper that often appears to suffer from such discoloration is the series of papers known as Creswick. It is often assumed that there were only one or two basic Creswick tones, and that the wide range of tones actually found are discolored versions of those two basic papers. In fact the range produced was enormous, colors and tones being continuously refined in response to the market by Thomas Creswick, the papermaker and pasteboard maker who operated Hatfield Mill in Hertfordshire and a manufactory at Islington in North London between 1803 and 1839. His papers and boards, which ranged in color from a warm white to a deep buff and in finish from a very high glaze to a very rugged, rough surface, were so renowned that after his retirement there was so much demand for this type of paper that William Balston, the producer of Whatman paper, began producing Whatman's Imitation Creswick. By the end of the nineteenth century, they had dropped the word "imitation" and the paper was just known as Creswick.

However good the products of the various suppliers, many artists preferred to prepare their paper to suit their particular ways of working by resizing sheets, tinting them with colored washes, stiffening them by laying them down onto other sheets, changing the surface by abrasion, rubbing with a rough cloth, polishing them to make them smoother, etc. Although colormen started to sell such prepared papers themselves to suit all sorts of tastes, many people developed their own methods of preparing papers for different media, sometimes evolving quite extraordinary systems of preparing their working surfaces. Samuel Palmer (1805–1881) wrote to Richard Redgrave describing some of his curious methods using:

> six-sheet "London Board" of best quality from Newman's, Soho Square ... I take off the gloss by rapidly sponging with water (a *little* alum dissolved in it might perhaps do no harm) rapidly lest the paste should soften underneath ... At first I felt like a baby on a slide, but found that this might be avoided by using colors less wet than on a simple paper ... a very thin wash of zinc white over the sky and distance part, before beginning, might facilitate.[33]

> These difficulties have their equivalent advantage in the ease with which a bit of sponge or even a wet brush will remove dried color. In this respect the surface is intermediate between paper and ivory. I have done my largest things upon it; getting it made on purpose rather than use paper. I fasten it around the edge with the smallest flat-headed brass pins (not passing them though it, which if you avoid working on parts already wet, prevents warping), upon a thin, light, drawing-board of seasoned pine ... a sheet of white paper is fastened on the front of the drawing-board, so that even if you have not much margin of the "London Board" you have a sort of white mount.[34]

According to Palmer, a London Board prepared in this manner

> reflects light through the pigments. It shows the full depth of the colors laid on, and as you retouch with more color the retouchings are still darker, as one would suppose they naturally must be. In the modern papers, you come to a full stop at a point far short of the depth of the pigment – in fact it is an ocular paradox.

Palmer had very distinct views on the kinds of papers he wanted depending upon what sort of work he was contemplating, and he hated some of the developments in artists' papers that the merchants and the papermakers were insisting on:

> Supposing there should be a lucid interval in this vile grey and blue paper system ... it is evident to me that something has been done either to blue or bleach the paper. If we asked the papermaker to make us a creamy tone we should rue the day. All I say is, after you have washed them, make up the rags, pure and simple or impure and simple, into paper, just as they are, and let them alone: no bleaching acid, please, no blue bag, no toning.[35]

Acknowledgments

I would like to thank Camilla Baskomb, Sally Bower, Robin Hamlyn, Richard Hills, Annie Lyles, Greg Rubinstein, Julien Stock, Rod Tidnam, John van Oosterom, Barry Watson, and Tony Woolrich for their advice, assistance, and practical help with different aspects of the research behind this paper.

Endnotes

1. D. C. Coleman, *The British Paper Industry* (Oxford, 1958), pp. 54–55.

2. The "Glorious Revolution" of 1688 that saw the accession to the throne of William and Mary.

3. *The British Merchant* (1713), quoted by Rhys Jenkins, "Paper-making in England 1495–1788," *Collected Papers* (Cambridge, 1936), p. 184.

4. Joseph Collier, *The Parent's and Guardian's Directory and the Youth's Guide in the Choice of a Profession or Trade* (London, 1761).

5. "Bastard" was not always a description of the quality of the paper. It was occasionally used to describe an odd size or quality of paper, a *nondescript* paper difficult for the Excise men to categorize in terms of the duty payable.

6. Bistre papers are generally white, usually laid papers prepared with a brownish-gray wash made from soot and water. Different soots from different woods or coals produced different colored washes.

7. As part of the colormen's marketing strategies, Balston and Hollingsworth were both making ranges of Whatman Antiquarian and smaller drawing papers for individual colormen. These were only available from particular shops and were identifiable for the customer by watermark details or by blind embossed stamps: e.g. a copperplate "N" was added to all the Whatman drawing papers produced for Newman in Soho Square and the Griffin stamp was on the Antiquarian made for Winsor & Newton. But the main feature of this production was that this was not the same paper, packaged up for different colormen: these were actually very different papers. The choice of fiber and its preparation, the beating of the rags, sizing, and finishing details all differed, leading to papers with very different characteristics: changing the color, tone, surface texture, stability, and strength, producing distinctly different papers.

8. John Houghton, *A Collection for the Improvement of Husbandry and*

Trade (London, 1727), Vol. 2, p. 411, letter 356, 19 May 1699.

9. Ibid., pp. 378–79.

10. Coleman 1958 (as in note 1), fig. 2, opposite p. 91.

11. Ibid., p. 215.

12. Gerard Meerman et al., *De chartae vulgaris seu linea origine*, ed. by Jacob van Vaasen (The Hague, 1767). Among the contributors were T. Chr. Gottschel, C. A. Bell, D. Gerdesius, D. Longolinus, G. Maiansius, A. C. Ducarel, J. S. Herringen, J. D. Michaelis, and J. Murray.

13. Gilpin MSS, Pennsylvania State Archives, Harrisburg, Penn., LVII. The quotations from this document are from A. P. Woolrich's transcription, first published in *The Quarterly*, no. 3 (July 1990), pp. 3–7. Joshua and Thomas Gilpin established the first mill in the state of Delaware, on the Brandywine River, in 1787. The Gilpins were later instrumental in introducing the paper machine to the U.S. in 1817. This machine was of their own design but owed much to John Dickinson's Cylinder Mould Machine.

14. Kenneth Garlick and Angus Macintyre (eds), *The Diary of Joseph Farrington* (New Haven, Conn., and London, 1979), Vol. 1, p. 209, 3 July 1794.

15. Fritz Lugt, *Les Marques de Collections de Dessins et d'Estampes* (Amsterdam, 1921), p. 109.

16. Patent No. 2428, 24 July 1800.

17. Garlick and Macintyre 1979 (as in note 14), 21 June 1808.

18. Hector Campbell, Patent No. 1922, 28 Nov. 1792: "An Improved method of Destroying and Taking away all the Carbonic, Oleaginous and Colouring Elements and Particles in Linen, Cotton, Hemp, and in Coloured Rags and other materials used in Making Paper, in a manner hitherto unattempted in the Kingdom, and at much less Expence than any Method now in Use."

19. Hector Campbell, "Remarks of the present state of paper-making in England and France," *Nicholson's Journal of Natural Philosophy*, Vol. 2 (London, 1802), p. 8.

20. Barry Watson, "Rags as raw materials for paper-making, Part 2," *The Quarterly* (September 1993), p. 7.

21. Paper made from straw with small amounts of linen can be found in Vol. 2 of Jacob Christian Schäffer, *Vesuche und Muster Ohne alle Lumpen oder mit einem geringen Zusatze derselben Papier zu machen* (Regensburg, 1765–71).

22. Some copies of Matthias Koops, *Historical Account of the Substances Which Have Been Used to Describe Events and Convey Ideas from the Earliest Date to the Introduction of Paper* (London, 1800), were printed on straw paper.

23. For further discussion of straw as a papermaking raw material, see Richard L. Hills, "The use of straw in papermaking," Peter Bower, "Straw in 19th-century papermaking," and James Brander, "Straw," all in *The Oxford Papers*, SBPH Vol. 1 (London, 1996), pp. 9–13, 15–22, and 35–42, respectively.

24. Ludwig Piette, *Die Fabrikation des Papier aus Stroh und vielen andern Substanzen in Grossen* (Cologne: Verlag von Dümont-Schauberg, 1838).

25. Ludwig Piette, *Handbuch der Papierfabrikation,* transl. from the original French by Carl Freidrich Alexander Hartmann (Leipzig: Von Gottfried Basse. 1833).

26. Piette's biographical details are taken from two sources: G. Ullmann and P. Bean of the East Lancashire Paper Mill, Ratcliffe, Lancashire, "Early Efforts in Papermaking from Straw," unpublished typescript, 1951, and unpublished letter, 14 February 1951, to G. Ullmann from the librarian of the Institute of Paper Chemistry, Appleton, Wisconsin.

27. Ullmann and Bean 1951 (as in note 26), p. 8, transl. by G. Ullmann.

28. John Balston, *The Whatmans and wove (velin) paper* (West Farleigh, 1998), p. 211.

29. Quoted in Sarah Vallance, "The Development and Application of Chromatographic Techniques in the Characterisation of Artists' Media" (Ph.D. diss., University of Northumbria at Newcastle, 1997), p. 151.

30. Minutes of the Council of the Society of Antiquaries, 3 Apr. 1772.

31. Thomas Balston, *James Whatman, Father and Son* (London, 1957), p. 59.

32. Houghton 1727 (as in note 8), pp. 417–18.

33. A. H. Palmer, *The Life and Letters of Samuel Palmer* (London, 1892; repr. 1972), letter 47 to Richard Redgrave, 14 July 1865. James Newman, one of the oldest established artists' colormen in London, specialized in selling papers made up to his own specifications.

34. There were many attempts at making boards that replicated the ivory tablets used by painters of miniatures and several recipes survive. Bryan Donkin, in his MS *Recipe Book* (c. 1820), records one such preparation for an "Ivory" Board: "Make a paste with fine flour or starch. When it is sufficiently boiled, add to it common turpentine in such quantity as shall be equal to about a twelfth part of the starch or paste. Stir them well together and let them simmer over the fire for 5 or 6 minutes. Use the paste hot." Among the large works Palmer referred to are his watercolors based on Milton's *L'Allegro* (three subjects) and *Il Pensero* (five subjects), dating from 1868 to 1870. The normal way of exhibiting "Exhibition" watercolors during this period was to mount them in a gold mount, right up to or overlapping the edge of the painted area. It is interesting that Palmer seems to be recommending a white mount: quite simply it allows the colors to stand on their own.

35. Palmer 1892 (as in note 33), letter 86 to Richard Seeley, 6 Nov. 1872.

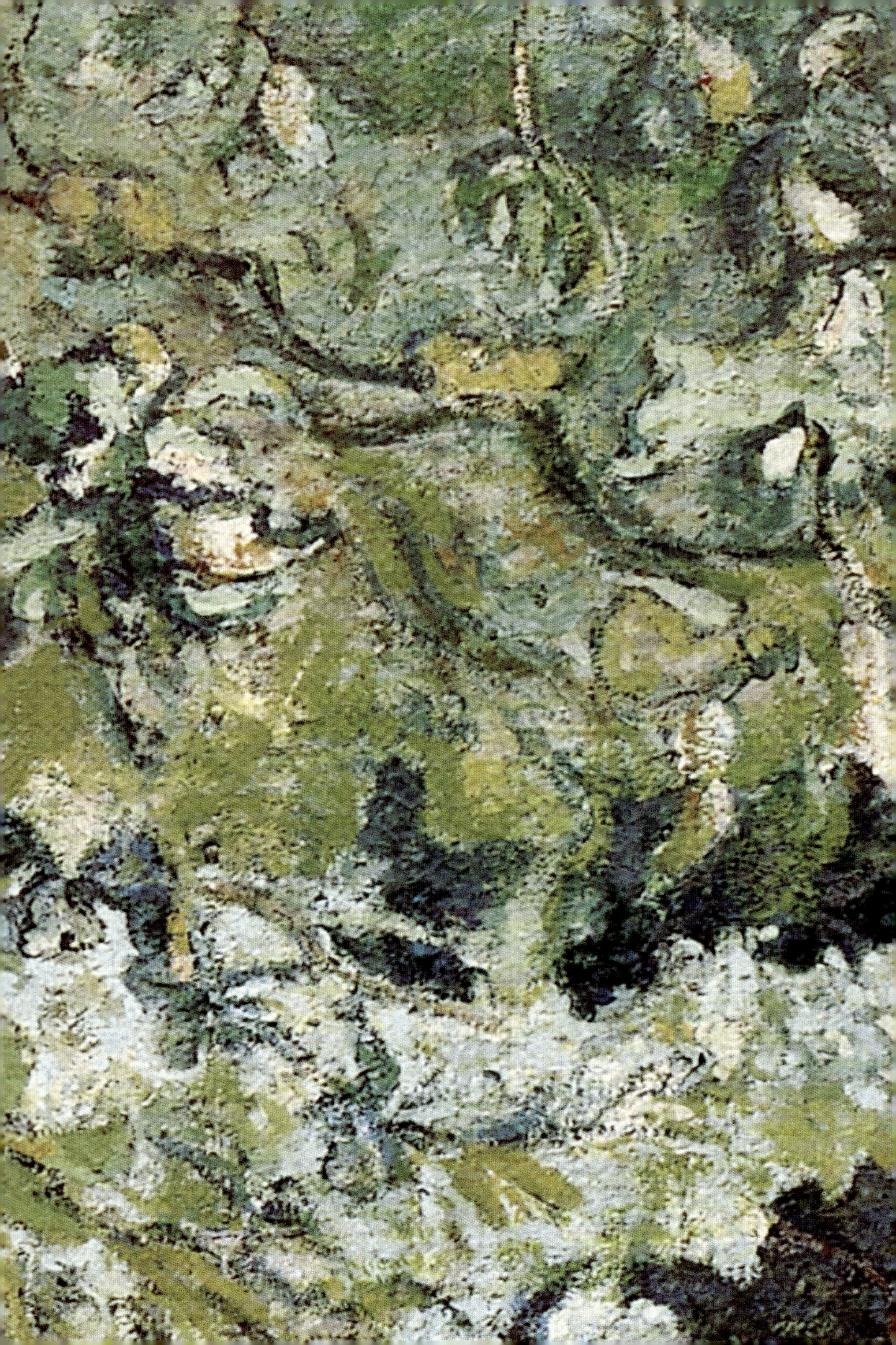

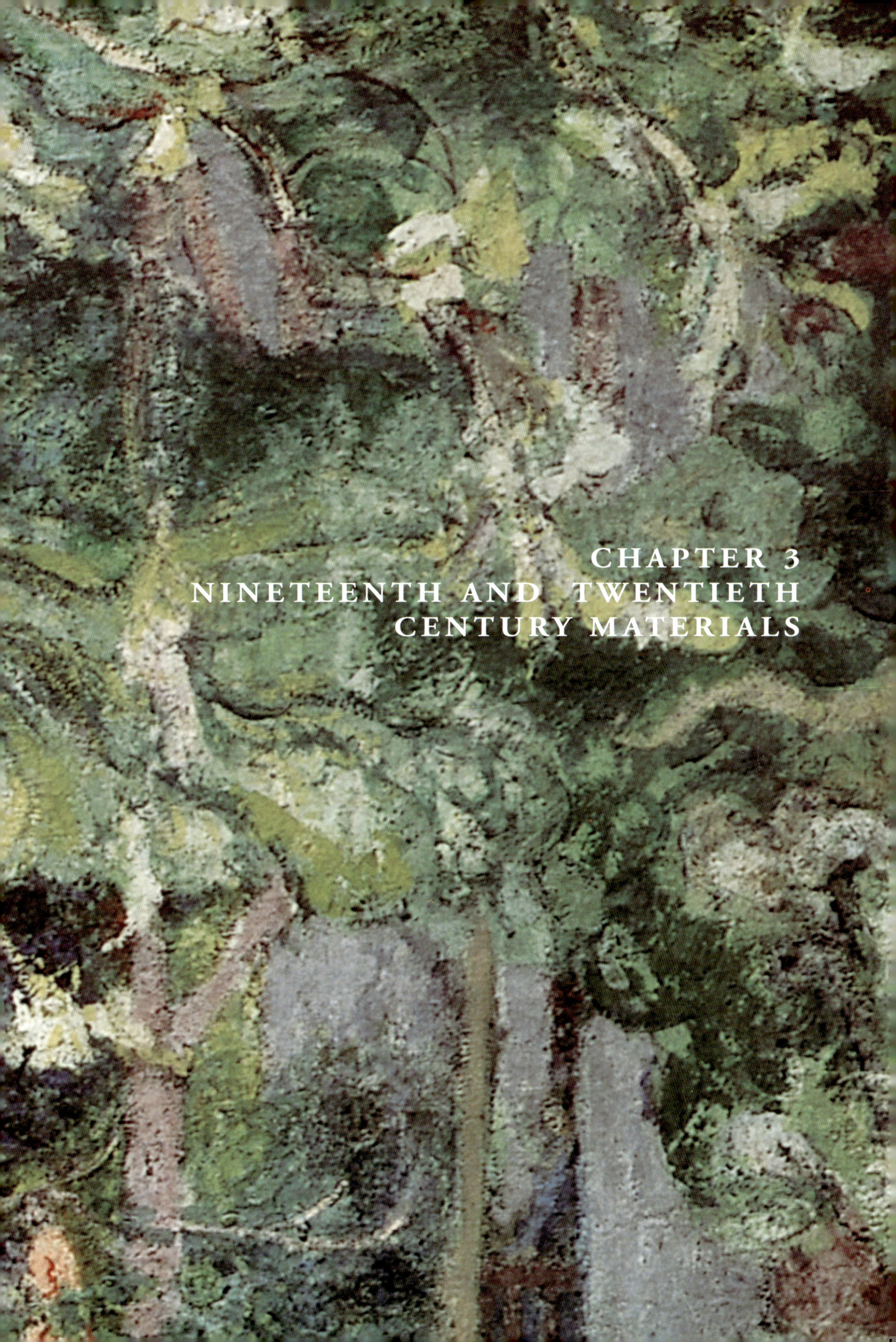

CHAPTER 3
NINETEENTH AND TWENTIETH CENTURY MATERIALS

In the Kitchen with Paul Gauguin: Devising Recipes for a Symbolist Graphic Aesthetic

Peter Kort Zegers

In 1893 Paul Gauguin (1848–1903) returned to France following a two-year stay in Tahiti. He brought with him the fruits of his time there – intensely colored, boldly decorative representations of an exotic culture. Gauguin's Tahitian vision was only partially shaped by his experiences in the South Pacific; it was, in fact, an amalgam of exotic fragments from different cultures, times, and places. These fragments included photographs of an Algerian woman by a contemporary Frenchman, Hippolyte Arnoux, and of a frieze at the Buddhist temple complex at Borobudur in central Java.[1] Gauguin had hoped and expected the body of work he had created in Tahiti to bring him fame and fortune back home. He was bitterly disappointed. An exhibition held at Durand-Ruel's in Paris during the fall of 1893 was not a success.[2] As a result, Gauguin began thinking of ways to publicize his work to a wider audience.

To this end, and as a means to widen his potential market, the artist began producing works on paper that derived from his paintings. These works exhibit a wide range of styles. Some have precise contours, intense colors, and the overall clarity of the paintings from which they derive.[3] Others reverse the paintings' main figure groups and have softer colors and contours, appearing, to

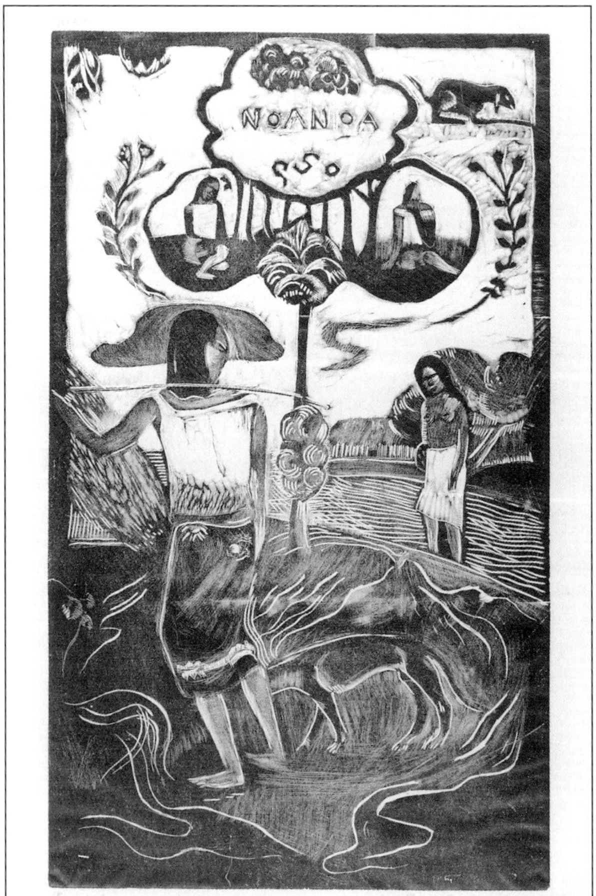

Figure 3.1 Paul Gauguin (French, 1848–1903), *Noa Noa* (*Perfume*), 1921. Woodcut printed posthumously and published by the artist's son Pola Gauguin. The Art Institute of Chicago, Gift of the Print and Drawing Club, 1924.1197.

the untrained eye, to be faded and curiously drained of the intensity and legibility one associates with the oils.[4] Such differences are of course related to the techniques Gauguin employed. Many of his gouaches are the result of a direct application of the medium. But those gouaches that are in reverse orientation from the painted source are clearly the result of a "printing" technique. Traditionally designated as "monotype," the true nature of Gauguin's working method in these works has remained until recently a "mystery" – this was the word used to describe them in 1894, when the artist exhibited them at his studio. But the reviewer gave insight into the mystery when he referred to Gauguin's "recent experiments" and his "procédé d'impression a l'eau" – a water-based printing process.[5]

Gauguin carried out his experimentation in a climate where printmakers were constantly seeking new means to achieve different effects. Their recipes, commonly referred to as *cuisine* (cookery, in English), were closely guarded secrets. We set ourselves the challenge of understanding why and how the artist pursued such different graphic ends. Not only did we hope to arrive at a better understanding of the artist's aims in his graphic work; questions of attribution were at issue as well. Less visually consistent than the paintings, Gauguin's graphic oeuvre is, in its diversity, perplexing – so much so that some (scholars and dealers alike) have been willing to purge the graphic oeuvre of inconsistency by removing disruptive works from the canon.[6] Unraveling the secrets of Gauguin's procedures – the *cuisine* that resulted in his somewhat baffling graphic work – involved trying to understand his aims and means from the inside out. This entailed rolling up our sleeves and, as it were, getting in the kitchen with Gauguin in the hope of re-creating his recipes. This was a process that started before the 1988 retrospective exhibition of the artist's work and continued for several years thereafter.

We began by trying to make sense of the work visually, starting with the woodcuts. A similar contrast to that between gouache and monotype is evident between two different impressions of the title page for the *Noa Noa* series of woodcuts that Gauguin made in 1894. One impression (Fig. 3.1) exhibits a crispness and legibility lacking in the other (Fig. 3.2). Gauguin conceived the woodcuts to accompany his literary account of his Tahitian experience, which he intended to both advertise and "explain" the work he had done there. The title page is loosely based on one of his paintings.[7] Given Gauguin's

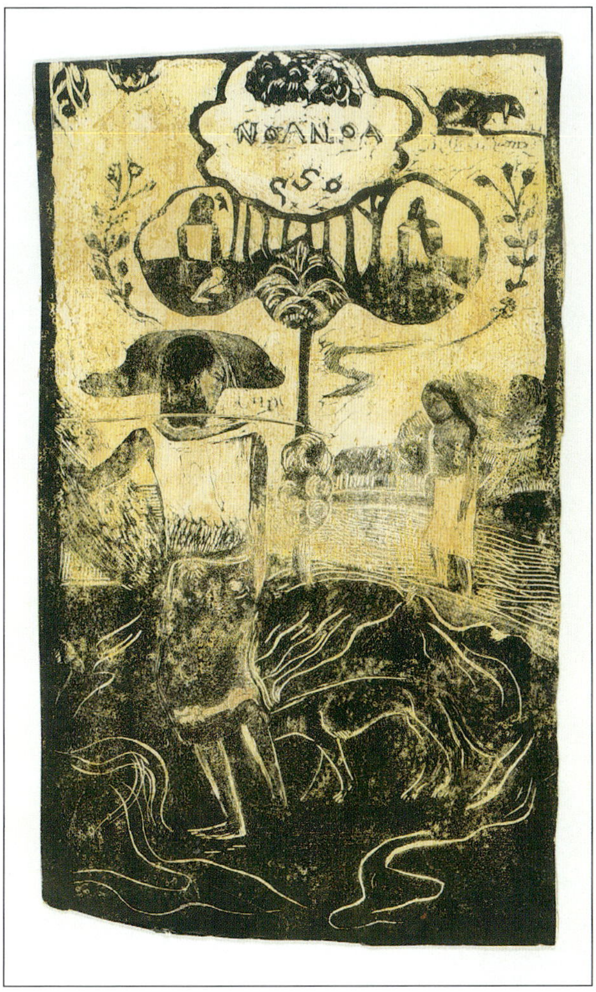

Figure 3.2 Paul Gauguin, *Noa Noa* (*Perfume*), 1894. Woodcut printed by Gauguin. The Art Institute of Chicago, Clarence Buckingham Collection, 1948.255.

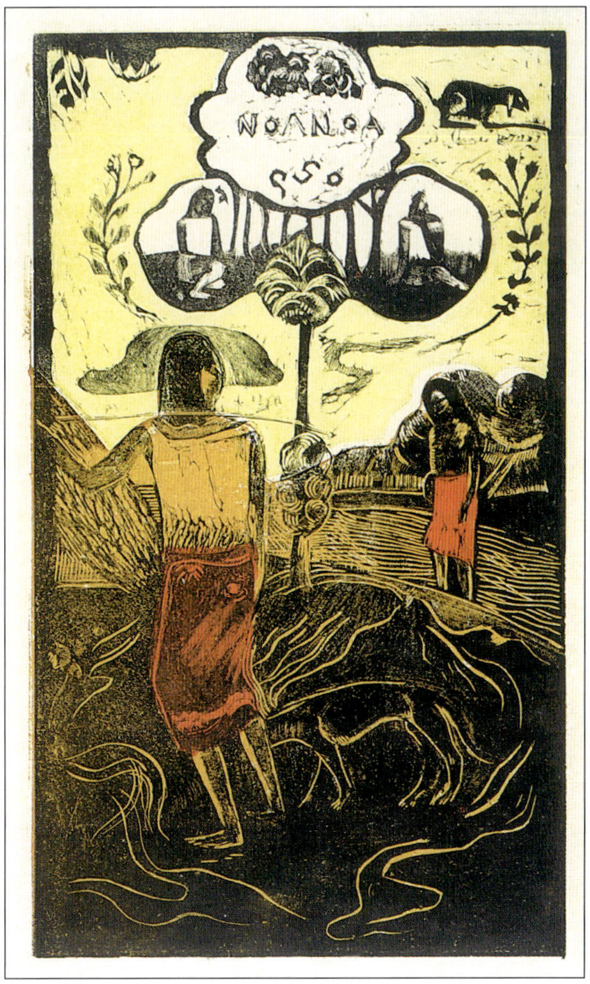

Figure 3.3 Louis Roy, *Noa Noa* (*Perfume*), 1894. Woodcut printed by Louis Roy. National Gallery of Art, Washington, D.C., Rosenwald Collection.

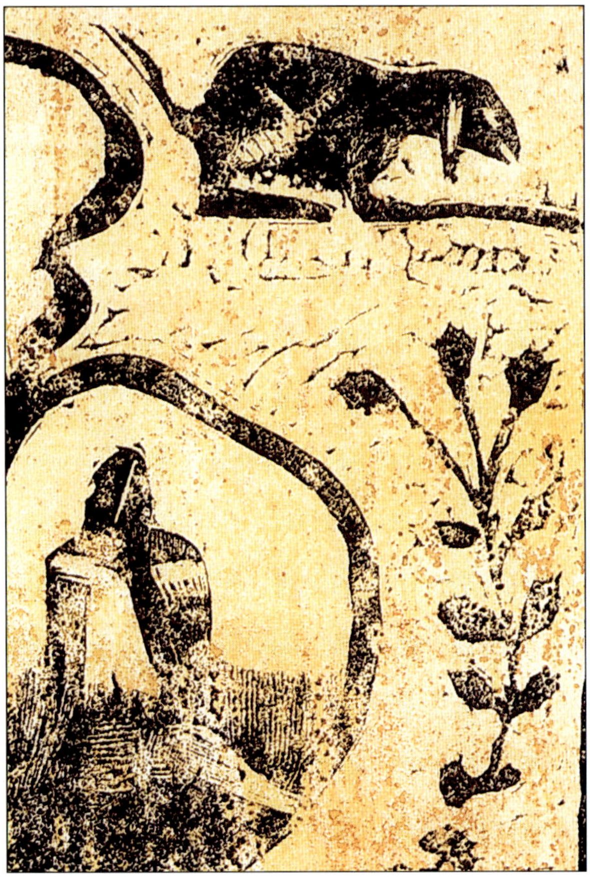

Figure 3.4 This detail of the Gauguin impression of *Noa Noa* (Fig. 3.2) illustrates that the black ink was printed over passages of colored media applied by hand.

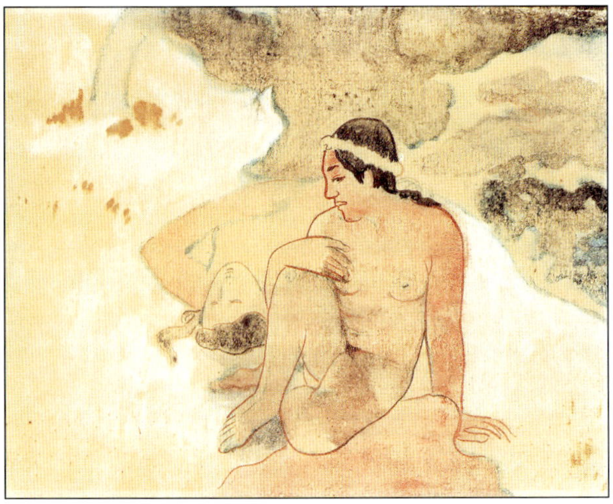

Figure 3.5 Paul Gauguin, *Aha Oe Feii?* (*What! Are You Jealous?*), 1894. Monotype selectively heightened with brush and water-based colors, brown ink, and white chalk on Japanese paper. Private collection.

aims, and following traditional guidelines of connoisseurship, one could imagine that the more legible impression is both earlier than and preferable to the other. It conveys more of the information contained in the printing matrix and transmits it more clearly. However, the traditional guidelines for assessing Old Master prints do not work in Gauguin's case. Quite the opposite – the more legible impression was pulled posthumously from the woodblock by Gauguin's son Pola in 1921, while the other is a rare example of an impression pulled by Gauguin himself in 1894. To complicate matters further, Gauguin asked his friend, the printmaker Louis Roy, to print an edition of the *Noa Noa* woodcuts that same year. Roy's impressions reflect the roughness – the "sloppiness" – seen in Gauguin's own impressions (see Fig. 3.3). For a professional printmaker to produce work like this clearly indicates that a kind of crudeness, a suggested "primitiveness," was Gauguin's intention.

The appearance of Gauguin's prints cannot be fully explained by the fact that he printed them himself using his bedpost and whatever furniture or implement he had at hand to exert the necessary pressure. As we discovered, it is not the unevenness of the printing alone that accounts for the fact that no two impressions he printed are alike. Microscopic study of the prints conducted by Mary Cropley and Harriet Stratis following the 1988 exhibition revealed a more willful procedure: it produced evidence that prior to printing his woodblocks, Gauguin applied color by a variety of means to the paper on which he would subsequently print. The printed lines are over the areas and particles of color (see Fig. 3.4). Microscopic examination of many of the Gauguin impressions in the *Noa Noa* series in The Art Institute of Chicago's collection revealed a comparable procedure. In *Auti Te Pape* (Fig. 3.6), it is clear that rather than applying color to black-and-white impressions, Gauguin applied color to the paper first, before printing his block.[8] This practice, plus his unorthodox, willfully "primitive" printing techniques, guaranteed that chance would play a role in the final result. Appearing out of register and blurry, Gauguin's woodcuts are suggestive in ways that the paintings from which they derive are not.[9] They have something of the appearance of fragments of ancient frescoes; their surfaces evoke wear, erosion, the passage of time, the fading of a vision.

The monotypes are similarly evocative. Indeed, the woodcuts exhibit characteristics that are visually and technically similar to those of the monotypes. There is also new evidence that Gauguin's work on the monotypes shaped his approach to the woodcuts, and that he was experimenting with both at the same time. In the monotype *Aha Oe Feii?* (Fig. 3.5), Gauguin repeated the pairing of two figures – one seated and one reclining – seen in his earlier painting of the same subject, although he has reversed their orientation.[10] In the woodcut *Auti Te Pape* (Fig. 3.6), the reclining figure is absent. Yet as we studied the print, we realized that the rocklike form to the left of the figure corresponds to the position of the head of the reclining figure in the monotype. When an outline tracing of the monotype was placed over the woodcut, an exact correspondence was revealed. It seems that Gauguin first made his monotype, then transferred its outlines to the surface of his wooden block. In subsequently cutting the block, he suppressed the reclining figure, transforming the head into an indistinct, rocklike form.

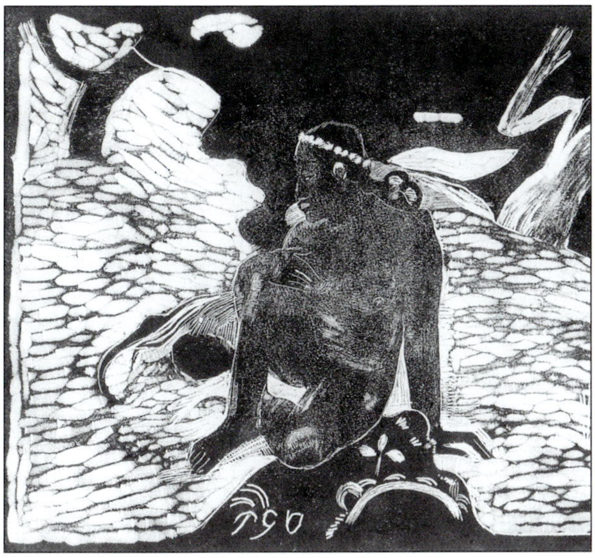

Figure 3.6 Paul Gauguin, *Auti Te Pape* (*The Fresh Water Is in Motion*), 1894 (detail). Woodcut. The Art Institute of Chicago, The Frank Hubachek Collection, 1954.1191.

To recap our findings thus far: back in Paris in 1894, Gauguin's exploration of graphic media revealed the possibility of an evocative visual style that uniquely suited the Symbolist aesthetic of mystery and suggestion. This was something that he had not attempted in his paintings in Tahiti. Nor did he attempt such effects in the Tahitian-inspired paintings, such as *Day of the God* of 1894 (The Art Institute of Chicago), which he was making at the exact same time as his work on paper. Yet they reappear in the woodcut version Gauguin made right after finishing this painting.[11]

However, in some of his works on paper, including a gouache in The Art Institute of Chicago (Fig. 3.7), Gauguin adopted a style more consonant with the paintings. Stylistically akin to *Day of the God* with its definite shapes and color areas, the gouache was likewise composed by piecing together motifs from various earlier works. The standing figure derives from a painting now in the State Hermitage Museum, St. Petersburg.[12] The crouching figure comes from a painting now in a private Swiss collection.[13] In the face, for example, the gouache is applied in a similar manner as in the painting. It is precisely because the gouache exhibits these stylistic and compositional strategies that some experts have been led to doubt its authenticity. This is despite the fact that the work has been known to be by Gauguin from the early years of the twentieth century – long before it would have been worthwhile to create a forgery. In a photograph of the Chicago installation of the 1913 Armory Show, the disputed gouache hangs together with other works by Gauguin now in the Art Institute's collection. Although we did not doubt the work's attribution, we did find the difference between the gouache and the monotypes puzzling. So we set about trying to account for the difference in appearance of the gouache on the one hand, from the woodcuts and monotypes on the other.

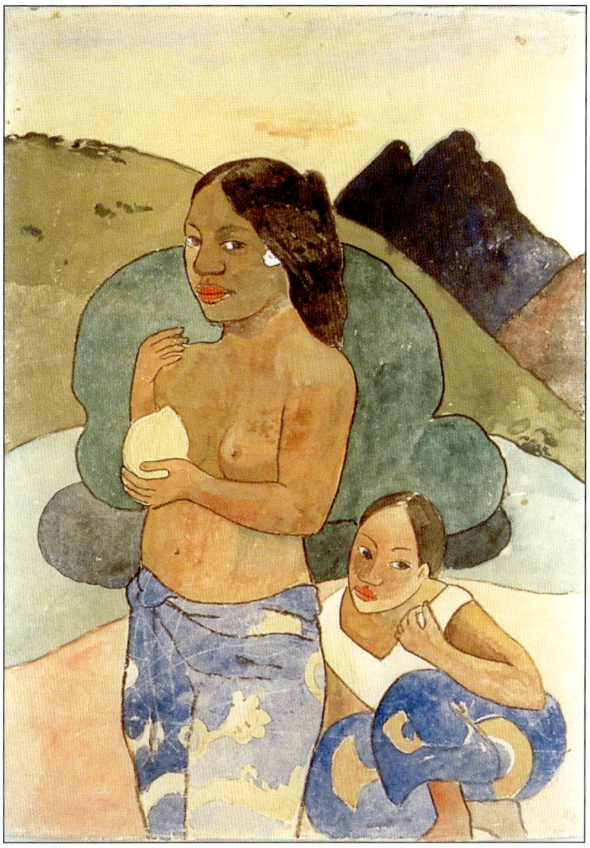

Figure 3.7 Paul Gauguin, *Two Tahitian Women in a Landscape*, 1892. Watercolor, gouache, and ink on Japanese paper. The Art Institute of Chicago, 1922.4795.

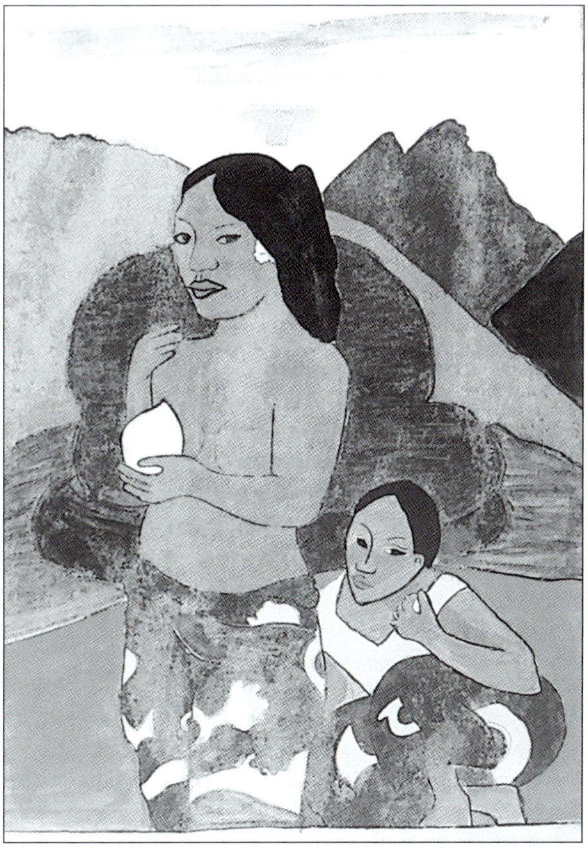

Figure 3.8 The gouache facsimile of Figure 3.7.

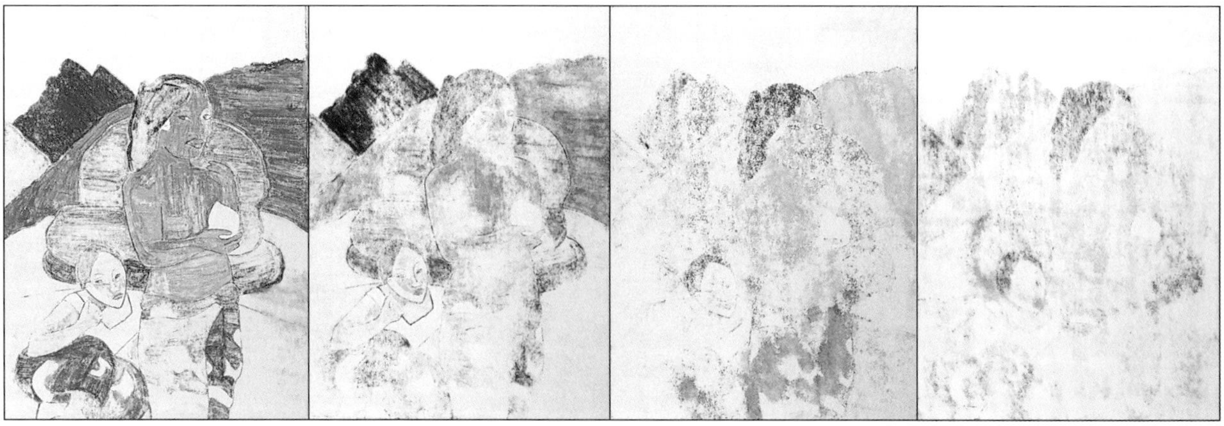

Figure 3.9 The four counterproofs sequentially pulled from the gouache facsimile (Fig. 3.8).

The fact that we had discovered that a number of Gauguin's so-called monotypes are in fact counterproofs was suggestive. Gauguin used this time-honored technique for duplicating works executed in friable media with some of his pastels.[14] We established that they are directly related rather than merely similar by using outline tracings made on Mylar®. Early on we had decided that we had to establish an object standard to test visual similarities. To this end, we began making tracings on Mylar® of the works on paper that came our way – from people walking into the print room from the street to requests for authentication from collectors, dealers, and auction houses. With these tracings in hand, we could test whether or not similar elements in different compositions actually shared the exact same source.

Gauguin took counterproofs from non-friable aqueous media (as well as from his pastels). We found that some of his gouaches could yield as many as four offspring.[15] To achieve this, Gauguin would have had to place a dampened piece of paper directly on the surface of the gouache and then apply pressure. As he did not have ready access to a printing press, he probably applied pressure using a kitchen implement – very likely a wooden spoon. In this process, however, he risked compromising the gouache.

This led us to wonder, could it be that Gauguin made the gouache like the one in the Art Institute (Fig. 3.7), with its simplified design and application, in order to function – perhaps primarily – as a printing matrix? To test this hypothesis, we created a gouache facsimile of our gouache (Fig. 3.8). From this we printed one impression using a dampened sheet of medium-weight newsprint and the back of a wooden spoon, the most utilitarian of kitchen utensils, to exert pressure. A second impression was taken, again using medium-weight dampened newsprint and spooning. A third impression was achieved with spooning on a rather wet sheet of newsprint. Finally, we took a fourth counterproof, again employing spooning and rather wet newsprint (see Fig. 3.9). Clearly, unlike traditional monotypes created with ink on non-absorbent surfaces, Gauguin's counterproof method using a gouache matrix did not produce, strictly speaking, monotypes or single impressions, but rather two or more viable counterproofed images. In turn, each of these could potentially be reworked to produce unique yet related compositions.

Tracings allowed us to establish that other works – previously considered to be different versions of the same subject – were in fact created from the same matrix.[16] Moreover, in at least one case, Gauguin mounted two counterproofs derived from the same matrix side by side on the same cardboard support. Two counterproofs of *Words of the Devil* appeared like this in the 1918 Degas sale. Subsequently separated, they ended up in different collections.[17]

Having established this relationship to a common matrix, we began to compare – by means of the outline tracings – works that on first glance seemed to be different versions of the same theme. In a number of

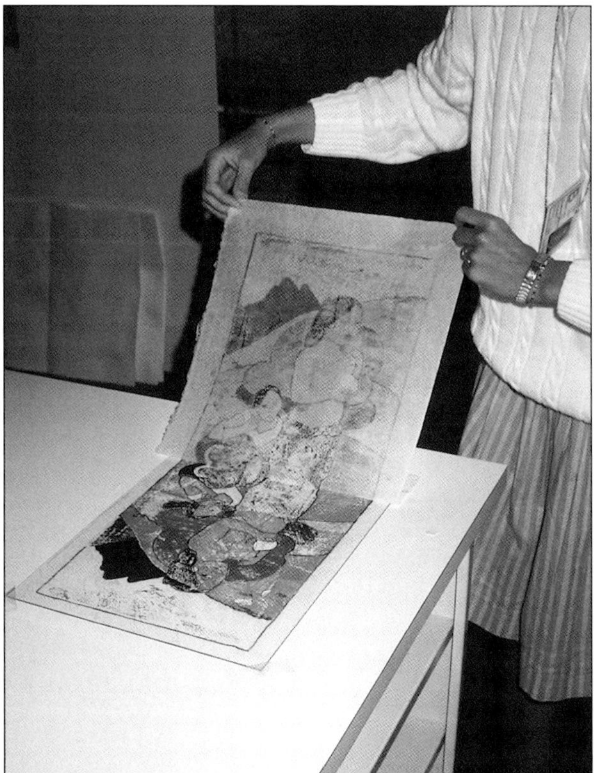

Figure 3.10 A counterproof being pulled from the gouache design on glass.

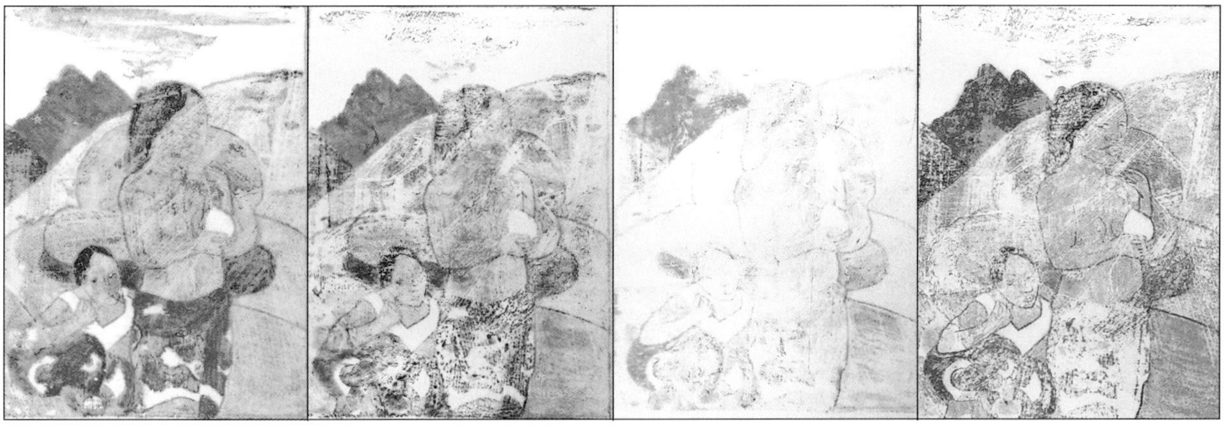

Figure 3.11 The four counterproofs sequentially pulled from the gouache design on glass.

cases, the tracings made from the gouaches established beyond doubt their direct relationship to the monotypes. This correspondence, in turn, called our initial hypothesis into question. For if a group of monotypes shared one matrix, a work in gouache, then how could one explain the many differences between these monotypes? This led to a new hypothesis: that a gouache matrix was only indirectly the source of the so-called monotypes, and that the actual matrix was a piece of glass that Gauguin placed over the gouaches, treating the glass like tracing paper, simplifying the gouache design as he copied it onto the smooth, non-absorbent surface.

Again, using a gouache facsimile and a sheet of glass, we put our hypothesis to the test (see Fig. 3.10).[18] The gouache design on the glass surface yielded more impressions than we had imagined it could. For the first, we used Japanese paper coated with diluted paste to replicate the streaky effect found in some of the "monotypes," and we employed only the pressure of the hand. As expected, subsequent impressions were lighter, but by varying the pressure and once again using the back of a wooden spoon, each successive impression differed considerably. For example, the impression realized on a slightly damp sheet of Japanese paper with light spooning was quite pale, while an impression on a moderately damp sheet of Japanese paper in conjunction with both hand pressure and spooning was darker. Applying conventional wisdom, it could be assumed that the darker impression was made prior to the paler impression. But here the greater richness was a function of the greater pressure exerted – hand pressure followed by vigorous spooning – and the use of a medium damp Japanese paper. Remarkably, after pulling several impressions, the design on the glass matrix was still legible. After four transfers the matrix could still yield distinctly visible results; however, we did not proceed further (Fig. 3.11).

The results derived from testing this hypothesis were promising. The surface textures of colored passages printed from a water-based medium on glass were close to those of the artist's surviving "monotypes." Moreover, the use of glass could explain the fact that works linked by identical figure groups nonetheless exhibit notable differences. But there remained the problem of testing this hypothesis against Gauguin's extant graphic oeuvre – a body of work that has by no means survived intact! Works in gouache are numerically fewer than the so-called monotypes. And there are relatively few instances where a gouache appears to be connected to color "prints" that are compositionally oriented in "reverse." A notable exception is a gouache in Belgrade, which clearly relates to a work that Gauguin pasted into the manuscript version of *Noa Noa*, now in the Louvre.[19] The exact nature of the relationship could not be tested until it was possible to visit Belgrade. When the occasion arose, the curators in Belgrade gave us permission to make an outline tracing on Mylar® of their gouache. Subsequent application of the tracing over the *Noa Noa* monotype in the Louvre both confirmed the relationship and revealed Gauguin's working procedure. Laying a piece of glass over the gouache, the artist traced a figure and tree just as they appear in the gouache. However, before copying the form of a dog, he shifted the position of the glass upward and to the right. As a result, the dog in the monotype is on a different plane and closer to the figure than it is in the gouache.

The ease with which it was possible, using this technique, to both replicate and alter elements of an existing composition suited an approach to image-making that relied on the reuse and reshuffling of a stock of pictorial motifs. Moreover, the element of chance inherent in the process, and the evocative blurring of the resulting images, clearly revealed new expressive possibilities to Gauguin.

In 1994, Gauguin scholar Victor Merhlès published a previously obscure article written in 1898 on the American artist Augustin Koopman in which Gauguin's prints are invoked.[20] The author of the Koopman review, a friend of Gauguin's, described how Gauguin had pioneered a technique that built upon the traditional approach to monotype. The reviewer wrote: "Gauguin painted with watercolor on a sheet of glass and pulled impressions on dampened paper." There it was – the confirmation of our research. In one sense, reading this earlier would have simplified our investigation considerably. But by proceeding as we did – by attempting to understand Gauguin's *cuisine* by re-creating his recipes – we came to an understanding of his aesthetic aims that we would not have reached otherwise. Simply reading a recipe is a vastly different experience from trying it out in the kitchen.

Endnotes

1. See *Ia Orana Maria* (*We Greet Thee, Mary*), c. 1891–92 (W 428), an oil painting in the collection of The Metropolitan Museum of Art, New York (Georges Wildenstein, *Gauguin*, ed. by Raymond Cogniat and Daniel Wildenstein, Vol. 1, *Catalogue* [Paris, 1964]; hereafter referred to as W). Gauguin based the figure of the Tahitian Mary carrying the Christ Child on an Arnoux photograph of a North African woman with a naked child on her shoulder, reproduced in Alain Fleig, *Rêves de Papier. La Photographie Orientaliste, 1860–1914* (Neuchâtel, 1997), p. 111. Hippolyte Arnoux was active c. 1860–90 with a studio in Port-Saïd, Egypt, Place Ferdinand de Lesseps. The figures at left derive from a photograph of the Borobudur frieze. Richard Brettell et al., *The Art of Paul Gauguin*, exh. cat. (Washington, D.C.: National Gallery of Art, 1988), p. 245.

2. *Exposition d'oeuvres récentes de Paul Gauguin*, exh. cat. (Paris: Galerie Durand-Ruel, 1893).

3. See for example, *Te Faruru* (*To Make Love*), 1892, a gouache in the collection of the Museum of Fine Arts, Springfield, Mass.: the figure in this gouache is based on the middle figure on the left in the painting *Ia Orana Maria* (*We Greet Thee, Mary*). Brettell et al. 1988 (as in note 1), p. 242.

4. See for example *Ia Orana Maria* (*We Greet Thee, Mary*), 1894, reproduced in Richard S. Field, *Paul Gauguin: Monotypes* (Philadelphia Museum of Art, 1973), no. 1, p. 53.

5. Julien Leclerq, "Petites Expositions. Augustin B. Koopman," in *Chronique des arts et de la curiosité*, 3 December 1898, p. 343, repr. in Victor Merlhès, "Notes sur les monotypes," in *Racontars de Rapin, Fac-similé du manuscrit de Paul Gauguin* (Taravao, Tahiti: Editions Avant et Après, 1994), pp. 115–16.

6. The Springfield gouache (as in note 3) was such a casualty. Although stylistically consistent with the painting, the gouache appears insistent – brittle and static – in comparison with the Boston monotype (as in note 4), which is certainly more typical numerically in Gauguin's graphic oeuvre.

7. In fact the entire series is loosely based on his paintings. Compare for example the title page (Fig. 3.1) with the oil painting *I Raro Te Oviri* (*Under the Pandanus Tree*) (*II*), 1891 (W 432), Minneapolis Institute of Art; reproduced in color in *Paul Gauguin, Das verlorene Paradies*, exh. cat. (Essen: Museum Folkwang, 1998), no. 10, n.p.

8. See for example *Auti Te Pape* (*The Fresh Water Is in Motion*) in the collection of The Art Institute of Chicago. Brettell et al. 1988 (as in note 1), p. 322.

9. See for example *Aha Oe Feii?* (*What! Are You Jealous?*) (W 461), an oil painting in the collection of the Pushkin State Museum of Fine Arts, Moscow. Brettell et al. 1988 (as in note 1), p. 278.

10. Ibid., p. 278.

11. Ibid., p. 364.

12. *Ea Haere Ia Oe* (*Where Are You Going?*) (*II*), 1893 (W 501), the State Hermitage Museum, St. Petersburg; reproduced in color in *Paul Gauguin, Das verlorene Paradies* (as in note 7), no. 31, n. p.

13. Brettell et al. 1988 (as in note 1), p. 264; *Nafea Faa Ipoipoi* (*When Will You Marry?*), 1893 (W 454); reproduced in color in *Paul Gauguin, Das verlorene Paradies* (as in note 7), no. 21, n. p.

14. See for example the pastel study of a reclining nude and its counterproof in Brettell et al. 1988 (as in note 1), p. 308.

15. See for example the gouache of a Tahitian girl and two of the counterproofs it yielded in ibid., pp. 356–57.

16. See for example two counterproofs created from the same matrix in ibid., p. 354.

17. Ibid., pp. 352–53.

18. The counterproofs were pulled with the assistance of intern Constance Coburn, to whom I extend my thanks.

19. *I Raro te Oviri* (*Under the Pandanus Tree*), 1894, gouache, Narodna Gallerija (National Gallery), Belgrade. I would like to thank Branka Nakanishi for making the tracing in Belgrade. For the frontispiece monotype pasted into *Noa Noa*, see Field 1973 (as in note 4), no. 124, p. 135.

20. Merlhès 1994 (as in note 5).

The Use and Misuse of Distemper in the Works of Édouard Vuillard: A Curator's View

Gloria Groom

This paper grew out of research toward an exhibition devoted to the decorative paintings of four artists: Pierre Bonnard, Édouard Vuillard, Maurice Denis, and Ker-Xavier Roussel.[1] It was the first exhibition to look seriously at the decorative achievement of these artists from their Nabi works of the 1890s to their late decorations of the 1920s. Though each was successful in receiving commissions for decorative panels, folding screens, and single-painting decorations for private interiors, only Vuillard used distemper – a medium traditionally known for its ability to cover large areas of wall cheaply and with quick drying time. It is a medium both generous and unforgiving, capable of incredibly rich and evocative surfaces. It is also one of the most revealing in the sense that it records for posterity the artist's hand – the workings and reworkings – and, as I will show, the ease or complete unease with which a project was realized.

My purpose is to introduce Vuillard and the distemper technique within the framework of avant-garde painting practices at the end of the nineteenth century in France, and to look specifically at how Vuillard experimented with the medium. Faye Wrubel, with whom I have had the privilege of working on this research, takes up the technical issues and discusses the practical problems involved with treating Vuillard's works in distemper in her paper "The Use and Misuse of Distemper in the Works of Édouard Vuillard: A Conservator's View" (this volume, pp. 151–54).

As a member of the Nabi group of artists, Vuillard believed Paul Gauguin's credo that all art is decoration.[2] For the Nabis – who sought to integrate art into the art of living – printmaking, applied arts, and mural painting were all worthwhile expressions and undertakings. With his friends, Vuillard made set flats for avant-garde theaters in the early 1890s. Quickly painted and just as quickly painted over or destroyed, the sets no longer exist, but contemporary accounts of the "purple and gold mist" or the "lamplit darkness" indicate the schematic and abstract nature of these large surfaces, which are habitually cited as the catalyst for the Nabis' subsequent work on large-scale decorative paintings.[3] Though the theater sets were influential for the Nabis' developing ideas about subject matter, format, and support, only Vuillard adopted the distemper medium.

Perhaps, as Jacques Salomon (Vuillard's nephew by marriage and the author of several books on the artist) has written, Vuillard used distemper initially because of its cheapness, though it is likely that aesthetics also played a role.[4] Like Bonnard, who wrote to Vuillard in 1892 that he was "tormented by oil painting,"[5] Vuillard was also tentative about painting in oil, for fear of falling into the trap of naturalism by using the medium as he had been instructed at the École des Beaux-Arts and at the Académie Julian. In the beginning, both artists worked in oil on modest-size cardboard, which they left unvarnished. Vuillard especially was drawn to this support, perhaps because he had a steady supply of boxes from his mother's dressmaking studio, but also because

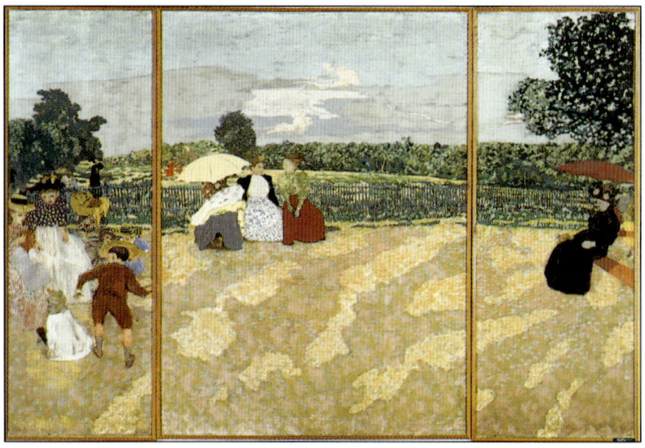

Figure 3.12 Édouard Vuillard (French, 1868–1940), triptych from *Public Gardens: Conversation, Nursemaids, Red Sunshade,* 1894. Distemper on canvas, 215.5 x 83 cm; 213 x 143 cm; and 214 x 81 cm. Musée d'Orsay, Paris. © Réunion des Musées Nationaux/Art Resource, N.Y.

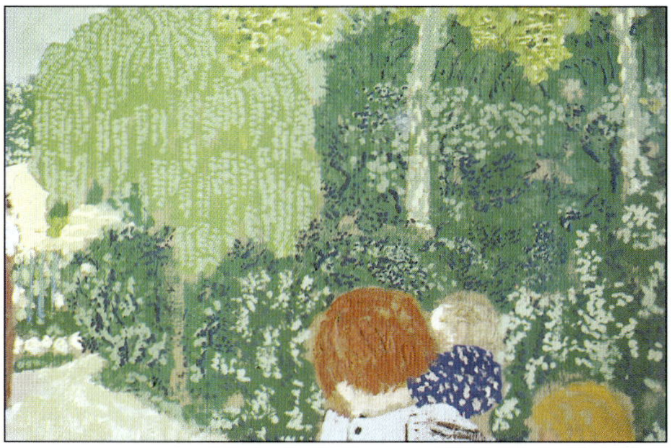

Figure 3.13 Édouard Vuillard, panel from *Public Gardens: Promenade* (detail), 1894. Distemper on canvas, 212 x 96 cm. The Museum of Fine Arts, Houston, Texas, the Robert Lee Blaffer Memorial Collection, gift of Mr. and Mrs. Kenneth Dale Owen.

of cardboard's absorbent quality and its ochre or gray tones, which created a neutral background for his color harmonies.

Vuillard's earliest works show that he was interested in creating flat areas of color silhouetted by the arabesque line. When he began painting pattern on pattern around 1893, he still relied on line to make the figure stand out from the ground. By around 1894–95, this goal gave way to paintings in which priority is given to neither ground nor figure; it is this dabbed, fuzzy, blurry style that Vuillard would eventually adopt for the distemper medium.

Distemper is an aqueous substance made by mixing dry pigment in hot animal glue. In 1894 Vuillard received his first major commission for a series of nine panels on the theme of public parks, which he executed in distemper. These paintings on canvas, five of which are in the Musée d'Orsay, Paris (see Fig. 3.12), are splendid examples of the medium used consistently to create an even pattern and overall harmony. Up to now, Vuillard had only produced fragments; small, highly original works that were coveted by private collectors. His choice of distemper allowed him to break dramatically from the smaller paintings (some of them no larger than postcards) and, at the same time, to reinforce the notion of a painting as expressive decoration. It was also a way of coming to grips with a theory and process that would fit his temperament. In his private journals (now housed at the Institut de France in Paris), Vuillard noted at this time the two ways he could approach art: by the study of exterior perception filled with "dangerous experiences" and thus "dangerous for his nervous personality," or by the study of pictorial decoration, which, while limiting, afforded him the "tranquility of a worker."[6] Like Delacroix, whose recently published journals he read that year with enthusiasm and whose advice was "to avoid that infernal facility of the brush" by choosing "a medium that is hard to work," Vuillard turned to distemper to slow down and control his too-rapid dexterity and casual working method.[7] As seen in the detail of one of the panels from the public parks series in the Houston Museum of Fine Arts, here is the work of a craftsman (Fig. 3.13). None of the paint strokes blend with another, but are either adjacent or laid down over the other, causing a tapestry effect that was not lost on Vuillard. He wrote in his journal, "Really for a decoration for an apartment, a subject that's objectively too precise could easily become unbearable. One would grow less quickly tired of a textile, of designs that don't have too much literary precision."[8]

Following the public parks commission, Vuillard received other projects for which he followed what I will term the "tapestry aesthetic" – creating painted textiles by juxtaposing large areas of pattern. As he experimented with the distemper medium, he used it much as he did oil. Compare, for example, one of five panels for a decorative ensemble called *The Album,* painted in oil in 1895 (Fig. 3.14), with one of four panels painted in distemper (Fig. 3.15). Even when painting in oil, Vuillard did not take advantage of its blending power – the ability to paint wet-on-wet and cover over. Instead, he treated both oil and distemper with the same respect and restraint. With distemper, mixing on the canvas is not possible. Dry pigments are added to the glue by the artist and applied to the support while warm. The paint dries quickly and can be painted over, but the sensuous blending and translucency possible in oils is absent. In these two works, both oil and distemper are dabbed on

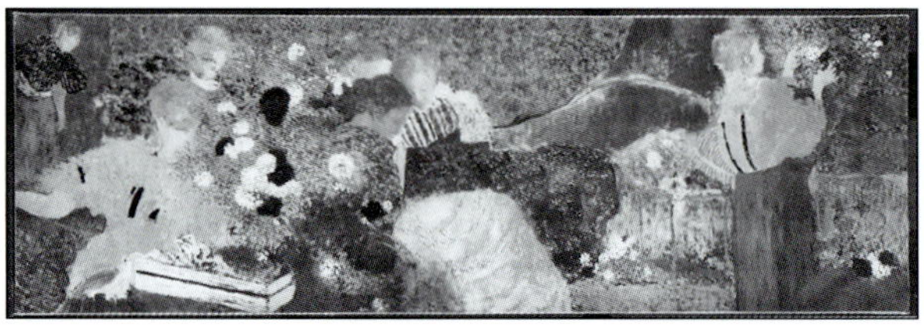

Figure 3.14 Édouard Vuillard, *The Album,* 1895. Oil on canvas, 67.6 x 204.5 cm. The Metropolitan Museum of Art, New York, the Walter H. and Leonore Annenberg Collection, partial gift of Walter H. and Leonore Annenberg, 2000.93.2.

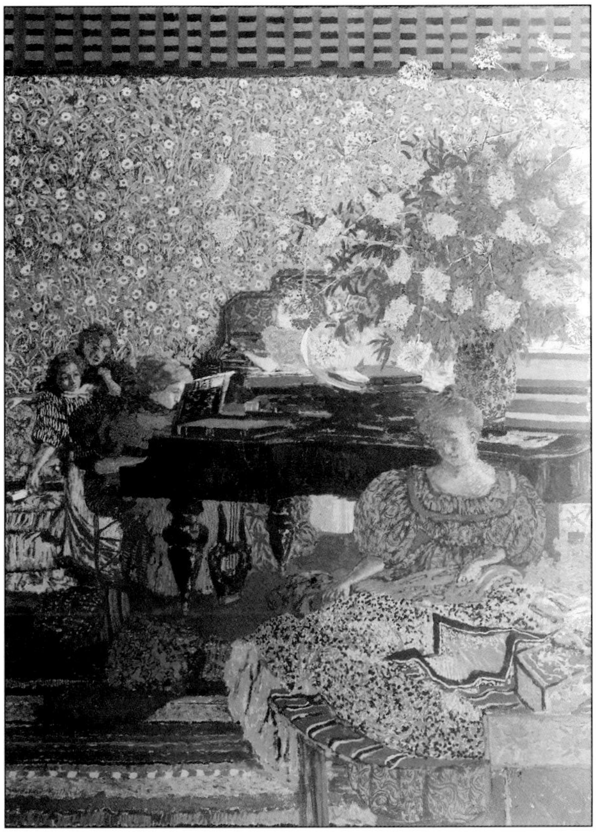

Figure 3.15 Édouard Vuillard, *The Piano*, 1896. Distemper on canvas, 210 x 153 cm. Musée du Petit Palais, Paris. © Giraudon/Art Resource, N.Y.

top of but not blended into one another. At other times, such as for the 12-foot-wide *Window Overlooking the Woods* (1899; The Art Institute of Chicago) painted in oil, he leeched the oil so that it became a drier medium and laid paint down in broad bands of undifferentiated color, as if he were thinking of filling large spaces (such as for scene designs) in distemper.

At this point I want to detour momentarily to bring in the one other artist in Vuillard's circle who painted in distemper, Odilon Redon, who seems to have, as one critic put it, "learned some technical tricks" from Vuillard.[9] Indeed, after 1900 Redon created numerous large-scale works in distemper, several of which, such as a decorative panel in the Rijksmuseum Twenthe in the Netherlands and *Andromeda* in the Arts Center in Little Rock, Arkansas, traveled successfully to Chicago, Amsterdam, and London during the 1994 retrospective on the artist. This led me to believe that borrowing large works by Vuillard in the same medium and from the same time period would be no big deal. There I was wrong – very wrong. For Redon, who had worked extensively in pastel, turned to distemper only later in life and worked with the medium in much the same manner as pastels, that is, in thin layers that adhered well to the glue-coated canvas.[10] Vuillard's use of distemper was much more complicated and varied from project to project. Moreover, as his painting evolved stylistically, he forced the medium to adapt as well, sometimes with disastrous results.

By 1905 Vuillard's social circle had radically changed from the middle-class bohemian crowd of the Nabis to the circles of dealers like the Bernheim-Jeunes and their even wealthier clients. As the scale of the interiors grew, so did Vuillard's decorative ambitions. But something else happened. Vuillard began a more "Impressionist" phase. While holding dear to the Nabi principles of suggestion and flatness, he more and more introduced real people and places into what had previously been decorative landscapes or interiors taken out of time.

Vuillard also began to use plain brown wrapping paper as the support that he would glue to canvas. This allowed him to figure out scale and compositional strategies, and also to anticipate certain vagaries of color inherent in distemper, before investing in large, custom-cut canvas. Paper supports were slightly better than canvas and cardboard since they absorbed the medium and many of the works on paper have remained in fairly good condition. Without a catalogue raisonné, it is difficult to assess Vuillard's working method comprehensively, but we know he did hundreds of studies for any given decorative commission, and many times full-size distemper studies immediately preceded the final version. A distemper study on paper mounted on canvas known as *Place Vintimille* (Fig. 3.16) has the identical subject and scale as the final version painted in distemper on canvas as part of a decorative triptych (Museum of Modern Art, New York).

Perhaps because he acknowledged the inherent fragility of paper, Vuillard tended to rework a painting in distemper on paper rather less than one on canvas. But paper can be problematic as well, and when the support moves or undulates, it can cause the glued paper layer to weaken or tear, and the entire work may be adversely affected, sometimes irreparably. Light exposure can also ruin the appearance and undermine the stability of a decoration on paper. Overexposure of the unprimed, unpainted paper to light readily darkens it, and can significantly alter the paper tone. This is the case with another image of *Place Vintimille* (Fig. 3.17). The paper, once meant to act as a neutral background, has turned a hot brownish orange, in contrast to the otherwise soft green and ochre palette. Though today the nearly 9-foot-tall paintings (dismantled from their original five-panel screen format) are hung responsibly behind glazing in low light, decades of use in a private interior have forever altered the balance between ground and paint and thus have undermined the artist's intent.

Impressionist subjects like the *Place Vintimille* panels required a relatively unprogrammatic conception of the "decorative, tapestry effect," since they were indeed based on nature. But Vuillard also painted his friends in their interiors. He developed a curious hybrid genre called a "decorative portrait," in which he sought to capture not only the character of the sitter but the milieu defining that character. Taking the realist program of the Impressionists to another level, Vuillard undertook a number of commissions for life-size portraits of friends and acquaintances with the unyielding distemper

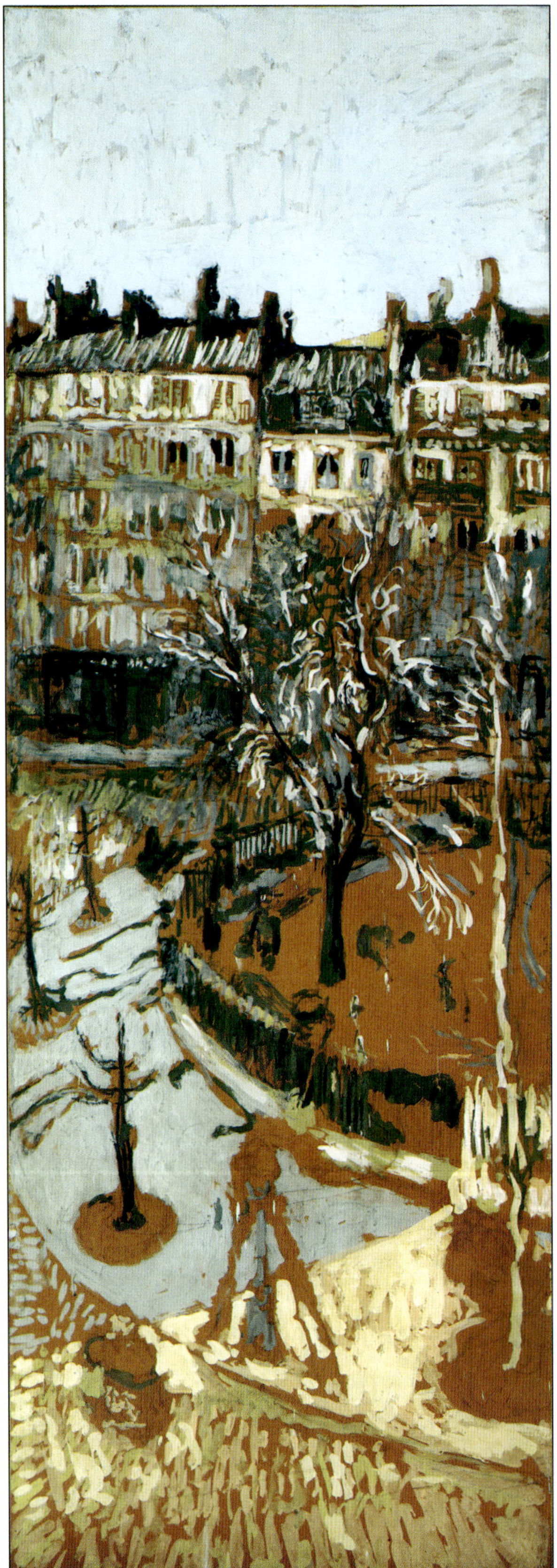

Figure 3.16 Édouard Vuillard, study for left side of *Place Vintimille* triptych, 1906–07. Distemper on paper, mounted on canvas, 196 x 69 cm. Kunsthandel Wolfgang Werner, Bremen/Berlin; Kunsthandel Sabine Helms, Munich.

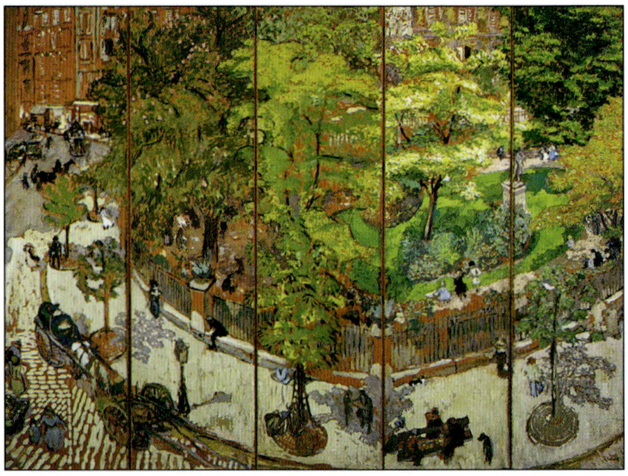

Figure 3.17 Édouard Vuillard, *Five-Panel Screen for Miss Marguerite Chapin: Place Vintimille*, 1911. Distemper on paper, mounted on canvas, each panel 230 x 60 cm. National Gallery of Art, Washington, D.C., gift of Enid A. Haupt, 1998.47.1.

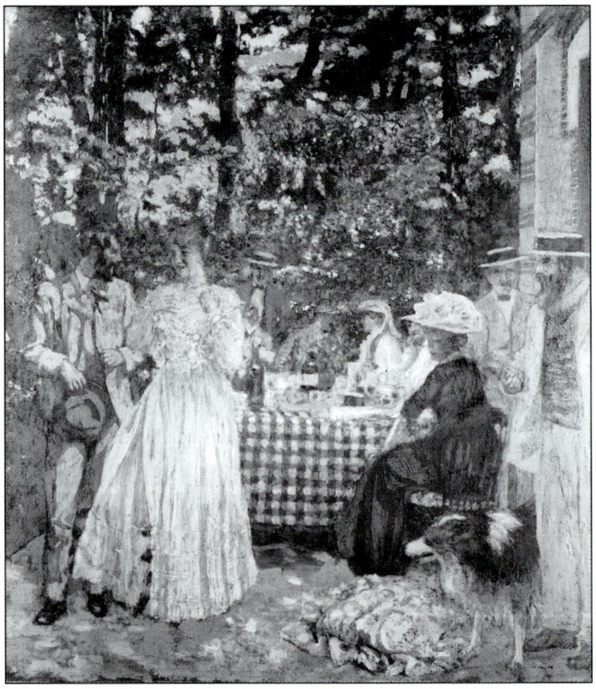

Figure 3.18 Édouard Vuillard, *Luncheon in Vasouy*, c. 1901 and 1936. Right half of a painting separated by 1931. Distemper on canvas, 218 x 183 cm. Tate Modern, London.

medium. Portraiture itself is a tricky business, and Vuillard's art was victim to the personalities and individual circumstances of the lives of his sitters. Such was the case with Vuillard's family luncheon scene of 1901 (Fig. 3.18). The paintings, which are now in the Tate Modern in London, originally formed one large panel. Thirty-five years later, Vuillard cut the canvas in half to make it more salable for the collector's widow. At that time, he reworked the surface so that certain people, including Léon Blum (1872–1950), who by 1936 had become the first Jewish and socialist prime minister of France, were painted out or turned into someone else. But distemper does not work like oils, and overpainting is never seamless. Instead, layering results in cracking, cupping, and a very noticeable change from the original surface. Aesthetically, it may change the entire balance of a work, since the layers become three-dimensional – forming a dry, craggy, cement-like structure.

On the one hand, Vuillard's method of working from tiny thumbnail sketches and photographs to finished images in distemper was well suited to Impressionist scenes that did not require a flattering likeness. But the properties of distemper are antithetical to portraiture. Jacques Salomon remembers watching in anguish as Vuillard painted again and again the same area of his portrait of the Comtesse de Polignac – perhaps the leg, which he adjusted multiple times in notebook sketches. Apparently, just as Salomon started to say something critical, Vuillard jumped up, brushed the area with hot water, and scratched off the paint layer with a penknife only to begin again.[11]

Vuillard's *Portrait of Madame Guérin* seems to have suffered a similar fate (Fig. 3.19). The painting entered The Art Institute of Chicago in 1985, and since that time has seldom been on view. From a distance, it is a wonderful example of this hybrid portrait-cum-decoration. Standing over 5 feet high, it shows Madame and her poofy Pomeranian in a luxuriously appointed boudoir that is reflected in the mirrored wardrobe. A virtuoso display of mauves, lime greens, and pinks, it appears brilliantly fresh – from a distance of 3 feet, that is. But advance any closer and you will see what distemper and the artist's change of mind 15 years later can do to this frail medium and genre. Like the previously considered large-scale decorative portraits, this likeness was also returned to the artist, and over time Vuillard reworked the subject's face, pearled neckline, and especially her coiffure, which he updated to a bouffant that, because it is layered over several other layers, sits ungraciously on her head like a football helmet.

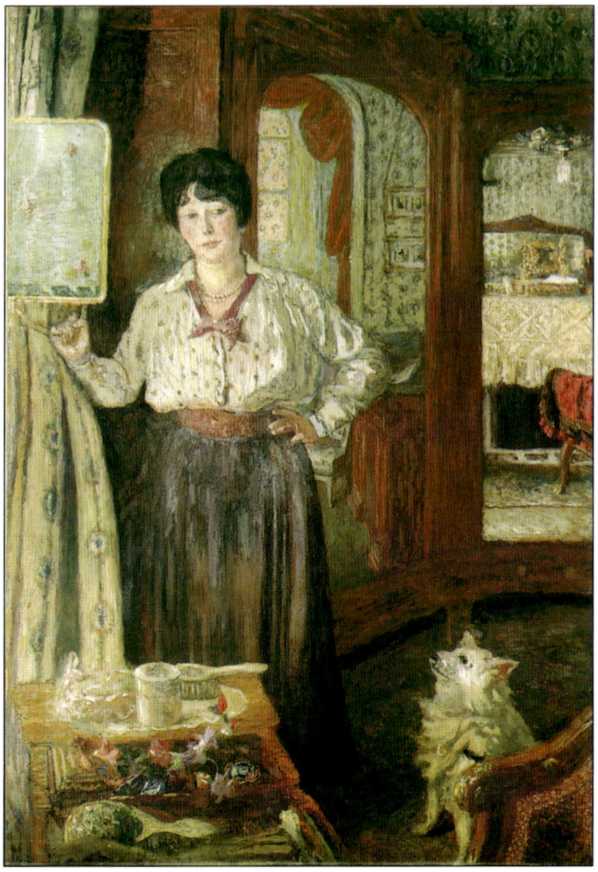

Figure 3.19 Édouard Vuillard, *Portrait of Madame Guérin*, 1916. Distemper on canvas, 200.7 x 121.9 cm. The Art Institute of Chicago, gift of Sara Lee Corporation, in memory of its founder, Nathan Cummings, 1986.868.

What can be concluded about Vuillard's use of distemper? Clearly he was a master in a difficult medium, and his oeuvre has more great successes than failures. In the more abstract, non-personalized patterned decorations of the 1890s, and even in the Impressionist scenes dating from the early twentieth century, Vuillard used distemper in the way it worked best, and in so doing came close to his aspiration of working tranquilly and like an artisan. With the portrait-decorations, however, he added a new, and not always successful, dimension to the medium and technique. For this reason, technical examinations of many works in distemper, and a conservator's experiments with the medium itself, have been invaluable from both a conservation as well as an art-historical standpoint as a means of getting inside the artist's working process and thus his art.

A portrait of the artist at age 57 shows Vuillard washing his hands in the studio, perhaps after or during a distemper session. Vuillard, known for his reticence, for his discreet and prudent character, once told a biographer that silence protected him. And so it is poignantly ironic that this most private of artists should have chosen not only a medium that prevents hiding or successful repainting, but, as in the case of the decorative portraits, a genre that makes him the center of attention.

Endnotes

1. *Beyond the Easel: Decorative Painting by Bonnard, Vuillard, Denis, and Roussel, 1890–1930*, was organized by The Art Institute of Chicago in collaboration with The Metropolitan Museum of Art, New York. The exhibition was held in Chicago, 25 February–16 May 2001, and in New York, 26 June–9 September 2001.

2. See P. L.-Maud [Maurice Denis, pseud.], "L'Influence de Paul Gauguin," *L'Occident* (Oct. 1903), p. 163.

3. See Geneviève Aitken, "Les Peintres et le théâtre autour de 1900 à Paris" (Ph.D. diss., École du Louvre, Paris, 1978), p. 87 passim.

4. Jacques Salomon, *Témoignage de Jacques Salomon* (Paris: Albin Michel, 1945), p. 124.

5. Letter from Pierre Bonnard to Vuillard, 16 August 1892, in Guy Cogeval, *Vuillard: Le Temps détourne* (Paris: Réunion des musées nationaux, 1993), p. 119.

6. Entry dated 16 July 1894 in Vuillard, Journal (Bibliothèque de l'Institut de France, Paris), MS 5396, carnet 2, p. 44 (recto), cited and transl. in Gloria Groom, *Édouard Vuillard, Painter-Decorator* (London: Yale University Press, 1993), p. 58.

7. Entry dated 20 July 1824 in *The Journal of Eugène Delacroix*, transl. by Walter Pach (New York: Crown Publishers, 1948), p. 99.

8. Entry dated 28 August 1894 in Vuillard, Journal (as in note 6), cited and transl. in Belinda Thomson, *Édouard Vuillard* (New York: Abbeville Press, 1988), p. 40.

9. François Thiébault-Sisson, "Salon d'Automne au Grand Palais," *Le Petit Temps* (14 Oct. 1904), p. 278. For a comparison of Redon's decorative aesthetic to Vuillard's use of distemper and decorative paintings, see Gloria Groom, "The emergence of a decorative aesthetic," in *Odilon Redon: Prince of Dreams*, exh. cat. (Chicago: The Art Institute of Chicago, 1994), pp. 306–07, 319.

10. For Redon's pastel technique, see Harriet K. Stratis, "Beneath the surface: Redon's methods and materials," in *Odilon Redon: Prince of Dreams* (as in note 9), especially pp. 368–73.

11. Jacques Salomon, *Auprès de Vuillard* (Paris: La Palme, 1953), p. 73.

The Use and Misuse of Distemper in the Works of Édouard Vuillard: A Conservator's View

Faye T. Wrubel

Distemper, also known as a *tempera*, *detrempe*, and *a la colle*, is one of the oldest known painting processes. Used in wall decoration by the Egyptians, Etruscans, Romans, and Greeks,[1] distemper – which calls for powdered pigment to be mixed with warmed glue – is different from gouache and other water-based media because it is temperature-dependent. The technique is demanding, the medium unpredictable, and, in many cases, the conservation problems are extremely complex.

In early American times, distemper was the wall covering of choice. It was cheap – made from readily accessible materials – and not as labor intensive to prepare and apply as oil-based paints. Interestingly, the job of applying the distemper belonged to plasterers, not decorators. Traditional lime plaster requires several years to cure, and applying oil paint would impede the curing process. To combat the problem, plasterers would put a coat of distemper on a new wall as a protective coating that could be replaced after the plaster had fully cured, usually after two years. The medium's lack of permanence may have been its most attractive feature. Distemper painting could hide flyspecks and candle-soot – two pressing problems for the colonial homemaker. But, unlike other wall coverings, it could easily be washed off with water and wet rags, and replaced every few years without any unsightly paint build-up. And the color could be changed at the same time to reflect changes in fashion or personal taste.[2]

By the nineteenth century, the fast-drying medium had found a new audience among the scenic painters and designers of theatrical sets. Édouard Vuillard himself was first exposed to the technique in the early 1890s, while painting theater sets for Aurélien Lugné-Poë's Théâtre de l'Oeuvre.[3] Although most fine artists preferred to use aqueous media only for smaller works, by the end of the century, several well-known European artists were experimenting with distemper for larger-scale decorative paintings.[4] They were drawn to the medium in an attempt to reproduce the soft pastel quality of fresco, a medium impossible to employ in France because of the damp climate.

According to Jacques Salomon, Vuillard's nephew by marriage and an artist himself, Vuillard first experimented with cardboard as a support thanks to a steady supply of box bottoms from his mother's dressmaking studio. Later, he became so enamored both of its absorbent quality and of its ochre or gray tone that he used tinted papers to achieve the same base for his color harmonies. Initially, Vuillard was drawn to distemper for reasons of economy, though he grew to love its end result, prizing the matte tones produced by this technique (see Fig. 3.20).[5] Ironically, in later years this matte quality has led to his paintings being misidentified as works in pastel or gouache. It is the very dry quality that gives it a resemblance to a chalky medium.[6]

Distemper demands a great deal from anyone brave enough to work with it. A late nineteenth-century dictionary of decorative arts defines distemper as

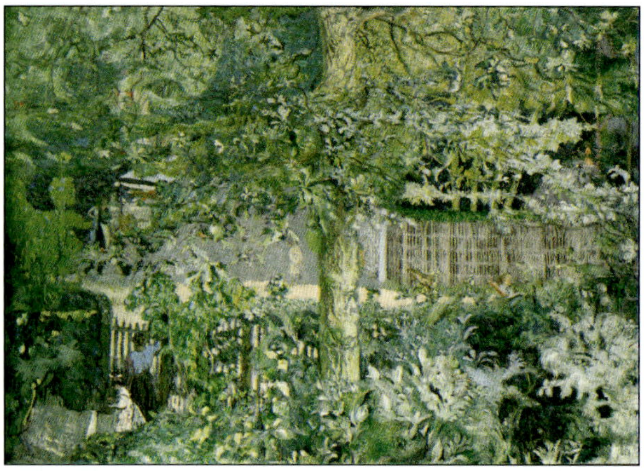

Figure 3.20 Édouard Vuillard, *Foliage: Oak Tree and Fruit Seller*, 1918. Distemper on canvas, 193 x 284.8 cm. The Art Institute of Chicago, Millennium Gift of Sara Lee Corporation, 1999.373.

requiring an enormous amount of practice.[7] It is no accident that, when describing the process, Salomon alludes to *cuisine.* Unlike oil painting, distemper requires an elaborate kitchen-studio: a spirit lamp or electric stove or burner, quantities of pots and pans, bags and boxes of powdered colors.[8] And of course the ubiquitous brown Tottin – hard, brittle sheets of rabbit glue to be soaked and boiled down into a foul-smelling, syrupy liquid.[9]

In Vuillard's time there were few published recipes for making distemper, and most contained vague recommendations to "take some thick glue size and thin with water to desired consistency."[10] There was considerable latitude given also with the ratio of medium to dry pigment. In earlier writings about egg tempera, Velásquez's teacher noted that fluffy pigments require more medium than do heavier ones; this seems to hold true for distemper as well.[11] With or without a recipe, the preparatory process was extremely elaborate and messy. Salomon, the only source who has written explicitly about Vuillard's technique, says that the artist obtained his Tottin from the corner druggist or from art-supply houses (it was originally a by-product of the French glove-making industry),[12] though he sometimes also used a strong cabinetmaker's glue made from animal hoofs.[13]

For his largest paintings, Vuillard prepared his colors in advance – soaking the glue sheets in four or five times their volume of water until they would swell and soften – a process that took 12 hours on average. He would then mix individual colors in separate pots and keep them warm in a double boiler over a stove or lamp to prevent them from thickening, taking care not to "overcook" or boil the mixture, which would cause the glue to spoil.[14]

Once the distemper was applied, the water evaporated in 5 to 15 minutes, depending on the temperature of the mixture, the ambient temperature in the artist's studio, and also on the thickness of the layers of paint already applied.[15] With the changes in temperature, the paint would exhibit a wide variance in viscosity, ranging all the way from runny to thick and gelatinous – unlike any other paint with the exception of encaustic.

The colors altered in appearance by lightening during the drying process, and Vuillard would often remix and correct them.[16] With such a rapid drying time, a sure, steady hand was imperative.[17] Salomon recounts how colors would often change right before the artist's eyes and acquire a nuance that appealed to him; he would then have to remix his colors in an attempt to "reconstitute that nuance."[18]

In addition, the pigments behaved differently from one another depending on their wetting properties. Some would dry smooth, and others rough and granular – so much so that in later years it was erroneously believed that Vuillard had added sand to his paint mixture. And, as if these properties weren't challenging enough, distemper also had a very short life, spoiling very quickly – actually turning rancid and odiferous – requiring new paint batches to be mixed fresh daily. This was especially true in the warm summer months.

As Vuillard worked, the glue would become sticky and thick, but, according to Salomon who watched in dismay, Vuillard would continue working it – soaking whole sections in hot water, scratching the surface with a penknife, overpainting to adjust color and tone, all the while layering new paint layers over old ones that had not yet dried.[19]

On drying, distemper is extremely brittle and has very poor elasticity. This in itself was troublesome, especially given the grand size and scale of the works Vuillard liked to paint. Because glues are both extremely non-elastic and adhesive, it is difficult to lay down more than two or three paint layers without experiencing cracking or peeling (see Fig. 3.21). Combating this process calls for reversing the traditional rule of "fat over lean," and instead painting the upper layers with paint that is progressively more dilute and weakly bound.[20]

Unlike commercially prepared paints, in distemper there is no consistency from color to color. Observing a large body of works painted in this process, one will find that nearly every picture has one or more colors that show damage – an indication that they were poorly mixed. In addition to cupping and cracking paint, distemper also exhibits air bubbles and craters where water molecules have evaporated, a property that is often used to distinguish distemper from oil paint (see Fig. 3.22).

There is evidence that Vuillard himself may have been aware of the instability of his medium. Salomon teased him about the encrustations of paint that might endanger the survival of his work. Vuillard protested that, in his estimation, his paintings would become "hard as a rock." Apparently, though, he and his contemporaries were aware that the paintings' textured, unvarnished surfaces could easily become magnets for moisture and dust.[21] As a precaution, Vuillard typically covered his pictures with a protective layer of glass. However, contemporary collectors of his works, who perhaps did not share Vuillard's love of the matte finish,

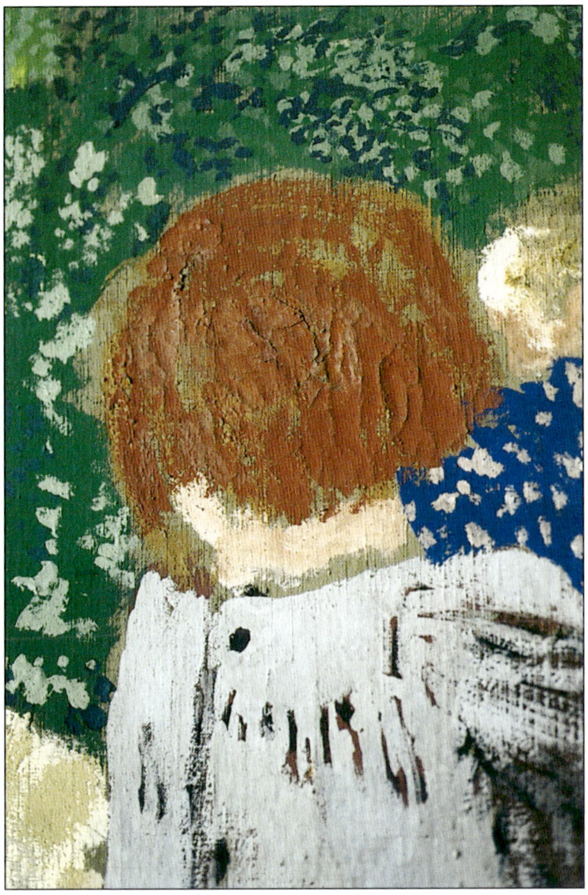

Figure 3.21 Édouard Vuillard, *Public Gardens: Promenade* (detail), 1894. Distemper on canvas, 212 x 96 cm. The Museum of Fine Arts, Houston, Texas, the Robert Lee Blaffer Memorial Collection, gift of Mr. and Mrs. Kenneth Dale Owen.

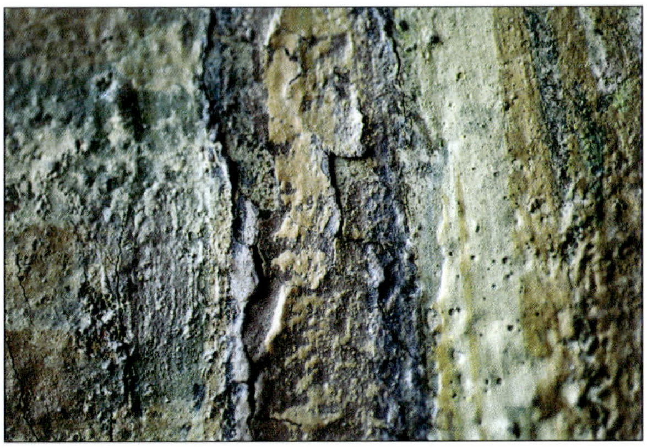

Figure 3.22 Édouard Vuillard, *Portrait of Madame Guérin* (detail of Fig. 3.19), 1916. Distemper on canvas, 200.7 x 121.9 cm. The Art Institute of Chicago, gift of Sara Lee Corporation, in memory of its founder, Nathan Cummings, 1986.868.

sometimes attempted to varnish his paintings, often with disastrous consequences.[22]

While Vuillard's glue-sized supports could be varnished safely, those that had been left "raw" could be irreparably damaged and distorted as dried varnish darkened and discolored the areas of the support that had been left unpainted. Even those paintings whose surfaces Vuillard had toned and sized with a layer of glue and powdered paint were problematic because the paint absorbed the varnish and darkened. Salomon recounts tales of "imprudent art lovers [who] often came shamefacedly to confess their [varnishing] misadventures to Vuillard"; the damage was irreparable.[23]

Today's conservator, encountering distemper paintings in museums and private collections, is faced with some complex issues. Perhaps the most daunting is the fact that the glue remains both water-soluble and hygroscopic, responding to environmental changes in humidity. Because the paint becomes brittle when dry, movement of the flexible supports underneath – be they canvas or paper glued to canvas – often leads to cracking and flaking. Finally, distemper is not particularly absorbed by the support. Instead, it rests on the surface and flakes off easily, a problem exacerbated in Vuillard's case by his fondness for painting in multiple layers (see Fig. 3.23).

Typically, successful treatments are those that have involved some form of humidification to soften the cupped or flaking paint, followed by the introduction of a consolidant. A very dilute gelatin is most commonly used. Because the dry, chalky appearance is intrinsic to the works, it is especially important during consolidation not to use an adhesive that will darken or change the saturation of the paint or become glossy.

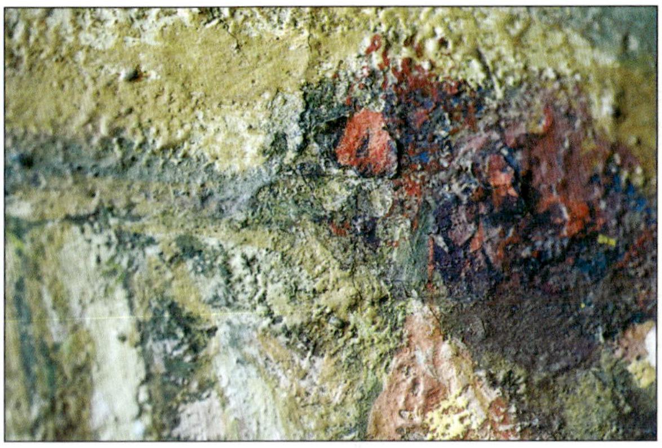

Figure 3.23 Édouard Vuillard, *Portrait of Madame Guérin* (detail of Fig. 3.19).

Depending on the extent of damage, treatment usually involves local humidification – possibly using Gore-Tex® and a damp blotter – or an overall treatment calling for humidifying the entire painting on a vacuum suction table. On completion, a panel back insert can be attached to the stretcher to reduce the risk of vibration to the support – a remedy that is non-invasive and completely reversible. The former technique was used in a recent treatment of a Vuillard painting at The Art Institute of Chicago to swell and soften an island of brittle, cupped paint, which was then consolidated with dilute gelatin, weighted, and pressed in place. Of interest was the fact that the softened paint became rubbery, unlike any other type of medium experienced by this conservator.

Vuillard did not invent the distemper process. His

originality consisted in his exceptional feat of using glue for works that seem better suited to straightforward oil painting.[24] Some art historians suggest that Vuillard's color sense – the juxtaposition of warm ochres and cool, vibrating blues and greens – and his disdain for blending made distemper an ideal, and enabling, medium for him to employ. Others believe that the interplay of colors came about directly as a result of the medium's unique constraints.[25]

While Édouard Vuillard himself believed his works to be enduring, time has shown them to be eminently changeable and fragile, subject to instability and requiring continuous monitoring. Vuillard was drawn to distemper for reasons of economy, though he grew to love its unique properties and the challenges it presented.[26] More than 60 years later, his choice of medium and method has left paper and painting conservators with a whole host of conservation problems that are as unusual as the works themselves.

Endnotes

1. Henry Harvard, *Dictionnaire de l'ameublement et de la décoration depuis le XIIe siècle à nos jours*, Vol. 2 (Paris, 1888), p. 102.

2. Christopher Ohrstron, "Some notes on distempers, calcimine and casein paints," *Traditional Paint News* 1, no. 2 (1996), pp. 24–25.

3. Charles H. Olin and Alexandra Riddleberger, "The special problems and treatment of a painting executed in hot glue medium, *The Public Garden* by Édouard Vuillard," *AIC Preprints* (Washington, D.C., 1983), p. 98.

4. Bernard Chaet, "Studio talk," *Arts* (Dec. 1956), p. 67.

5. John Russell, *Vuillard* (Greenwich, Conn.: New York Graphic Society, 1971), pp. 137–38.

6. Pamela Randolph Young, "History, analysis and treatment of *La Salle à Manger au Château de Clayes*, 1938, by Édouard Vuillard," *AIC Preprints* 1983 (as in note 3), p. 113.

7. Harvard 1888 (as in note 1), p. 102.

8. Russell 1971 (as in note 5), p. 67.

9. Olin and Riddleberger 1983 (as in note 3), p. 98.

10. Chaet 1956 (as in note 4), p. 67.

11. Ibid.

12. "Glues and size painting," *Artist* (1984), p. 17.

13. Olin and Riddleberger 1983 (as in note 3), p. 98.

14. Ibid., p. 99; and *Artist* 1984 (as in note 12), p. 17.

15. Olin and Riddleberger 1983 (as in note 3), p. 99; and Russell 1971 (as in note 5), p. 139.

16. Olin and Riddleberger 1983 (as in note 3), p. 99.

17. Harvard 1888 (as in note 1), p. 102.

18. Russell 1971 (as in note 5), p. 139.

19. Olin and Riddleberger 1983 (as in note 3), p. 99.

20. *Artist* 1984 (as in note 12), p. 29.

21. Olin and Riddleberger 1983 (as in note 3), p. 99.

22. Russell 1971 (as in note 5), p. 138.

23. Ibid.

24. Ibid.

25. Chaet 1956 (as in note 4), p. 67.

26. Russell 1971 (as in note 5), p. 138.

The Treatment of Two Karel Appel Gouaches

Piet van Dalen and
Gabriëlle Beentjes

In November 1948, a number of artists formed a movement that became known for its unorthodox techniques and imagery: CoBrA. One of the founders was the Dutchman Karel Appel; fellow group members included Corneille and Constant, also Dutch, Christian Dotremont and Joseph Noiret from Belgium, and the Dane Asger Jorn. The hometowns of the group members – Copenhagen, Brussels, and Amsterdam – formed the basis for the name CoBrA. The group was inspired by primitive and folk art and, particularly in the case of the Dutch members, children's drawings. They used primary colors combined with strong, dynamic, independent lines to give their works a spontaneous, almost nonchalant appearance.

During their early years, CoBrA members faced uneasy financial situations. Their work differed greatly from the acceptable norm of what "art" should be and was therefore not easy to sell. They chose to use gouache primarily because it was cheap. Old papers were often used as supports, varying from old newspapers to previously painted sheets. CoBrA members also often used large cheap pots of Eta-paint, a decorative paint made by Talens. Appel's creative process is typical of the CoBrA group. While painting and drawing, he struggled to find a form that suited him. He overpainted some areas, sometimes scratched wet paint layers away to reveal underlying colors, and combined a variety of incompatible materials such as chalk and paint. Moreover, Appel was not at all economical with his paint, and it is often layered very thickly on the paper supports.

The Stedelijk Museum in Schiedam, the Netherlands, owns 29 Appel gouaches on paper, all made in 1948. Of these, 20 are so deteriorated that they can only be stored flat and any exhibition of them is completely out of the question. Since these are important works representative of the artist's early oeuvre, the museum decided to conserve two of them as part of a pilot project to find a suitable method for treatment of the entire group. This paper discusses the treatment of *Three Heads* and *Standing Figure* and makes clear that the artist's creative process was directly responsible in part for the damage to the works; this informed the course of conservation treatment undertaken to stabilize them.

The fact that 1948 was not a financially stable time for the artist is clearly reflected in his choice of materials. Appel used inexpensive gouache on sheets of strongly calendered, machine-made, wove paper, many of which had been previously painted. Paintings are found on the backs of the supports as well as underneath his final renderings. It is possible that the bottom-most layers were painted much earlier than the upper layers. The resultant paint layers are very thick, probably leading to the current adhesion problems between paint and paper, and between the paint layers themselves. The upper layers of gouache are often cracked and locally flaking, especially where the uppermost paint layer is particularly thick. Under magnification, the uppermost paint layer looks like a sun-dried mudflat (see Fig. 3.24). In addition

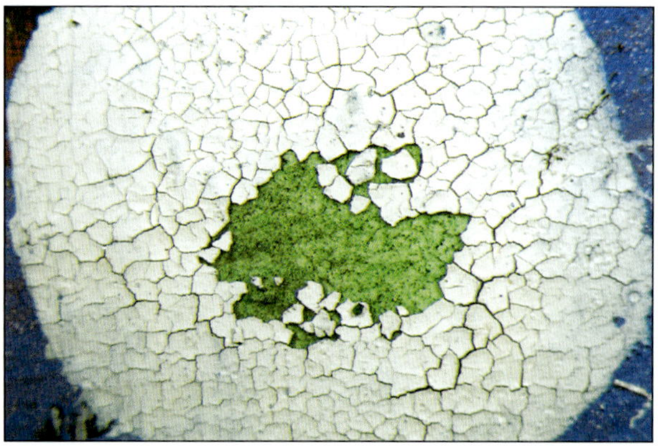

Figure 3.24 Karel Appel (Dutch, b. 1921), *Three Heads* (detail of Fig. 3.27), 1948. Cracks visible at 10x magnification through the binocular microscope.

to pigment, gouache contains a large quantity of filler, such as barium sulfate or chalk; a little binding medium, such as gum arabic; and a softener in the form of glycerine. The amounts of filler and binder influence the behavior of the paint over time. A high percentage of filler and a low percentage of binder can cause cracks to form in the paint layer as it dries because cohesion within the paint is not sufficient to form a uniform film. Both *Three Heads* and *Standing Figure* have cracks that appear to have formed shortly after completion of the works. An important observation supporting this hypothesis is that in *Standing Figure*, some of the later applications of paint have run into these cracks (see Fig. 3.25). Around 1985, the works were mounted and framed. The fact that no dislodged paint fragments were discovered in the frames when they were opened over a decade later for this project suggests that the damage occurred prior to this time.

In 1997 and 1998, the relative humidity (RH) and temperature where the works were stored were monitored for a period of nine months. It appears that the RH varied between 36.8% and 53.7% and the temperature between 14.5°C and 30.3°C. Assuming that these fluctuations were not restricted to this particular time period, they certainly would have contributed to the desiccation of the paint and subsequent formation of cracks. Furthermore, these unfavorable conditions probably caused tension between the paper supports and the bottom-most paint layers, as well as between the intermediary paint layers as each responded differently to changes in temperature and RH.

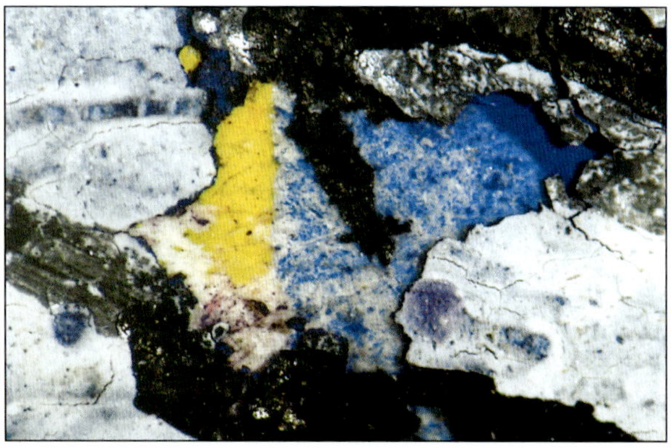

Figure 3.25 Karel Appel, *Standing Figure* (detail), 1948. Cracks visible at 25x magnification through the binocular microscope.

Since the choice of consolidant is strongly dependent on the constituents of the paint, and the course of treatment would in turn be dependent on this information, a number of pigment samples from *Three Heads* were sent to the laboratories of the Instituut Collectie Nederland for analysis. The research was aimed at establishing what binder was present in the white paint used in the uppermost image layer and in the underlying green paint layer. The investigation was restricted by the quantity of material available for analysis. The primary results of the analysis revealed the presence of gum and significant quantities of linseed oil in both the white and green paint. The quantities of linseed oil were not sufficient enough to call the medium oil paint rather than gouache. However, the presence of linseed oil could well explain the poor adhesion of some of the paint layers. With the aid of a binocular microscope, we were able to establish that the white and green paints in *Three Heads* had a characteristic glossiness comparable to that of the purple and green paint in *Standing Figure*. This gave us reason to believe that these paints contained linseed oil as well, though they were not analyzed.

Traditionally, linseed oil is used as a binding agent in oil paints because it is the fastest drying oil. During the drying process, the oil gradually air-dries and becomes thick and hard. Raw linseed oil, on the other hand, dries slowly and is used to store paint brushes to prevent them from hardening and drying out. When smeared thinly, raw linseed oil dries to an elastic, glossy, transparent layer in three to five days. Appel's reasons for using gouache containing raw linseed oil can only be surmised. It could be that he always stored his brushes in raw linseed oil and used them without wiping the oil from them – this would indeed explain the small quantities in which the oil is present. It is also possible that he added raw linseed oil to his paint to make it more smooth and spreadable. Since gouache contains a gum emulsifier, small quantities of oil can be mixed into it despite the fact that it is a water-based paint. A further possibility is that Appel added any leftover oil paint on his palette to his gouache colors.

The treatment of both paintings centered on the consolidation of lifting and loose fragments of paint. It was felt that the traditional method of using a fixative would not prevent further damage to the paintings, as it was unlikely that a fixative would penetrate the paint layers sufficiently to restore the bond to the support. Indeed, it was more likely that a fixative sitting on the surface would increase tension and cause the lifting paint to distort even more. Instead, a consolidant was sought that could be introduced under the flaking paint. Based on a literature survey and previous experience, methyl cellulose was considered. The long-term stability of methyl cellulose is well known in paper conservation practice. It is flexible, and from a chemical standpoint is compatible with the constituents of gouache. The literature on similar projects confirms the success of using methyl cellulose: a good bond is formed without imparting glossiness or discoloring the paint, and subsequent applications of methyl cellulose could be made to enhance the effects of the first.

Several factors were taken into consideration when deciding on a specific consolidant. First, the consolidant should not dissolve the paint or pigment or change the image or structure of the paint. This is a possibility when the solvent used is either water or alcohol (or both). Second, the color should not change. If the paint contains a lot of pigment and little binder or filler, the addition of a consolidant would fill the large spaces between the pigment particles and make the color appear darker, which would be unacceptable. Third, no gloss should form on the matte surface. The difference between a glossy and matte surface is an optical effect caused by the quantity of binder present in the paint. The more binder, the smoother the surface, and therefore the glossier the appearance. If the consolidant were to end up on and around the flaking paint, the matte surface could easily become glossy and discolored. Thinner, more finely applied consolidants reduce the chances that this phenomenon will occur.

A low-viscosity consolidant, a means to introduce a very fine spray of consolidant, and a climate-controlled space were required for treatment to proceed (see Fig. 3.26). Three methyl celluloses were purchased from Dow Chemicals, varying from low viscosity (Methocel A-15), through medium viscosity (Methocel A-4–C 400 cps), to high viscosity (Methocel E-4–M premium 4000 cps). All are colorless, even after aging, and all are soluble in water and/or alcohol. To finely spray the consolidant, an ultrasonic mister based on the ultrasonic mister designed by the Canadian Conservation Institute was purchased. The mister enabled ultrasonic movement of the consolidant. Using a flexible hose and a nozzle, it was possible to direct the consolidant precisely to the right location. The fine dispersion of the consolidant prevented shine, unless applied in very large quantities. Furthermore, in this form the consolidant could be easily sprayed in the cracks between the flaking pigment so that it would spread out beneath them.

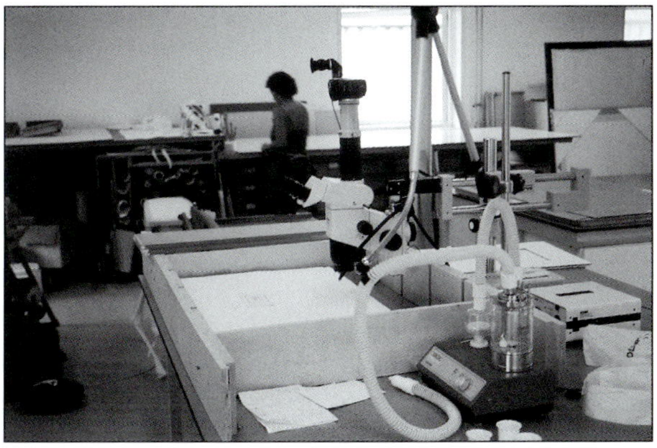

Figure 3.26 The ultrasonic mister and humidity chamber.

The literature indicates that it is advisable to work in a climate-controlled space. In a space where the RH is high, the consolidant will evaporate very slowly and form an even layer. In a very dry atmosphere, the consolidant will evaporate too quickly and be drawn back to the surface. In addition to high RH, the object should also be damp to encourage capillary penetration of the consolidant. Unfortunately, the studio at Art Conservation B.V. did not have a separate humidity chamber. It was therefore necessary to build one in box form, covered by two glass plates. The set-up was slightly larger than the artworks. The air within the chamber was humidified with mist. A major disadvantage of this method was that the conservator had to work from outside the climate-controlled space.

Three Heads was the first gouache to be treated (see Fig. 3.27). After being cleaned with a moist cotton swab to remove loose surface dirt, the gouache was dampened and placed in the humidity chamber on top of Gore-Tex® over a wet blotter. The RH within the chamber was then raised to 85% by means of mist. At this point the first problem arose: to treat a corner of the gouache, one of the two glass plates covering the chamber would have to be opened quite wide, causing the RH to drop sharply. Moreover, the conservator and the humidity chamber were covered under a large sheet of clear plastic to maintain the proper RH level. Though this proved effective for humidification purposes, it was not particularly comfortable for the conservator. A more pleasant solution was found by using several smaller glass plates in place of just two large ones. In this way, an opening could be made in just the right place for localized treatment. When necessary, the opening was made even smaller with the aid of polyester film.

Once this initial problem was solved, the application of the Methocel could begin with the aid of the ultrasonic mister (see Fig. 3.28). The first application was of a 0.25% solution of the medium-viscosity Methocel, which we believed would penetrate deeply into the various layers until they became saturated and absorbed no further consolidant. Theoretically, this would have solved the cohesion problem: the bond between the paint

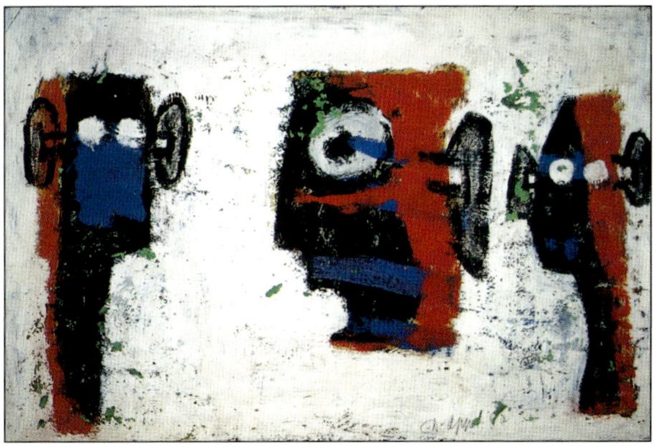

Figure 3.27 Karel Appel, *Three Heads*, 1948 (before treatment). Gouache on paper, 34.3 x 50 cm. Stedelijk Museum Schiedam, The Netherlands.

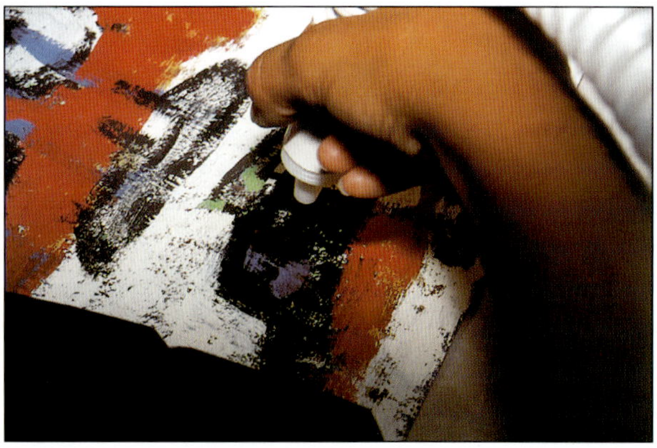

Figure 3.28 Applying the Methocel with the ultrasonic mister.

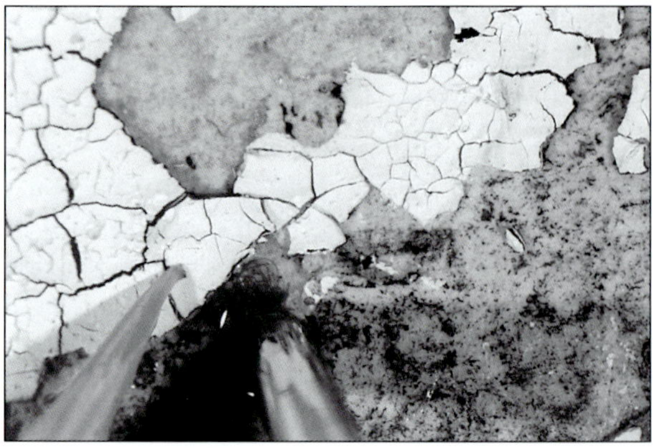

Figure 3.29 Applying the Methocel with a brush and wooden skewer.

layers would be restored. After several applications of the 0.25% solution, a 2% solution of low-viscosity Methocel was applied to consolidate the flaking paint. It became readily apparent that the low-viscosity consolidant was too thin; the paint layer became wet, but there was not enough tack to adhere the lifting paint. This technique, which in the literature had seemed ideal for our purposes, did not appear to have any practical effect. The adhesion between the paint layers was insufficient, and the paint layers remained loose despite many applications of Methocel. It is possible that the linseed oil present in the paint interfered with adhesion, the methyl cellulose chosen was not strong enough, or the solutions used were too dilute to form an adequate bond. In an attempt to rectify this, the ultrasonic mister was replaced with a pressurized syringe to apply the methyl cellulose in carefully controlled doses to specific areas. A stronger Methocel solution, 1% instead of 0.25%, was then tried, but it still failed to have the desired effect. Examination under a binocular microscope at this point showed that large droplets of Methocel were being formed which, though not visible to the naked eye, were larger than those formed by the mister. Unfortunately, in this quantity the consolidant did not spread adequately over the surface.

In preliminary studies it was noted that a combination of alcohol and methyl cellulose could produce good results. So a new approach was tried. A solution of high-viscosity Methocel mixed with 20% alcohol was applied with the ultrasonic mister. The presence of the alcohol enabled the Methocel to penetrate more deeply and work effectively to consolidate the lifting paint. This approach was only partially successful. From our various attempts to consolidate the painting, we were able to draw a number of conclusions as to the effectiveness of our methods. First, the bottom-most paint layer appeared to absorb all of the methyl cellulose applied, so no adhesion could take place between the paint layers themselves. The use of a more viscous solution caused the consolidant to rest on the surface, eventually providing adhesion, but only between the uppermost paint layer and the loose, flaking paint on the surface. We finally concluded that the loose paint particles could be consolidated with a 2% solution of medium-viscosity Methocel applied with a brush and wooden skewer (see Fig. 3.29). With the help of a microscope, which was primarily used to check that no unexpected changes had occurred, the loose pigment particles were set back down in their original positions. Small fractures in the paint layer were treated with a low- or medium-viscosity Methocel, applied with the ultrasonic mister.

As the project continued, the means of humidification were modified. For the treatment of *Three Heads*, moisture was introduced via the verso of the painting, as this seemed the safest method with respect to the transport of dust and dirt. In practice, this had more disadvantages than advantages. When weights were applied locally after treatment with the consolidant, the object absorbed more moisture from the damp blotting paper in the weighted areas, which led to uneven tensions in the paper and image layer. Some capillary action did occur, but it was unclear whether this was caused by the added moisture or because the underlying blotter was acting as a sort of sponge. Fortunately, the migration of dirt was prevented, since the image layer was barely absorbing moisture.

Armed with the knowledge and experience of the first treatment which, though ineffective as far as our original plan was concerned, was ultimately considered a success after much modification, we turned our attention to the treatment of *Standing Figure*. The same set-up was used, though this time humidification occurred only through conditioning in the humidity chamber. In this way moisture was evenly introduced to all areas of the object, minimizing the risks of tension through uneven absorption. 0.25% and 2% solutions of Methocel A-4–C were alternately applied with the finest nozzle of the ultrasonic mister. The weaker solution was used to saturate the bottom-most paint layers, and the stronger solution was applied in the cracks to restore the adhesion between the paint layers. These areas were then weighted down with a sandwich of Hollytex, wool felt, and glass plates. During treatment we observed to our surprise that the underlying green paint was showing through the white areas. Fortunately, once the Methocel was dry it was no longer visible. This did, however, emphasize the sensitivity of the gouache to moisture, and it is advisable to proceed with great caution when using water-based consolidants on gouache.

The alternating use of 0.25% and 2% solutions for the purpose of saturating the underlayer and adhering the top layer was still not satisfactory. By using a blotter instead of felt when weighting the object, we hoped to increase the pressure placed directly on the paint layers. Eventually it became apparent that the bottom-most paint layer needed to be completely saturated with consolidant before the loose paint would adhere to the substrate. As a consequence of this, however, the bottom-most layer did acquire a glossiness, but this did not detract from the results as it was only visible through the microscope. We were, after all, dealing with underlayers of paint, which the artist did not intend to be visible. Furthermore, the saturation of the underlayers with consolidant was necessary to allow adhesion to take place.

Appel's superimposition of different paint layers of varying thicknesses on previously painted paper was a major contributing factor to the current state of the works. His technique certainly influenced our approach to the conservation of the paint layers. The use of the humidity chamber in the treatment of the paintings was of utmost importance. The objects relaxed, as did the flaking and lifting paint. The ultrasonic mister proved useful for the application of Methocel. By using this apparatus, a great deal of Methocel was introduced onto the surface and between the paint layers before any glossiness appeared. Using this technique, cohesion between the paint layers was somewhat improved. To adhere the lifting paint to the bottom-most layer, a small brush and a wooden skewer gave the best results. The use of different percentages (0.25% and 2% solutions) of the same medium-viscosity Methocel worked well. The low-viscosity Methocel was too thin and did not provide good adhesion; the high-viscosity Methocel formed droplets that were too large. In the final assessment, it is clear to us that a conservation plan, even when preceded by an extensive literature survey of comparable treatments supplemented by pigment analysis, still cannot predict the behavior of a specific object during treatment. It is therefore important that the conservator be flexible enough to adapt and change the plan where necessary, and wherever possible, to meet the demands made by the object.

References

Cumming, L., and J. Colbourne 1998. "The conservation of Mrs. Marton, an eighteenth-century pastel and gouache portrait by Daniel Gardner," *The Paper Conservator* 22, pp. 38–47.

Dierks-Staiger, R. 1996. "Anwendung von Aerosolen zur Konsolidierung von Gouachemalerei auf Papier." Ph.D. diss., Staatlichen Akademie der Bildenden Künste, Stuttgart. In *Schriftenreihe des Instituts für Museumskunde an der Staatlichen Akademie der Bildenden Künste,* Vol. 14 (Stuttgart: K. W. Bachmann Hsg., 1997).

Dignard, C., et al. 1997. "Ultrasonic misting, part 2: Treatment applications," *Journal of the American Institute for Conservation* 35–36 (spring 1996–winter 1997), pp. 127–41.

Hansen, E. F., and R. Iowinger 1990. "Investigations into techniques for the consolidation of high pigment volume concentration paint at the Getty Conservation Institute," *WAAC Newsletter* 3, no. 12.

Hansen, E. F., and P. Volent 1994. "Solvent sensitivity testing of objects for treatment in a vapor-saturated atmosphere," *Journal of the American Institute for Conservation* 33–34 (spring 1994–winter 1995), pp. 315–16.

Kaltenegger, U. 1994. "Conservering van afbladderende gouaches van Charlotte Salomon," *Restaurator* 15 (1994), pp. 14–23.

Kessler, A. 1997. "Anwendung von Aerosolen zur Konsolidierung matter Farbschichten unterschiedlicher Schadensbilder unter Berücksichtigung des Eindringverhaltens des Kosnsolidierungsmittels." Ph.D. diss., Staatlichen Akademie der Bildenden Künste, Stuttgart.

Michalski, S., and C. Dignard 1993. "Consolidation of powdery paint using the ultrasonic mister," *ICOM Committee for Conservation, 10th Triennial Meeting, Washington, D.C., Preprints, Poster Abstracts.* London: James & James.

Stokvis, W. 1990. *Cobra.* Ph.D. diss., Vrije Universiteit, Amsterdam.

Daylight Fluorescent Colors as Artistic Media

Margaret Holben Ellis, Christopher W. McGlinchey, and Esther Chao

Once limited to industrial usage, daylight fluorescent colors, known to most readers as Day-Glo®,[1] are now incorporated into a vast array of everyday products, ranging from packaging, clothing, jewelry and accessories to sports equipment and works of art.[2] The widespread commercialization of the colors led to their induction into the Pop Art movement of the 1960s. Used ever since as artistic media, the colors' direct, almost impudent, aesthetic message demands both a psychological and physiological response.

As they age, daylight fluorescent colors first darken and then lighten. While it can appear that two separate reactions are occurring, in reality they are taking place simultaneously – that is, the reduction of fluorescence and the fading of the actual chromophore. The most notable reaction appears to be reduction of fluorescence, which proceeds upon exposure to both light and heat. As further aging takes place, an overall lightening of the daylight fluorescent pigment occurs, and fading becomes more apparent. This mechanism makes accurate color matching for inpainting procedures difficult, if not impossible, to achieve.

Complicating efforts to retard or at least monitor the colors' inevitable shift in appearance is the difficulty of documenting them accurately. In many museums, visual recording systems (photographic and digital imaging) are used to record or track color changes. However, no color-reproduction process is capable of replicating the phenomenon of daylight fluorescence. The use of instrumental methods for colorimetric measurements of daylight fluorescent colors is likewise problematic because of the specialized equipment required. As a consequence, different parameters for "measuring" daylight fluorescence have been established by the paper and textile industries, among others, to fit individual applications, with the result that comparison and reproduction of data is difficult.

When they are used to create works of art, daylight fluorescent colors pose unique conservation problems for conservators and curators. Such artworks are best viewed under ultraviolet (UV) light exclusively or full-spectrum visible light, the very agents that will accelerate the colors' deterioration. In the interests of preservation, compromises in exhibition are, therefore, often recommended.

History

During the 1930s, Robert Switzer was attending the University of California at Berkeley on a scholarship and had a job unloading crates from railway cars for the Safeway grocery chain. He fell and hit his head on a rail. Months later he awoke from a coma, his vision blurred. He had suffered a fractured skull, a partially severed optic nerve, heavy internal bleeding, and extensive memory loss. The family doctor recommended that he recuperate in a darkened room, which, together with his younger brother Joseph's hobby of magic, led to an interest in fluorescence. The brothers discovered that certain organic compounds, found in the storeroom of their father's pharmacy, fluoresced under UV or "black light." Soon they were brightening up Joseph's amateur magic act by using homemade fluorescent paints. Their most popular act involved applying the paint to a mask and headdress depicting a Balinese dancer which, under UV illumination, created the illusion of an exotic woman going one way while her head floated off in another. In 1934, with a pooled investment of $220, the Switzer brothers established the Fluor-S-Art Company in their mother's kitchen, using her Mix-Master as their primary workhorse.

By 1940, in addition to creating pigments that fluoresce only under UV light, Bob and Joe Switzer had developed the first so-called daylight fluorescent or "coldfire" pigments, colors that glow in visible light with or without the aid of a black light. Their first commercially successful products were brilliant daylight fluorescent panels of cellulose acetate fabric, used in World War II by Allied ground troops in North Africa to signal their presence to Allied bombers. Because some ordinary colors cannot be distinguished at extreme distances and in low light conditions, such as at dusk and dawn, the troops would lay out the Switzers' vividly colored strips in coded sequence to identify their units and send simple messages.[3] Another early military application of the colors was on aircraft training missions. By using the colors to mark the planes of student pilots, mid-air collisions were significantly reduced. By 1944 mention of the colors' potential usage by artists had already appeared in *American Artist* magazine, where their suitability for special effects was noted, as well as the eventuality that one day entire rooms would be illuminated by painted walls and ceilings rather than fixtures, bulbs, and tubes.[4]

In 1946 the two brothers created Switzer Brothers, Inc., in Cleveland, Ohio (renamed the Day-Glo Color Corporation in the late 1960s) to meet the growing demand for their daylight fluorescent products, which were still largely reserved for hazard warnings and industrial applications. By the late 1950s, daylight fluorescent printing inks for both silk screen and offset lithography had been developed. In 1958 Proctor and Gamble came out with the now familiar Tide box, touted to "jump off the shelf" and into the hands of the unsuspecting consumer.[5]

Packaging applications proliferated in the 1960s, and daylight fluorescent colors became synonymous with the American graphics scene. Today these pigments are added to everything from molded plastic detergent containers and hockey pucks to Hula-Hoops. Like the brand names Xerox, Scotch Tape, and Magic Marker, Day-Glo has become a generic term and has entered everyday vocabulary.[6]

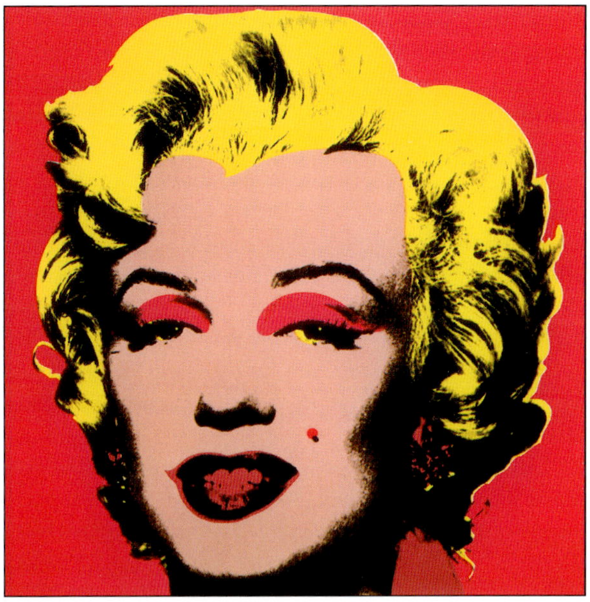

Figure 3.30 Andy Warhol (American, 1928–1987), *Marilyn*, 1967. Screen print with Day-Glo® colors on white card. The Art Institute of Chicago, Margaret Fisher Endowment, 1998.442.10. This is a later printed version of a 1962 painting of the same subject employing a Day-Glo palette.

The psychological response to daylight fluorescent colors

Daylight fluorescent colors quickly became associated with the freewheeling hippie lifestyle of the 1960s. At a 1967 Central Park "happening" in New York, for example, the period's flower children adorned their bodies and faces with Day-Glo paints.[7] Commercial artists were quick to exploit the "psychedelic" colors, as they became known.[8] Many of the designs they created intentionally mimic or were meant to enhance the mind-altering effects of lysergic acid diethylamide (LSD) and mescaline on human color perception.[9]

The Fillmore West concerts in San Francisco left behind a legacy of graphic rock-and-roll memorabilia. For each weekly concert from 1966 to 1971, posters, handbills, and tickets, many in daylight fluorescent colors, were designed by a group of artists who became known for their idiosyncratic style.[10] It has been said that the posters of the Fillmore are "as much about art as … about one of the most profound periods in popular music, and ultimately what was and will remain relevant to people's lives."[11] Some of the era's rock-and-roll LP album covers were memorable as well. The artist Peter Max will forever be associated with the cosmic colors and imagery he used for the Beatles' *Yellow Submarine* and *Magical Mystery Tour* album covers.[12] Mere ephemera then, first-edition Fillmore posters and rock-and-roll memorabilia have become valuable commodities.

A genre of graffiti artists, specializing in covering New York subway cars with their signature spray-painted designs, shunned the dazzling daylight fluorescent colors due to their poor light-fastness and difficult handling properties. Because the pigments are essentially transparent, they have poor covering power, and used as spray paints, they tend to drip and run easily.[13] Given the considerable risks, both legal and physical, entailed in creating their works, the artists' preference for more permanent materials is understandable. Keith Haring, who started out as a graffiti artist, seems never to have used daylight fluorescent colors outdoors, growing fond of them only after becoming an established indoor artist. Conversely, the graffiti artist Peter Mayer, whose slapdash schnauzers popped up across New York City in the late 1980s and are re-emerging today, used whatever "real-time" colors were at hand, including Day-Glo.[14]

Daylight fluorescent colors naturally appealed to Pop artists of the 1960s, who transformed commonplace images into gaudy pseudo icons. Examples include Andy Warhol's *Marilyn* (Fig. 3.30), whose face "seems perpetually illuminated by the afterimage of a flashbulb,"[15] and his *Flowers* of 1964, whose "garish Day-Glo hues were meant to clash with tasteful interior decoration."[16] The trend continued after Abstract Expressionism with the artist Frank Stella, whose intentional use of man-made colors prevented any degree of illusionism. In Stella's lithographs, the alien colors are combined with glitter for even higher voltage.[17] More recently, the paintings of Peter Halley have been said to convey a sense of "urban alienation"[18] through his use of daylight fluorescent colors, described as "unlike anything in nature."[19] Thus, despite their blue-collar pedigree, daylight fluorescent colors were ushered into the elegant precincts of galleries and museums.

The physiological response to daylight fluorescent colors

In addition to their unequivocal aesthetic of brash confrontation, daylight fluorescent colors also evoke a reflexive physical reaction because of their greater perceptibility by the human eye. A daylight fluorescent color not only absorbs colors the way a conventional one does, it also takes in the absorptions that are occurring at higher energies and converts them to luminescent light whose emission wavelengths, the perceived color, and the conventional color overlap. The peak wavelength that corresponds to this color is referred to as the dominant wavelength of the color. The energy available from solar radiation capable of promoting excitation includes UV (c. 300–400 nm) and up to the dominant wavelength of the fluorescent color (Fig. 3.31). Fluorescent color appears to be literally making its own light, because the eye cannot justify the amount of color being produced with its light source; that is, the color "glows." Because fluorescence emission and conventional reflectance are superimposed, a brilliant color of high purity or luminosity is produced.

Under optimal lighting a clean, bright conventional color is able to reflect 80–90% of its color; a daylight fluorescent color can appear to reflect up to three times as much. Thus it is that Blaze Orange™, which reflects in the neighborhood of 300%, is used for roadside safety signage, life preservers, and airport directional signals, potentially life-saving markers that are more easily detectable from long distances and under poor weather conditions than conventional colors. Reflectance data for Day-Glo colors indicate that all, with the exception of blue, re-emit well over 100% of incident light.[20]

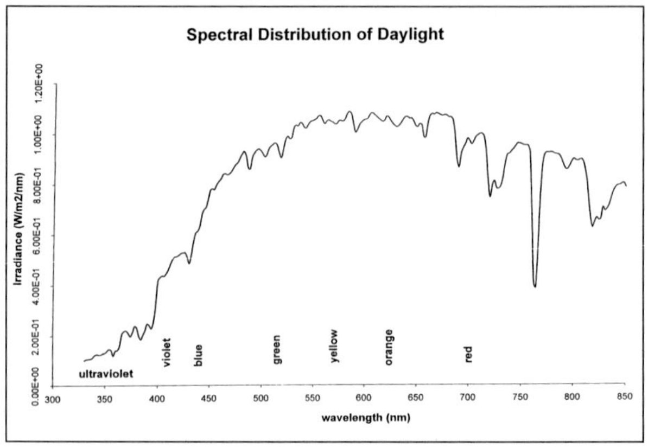

Figure 3.31 Spectral irradiance of daylight. Ultraviolet radiation and wavelengths of visible light below the dominant wavelength of the daylight fluorescent color are capable of promoting fluorescence.

The physics of daylight fluorescent colors have been exploited in the creation of works of art and can lead to conservation problems. For example, a dab of daylight fluorescent color can be used to "juice up" muddied printing inks, a trick of the printmaking trade used to salvage over-mixed inks. Additionally, in screen prints one daylight fluorescent color may be included among a host of individually applied conventional colors in order to create the resonance among them. Some portraits by Chuck Close, made from over 100 separately screened colors, are examples of this practice.[21]

The recommended practice of using optically brightened paper below daylight fluorescent media also takes advantage of their super-reflectance. This type of support serves to make the colors reverberate even more, but of course compounds conservation problems, since optical brighteners, being fluorescent, appear to follow a similar degradation pattern.[22]

The most common artificial source used for illuminating daylight fluorescent colors is the fluorescent black-light fixture having a spectral peak at 365 nm. These lamps are used to illuminate works of art meant to be viewed under UV light as well as visible light. Examples of such works range from inexpensive posters available in local headshops to Warhol's 1986 *Christ* paintings.

Daylight fluorescent colors differ from phosphorescent or luminous pigments. They are similar in terms of the re-emission of light but differ in their time of afterglow. Phosphorescent substances continue to emit light long after the input of energy ceases. Artists such as Warhol and Nancy Haynes have used phosphorescent pigments to create phantasmal images intended to be displayed in a darkened gallery.

Daylight fluorescent pigments

Daylight fluorescent pigments are fluorescent dyes that have been dissolved in thermoplastic resins.[23] The solubility properties of the original thermoplastic resin components contributed to the poor solvent resistance of early daylight fluorescent paints. These resins, originally polycondensation resins based on a sulfonamide structure, typically require less than 5% dye. At higher concentrations quenching of the dye occurs, and the fluorescence appears to diminish in brightness. The brittle product consists of a solid solution of resin and dye that can be easily ground into a powder and called a pigment, in spite of the fact that a binder is present. Additional binders are required to make this composite "pigment" into a bona fide paint or ink.

Advancements have been made by copolymerizing the fluorescent dye with binders to produce a fluorescent pigment more resistant to solvents.[24] Because the binders polymerize by a condensation mechanism, the procedure is a co-condensation reaction. As with the earlier mixtures of dyes and binders, these co-condensates can shift the luminescence of the dye by forming intermolecular hydrogen bonds. The increased solvent or bleed resistance of these materials has allowed fluorescent pigments to be prepared in a wider range of vehicles.

Daylight fluorescent paints covered with a clear coat containing UV absorbers offer vastly improved outdoor light-fastness. The UV-absorbing coatings immediately darken the color slightly because of the decreased amount of UV radiation reaching the dye. Rudimentary tests, however, indicate that the UV absorbers in these coatings mainly absorb the shorter wavelength of UV, leaving the longer wavelengths available to promote fluorescence. Today, additives such as hindered amine light stabilizers (HALS) are also used to retard decomposition, and the search is on for more stable dyes.[25]

Conservation problems of daylight fluorescent colors

Both empirical and scientific evidence indicates that daylight fluorescent colors eventually lose their ability to become excited by irradiance.[26] The first visible change is a darkening, followed by the more familiar lightening or fading.[27] While it can appear that two separate reactions are occurring, the reduction of fluorescence and the fading of the actual chromophore are actually taking place simultaneously.

Five different printed lithographic colors from a Funky Enterprises poster and one fine-art silkscreen ink, Handschy Aurora Red™ dating from 1974, were tested using a fadeometer for light-fastness evaluation. The Handschy ink is the same used in the creation of James Rosenquist's *F-111*, printed by Petersburg Press.[28] The Handschy ink was further studied under "natural" light conditions and separate heat aging tests. Both aging methods were conducted under ambient and anoxic conditions.

The samples were evaluated under two different lighting conditions: UV fluorescent and UV-filtered xenon (see Fig. 3.32). Examination of one color's spectral reflectance under either of these conditions indicates that the most significant change initially occurs at the dominant wavelength of fluorescence (see Figs. 3.33 and 3.34). This initial drop corresponds to the darkening. In addition to the reduction in the reflectance at the dominant wavelength, upon further aging a broadening of reflectance occurs at colors adjacent to the dominant wavelength that corresponds to fading. This development affects what is referred to as the "purity" of the fluorescent color. The loss of chromatic purity results in subtle dulling. The darkening was observed for both

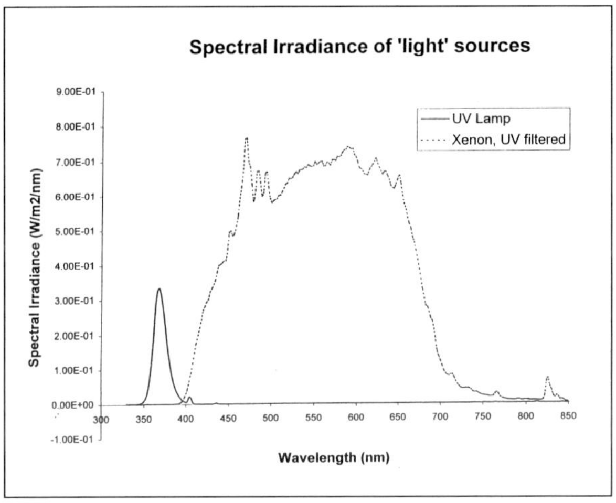

Figure 3.32 Spectral irradiance of light sources used in this study. The irradiance values shown correspond to the distance of the light source to the image surface for the data collection in Figures 3.33 and 3.34.

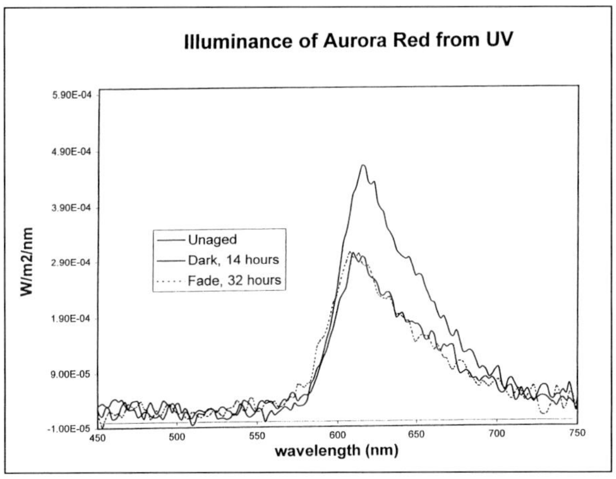

Figure 3.33 Combined reflectance and illuminance of Aurora Red™ from ultraviolet-filtered xenon arc lamp. Dominant wavelength is 622 nm for unaged sample. Measurements were taken in the diffuse reflectance region.

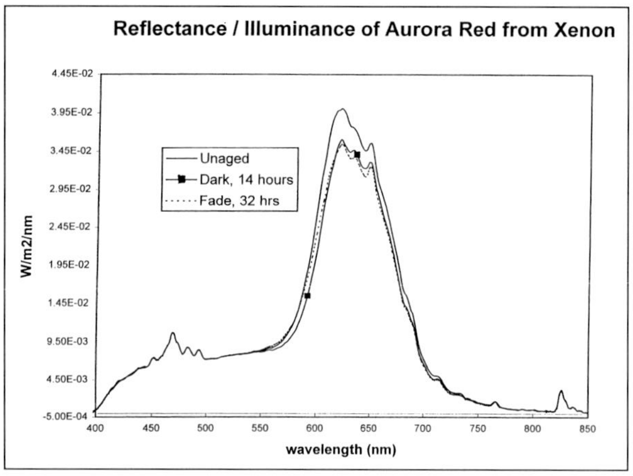

Figure 3.34 Illuminance of Aurora Red™ from black light fluorescent light. Dominant wavelength is 622 nm for unaged sample. Measurements were taken in the diffuse reflectance region.

natural and accelerated aging and heat aging. Anoxic conditions did not appear to have an effect in either case. Furthermore, this trend of darkening and fading occurred for all materials tested. (It should be noted that there is no point at which the aged material has the same appearance as the unaged one under identical viewing conditions. What occurs is a fading of the slightly muddied version of the color.)

This mechanism of aging has practical implications for color matching in the paint industry and, by extension, for conservation procedures involving color matching, such as inpainting. If the original daylight fluorescent medium has aged beyond the initial darkening point and lightening is now proceeding, any fluorescent inpainting medium will first darken, thereby diverging asynchronously with continued visual changes in the original. Not knowing where an artwork is in its life cycle of aging makes accurate color matching difficult, if not impossible. Additionally, metamerism, whereby colors match under one light source but not under a different one, can occur if the color's appearance must match under black light, as well as in visual light.

Whitmore's recently developed device for making micro-scale light-fastness predictions, with UV not filtered from its xenon light source, might prove helpful in predicting this problem.[29] Samples whose "aging history" is unknown, i.e., those that do not go through the initial darkening stage when tested with the micro-fading equipment, may have already undergone overexposure to UV-rich light or, perhaps, excessive heat. The conservator, therefore, can be less concerned with matching fluorescence and can concentrate instead on matching color. The short time required for microfading and the moderately elevated temperatures from the light beam in the device suggest that darkening would be caused not by thermal degradation, but by low-wavelength visible light and whatever UV is present.

The difficult task of matching both color and fluorescence is not restricted to conservators. The paint industry admits to using the "lumberjack" style of calculating fluorescence in color matching: by "chopping off" the fluorescent peaks of a color's spectrum and relying instead on instinct and a good "eye."[30]

Documentation

Documentation poses a further challenge in the effort to retard, or at least monitor, the colors' inevitable shift in appearance. The photographic and digital imaging systems typically used in museums to record or track color changes are incapable of replicating the phenomenon of daylight fluorescence. Instrumental methods for colorimetric measurement of daylight fluorescence are likewise problematic, requiring specialized equipment. The paper, textile, and signage industries, among others, have developed different parameters for "measuring" daylight fluorescence to fit their particular needs. Consequently, comparison of materials and data is difficult.[31]

Also, the color and degree of fluorescence produced by daylight fluorescent colors is dependent upon the type of illuminating radiation used. In fact, though the American Society for Testing and Materials (ASTM) has created guidelines for the measurement of daylight fluorescent colorants, the committee encourages their evaluation under lighting conditions that "duplicate those of anticipated use."[32]

Instrumental analysis of daylight fluorescent colors

The spectral reflectance readings of Figures 3.31–3.34 were taken using a radiometer capable of measuring radiation from 330 nm to 850 nm. The values measured show the total reflected energy from a particular source, in this case UV fluorescent and xenon arc light that has been UV filtered. The radiometer is also capable of measuring the spectral irradiance of these light sources. The reflectance values from the UV source correspond closely to luminescence because the UV source contains little visible light. The xenon arc does not correspond exclusively to reflectance because, as their name implies, daylight fluorescents are capable of fluorescing in the presence of visible as well as UV light.

Recommendations

The chief concerns for the preservation of daylight fluorescent colors are to minimize exposure to UV and short-wavelength radiation during exhibition, and to avoid excessive heat when in storage. For the former, three factors – type of light, intensity of light, and duration of exposure – need to be addressed.

As stated earlier, the fluorescent black lights available over-the-counter emit at 365 nm; this value corresponds to the radiation spike from mercury in the fluorescent tube. Lamps with an emission peak below 365 nm are not commercially available. Using a 15-watt fixture, the intensity of fluorescence was evaluated at various distances to determine how far the light source used in this study could be placed away from the artwork and still maintain an appreciable level of illumination. For the material examined in this study, the distance was between 2 and 3 meters. This corresponds to approximately 0.02 watts of UV at the image surface. Correlation of this value with the UV aging tests carried out at 5 watts of UV at the image surface indicates that, assuming reciprocity, significant darkening would occur after 1,500 hours of continuous exposure at 0.02 watts. A value of 1,500 hours can obviously be extended if UV illumination is more strictly controlled by timers activated either by viewers or by motion detectors, as has been done in several exhibitions. Even so, under the most stringent controls, significant darkening will occur after say less than five years under regular but intermittent illumination. It is possible that such black-light conditions may be suggested by darkened gallery walls and more focussed lighting, though the luminescence would only be suggested and may prove more fatiguing for the viewer than authentic.

UV-filtering acrylic sheets or UV-free lights are typically used to protect works of art from excessive exposure to the damaging wavelengths of visible light. Such measures, however, affect the way daylight fluorescent colors appear to the viewer, a critical factor if their UV-responsiveness is part of the artwork's aesthetic. Obviously, black-light posters cannot glow if they are glazed behind UV-filtering acrylic. Thus, under "safe" lighting conditions, the vibrant psychological message of daylight fluorescent colors is gradually silenced in proportion to decreasing luminescence. Non-filtering acrylic likewise reduces fluorescence, though not as dramatically.

Similar calculations were carried out for works of art meant to be viewed in visible light. It is predicted that under 100 lux or 10 footcandles, darkening will occur within 13,000 hours of total exposure, assuming reciprocity. These odds are not much better, however, because illumination under these circumstances tends to be more continuous; at eight hours per day this corresponds to about five years of exposure. For artworks not intended to be viewed in black light, clearly it is prudent to eliminate the UV component of any illumination source.

Conclusion

The paradox of daylight fluorescent colors is that, when viewed as they were intended to be seen, they elicit not only a psychological response to their brash twentieth-century aesthetic, but also a genuine physical reaction. Just as a hunter's Day-Glo safety vest is automatically detected by the human eye, so is a Day-Glo painting by Frank Stella reflexively registered. In the latter case, mental note is simultaneously taken of the artwork's aura of artificiality because the colors are not "natural." To sacrifice fluorescence, via the conservator's typical arsenal of UV-reducing tactics, is potentially to mute this two-fisted message. Ironically, for some artworks, especially those meant to be seen under UV light exclusively, an authentic viewing experience may come at the price of preservation.

Appendix

Instrumentation

The radiometer used in this study was a Li-Cor radiometer (Model LI–1800) fitted with a remote cosine receptor capable of registering UV and visible radiation from 330 nm to 850 nm in 2 nm increments with a bandwidth of 4 nm. Scans were averaged from five sequential readings. (Li-Cor, Lincoln, Nebraska)

Aging conditions

Accelerated aging was conducted with an Atlas Ci35F Fadeometer. The xenon arc was fitted with borosilicate and sodalime filters to remove shortwave UV. It was set to produce approximately 100,000 lux with a black panel temperature of 50°C giving an RH of about 40%. (Atlas, Chicago)

UV aging was conducted using a 15-watt fluorescent black light (General Electric F15T8.BLB). Wattage in the UV region (peak 365 nm) measured 5 watts on the image surface.

Indoor "natural" light aging was carried out behind window glass facing east. The window glazing absorbed UV up to 390 nm. Samples received approximately 250,000 lux hours per day on cloud-free days.

Heat-aging studies were conducted at 50°C under ambient and anoxic (N_2) conditions. Samples were placed in vessels with a Teflon® gasket and purged with nitrogen four times in a glove box, sealed, and wrapped with Teflon® tape.

Acknowledgments

The authors would like to extend their deepest appreciation to Fred Switzer, brother of Robert and Joseph, the founders of Day-Glo Color Corporation, and Gordon Russell, its former Vice-President of Marketing, and to the following friends and colleagues: Sandra Amann, Liesa Bing, Robert Blanton, Bob Brumfeld, Karl Buchberg, John Burke, Kevin Carr, David Burnes, Lee Ann Daffner, Elise Effmann, Elizabeth Estabrook, Alan Farancz, Anne Fitzgerald, Gillian Forrester, Michele Frazier, Elizabeth Freeman, Michael Geisert, David Kiehl, Margaret Livingstone, Alison Luxner, Susan Mathisen, Robert McKeever, Winline Mei, Chuck Moore, Josh Nefsky, Chris Paulocik, Karen Pavelka, Roy Perkinson, Andrea Pitsch, Neil Printz, Will Real, Nita Roberts, Shelley Sass, Jerry Shiner, Judith Solodkin, Robert Stacy, Serena Urry, and William Walmsley.

Endnotes

1. Day-Glo® is a registered trademark of Day-Glo Color Corporation, 4515 St. Clair Avenue, Cleveland, OH 44103; Rad-Glo® is manufactured by Radiant Color N.V., Europark 1046 – B-3530 Houthalen.

2. L. Bing, *The Story of Day-Glo* (Old Lyme, Conn.: Day-Glo Color Corporation, 1991), pp. 4–6.

3. The way various colors are perceived under low light conditions is also influenced by the absorption peaks of the rods and cones found in the retina of the human eye (M. Livingstone, Boston, personal communication, 1999).

4. "The art mart," *American Artist* 8, no. 8 (1944), p. 34.

5. Day-Glo Color Corporation, *The Day-Glo Designer's Guide* (Cleveland, Ohio: Day-Glo Color Corporation, 1969), p. 5. The product name Day-Glo was trademarked in Canada in 1944 and in the U.S. in 1947.

6. Daylight fluorescent colors are sometimes called "neon," probably in reference to the red/orange glow produced by elemental neon gas.

7. A. Warhol, *POPism: The Warhol '60s* (New York: Harcourt Brace Jovanovich, 1980), p. 207.

8. Coined by Humphrey Osmond, the word was originally "psychodelic," but was changed because of its negative allusions to mental illness (C. Pringle, "The language of hip," http://www.halcyon.com/colinp/hipterms.htm, 1995–98, p. 9).

9. In describing an experience with mescaline, the English writer Aldous Huxley (1894–1963) noted that "it raises all colors to a higher power and makes the percipient aware of innumerable shades of difference, to which, at ordinary times, he is completely blind." A. Hope and M. Walsh, *The Color Compendium* (New York: Van Nostrand Reinhold, 1990), p. 25.

10. Bill Graham acquired a lease in 1966 on the original Fillmore Auditorium in San Francisco and later moved to a larger space called Fillmore West, where he continued to present weekly rock-and-roll concerts until 1971. Fillmore artists included Stanley Mouse, Wes Wilson, Victor Moscovo, Alton Kelley, and Randy Tuten, among others. D. Henkel, *The Official Price Guide to Rock and Roll Magazines, Posters, and Memorabilia* (New York: House of Collectibles, 1992), p. 403.

11. G. Lemke, *The Art of the Fillmore 1966–1971* (Petaluma, Calif.: Acid Test Productions, 1997), p. 8.

12. http://www.microtimes.com/141/max.html

13. Anon., @149st., http://www.@149st.com, personal communication, 1999: "The only work rendered in fluorescent colors that comes to recollection was 'No Nukes' by the artists ZEPHRY and MIN. This painting was done in protest to nuclear arms in 1979 and appeared on the #1 IRT Broadway Local. There are some bright colors of spray paint that are frequently used that are very bright and on first glance may appear fluorescent. Examples are Jungle Green and Federal Safety Purple manufactured by Krylon and Mad About Pink manufactured by WetLook."

14. K. Littmann, ed., *Keith Haring: Editions on Paper, 1982–1990* (Stuttgart: Editions Cantz, 1993). D. Magidson, Magidson Fine Art, New York, personal communication regarding Peter Mayer, 1999.

15. R. Rosenblum, "Andy Warhol: court painter to the '70s," in D. Whitney (ed.), *Andy Warhol: Portraits of the 70s* (New York: The Whitney Museum of American Art, 1979), p. 14.

16. F. Feldman and J. Schnellmann, *Andy Warhol Prints: A Catalogue Raisonné* (New York: Ronald Feldman Fine Arts, 1985), p. 14. An invoice from Concord Artists Materials on Lexington Avenue, New York City, confirms "1 set Dayglo Colors (jars sold for $1.95 plus tax) to A. Warhol" on January 12, 1959. B. Richardson, *Andy Warhol Camouflage* (New York: Gagosian Gallery, 1998), p. 23.

17. R. Axsom, *The Prints of Frank Stella: A Catalogue Raisonné 1967–1982* (New York: Hudson Hills Press, 1983).

18. Eli Broad Foundation, "Peter Halley," http://www.broadartfdn.org/bio-halley.html (1999), p. 1.

19. A. Mac Adam, "Peter Halley," *Art News* 98, no. 7 (1999), p. 153.

20. A brilliant blue fluorescence is difficult to achieve because, aside from violet, which is the only visible light more energetic than blue, it must rely on UV for excitation. Daylight fluorescent blue is utilized in fluorescent whitening agents (FWAs) or optical brighteners, often added to papers and clothing detergents.

21. B. Blandon, Brand X, New York, personal communication, 1999.

22. F. Leclerc and F. Flieder, "Influence of optical brighteners on paper permanence," in S. Fairbrass (ed.), *Conference Papers Manchester 1992* (London: The Institute of Paper Conservation, 1992), pp. 257–63.

23. Important dyestuffs used for daylight fluorescent pigments include Rhodamine B, Rhodamine F5G (BASF), Xylene Red B (Sandoz Chemical), Fluorescent Yellow Y (L. B. Holliday), Maxilon Brilliant Flavine 10Gff (CIBA-GEIGY), Alberta Yellow, Potomac Yellow (Day-Glo Color), and Macrolex Fluorescent Yellow 10GN (Bayer). R. Ward and E. Kimmel, "Fluorescent daylight," in M. Grayson (ed.), *Kirk-Othmer Concise Encyclopedia of Chemical Technology* (New York: John Wiley and Sons, 1985), pp. 712–13 (see p. 713).

24. B. Krasovitskii and B. Bolotin, *Organic Luminescent Materials*, transl. by V. G. Vopian (New York: VCH, 1988), p. 171.

25. C. Moore, chemist, Day-Glo Color Corporation, Old Lyme, Conn., personal communication, 1999.

26. It should be pointed out that the authors examined only a daylight fluorescent ink and not other media, such as acrylics. The manufacturer of one popular brand of acrylic, Golden Artist Colors, emphatically states that its fluorescent colors are not light-fast. Golden Artist Colors, "Fluorescent colors," http://www.goldenpaints.com/fluores.htm (1998), p. 1.

27. In interviews, conservators frequently mentioned fading; however, many also described the colors as becoming "tired," probably referring to the initial loss of fluorescence.

28. J. Solodkin, Solo Press, New York, personal communication, 1999.

29. P. Whitmore, X. Pan, and C. Bailie, "Predicting the fading of objects: Identification of fugitive colorants through direct nondestructive lightfastness measurements," *Journal of the American Institute for Conservation* 38, no. 3 (1999), pp. 395–409 (see p. 396).

30. E. Ganz, "Problems of fluorescence in colorant formulation," *Color Research and Application* 2, no. 2 (1977), pp. 81–84 (see p. 84). F. Switzer, former president, Day-Glo Color Corporation, personal communication, 1999.

31. ASTM Committee E12.05 on Fluorescence Measurement, David Burns, Chair.

32. ASTM Standard E991–90, published 1991.

Assessing the Impact of Storage Environments on the Color Transfer of Felt-Tipped Pen on Paper

Barbara Rosenberg

The felt-tipped pen, generically known as the porous pointed pen, is an unstable dye-based medium that not only fades, but transfers color onto adjacent paper surfaces. This study assesses the impact of storage environments on the color transfer of felt-tipped pen on paper. There are documented occurrences of color transfer in works by contemporary artists including the drawings of Inuit artists.[1] In the 1960s, the felt-tipped pen marker entered the artist's palette, and the fugitive nature of the dyes soon became obvious. In response, conservation scientists conducted experiments on various pens and markers, investigating the effects of light and various solvents on a wide range of porous pointed pens.[2] Their research helped to define standards and stimulated an awareness of problems related to the medium's permanence, light-fastness, and water-fastness.[3]

In 1990, the McMichael Canadian Art Collection received a long-term loan from the West Baffin Eskimo Co-operative's Cape Dorset archival collection, consisting of over 100,000 works on paper, including approximately 10,000 felt-tipped pen drawings. Soon after the collection arrived at the McMichael, conservators noted that the drawings with felt-tipped pen medium showed color transfer during storage (see Figs. 3.35 and 3.36). To stem this problem, Canadian Conservation Institute scientists developed a housing system of interleaving with unbuffered Silversafe tissue (Atlantis™ 40g/m^2 Silversafe) and Mylar® folders (Dupont Mylar® Type D-3mil).[4] However, in the drawings with the more heavily applied felt-tipped pen medium, color transfer was still actively occurring onto both the tissue and Mylar®.

This study investigated the color transfer of felt-tipped medium on paper. Gathering experimental data based on three colors (red, blue, and yellow) in five storage conditions, the color changes were tracked through colorimeter readings using the CIE 1976 L*a*b* system. The final results indicate that low storage temperatures inhibited the color transfer in the samples, while high temperatures promoted color transfer. The preconditioned moisture content of the samples was not a significant factor in the transfer of color.

Methodology

A large number of variables can contribute to color transfer. For the purposes of this study, data is standardized using one type of pen and uniform paper components, with experimental environments reflecting actual storage conditions (see Table 3.1). Prior to the study, 29 Magic Marker™ Studio felt-tipped pen colors were tested on a paper substrate and artificially aged for a month in a humidity chamber at 60% relative humidity (RH) and 60°C. Nine colors indicated significant color transfer, and three were chosen for further research. These colorants were typical of the color transfer observed in the Dorset drawings. Also, the markers date from the 1970s and roughly correspond to the age of felt-tipped pens used by the Inuit.

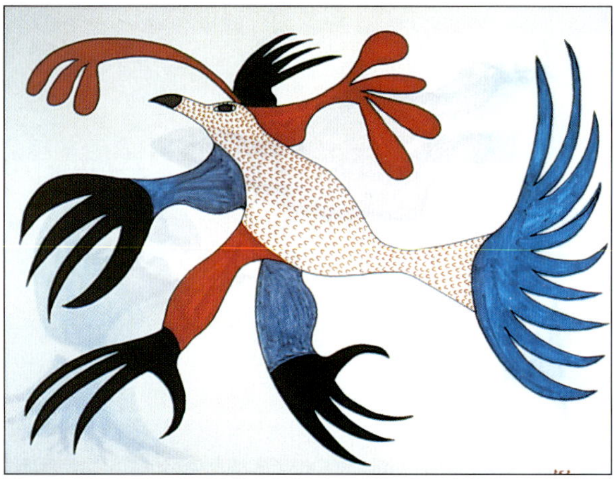

Figure 3.35 Johnniebo Ashevak (Inuit, 1923–1972), *Untitled*, 1960–65. Felt-tipped pen on paper, 50.9 x 66.6 cm. Collection of the West Baffin Eskimo Co-operative Ltd., Cape Dorset, Nunavat, on loan to the McMichael Canadian Art Collection. (Photograph courtesy of the McMichael Canadian Art Collection.)

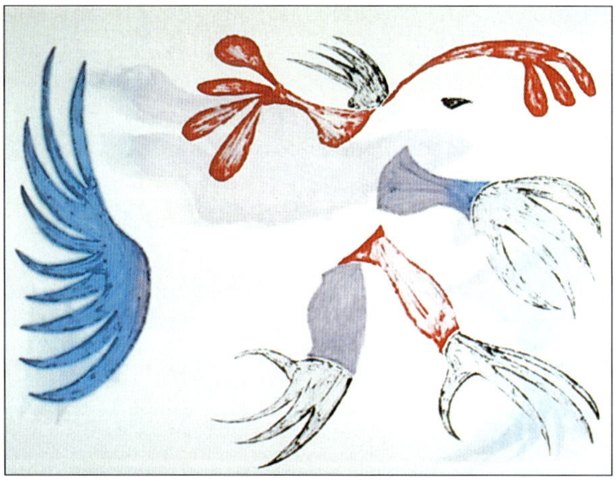

Figure 3.36 Johnniebo Ashevak, *Untitled* (verso). (Photograph courtesy of the McMichael Canadian Art Collection.)

Sample preparation

The experiment tested three Magic Marker™ colorants – Chinese Red A371, Mariner Blue 490, and Canary Yellow A713 – in five different storage environments: 50% RH 20°C, 70% RH 40°C, 70% RH 4°C, 25% RH 40°C, and 25% RH 4°C. Each sample consisted of the paper substrate with media sandwiched between two Silversafe tissues (unbuffered Atlantis™ 40g/m^2 Silversafe paper), which in turn was sandwiched between two pieces of buffered 100% rag board (see Fig. 3.37). The paper substrate was a buff-colored, 115 gsm machine-made paper with a pH of 5 and no lignin or alum content.

The markers were reactivated with xylene and the medium was applied to the paper substrate in three vertical strokes slightly overlapping one other. Five duplicate samples of each of the three colors and one control were placed in each of the five storage environments. Thus, there were 25 samples of each color for a total of 75 with medium and 15 samples without medium.

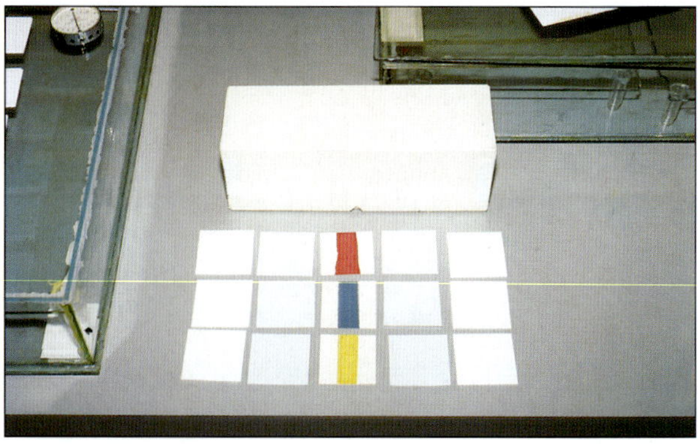

Figure 3.37 Red, blue, and yellow sample components of medium on a paper substrate sandwiched between unbuffered tissue, which in turn is sandwiched between buffered 100% rag board.

Temperature	Relative Humidity 25%	Relative Humidity 50%	Relative Humidity 70%
40°C	X		X
20°C		X	
4°C	X		X

Table 3.1 The five experimental storage environments.

Initial colorimeter readings were taken of the tissue prior to contact with the medium. The felt-tipped pen marker was then applied to the recto of the paper substrate and all samples were preconditioned at 20°C in their respective RH. Samples requiring 70% RH were preconditioned for 68 hours in a glass box with sodium chloride crystals. Samples requiring 25% RH were preconditioned for 68 hours in a second glass box with potassium acetate.[5] The samples requiring 50% RH were left in ambient room conditions in a glass box. The samples were tightly wrapped in foil and heavy-duty polyethylene, thereby creating a microclimate in which the preconditioned moisture equilibrium was maintained.[6]

Color measurement

The experimental data were gathered from initial and final colorimeter readings, using a Minolta CR-300 Chroma Meter and the CIE 1976 L*a*b* color system. Readings were taken from each tissue side that was in contact with the recto and verso of the paper substrate. For consistency, a Mylar® template was positioned on top of the tissue and three replicate readings were taken (see Fig. 3.38).

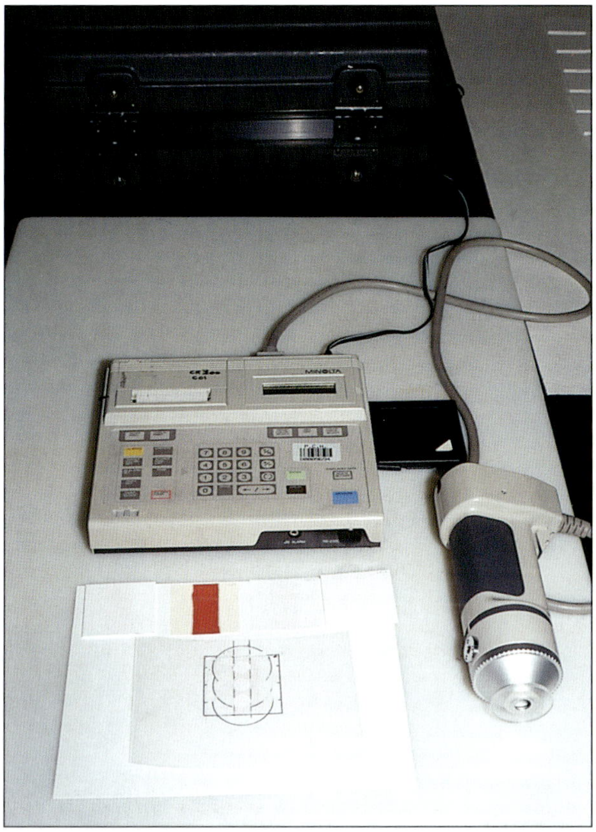

Figure 3.38 Mylar® template, red medium sample from the 70% RH 40°C storage environment, and Minolta CR-300 Chroma Meter. The template is on the recto tissue. Color transfer has occurred through the tissue and onto the mat board.

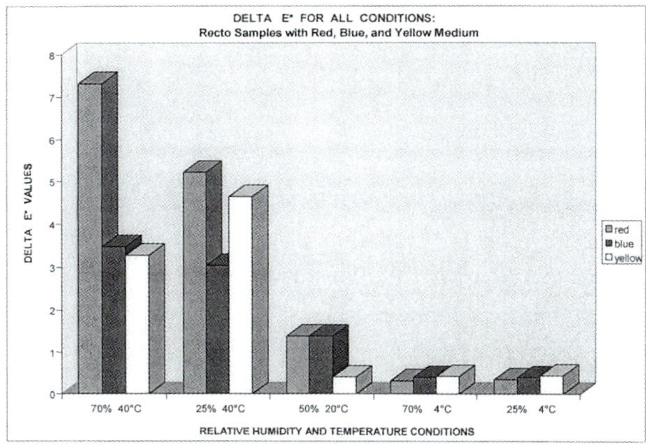

Figure 3.39 A comparison of delta E* values of recto red, blue, and yellow samples in five storage environments.

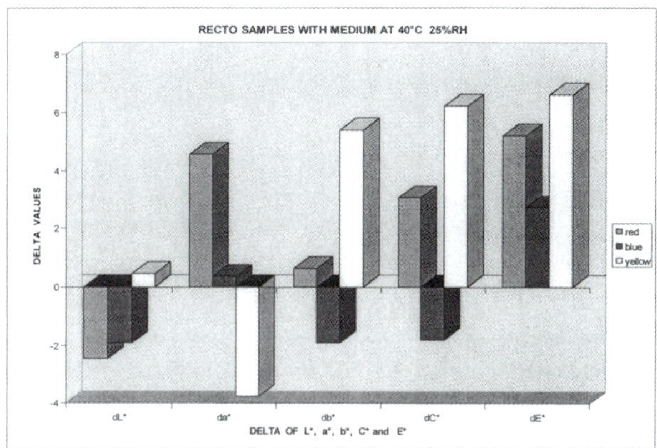

Figure 3.40 The delta values L*, a*, b*, C*, and E* of recto red, blue, and yellow samples in the 25% RH 40°C storage environment.

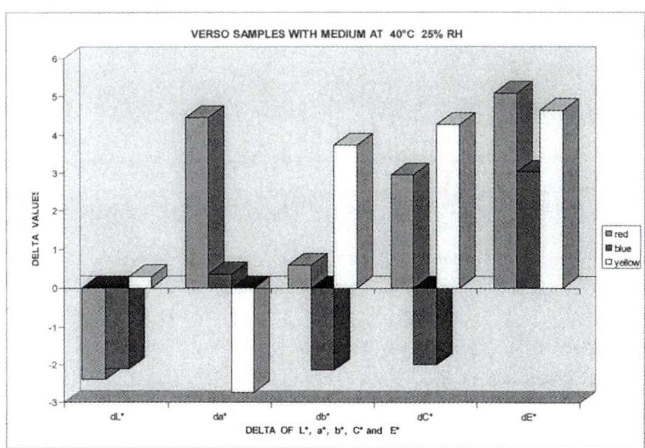

Figure 3.41 The delta values L*, a*, b*, C*, and E* of verso red, blue, and yellow samples in the 25% RH 40°C storage environment.

The experimental data were calculated according to the CIE 1976 L*a*b* color coordinates. Color transfer was measured by the total color difference (ΔE^*_{ab} and was defined by the equation:[7]

$$\{(\Delta L^*)^2 + (\Delta a^*)^2 + (\Delta b^*)^2\}^{1/2} = \Delta E^*_{ab}$$

A change greater than 0.2–0.5 units was considered significant; a change in the range of 0.5–1.0 units could be observed visually as color transfer to the tissue.[8] Initial data for the tissue, obtained prior to contact with the medium, showed a standard deviation of 0.01–0.05 units. After color transfer occurred, the standard deviation varied greatly. The higher standard deviations suggested that there was a degree of inconsistency in the deposition of the medium onto the paper substrate.

Discussion

The results indicated that at 40°C there was color transfer, at 20°C there was slight color transfer, and at 4°C there was no color transfer, although a longer experimental time could have resulted in color transfer occurring in cold storage (see Fig. 3.39). The red medium samples exhibited more color transfer than blue and yellow in high temperatures and at ambient conditions of 20°C and 50% RH. The yellow medium data indicated a significant ΔE*, but color change was less apparent due to the limitations of visual perception when observing yellow on a white surface. This study clearly demonstrated that color transfer occurred from both the recto and verso of the substrate (see Figs. 3.40 and 3.41). Color transfer occurred when the paper substrates were in direct contact with each other.

The color measurements for each color transfer on the recto and verso tissue in a specific condition were summarized and compared (see Table 3.2). All samples

COMPARISON OF COLOR MEASUREMENTS														
		40±5°C	25±5%						change from initial values					visual change
Color	Verso/Recto	Temp.	RH		L*	a*	b*	C*	dL*	da*	db*	dC*	dE*	none/slight/transfer
no medium	recto	40°C	25%	initial	97.56	0.1	2.45	2.5						
				final	97.26	−0.19	2.53	2.5	−0.3	−0.29	0.08	0.1	0.425	none
	verso	40°C	25%	initial	97.6	0.09	2.44	2.4						
				final	97.34	−0.21	2.5	2.5	−0.26	−0.3	0.08	0.1	0.401	none
red	recto	40°C	25%	initial	97.59	0.05	2.52	2.5						
				final	95.39	4.24	3.43	5.5	−2.2	4.19	0.91	2.9	4.819	transfer
	verso	40°C	25%	initial	97.61	0.07	2.51	2.5						
				final	95.11	4.84	3.38	5.9	−2.5	4.77	0.87	3.4	5.455	transfer
blue	recto	40°C	25%	initial	97.6	0.03	2.52	2.5						
				final	95.69	0.39	0.63	0.7	−1.91	0.36	−1.89	−2	2.711	transfer
	verso	40°C	25%	initial	97.62	0.02	2.52	2.5						
				final	95.51	0.38	0.38	0.5	−2.11	0.36	−2.14	−2	3.027	transfer
yellow	recto	40°C	25%	initial	97.62	0.11	2.48	2.5						
				final	97.7	−1.97	5.16	5.5	0.08	−2.08	2.68	3	3.393	slight
	verso	40°C	25%	initial	97.63	0.09	2.49	2.5						
				final	97.81	−2.49	5.95	6.5	0.18	−2.58	3.46	4	4.32	slight

Table 3.2 A comparison of the color measurements of one red, blue, yellow, and no-medium sample in the 40 ± 5°C and 25 ± 5% RH storage environment.

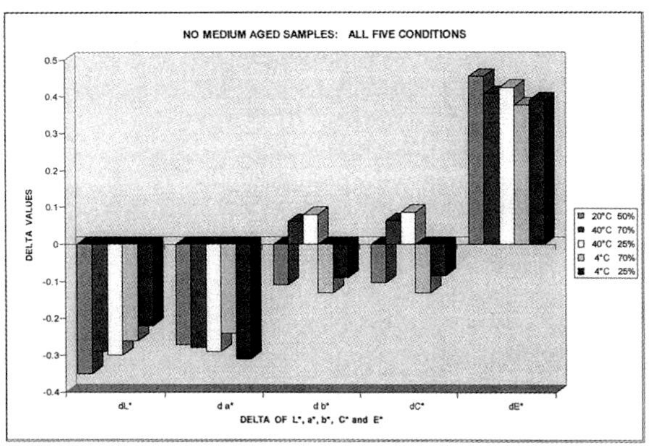

Figure 3.42 A comparison of delta values L*, a*, b*, C*, and E* of no-medium aged samples in five storage environments.

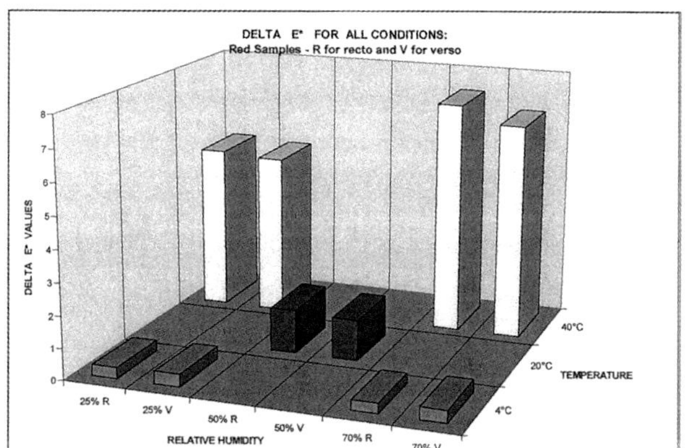

Figure 3.43 A comparison of delta E* values of the red recto and red verso samples in five storage environments.

without medium aged similarly, regardless of their storage condition. Also the tissue had some color change, that is less than 0.5 units (see Fig. 3.42). When color was transferred, it occurred as readily to the recto of the tissue, which was in direct contact with the medium on the paper substrate, as to the verso of the tissue, which was adjacent to the verso of the paper substrate (see Fig. 3.43). There appeared to be no significant difference between the recto and verso ΔE^*_{ab} for all colors in all conditions. The total color change (ΔE^*_{ab}) indicated the amount of color transfer, and as ΔE^*_{ab} increased, the magnitude of the color transfer increased.

There are many factors to consider when asking the question, "Why does color transfer occur?" Among these are solvents. A variety of solvents has been used in these pens, and it is probable that after application, the solvents have evaporated off. However, the resin and the surfactant present in the pens most likely remain with the dyes on the paper substrate. Since the study has shown that color transfer occurs in high temperature conditions, color transfer is probably related to the thermodynamics of the organic synthetic dyes. If this is the case, then color transfer would be determined by the physical and chemical nature of *both* the substrate and the felt-tipped pen media. Additional factors, clearly, are the application and actual deposit of the medium on the paper. A future study could investigate the type of bond occurring between the dye and fiber.[9] Added

Temperature	Relative Humidity		
	25%	50%	70%
40°C	**Color Transfer**		**Color Transfer**
20°C		**Slight**	
4°C	**None**		**None**

Table 3.3 Summary of visual color transfer in the five storage environments.

investigation could yield a greater understanding of this problem by focussing on dye diffusion, and the chemical and physical adsorption of the dye, with an emphasis on surface and mobile adsorption.[10]

Conclusion

Low storage temperatures inhibited the color transfer in the samples, while high temperatures promoted color transfer. The preconditioned moisture content of the samples is not a significant factor in the color transfer (see Table 3.3). A suitable storage environment is necessary to prevent color transfer of felt-tipped pen on paper. If objects are in contact with one another, color transfer may occur, and this must be considered when designing storage containers. A window-mat on *both* sides of the felt-tipped drawing will limit contact with paper substrates and inhibit this transfer. When assessing the impact of storage environments on the color transfer of felt-tipped pen on paper, it was found that the color transfer onto adjacent paper surfaces was minimized in cold storage conditions. Thus, if cold storage is economically feasible, then it could be a long-term solution to prevent, or at least retard, the color transfer that occurs in felt-tipped pen drawings.

Acknowledgments

I thank Linda Sutherland and Katherine Stewart for suggesting this area of research and John O'Neill and Professor A. Murray for their assistance in carrying out the study. I extend a further thank you to Charles Costain, Nancy Binnie, and Season Tse of the Canadian Conservation Institute for their assistance and expertise in the experimental design and planning of the testing procedures.

Endnotes

1. Margaret Holben Ellis, "The porous pointed pen as artistic medium," in S. Fairbrass (ed.), *Conference Papers Manchester 1992* (London: The Institute of Paper Conservation, 1992), pp. 11–18. On the Inuit artists in particular, see Linda Sutherland, "North moves South: Conservation solutions for a large collection of Inuit prints and drawings," in A. Richmond (ed.), *Modern Works – Modern Problems? Conference Papers* (London: The Institute of Paper Conservation, 1994), pp. 122–27.

2. See Raymond H. Lafontaine, "The lightfastness of felt-tip pens," *Journal of the International Institute of Conservation – Canada Group* 4, no. 1 (1976), pp. 9–16; V. Horie and J. Barry, "Solvent resistance of marking inks," *Conservation News* 41 (1990), pp. 11–12; A. K. G. Jones, J. J. M. Jones, and J. A. Spriggs, "Results of a marker trial," *Conservation News* 11 (1980), pp. 6–7; idem, "Results of a second marker trial," *Conservation News* 27 (1985), pp. 37–38; and J. H. Townsend, "Lightfastness of marker pens," *Conservation News* 30 (1986), pp. 6–7.

3. Robert L. Feller, "Felt-tipped markers and the need for standards of lightfastness for artists' colorants," *Bulletin of the American Group – The International Institute for Conservation of Historic and Artistic Works* 8, no. 1 (1967), pp. 24–26.

4. Sutherland 1994 (as in note 1), p. 126.

5. Lewis Greenspan, "Humidity fixed points of binary saturated aqueous solutions," *Journal of Research Physics and Chemistry* 81A, no.1 (1977), pp. 89–96.

6. Mervin Richard, "Control of temperature and relative humidity in packing cases," in M. Mecklenberg (ed.), *Art in Transit: Studies in the Transport of Paintings* (Washington, D.C.: National Gallery of Art, 1991), pp. 279–98.

7. Minolta Camera Co., Chroma Meter CR-300/CR-310/CR-321/CR-331/CR-331C, Instruction Manual, 1993, pp. 78–79.

8. Nancy Binnie, Canadian Conservation Institute, Ottawa, personal communication, 1997.

9. J. Richard Aspland, "What are dyes? What is dyeing?" *Dyeing Primer: Part 1* (Research Triangle Park, N.C.: American Association of Textile Chemists and Colorists, 1981), pp. 4–6.

10. Alan Johnson, *The Theory of Coloration of Textiles* (Bradford, U.K.: Society of Dyers and Colourists on behalf of the Dyers' Company Publication Trust, 1989).

The Preservation and Conservation of Ink Jet and Electrophotographic Printed Materials

Debbie Glynn

Digital printers have become an established tool for the production of fine-art and photographic prints. Artists such as Robert Rauschenberg, Chuck Close, Jim Dine, and David Hockney have all used and endorsed the medium. There is now an urgent requirement for research into the stability of this new form of artwork. The aims of my research are: 1) to establish the archival quality of computer-generated materials; 2) to identify possible conservation problems associated with this type of material, and to propose treatments to deal with these. Since artists have mostly exploited the use of ink jet and electrophotographic printers, the research has focussed on these areas.

The use of computer printers to produce fine-art prints really became established after Graham Nash, the rock musician, realized the potential of large-format ink jet printers in 1989. Being a keen photographer, his interest led him to investigate methods of high-quality digital output. He discovered the IRIS 3047, a large-format ink jet printer manufactured by Scitex, which was designed for offset proofing.[1] The printer was revolutionary, as it was the first digital printer capable of producing high-resolution, continuous-tone images on a variety of substrates. Nash started to use the machine to print his photographs and then opened up a business called Nash Editions, in California, to offer the services of the printers to other artists.[2]

Electrophotographic printers have been exploited by many artists, one of the most famous examples being David Hockney and his "Home Made Prints," made in the late 1980s. Hockney experimented with Xerox color photocopiers and fax machines to produce prints of his designs. He used the color photocopiers to layer various colors and textures on the paper surface by feeding the paper through the copier machine again and again – up to 12 times – for each color change.[3]

Concern for the stability of digital prints arose after it was found that the inks used with the IRIS ink jet printers had poor light-fastness characteristics, and were noticeably fading within only a few months. The archival quality of the inks and papers had originally not been an issue for the manufacturers, because the printers were intended to produce ephemeral graphics and proofs. Research was then undertaken by companies such as IRIS to develop better-quality inks, papers, and printers. This research has led to the availability of consistently higher standards of media and hardware.

Both ink jet and electrophotographic printers are categorized as non-impact printing (NIP) systems, as they produce images and text on paper without striking the page. NIP is the term given to a number of computer printers including thermal wax transfer, dye sublimation, photochemical systems, etc. The quality of the images produced by ink jet and electrophotographic printers is determined by several different variables, but the most important of these is the printer's resolution.

The resolution refers to the number of either the series of dots per inch for ink jet (dpi) or pixels per inch for photocopiers and laser printers (ppi) (see Figs. 3.44

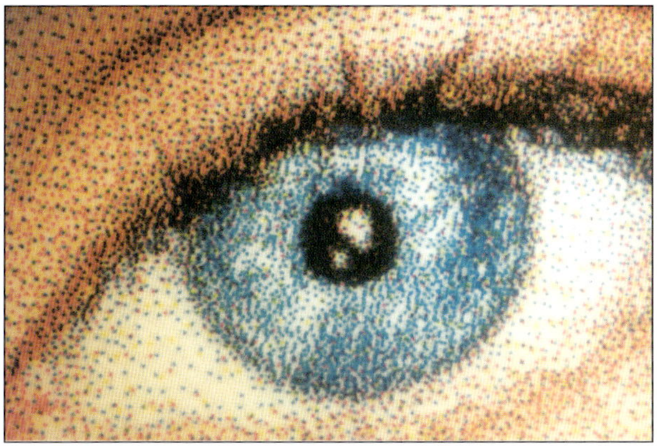

Figure 3.44 Magnification (1.6x) of a Xerox VivagrafX Xpress ink jet print, 720 x 720 dpi.

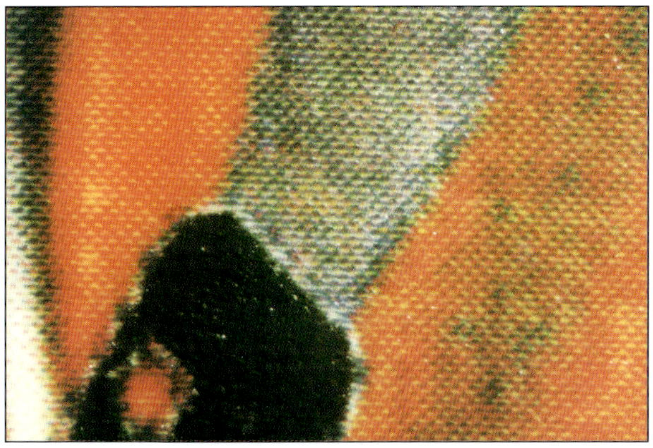

Figure 3.45 Magnification (1.6x) of an Apple Color LaserWriter 12/600 print, 600 x 600 ppi.

and 3.45 respectively). Ink jet resolution can range from 300 to 1,800 dpi, depending on the printer. A dpi of 1,800 is achieved through an apparent resolution value, where the individual dot can be varied from 0 to 32 droplets per dot. Electrophotographic systems tend to have a resolution between 300 and 600 ppi, depending on the printer. In both cases, resolution is variable depending on the configuration of the dots that make up an image, known as the "dithering pattern" for ink jet printers.

There are four different types of ink jet printers on the market today: continuous ink jet (CIJ), drop on demand (DOD), bubble/thermal ink jet, and phase change ink jet. Two classifications of electrophotographic printers exist: photocopiers and laser printers. Both systems use four colored inks – cyan, magenta, yellow, and black (CMYK) – to produce all the colors. The two printing devices are dependent on specialized inks and paper for high-quality output. Of all of the NIP technologies, ink jet systems are the most dependent on the type of ink and paper used with the printer for good-quality imaging.

Ink jet printers

Ink jet printers require a minute orifice or nozzle from which a tiny stream of uniformly shaped droplets of ink spurts out directly onto a paper surface. Ink must be very fine and fluid so that it may pass through the orifice. For CIJ systems, water-based inks colored with organic soluble dyes are used. The dyes are modified from textile dyes because of their color reliability, solubility in an aqueous medium, and thermal stability. Other ingredients such as hygroscopic solvents, typically a glycol, are added to the inks to lower the evaporation rate and to allow the ink to penetrate more rapidly into the paper. The DOD and bubble jet printers can utilize a larger orifice. Therefore, they can accommodate solvent-based pigmented inks, but these do not tend to have the vibrancy or large color gamuts of the dyes.

The liquid ink requires specially coated papers, so that the ink is not absorbed too far into the substrate. These coatings are composed of pigments, latex, fluorescent agents, and binders, and they are designed to have a very porous structure so that most of the ink is absorbed near to the surface of the paper. However, artists are also using good-quality watercolor and printmaking papers because they like the softening effect that results from the ink flowing along the fibers. These fine-art papers can considerably improve the light-fastness of an ink jet image because absorption of ink by the paper provides some protection against light damage and permits hydrogen bonding between the inks and the cellulose substrate.

Initially, the IRIS printer (which uses CIJ technology) became the preferred system to produce high-quality, large-format fine-art prints. These particular prints have been given the name Glicées, derived from the French word for "spit" or "spurt." However, recently printers have been introduced onto the market that can produce images almost equal to the quality of the IRIS but much more efficiently, such as the Epson Pro Series (DOD) printers, which have a resolution of 1,440 dpi.[4]

The required composition of the inks and paper is not of archival quality (see Fig. 3.46). The presence of dyes in the inks and the fluorescent agents and latexes in the paper coatings make the prints very unstable and are a cause for concern.[5]

Electrophotographic printers

Electrophotographic printers are based on forming electrostatic images by photoconductive discharge of an electrically charged surface. The charged surface is selectively discharged by exposure to light. The remaining charge is used to attract a colorant, referred to as a toner, which is then fused onto a sheet of paper by heat and/or pressure.

Electrophotographic toners and papers do not have as many limitations as the ink jet media. The toners are made up of a solid mixture of fine resin-coated pigments. The paper does require the addition of conductive polymers to prevent the moisture content of the paper becoming too high, as this can prevent proper fusion of the toners to the paper substrate.

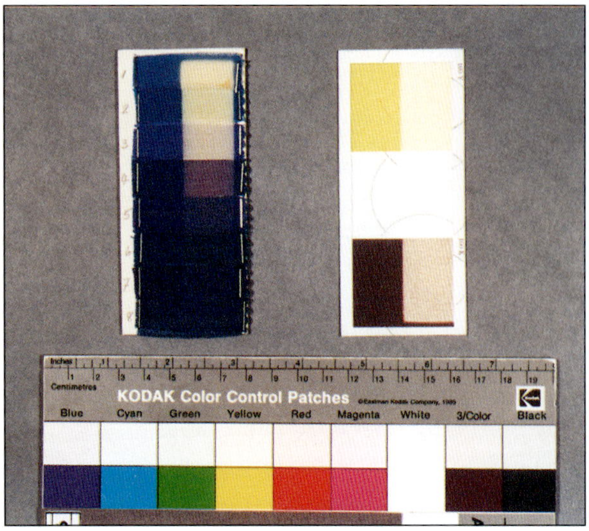

Figure 3.46 A sample of an IRIS 3047 ink jet print reproducing the format developed for light-fast experiments for this project and a blue wool scale. The sample consists of 100% printed yellow and black ink using one of the older IRIS inks, the Morgan FA ink set, on Somerset VRW paper (280 gsm). The sample and blue wool scale were exposed to a Microscal light tester for 176 hours (approximately 1 week). The yellow ink was found to have a blue wool scale rating of 4 and the black ink a rating of 1.

Research project

Research at Camberwell College of Arts has investigated, in particular, the damaging effect light has on the new media with the conduction of accelerated light-fast and natural aging tests. A colorimeter has been employed to plot the fading rates of different printing types, and comparisons of the results from the two methods of light exposure have been made.

The project is also investigating the effect different spectral distributions have on light-fast stability, following the research by conservation scientists David Saunders and Jo Kirby.[6] Their research has found that the amount of lux hours is not necessarily a good method of assessing and stipulating lighting display conditions. It was found that certain wavelengths of light were responsible for photochemical reactions, and the UV part of the spectrum was not necessarily responsible for catalyzing reactions of very sensitive materials. Sensitive colorants such as textile dyes were found to be unaffected by UV light, but wavelengths in the violet, blue, and green areas of the visible part of the spectrum were responsible for fading reactions. Therefore, colored filters have been employed with the light-fast tests to research the effect different wave bands of light have on the printing types.

The second stage of the project has focussed on the conservation of this type of printed material. Basic treatments have been reviewed in paper conservation and their application evaluated. Treatments tested include: mechanical dry cleaning, humidification, washing, deacidification, solvent application, pressing, and paper repair. Testing has been concerned with observations of any detrimental effects that may occur with these treatments.

Standard sample formats, contained on a computer file, have been developed for the experiments. Samples have been obtained from the following ink jet and electrophotographic manufacturers currently being used by artists and photographers: Canon, Encad, Epson, Hewlett Packard, IRIS, Xerox, and the ink manufacturer Lyson. (The research was submitted for a Ph.D. in September 2001, and the results of this project will be published in the near future.)

Acknowledgments

I would like to thank Epson (U.K.), Lyson, and Visualeyes for their support of this research.

Endnotes

1. Scitex America Corporation, 8 Oak Park, Bedford, MA 01730, U.S. www.scitex.com.

2. Nash Editions has a website that gives an introduction to how the business was established: www.nasheditions.com.

3. The stability of David Hockney's "Home Made Prints" has been investigated by Heather Norville-Day. Her research found that the photocopy toner underwent color changes during light-fast aging, except the black toner. A pronounced crazing of the coloring layer also occurred, which was more severe where two or more layers of toner were overlaid on top of one another. H. Norville-Day, "The conservation of faxes and colour photocopies, with special reference to David Hockney's 'Home Made Prints,'" in A. Richmond (ed.), *Modern Works – Modern Problems? Conference Papers* (London: The Institute of Paper Conservation, 1994), pp. 66–72.

4. Epson (U.K.) Ltd., Campus 100, Maylands Avenue, Hemel Hempstead, Hertfordshire, HP2 7TJ, U.K. www.epson.co.uk.

5. Research into the effect that various environments have on coated papers found that yellowing of the coatings occurred when nitrogen oxide reacted with the antioxidants present in the latexes and fluorescent agents. Ammonia intensified this reaction caused by the nitrogen oxide on the latexes. V. Mailly, J. Le Nest, J. Serra Tosio, and J. Silvy, "Yellowing of coated papers under the action of heat, daylight radiation, and nitrogen oxide gas," *TAPPI Journal* 80, no. 5 (1977), pp. 176–83.

6. D. Saunders and J. Kirby, "Wavelength-dependent fading of artists' pigments," *IIC Preprints to the Ottawa Congress, Preventive Conservation Practice, Theory and Research* (Ottawa, 1994), pp. 190–94.

On Material(ism) or Why Painting Conservators Have All the Fun

Daria Keynan

In dealing with contemporary art, we often come across the blurred boundaries between easel paintings and works on paper. Artists no longer feel restricted by traditional material combinations, techniques, or presentations. These fluid movements between and within the once traditional boundaries of material and technique have allowed artists, and consequently curators and collectors, to define works in new ways. The guidelines developed by the conservation tradition for the treatment of works from different art disciplines become more and more difficult to apply. In the following discussion, I should like to touch upon some of the complexities of the decision-making process regarding compensation for loss – inpainting – using modern and contemporary paintings on paper as examples.

My conservation training took place in the early 1980s and was dominated by two influential conservation teachers: Antoinette King, who taught paper conservation at the Conservation Center at New York University; and Konstanze Bachmann of the Cooper-Hewitt Museum in New York, with whom I did a two-year internship. Bachmann and King are in many ways very different conservators, but in one thing, at least, my education was consistent – I was taught that an ethical paper conservator never inpainted directly on the primary support. There were minor exceptions, allowing a little pastel to be dabbed on occasionally, and the rules for wallpapers were also somewhat different. But the basic rule held and I started my professional life believing that it was so.

This approach was based primarily on the need for reversibility in treatment, but other factors such as potential devaluation of a work (especially a print) also came into play. It came about in part because of the attitudes of curators, dealers, and collectors who often had lower expectations of "perfection" or "finish" from a work on paper than from a painting. They believed works on paper to be more vulnerable to change over time; acceptable, even perhaps a little more authentic, with a few stains and a bit of discoloration. To a certain extent, this still appears to hold true for older works on paper, but when we come to modern and especially contemporary art, the expectations from the works are completely different.

Over the years, my conservation practice has concentrated increasingly on modern and contemporary art, which comprises many paintings on paper. Untraditional, not to say unusual, use of media and media combinations and the resulting problems of flaking paint, interlayer cleavage, rejection between incompatible media layers, inclusions and collage elements that fall or move about – all of these and more often result in media loss or considerable changes in the artworks' appearance. To a certain extent, these problems, now common in paper conservation, were in the past more often the domain of painting rather than paper conservators. Because of traditional use of paper versus canvas or panel, and because of the nature of the material itself, paper conservation training and practice concentrated until fairly recently very much on the support, while painting conservation concentrated much more on media.

The problems presented to a paper conservator working in modern and contemporary art were, therefore, somewhat uncharted territory. One consequence of this is that for a while, and still to an extent, many painting conservators work on paintings on paper, and relatively few paper conservators want to deal with these works. Traditionally, paper conservators have shied away from loss compensation directly on the original support, while painting conservators have had that option. Our reluctance to inpaint directly on the support has caused several problems. At times we have developed elaborate justifications for leaving an image in a far less finished state than would a painting conservator. The different approaches to loss compensation developed by paper and painting conservators have often resulted in an aesthetic discrepancy between the treated works on paper done by a given artist, and those done by the same artist on canvas. Inadvertently, paper conservators have contributed to the expectation that a work on paper will have a look of more "immediacy" and less finish than a painting on canvas. This fact, in combination with other factors I will touch upon, has contributed to a small but poignant crisis in paper conservation in general, and in my treatment approach in particular.

It became increasingly clear to me that if one wanted to work with contemporary art, it would be virtually impossible to adhere to professional traditions that relate to another era, when material use and conceptual frameworks were completely different. How to resolve the basic principles of a traditional paper conservation approach with art that totally departs from the traditions on which these principles were based is an ongoing struggle in my practice. When faced with the possibility of inpainting, I go through a short checklist of basic considerations that I have created. These are very general and certainly do not cover all the issues that one must consider in any given treatment, but I have found them useful as general guidelines.

1. As with any treatment, the first step (and a less obvious one than it seems) is to ascertain whether you agree that there actually is damage. In other words, not all loss/change can be, in my opinion, equated with damage.
2. If there is damage or loss, determine its cause. Has it been caused by accident, wear and tear, inherent vice, or just plain aging?
3. Identify the location and type of loss. Is it loss of a top layer of media with more media under it; does the loss go through to the support; or is it on the support only?
4. Define the type of work – print, drawing, painting on paper, collage, etc.
5. Consider how the work, with its losses, relates to other works by the artist – the body of work, not just work on paper.

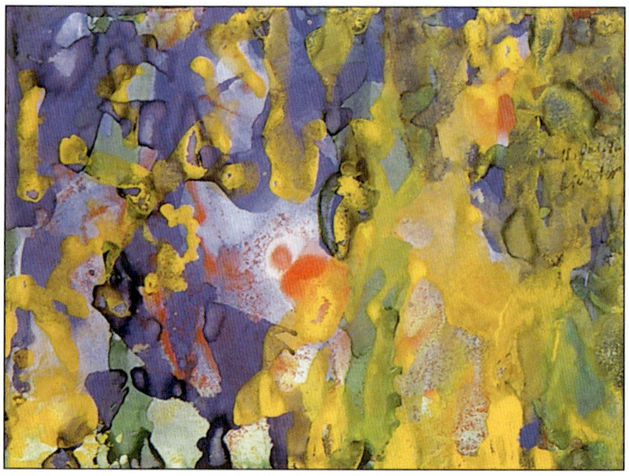

Figure 3.47 Gerhard Richter (German, b. 1932), *18 oct 92.* Watercolor on wove paper, 29.9 x 40 cm. Collection PaineWebber Group.

I should further like to examine some of these considerations using examples of modern and contemporary work on paper. The first issue must be to consider the change in a work and decide if it should be thought of as damage. To illustrate this issue I have chosen a Gerhard Richter watercolor, *18 oct 92* (Fig. 3.47), which exemplifies the extent of difficulty in evaluating the condition of some contemporary work. The watercolor exhibits extensive flaking, cracking, peeling, powdering, and color loss (see Fig. 3.48), but is it damaged or just altered? This is a difficult call. I do not believe it could be inpainted even if it were determined to be appropriate, but setting that aside for the moment, let us concentrate on the question at hand. What is this work? It is referred to as a "contemplation" work of sorts, created over a period of time, in several layers, with unusual use of Smincke watercolors. The artist employed wet-on-wet techniques, making pools of water on the paper and then dropping viscous watercolor into them; he could then drop water on to that to create a dispersion pattern. The enamel-like passages were created by placing the watercolor cakes under the faucet and letting them sit until they dissolved, then pouring them onto the paper. When I first examined such objects, I was baffled. Both the material and technique seemed unusual, and more information was needed before a treatment approach could be devised. I was fortunate to get a detailed first-hand account of the making and the selling of these works. Before dealers come to the studio to select work for sale, Richter takes out the watercolors and proceeds to "clean" them up. He does this by brushing the surface to remove what is already loose and then picks off flaking areas with his nail. As a finishing touch, he gives each work a good shake to make sure no loose bits remain on the surface (and to set in motion new flaking and cracking). The work is then ready for sale. Hearing this, my initial reaction was that if the artist is fully aware of the problem and chooses to deal with it

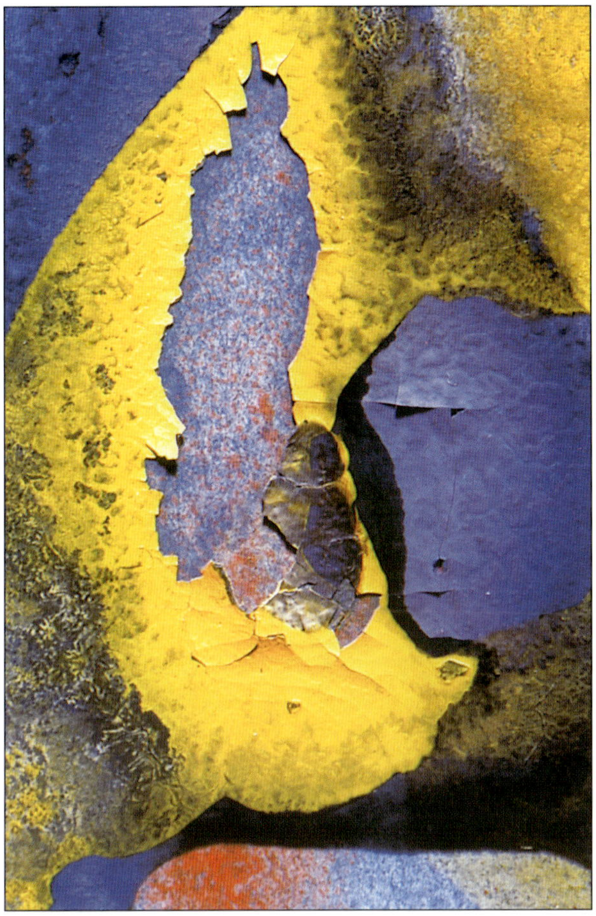

Figure 3.48 Gerhard Richter, *18 oct 92* (detail of Fig. 3.47, top left center).

in this way, then, at best, the objects should be considered changing art and the consequences accepted. However, every once in a while – rarely, I am told, but still it does happen – Richter will look at a work and determine that it has altered in a way that does not please him, and he will proceed to make additions or changes to correct it. This, of course, changes everything. Or does it? We ourselves cannot make the decision when a change has gone too far. We do not know what changes are or are not acceptable, nor can we accurately chart those changes in order to make an educated guess. Therefore, as a rule I do not inpaint these works and make a point of always explaining the process by which they were made, including the artist's "cleaning" phase, to the clients, letting them decide if even consolidation is called for, or if they can live with the idea that the work will continue to change.[1]

If an alteration in a work appears to be a damage, the cause of damage must be examined. Often the approach to treatment will be determined not just by the type of damage but by its cause. In modern and contemporary art on paper, inherent vice is not uncommon and remains one of the most difficult problems to deal with. The experimental nature of much of the work, and the frequent use of incompatible material combinations, make for unpredictable and, at times, undesirable aging characteristics. As an example I have chosen a Ben Shahn mixed-media work on paper entitled *The Harpie* from

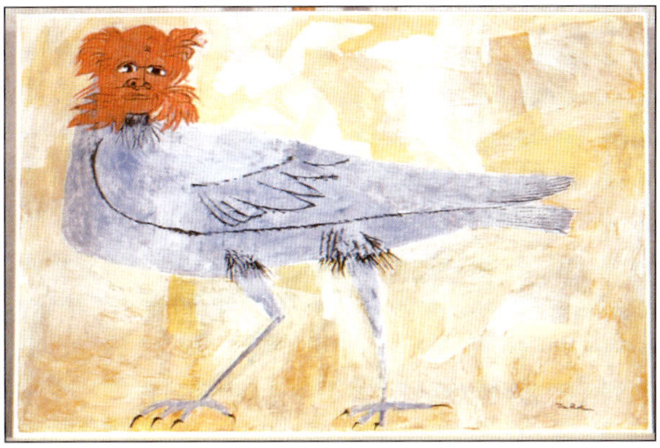

Figure 3.49 Ben Shahn (American, 1898–1969), *The Harpie*, 1951. Mixed media on wove paper, 63.5 x 96.5 cm. Collection of Mrs. Robert B. Mayer, Chicago.

1951 (Fig. 3.49). Shahn was intensely interested in materials and techniques, seeking particular effects and employing what he believed to be old recipes in order to achieve them. Apart from his rather broad interpretations of the recipes he employed, he also appears to have given rather little consideration to the appropriateness of their combinations. *The Harpie* features a very common Shahn combination: gouache, ink, and tempera. The tempera can be from any number of recipes. The artist liked the effect created by accenting areas of the gouache by applying tempera over it; he found it gave a depth and richness to the surface. Very often, incompatibility between the gouache and the tempera, as well as between the ink and gouache, causes extensive cracking and loss in the top layer, so that it sheers from the gouache underneath.

To date I have not found reason or justification to inpaint these works. There are several reasons for this. First, I have yet to come across an inpainting system that can be employed over an existing gouache layer without altering and/or harming that layer, which to me seems an impossible choice. Second, the tempera accenting layer itself has often altered visually, so that it is no longer creating the originally desired effect (changes in opacity and transparency are the most common). In the case of *The Harpie*, it was rather easy to justify my decision not to inpaint because the painting, while clearly showing paint losses, does not appear to me to suffer in any significant way from them. This is partly due to the image itself and partly due to the changes mentioned above. The tempera layer has become "cloudy." Inpainting the losses in this paint layer would have accented this alteration rather than minimized it.

Another common cause of damage is that of wear and tear. In today's international and very active art market, works of art are subjected to rigorous exhibition schedules. Modern and contemporary art on paper is often forced to roam the earth, unglazed and often unframed. Very often, the works are large and difficult to handle even under the best of circumstances. Not surprisingly, wear and tear in such works is a common occurrence. Traveling exhibitions, viewing schedules, trial periods at various potential buyers' homes, not to mention the difficulties of travel and storage, all leave their marks – insignificant perhaps in themselves, but they do add up after a time.

A group of such pieces, from the *Anthropometries* series by Yves Klein, were examined in order to evaluate the accumulated damage from such wear and tear. The works examined are fairly large, averaging 1.5–2 m; the medium is IKB in Rhodapast M, on a medium-thick wove paper from rolls, mounted to canvas, stretched on wooden stretchers, framed with wood strip frames, and not glazed. Upon examination, some of these works exhibited minor scratches, indentations, and losses in the paint layer and the paper support. Almost all the damage that I could see appeared related to handling, and the variety of fingerprints scattered about confirmed that the pieces had led an active life.

One cannot judge the conditions of works like these without considering what they are and how they were created. The *Anthropometries* date from 1960. They were created by painting the bodies of naked women and guiding them over the paper or canvas in an environment that included an orchestra and an audience. Mrs. Klein-Moquay, the artist's widow, described these works in an interview as follows:

> For Yves, the naked women were "living brushes" as he called them, a means of spontaneous energy transmission and a highly intense presentation of sensitive artistic collaboration for a coming art that does nothing in secret, has nothing to hide, but also integrates being-there, the present spectator as an active and creative participant in order to experience and witness the degree of immaterial cosmic presence together.[2]

There was no doubt in my mind that the works had altered over time, but considering what they are and how they were created, I believe that they are the type of works that can sustain all sorts of changes without in the least losing their visual integrity. The question of inpainting such works must be considered in relation to the context of their making. The live music played during the making of the works is very important; the music is in the work and part of it. If the disruption of the color had occurred in a static, color-field area (the equivalent of a static, still, or hovering note in the music), then one would have little choice but to inpaint. But when the losses occur at the edges of strong movements, there is very little disruption to the image or its rhythm. My inclination was, therefore, not to inpaint. Before making a final decision, I talked to several conservators, mostly painting conservators, about the approach I was taking. With a single exception, I received a negative response. It appeared that, with regard to Klein, the rule was to inpaint regardless of the location of the damage. This approach created a problem I had not anticipated. It is imperative, in my understanding, that a body of work by an artist should maintain some consistency of appearance. By proposing to depart from the norm of an accepted treatment approach, I fear that while I believe the integrity of the individual work to be preserved without inpainting, its relationship to and place in the body of work may be altered if it remains untreated.

Figure 3.50 Sam Francis (American, 1923–1994), *Gray Fields*, 1949 (after treatment). Tempera on paper, 37.5 x 27.3 cm. © 1999, the Estate of Sam Francis/Artists Rights Society (ARS), New York.

Accidents to the work also result from extensive travel and exhibition schedules. Often the repair of such damage involves the least theoretical and the most technical difficulty. One example is the Sam Francis drawing *Gray Fields*, of 1949 (Fig. 3.50). This is an early drawing, executed in tempera on paper. It was traveling in its frame, under glass; the glass broke, and the damage occurred (see Fig. 3.51). In my opinion – shared by all others whom I consulted at the time – this work could not exist in this condition. If the damage was not repaired, it could perhaps have been donated to a study collection, but it would have had no life as Francis's *Gray Fields*. After some consideration, I further concluded that doing a general toning rather than complete inpainting would create a false reading of the work, making it seem as if splashes and stains of paint, so typical of the artist's later, better-known work, were already implied here in this 1949 drawing. This is always a significant consideration, but if one looks at Francis's drawings between 1947 and 1950, the significance of this point is magnified, for this is the period when his signature visual vocabulary was forming. The treatment selected was,

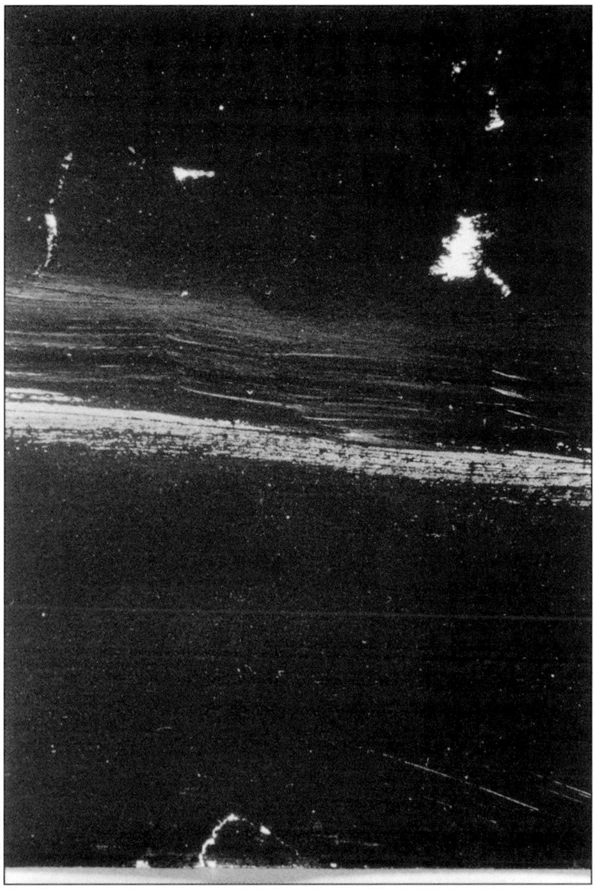

Figure 3.51 Sam Francis, *Gray Fields*, 1949 (detail before treatment, showing damage in center of lower section).

therefore, to do the most complete and accurate inpainting possible.

The inpainting imitated the artist's paint structure, which included a red ground and matched up the brush stroke direction. The torn and damaged paper support was prepared for inpainting, and evened out and smoothed with thick methyl cellulose where possible (it was not possible in the small scratches). The methyl cellulose was used in order to create a smooth surface for inpainting as well as an isolating layer between the paper and the inpainting. Its use as an isolating layer was minimal, since the inpainting itself was done with pastel in methyl cellulose as well. The question of reversibility is, in theory, an important one, since in principle this technique cannot be considered reversible. In practical terms, however, should the need arise, it can be mechanically reduced to make room for new inpainting, though never reduced enough to leave clean, exposed, untinted paper.

Another example of damage caused by an accident can be seen in Tom Wesselmann's *Still Life #33*, a 1963 oil-and-collage on canvas (Fig. 3.52). A forklift drove through the lower right side of the third panel, puncturing both canvas and paper collage (see Fig. 3.53). The treatment was a joint project with Amann Conservation Associates, a painting conservation studio in New York. They repaired the tears in the canvas, and I was to lay the paper collage back in place and inpaint if

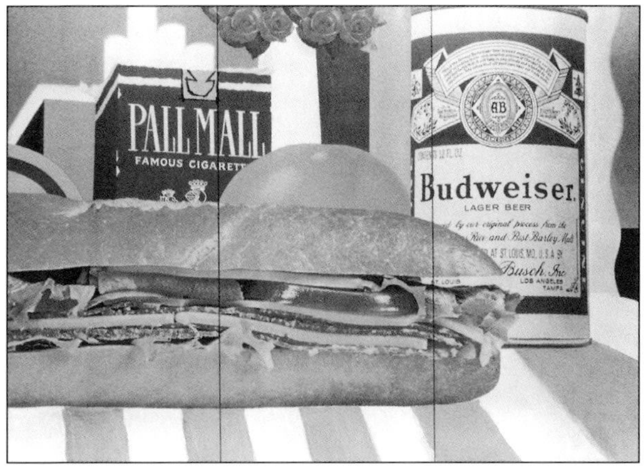

Figure 3.52 Tom Wesselmann (American, b. 1931), *Still Life #33*, 1963 (after treatment). Oil and paper collage on canvas in three sections, overall 335.3 x 457.2 cm. © Tom Wesselmann, N.Y.

needed. Sadly, it was needed. The collage elements are halftone screen-printed papers. In general, when I do inpainting on such material, I do not blend my colors but, under magnification, sort out the dot color and layer pattern and imitate it in order to get a better match. In this case, the decision to inpaint was easy because Wesselmann habitually coated the paper elements in these collages with Liquitex varnish, which allows for easy reversibility of the inpainting. Inpainting was needed because it was not possible, despite my best efforts, to completely flatten out the deformations in the paper. As a result the dot pattern did not match up perfectly when I laid down the paper and closed the tear. The point of the inpainting was to even out the look of these seam areas by creating, with inpainting, a slight shift in the dots' position. This was done using watercolor and Aquarelle pencils doctored with additional gum and/or methyl cellulose. A little very dilute Soluvar was the finishing layer to match the Liquitex shine. The Soluvar was dotted on in a tiny scattered pattern in order to break up the comparative matte look of the watercolor (see Fig. 3.54). In case the inpainting needs to be reversed in the future, a damp swab will do the trick because the inpainting media is not completely coated.

I would now like to turn to the murky waters of an artist's body of work, and how one's treatment fits into it. I have chosen to make my point using examples of the work of Robert Motherwell, since he is so well known. Motherwell, like many other modern artists, worked in large series, developing an image and repeating it in several variations over many years, using materials and material combinations interchangeably to create essentially the same type of image. The *Spanish Elegy* image, for example, was developed from 1948 on. It was done in drawing, painting on paper and board, as well as on canvas and in print. These images are not evolved as paper studies, oil sketches, and eventually the painting, but more as a fluid exchange of image and material, introducing new elements from one material to another. This is not to say that the artist did not do sketches or

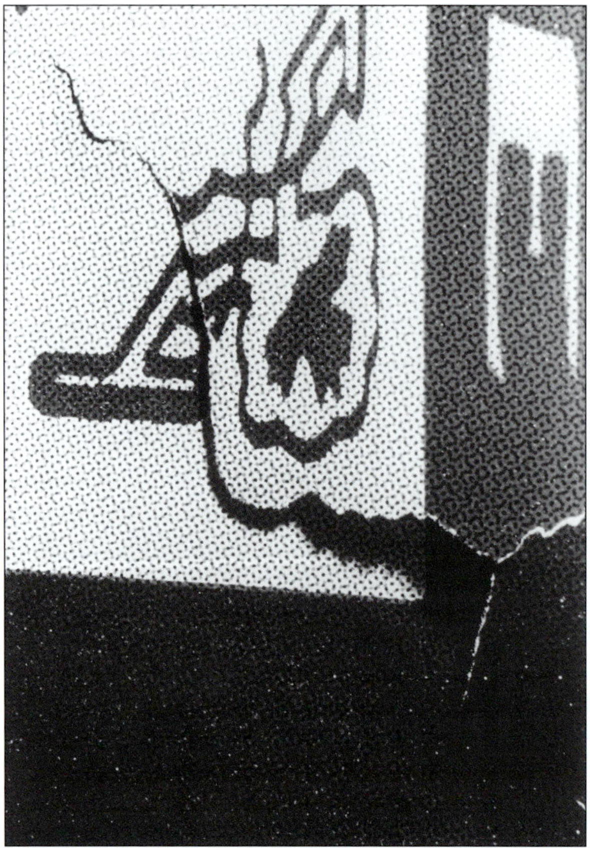

Figure 3.53 Tom Wesselmann, *Still Life #33* (detail before treatment of lower right section of panel #3).

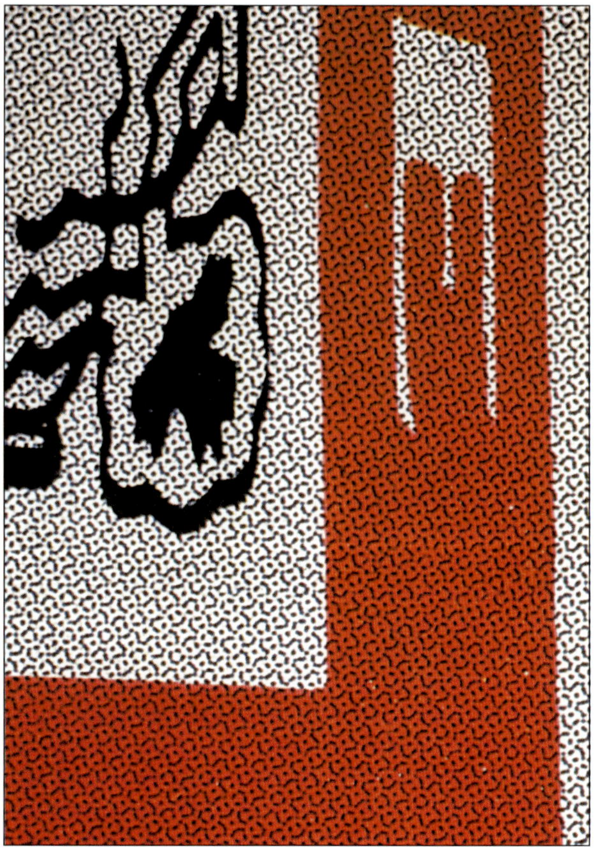

Figure 3.54 Tom Wesselmann, *Still Life #33* (detail of same section after treatment).

studies, but rather to illustrate that in the working methods of artists such as Motherwell, one will find a variety of approaches to developing images over time.

Very often when an artist's body of work needs conservation, the paintings are the first focus for curators, dealers, or artists' estates. By the time the works on paper get attention, the basic aesthetic approach to the final appearance after treatment has been hammered out with the painting conservator. We then find ourselves needing to take into consideration not only what we believe to be the needs of a particular work or group of works, but also what the body of work as a whole looks like. In other words we need to look at the treated paintings. I have been told by curators, "The paintings are already treated, match your work to them." When I first heard this, I had a bit of difficulty with the concept, but eventually I came to realize that in many cases it was indeed my only option. The traditional approach of the paper conservator is often to stabilize the work as needed and do no loss compensation, or alternately do a half-hearted toning that would serve as a compromise between our code of ethics and the demands of the image treated. Inadvertently, we create an alternate art history for such works, establishing the visual difference between the fully treated paintings on canvas and the unfinished-looking paintings on paper, thus creating an impression of preparatory sketches (paper-based images) and complete paintings (canvas). In time, this distorted view of an artist's body of work may completely alter the understanding of an individual artwork, of an artist's working method, and even of the artist's philosophy.

We are now confronted with the question of whether it is appropriate for us as conservators to be guided in our treatment choices by the material of the artwork (paper or canvas), by the visual requirement of a single work, or by the aesthetic of an artist's body of work. For example, in examining Motherwell's 1962 oil painting on paper *Beside the Sea #27* (Fig. 3.55), there were several areas of loss (see Fig. 3.56). This type of loss is not uncommon and is caused by a repeated problem related to framing. (This is also a common problem in Willem de Kooning's oils on paper.) The pieces are framed before the oil is completely dry, or in some cases it simply does not dry. As a result, oil paint sticks to the glazing, and with travel and/or climatic fluctuation, the paint often detaches from its support onto the glazing.

Many paper conservators, myself included, are trained to have a rather high tolerance for visual alterations in a work. If these oils on paper related to nothing at all in the rest of Motherwell's oeuvre, I could probably live with the losses, but they are part of a body of work and do relate to it. My treatment, I believe, needed to take this fact into consideration. Pastel in methyl cellulose gave a beautiful match in density of paint and in surface sheen and finish, making the inpainting reversible and rather easy to carry out. The paper is oil saturated so that the methyl cellulose stays on

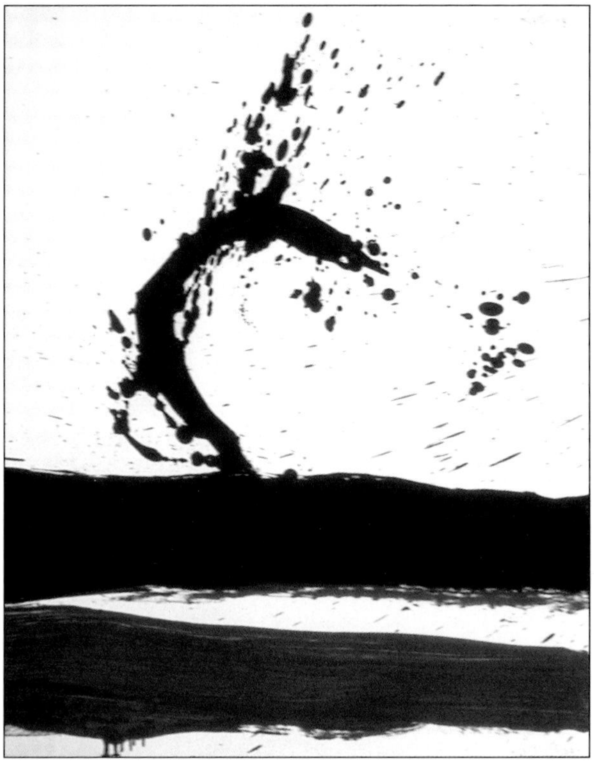

Figure 3.55 Robert Motherwell (American, 1915–1991), *Beside the Sea #27*, 1962 (after treatment). Oil on paper, 73.7 x 58.4 cm. The Dedalus Foundation, New York.

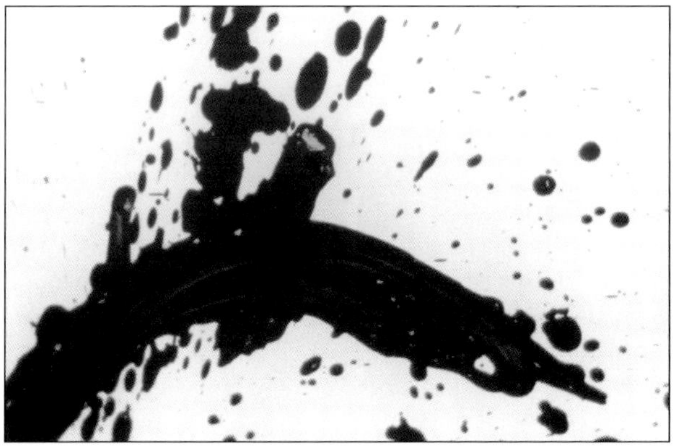

Figure 3.56 Robert Motherwell, *Beside the Sea #27* (detail of paint loss at top center).

the surface, but if necessary another barrier layer can be added – often the consolidant used at the edges of the loss can be extended to coat the paper if desired (see Fig. 3.56). This example is typical of my treatment approach to Motherwell's paintings on paper. I usually do inpaint them, since I have little doubt that a visual gap between the various types of painting creates a false differentiation between them and (unjustifiably) turns the material of a given work into a major factor in its visual perception.

In conclusion, when dealing with modern and contemporary art, we rarely have an individual isolated image. What we see in our studios most often is just a small aspect of a whole. I do not mean to say that a work in itself does not have an existence on its own, but that it does not have an isolated one. A conservator can do a beautiful job on a given piece, but if the work then stands apart visually from the body of work, is it an appropriate treatment? We must learn to examine how the appearance of a body of work evolves and what factors go into its making, as well as understand our own contribution to the perception of an artwork. The overall appearance, or the look, is created by a complex combination of factors, mostly out of our control and often out of the range of our awareness. It is made up of the aesthetic of our time, the scholarly readings and interpretation of a work, and the period in which it was created, as well as of our own culture, the market, and fashion, to name but a few factors. These are around and beyond us. We can only attempt to be aware of them and remember to try, to the best of our ability, to consider them when approaching the conservation of a work.

Endnotes

1. In a recently published exhibition catalogue, Richter makes reference to the fragility of his watercolor paintings. He states that he "had to accept that when the pools of color dried out, they could later break away. If one stands a sheet firmly on edge, it can trickle down, sometimes the loose bits of paint have to be scratched away." Though it can be inferred from his statement that the changes are a part of the technique, he never states this explicitly. He does however make it clear that he is fully aware of the changes. Dieter Schwarz, ed., *Gerhard Richter Aquarelle 1964–97*, exh. cat. (Düsseldorf: Kunstmuseum Winterthur, Richter Verlag, 1999), p. 22.

2. G. Tullmann and H. Weitemier, "An interview with Rouraut Klein-Moquay, September 1994," in *Yves Klein: The Anthropometries*, exh. cat. (New York/Cologne: Danese/Galerie Gmurzynska, 1997), p. 8.

CHAPTER 4
THE MATERIALS OF ASIAN ART

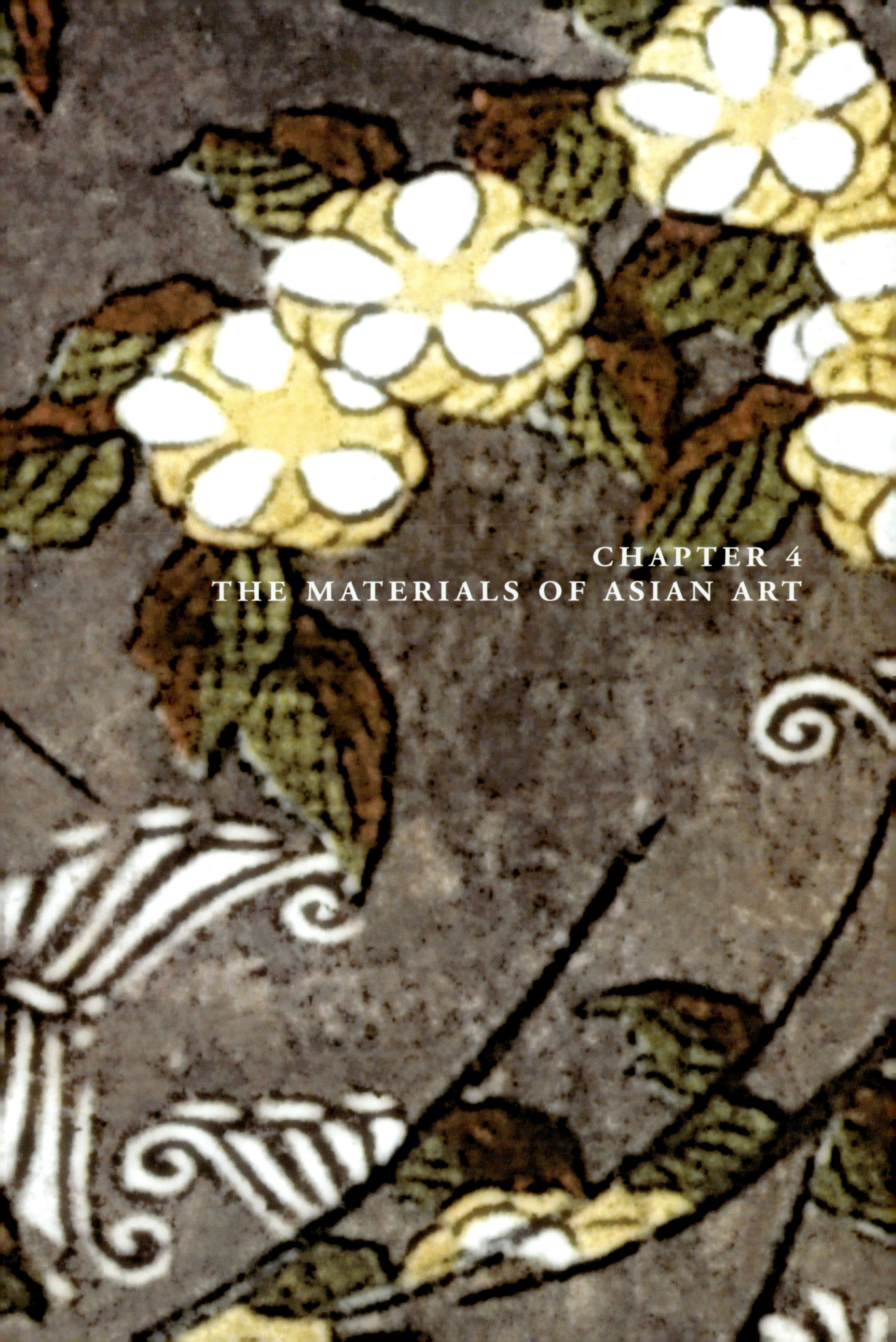

Color as Language in Traditional Japanese Prints

Roger S. Keyes and
Elizabeth I. Coombs

Color is the heart of most Japanese prints published between the 1680s and the early 1900s. It is the primary appeal of these prints for many people today. But printers used many pigments that change in hue or value when exposed to light, air, and moisture, so the prints depend on color effects that are inherently unstable.

Conservators need to assess color stability when choosing options for treatment. They often need to determine whether or not a color has faded to establish guidelines for preservation. They also need an objective language to discuss these matters with curators, museum administrators, and clients. But conservators face two challenges. First, there is still disagreement about the optical identification of many pigments that are well known by name. There is some consensus about dayflower blue, safflower red, Prussian blue, and the aniline dyes, but none, for example, for the ten yellows named in the literature. Second, without a prior instrumental reading, there is still no objective way to determine the degree to which a color has faded. Some colors were mixed; some were printed at different strengths in different periods, in different editions of the print, or even in the same picture. Also, pigments may have been prepared in different ways. Printers used distinctly different hues of safflower red (*beni*) in typical prints of the 1740s and the 1750s. Dayflower (*aigami*) ranges from a grayish but distinct blue to a warm but not-at-all-blue gray to a pale tan.

One solution to these difficulties is to consider colors in context. Printers used limited combinations of color during different periods and for different audiences. We will call these standard combinations of color "palettes." Changes of palette reflected changes in taste. They were sometimes caused by the accidental discovery of new color properties. Occasionally, they were catalyzed by the introduction of new pigments.

Here we introduce a number of palettes and ask that they be considered as components of the visual language of Japanese prints. This will enable recognition of certain visual effects that are characteristic of unfaded prints of a particular period, style, or audience. We approach the subject this way because there can be no absolute standards of color preservation. For example, the dilute red in an unfaded three-color print of the 1760s may be lighter and paler than the same red, originally saturated, in a light-struck, mid-nineteenth-century print. It is misleading to apply the color standard of one period to prints of the other, and it is inappropriate to apply a personal preference for colors of one palette to prints of another.

We have chosen eight palettes that we describe as warm or cool, wide or narrow, and dilute or saturated, depending on their color tonality, strength of color, and range of hue. *Warm* and *cool* refer to color tonality: *warm* colors are in the range from yellow and orange to red; *cool* colors are in the range from purple and blue to green. Browns in Japanese prints tend to be warm, and grays tend to be cool. A *wide* palette includes all three primary colors (red, yellow, and blue) and/or their derivatives. The colors in a *narrow* palette cluster around one family, like warm and cool, or one hue. *Saturated* colors are chromatically pure. These colors become *dilute* when mixed with white. Japanese printers diluted colors in two ways: by adding an opaque white pigment, either shell white (*gofun*) or lead white (*empaku*), or, more commonly, by printing pigments lightly so the white of the paper shows through.

We further characterize the palettes, as appropriate, by their principal color contrast: contrast of hue, warm/cool contrast, contrast of saturation, and light/dark contrast. Palettes with a full complement of primary colors in equal saturation achieve their effect by contrast of hue. This contrast is bright, energetic, and cheerful, and was used from the 1680s through the end of the nineteenth century. Red and green was the earliest warm/cool contrast. This was the palette for many hand-colored book illustrations in block-printed books in the seventeenth century, and the palette for the first color prints in the 1740s and 1750s. Pictures with hues of different intensities achieve their often subtle effects by contrast of saturation. Such palettes are often restricted to warm or cool tonalities, and even to single pigments, as for example, pictures printed entirely in shades of *beni*, or safflower red, which were used as talismans against illness in the nineteenth century. Black-and-white prints depend on light/dark contrast.

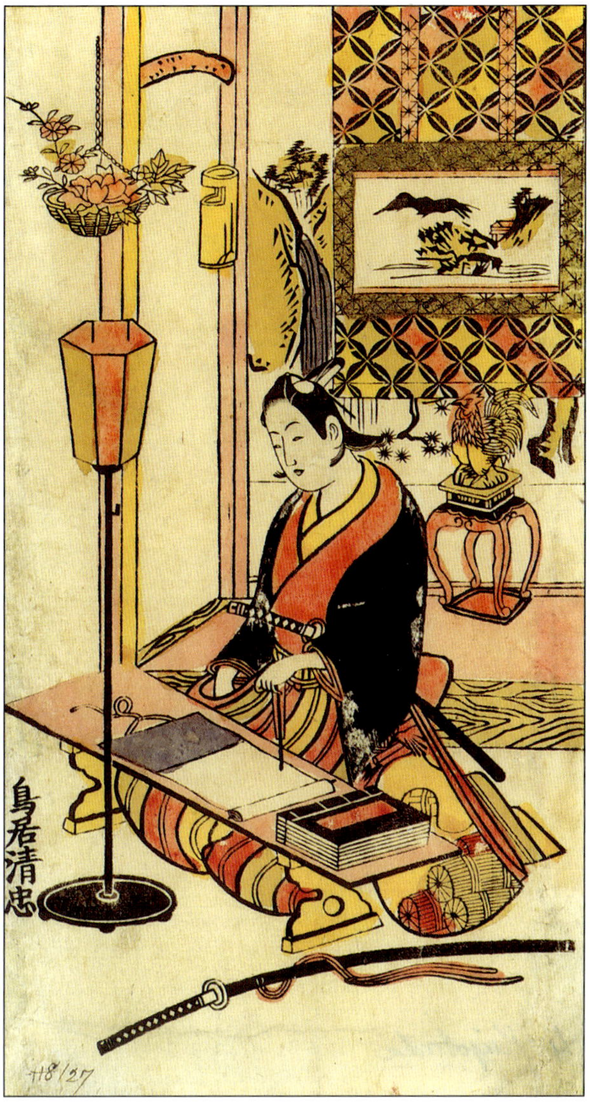

Figure 4.1 Torii Kiyotada (Japanese, active 1720s), *Young Man Seated at a Desk*, mid-1720s. *Beni-e*, hand-colored, *hosoban* format. A wide, saturated palette with contrast of hue. Herbert Egenolf Japanese Prints.

Basic palettes

Pictures printed from one block with black ink (*sumi*) are called *sumizuri-e*. They are printed with a *narrow, saturated palette with light/dark contrast*. The first dated pictures of this type were printed in 1162, although printed Buddhist images were certainly produced much earlier. *Sumizuri-e* were prevalent both as single-sheet prints and book illustrations through the end of the nineteenth century. Some examples were hand-colored at the time of publication; others were colored by later owners.

Tan-e, or hand-colored pictures with red lead, use a *narrow, warm palette with contrast of saturation*, and were prevalent between 1700 and the late 1710s. Red lead is saturated and opaque. The vivid yellow in these prints is transparent, so it is diluted by the white of the paper; the pink is so dilute as often to be barely visible. The colors in this palette vibrate and convey energy, which matches the vigorous style of drawing. Prints of this type were often sold without color and typically have a strong *light/dark contrast* as well.

Beni-e are hand-colored pictures in which safflower red (*Carthamus tinctorius*) has replaced red lead as the most conspicuous color (see Fig. 4.1). These prints, which were prevalent between the 1720s and 1740s, use a *wide, saturated palette based on contrast of hue*. Unfaded *beni-e* are lively and cheerful, and include sparkly and glossy effects that enliven their surfaces. Animal glue, or *nikawa*, brushed over *sumi* black gives a shiny lacquer-like surface – these prints are sometimes called "lacquer prints," or *urushi-e*, in the literature. Glue also may have been added to the red to give it a characteristic sheen. Finely pulverized brass filings were often brushed on for a glittering effect.

The first commercial color prints were produced in the early 1740s. Called *benizuri-e*, or printed *beni* pictures, they use two colors – organic safflower red and a mineral green – in addition to the black keyblock: a *narrow palette with warm/cool contrast* (see Fig. 4.2). This palette provided a satisfying analogue for the Buddhist notions of mutual interpenetration and non-separation. This may be why such a limited palette exclusively dominated the print world for over a decade, between the mid-1740s and the mid-1750s, even though printers were capable of printing more colors. The palette probably derived from hand-colored illustrations in seventeenth-century books that were called *tanroku-bon* because their principal colors were *tan* (red lead) and *roku* (copper green). The warm/cool contrast recurred in

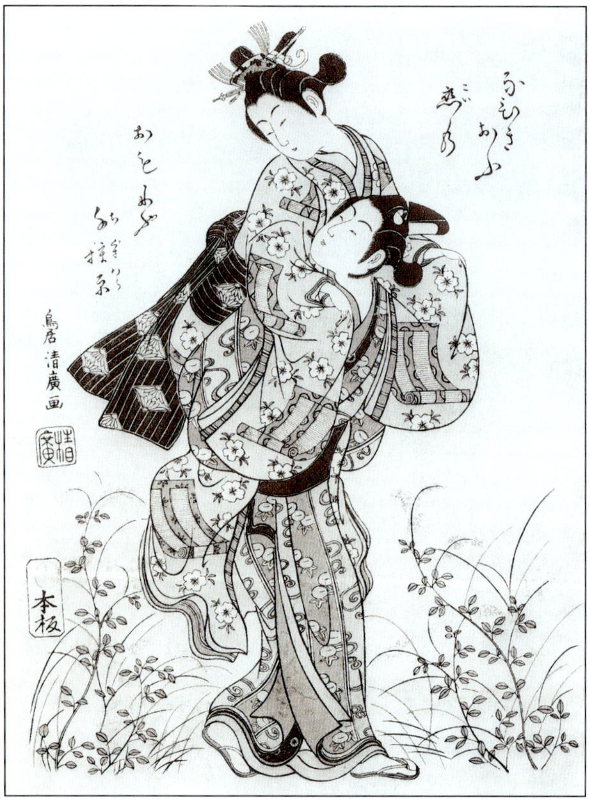

Figure 4.2 Torii Kiyohiro (Japanese, active 1750s), *Nakamura Shichisaburô II as Sukeroku and Nakamura Tomijûrô I as Agemaki*, 1753. *Benizuri-e*, printed in red and green, large *ôban* format. A narrow, saturated palette with warm/cool contrast. The Art Institute of Chicago, 1936.84.

later works by Toyokuni, Hokusai, and other nineteenth-century artists.

Transitional pictures printed with three or four colors, before the advent of full-color printing in 1765, were also called *benizuri-e.* Some of these were printed with a *wide, dilute palette with contrast of hue.* Artists like Kiyonaga and Hokusai used this palette for inexpensive prints on thin papers, notably actor portraits, between the late 1760s and the 1810s, during the period of full-color printing.

More affluent purchasers during this period preferred *nishiki-e*, or "brocade pictures," a term in use as early as the late 1760s to describe pictures printed with a *wide, saturated palette with contrast of hue* for the figures. These pictures were prevalent between 1765 and the 1880s. They often feature an additional *contrast of saturation* between densely colored figures and a lightly colored background, which gives the portrait figures a strong quality of presence.

In the middle of the nineteenth century, artists like Kunisada, Hiroshige, and Kuniyoshi in Edo, and Hirosada in Osaka, entirely covered the surfaces of their prints with saturated colors, giving a very dense effect in keeping with the complex narratives and iconography of the prints. This *wide, super-saturated palette with contrast of hue* was prevalent from the late 1840s through the early 1880s.

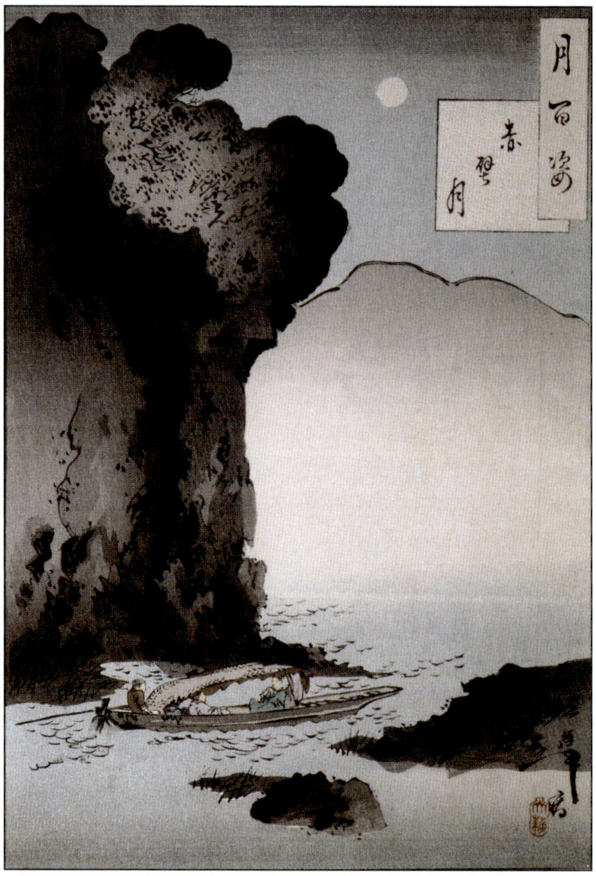

Figure 4.3 Tsukioka Yoshitoshi (Japanese, 1839–1892), *Moon over the Red Cliffs*, 1886, from *One Hundred Aspects of the Moon. Nishiki-e, ôban* format. A narrow palette with contrasts of light/dark, saturation, and warm/cool. Private collection.

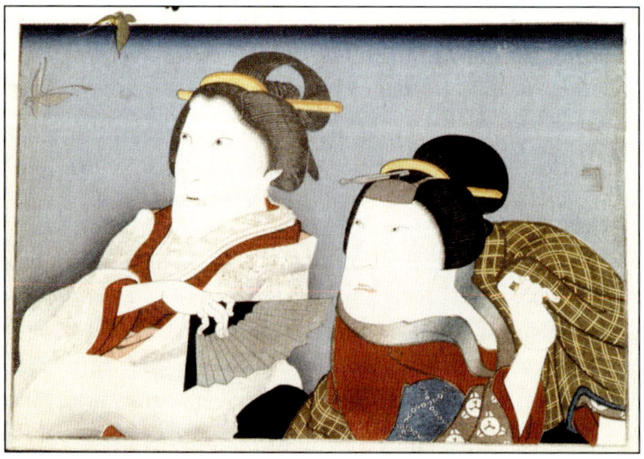

Figure 4.4 Konishi Hirosada (Japanese, c. 1810–1864), *Onoe Kikugorô III as the Ghost of Iwafuji and Arashi Rikan III as the Second Onoe*, 1849. *Nishiki-e.* A contrast between a narrow, dilute palette and a wide, saturated palette. Private collection.

Artists often combined palettes to create more complex color effects. Yoshitoshi (1839–1892) used a narrow palette with three different contrasts – of *light/dark, saturation*, and *warm/cool* – in *Moon over the Red Cliffs* of 1886 (Fig. 4.3). In an image of actors Onoe Kikugorô III and Arashi Rikan III of 1849 (Fig. 4.4), Hirosada (c. 1810–1864) employed a *narrow palette with*

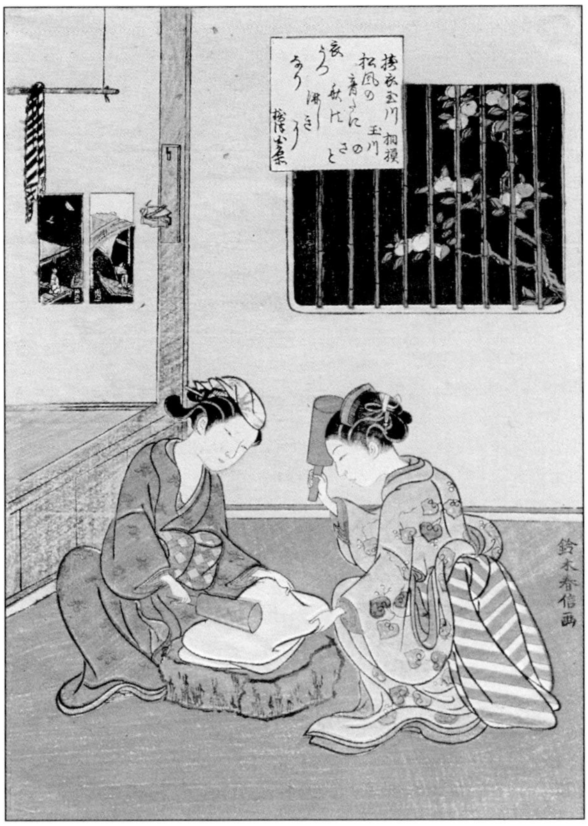

Figure 4.5 Suzuki Harunobu (Japanese, active late 1750s–1770), *Tôi Tama River* (night version), 1766, from an untitled series of six Tama rivers. *Nishiki-e, chûban* format. A wide palette with contrast of saturation. Raymond A. Bidwell Collection, Museum of Fine Arts, Springfield, Mass.

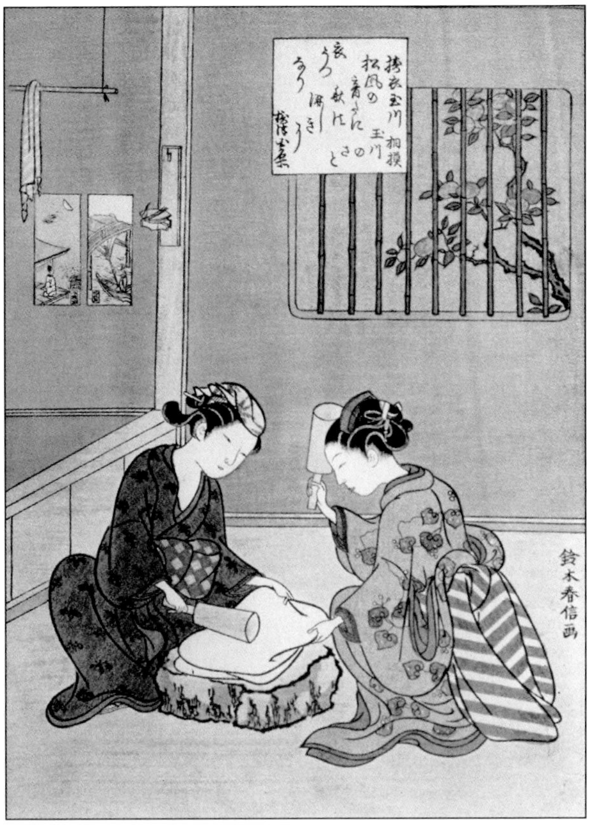

Figure 4.6 Suzuki Harunobu, *Tôi Tama River* (daylight version). A wide palette with contrast of hue. Private collection.

contrast of saturation for the malevolent ghost on the left, who is fading back to the spirit world, and a *wide, saturated palette with contrast of hue* for the woman on the right whom she has unsuccessfully tried to kill.

The history of blue

The organic blues used up until the 1820s are touchstone colors for assessing color preservation because they are so fugitive. If an organic blue is unfaded, the rest of the print is unfaded. A short history of the color blue will put this into perspective.

From the 1680s through the 1740s, *aigami*, the blue derived from the dayflower (*Commelina communis*), was apparently the only blue pigment used on hand-colored prints. From the mid-1750s through the beginning of the nineteenth century, it was the prevalent blue in both *benizuri-e* and *nishiki-e*, and it was also the blue commonly mixed with *beni* to produce a range of purples.

Aigami blue, as stated earlier, ranges from a grayish but distinct blue to a barely blue gray to a pale tan. The blue form is highly moisture sensitive and relatively light stable, whereas the gray form is relatively moisture stable and highly light sensitive. The actual hue seen on a print depends on the pigment's preparation, original saturation, and degree of subsequent fading. Different saturations of blue fade to subtly different shades of tan.

Aigami blue was frequently used as a background for actor portraits of the late 1760s and early 1770s. Alteration of this pigment may have been the inspiration for pictures with yellow backgrounds by Eishi, Utamaro, and others which were so popular during the 1790s. A convention arose that prints of beautiful women almost exclusively feature yellow backgrounds and a predominately warm palette. Simultaneously, *sumi* gray became the conventional background color for portraits of male kabuki actors.

Ai, the blue derived from indigo (*Polygonum tinctorium*), was the principal blue dye for fabrics in the eighteenth and nineteenth centuries. The extent of its use in prints of the eighteenth century has not yet been determined, but it was certainly used for the dark, saturated blue on some *surimono* of the 1810s and 1820s, and on landscapes of the same period by Hokuju. Indigo blue also fades to tan and is optically indistinguishable from *aigami* when it has faded.

An entirely new, much more stable blue was introduced perhaps as early as the 1790s. This was Prussian blue (ferric-ferrocyanide), an imported synthetic pigment known as *bero* or *bero-ai* in Japan, an abbreviation for *Berorin*, or Berlin. Originally available only in small quantities at very high prices, it was used sparingly on private publications.

Prussian blue became abundantly available in the late

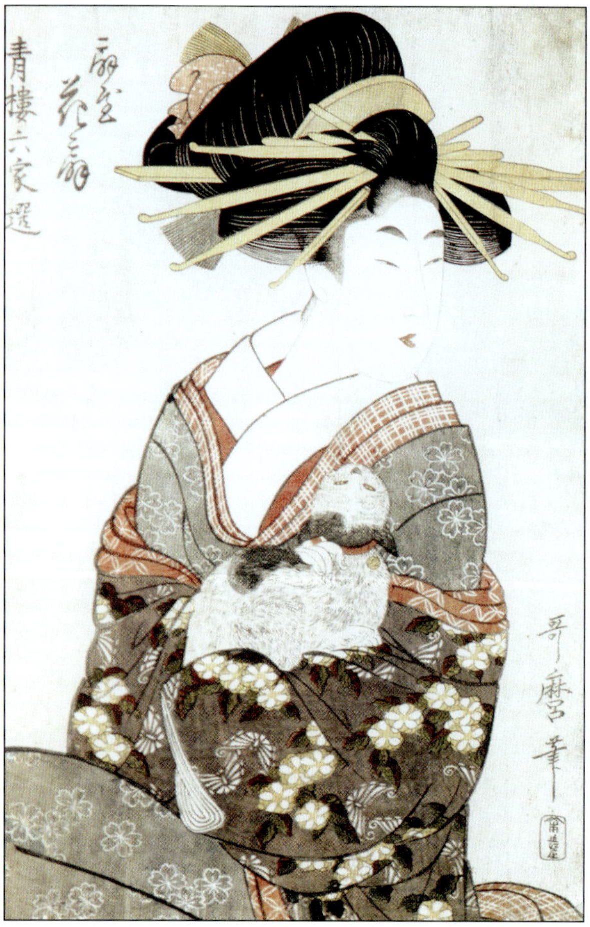

Figure 4.7 Kitagawa Utamaro (Japanese, 1735–1806), *The Courtesan Hanaôgi Holding a Cat*, c. 1801–02, from *Six Houses in the Yoshiwara. Nishiki-e, ôban* format. An unfaded impression; a wide, saturated palette with contrast of hue. Museum of Fine Arts, Boston.

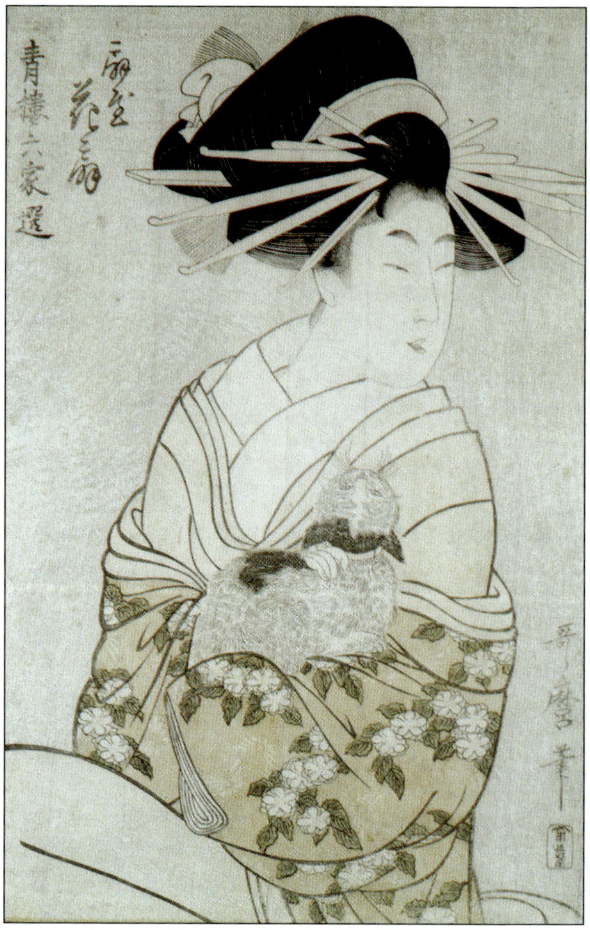

Figure 4.8 Kitagawa Utamaro, *The Courtesan Hanaôgi Holding a Cat.* A faded impression changed to a narrow, dilute palette with light/dark contrast. Formerly Anders Rikardson Collection.

1820s and was used at full strength thereafter. *Aizuri-e*, or "blue prints," were printed with Prussian blue as the primary color. This is the final palette we address here, a *narrow, cool palette with contrast of saturation.* Some artists enlivened their *aizuri-e* with touches of red for an additional *warm/cool contrast. Aizuri-e* may have owed some of their popularity in the late 1820s and early 1830s to the prevalence of blue-and-white porcelain. This electrifyingly brilliant pigment challenged mid-nineteenth-century artists to find more saturated hues of other pigments to match its intensity, and catalyzed the development of what we have called the super-saturated palette. By providing a relatively stable, brilliant color to represent sky and water, Prussian blue also revolutionized the genre of landscape prints.

Color variation

Palettes in different impressions of the same print can vary for several reasons. Common organic pigments like safflower red and dayflower blue may vary in color depending on how they were prepared. Different printers might choose different colors to print the same set of blocks. Printers in the late 1760s sometimes experimented with strikingly different combinations of color. One early impression of Harunobu's (active late 1750s–1770) *Tôi Tama River* is depicted as a warmly lit night scene (Fig. 4.5); another shows daylight (Fig. 4.6).

But most examples of extreme color difference and loss of contrast are accidental. If the originally cream-colored paper darkens, the resulting brown tone both warms the palette and reduces contrast. And many organic pigments are so fugitive that they can fade beyond recognition and sometimes disappear altogether, drastically changing the color balance and effectiveness of the print (see Figs. 4.7 and 4.8). A cool palette can become warm; contrast of hue can change to contrast of saturation. If much color is lost, a warm/cool contrast can change to a dark/light contrast.

With the widespread adoption of a relatively unfading blue in the late 1820s, we must look for other fugitive colors to tell us whether or not a print from after this time has faded. Organic yellows and reds, for example, were used separately and mixed with other pigments. Many mid-nineteenth-century prints have strongly colored yellow and red cartouches, which can fade dramatically. The purples fade to various shades of brown. The green in landscape prints is often a mixture of Prussian blue and a fugitive organic yellow. The effectiveness of many prints by Hiroshige and Hokusai depends on the contrast of a cool blue with warm yellows and greens. When light-struck, the yellow fades and leaches out of the green, making the prints look unpleasantly cold.

Finally, faded color may be deliberately restored. One form of restoration is hand-coloring. When color prints are exposed to light or moisture, the inorganic or stable pigments remain, but the organic pigments fade. It is therefore strange to see, on otherwise faded prints of the late eighteenth century, an apparently unfaded but normally fugitive color. On closer inspection we usually find that such areas have been restored by hand. Uneven application of color and irregular edges are characteristic of these overpainted areas.

The alternative to hand-coloring is block-printed color restoration, a common practice in Japan in the 1910s. The residual organic colors of genuine but partially faded prints of the late eighteenth century were intentionally bleached with sunlight, and then new color blocks were carved and printed for these bleached areas. The four clearest indications of this practice are: atypical color hues (a brighter blue, a cooler red, and a darker purple), since the restorers did not use organic pigments; traces of previously printed patterns visible as anomalous embossing in newly reprinted areas; certain faded areas left uncolored, apparently to avert suspicion; and pinpricks used to register the new color blocks, often visible in transmitted light.

"The colors of the flower fade," wrote Ono no Komachi in the ninth century. Japanese prints are a precious and fragile heritage. They depend on color for the many effects we have described, including presence, atmosphere, and mood. We hope our descriptions of the various palettes have contributed to a better understanding of how color works in this art form, that they will assist in helping to recognize when prints have changed from their original state, and that they will provide a vocabulary with which to discuss these issues. There is no substitute for looking at prints, and we encourage looking at prints of all types from all periods. Color is a language of the heart, of nuanced sensibility, of degrees and shades of feeling. And it is through the language of color, particularly through the refined and changing use of color palettes, that *ukiyo-e* prints speak to us so clearly and forcefully today.

Acknowledgment

We thank Shiho Sasaki for sharing with us her ongoing research into the characteristics of dayflower blue.

Aqueous Treatment of *Ukiyo-e* Prints of the Edo Period: Three Case Studies

Pamela de Tristan

"Pictures of the floating world" – *ukiyo-e* – evolved in Japan during the Edo period (1603–1868), so named because the feudal capital was moved from Kyoto to Edo (modern Tokyo) in the early seventeenth century. Lasting roughly 250 years, during an era when Japan was relatively isolated from Western influences, the Edo period was a time of invention and intense creativity. Woodblock printing, first accomplished in Japan in A.D. 764 with Buddhist devotional texts, was employed by the mid-seventeenth century to produce secular books and hand-colored prints, including those made so popular by artists of the *ukiyo-e* school during the late seventeenth and early eighteenth centuries.[1]

Many of the earliest prints of the period, the *sumizuri-e*, were printed only in *sumi* on *kozo* paper. Among the first-generation *ukiyo-e* artists was Hishikawa Moronobu (d. 1694), who is perhaps best known for his "spring pictures" (*shunga*), explicitly erotic works that appear in illustrated books and hand-scrolls. The *sumi* used for these prints remains stable and unaffected by light to this day. The black pine soot ink is comprised of pigment particles so small that they readily penetrate a paper matrix and integrate with the paper fibers. Because of this, *sumi* does not flake from its support and stands up to aqueous treatment.

The colors added by hand to prints in the early years of the eighteenth century most commonly include the red lead pigment *tan*, as well as a bright yellow, a very pale pink, and blue.[2] *Tan-e*, as these hand-colored prints are known, were popularized by artists such as Torii Kiyomasu I (active c. 1704–18?), a specialist in the depiction of beauties and kabuki actors. *Tan-e* must always be tested for potential solubility problems. Although it is often not known what medium was chosen to mix with the *tan*, rice starch, which can be quite soluble, is likely.

Prints that are hand-colored with *beni*, rather than *tan*, are known as *beni-e*. These were often embellished with metallic powders and glues to heighten the colors and provide selectively glossy surfaces. When animal glue was applied over *sumi* or mixed with it and applied to the surface of a print, the black medium resembles shiny lacquer. Hence these prints are known as *urushi-e*, or "lacquer prints."[3] Okamura Masanobu (1686–1764) is credited with introducing this technique in the middle of his working career, around 1720. He is also well known for his additions of glue over blind printing, or *gauffrage*, to create a lacquered appearance in dimensional passages. The use of animal glue makes these "mixed-media" prints vulnerable to damage with changes in humidity. Lacquered areas easily crack, and contraction, if it occurs, can readily cockle surrounding areas. Metals dusted over the surface of the prints can sometimes bite through the paper substrate if they have a tendency to be corrosive. This phenomenon is quite rare, as the adhesive between the metallic dust and the paper tends to act as a buffer, but when it does occur, areas of damage should be reinforced with Japanese paper repairs on the verso. Moisture must be used most judiciously in the conservation of these works.

In the early 1740s, Japanese printing underwent a revolution with the invention of the *kento* technique for registering wood blocks. In this technique registration marks were cut into the long side of each block, allowing for accurate alignment of the paper with multiple blocks. A keyblock printed the outlines of the composition, and individual blocks were used for each color. The earliest prints produced in this manner, *benizuri-e*, were printed in two colors: *beni*, the organic safflower red (*Carthamus tinctorius*), and a mineral green. The keyblock was customarily printed with *sumi*. By the early 1750s, two additional colors, *aigami*, the organic dayflower blue (*Commelina communis*) and *ukon*, the organic turmeric yellow (*Curcuma longo*), were prevalent. By 1765 a full range of organic and inorganic colorants came into use.[4] Overprinting of several primary colors could produce an array of greens, oranges, and purples as well.

Highly sophisticated color effects were achieved in the prints known as *nishiki-e*, or "brocade pictures," which appeared in 1765 and remained popular well into the nineteenth century. Suzuki Harunobu (d. 1770), who may be considered one of the greatest masters of the brocade picture, often used as many as ten blocks to achieve his color harmonies. He is perhaps best known for the "picture calendars" that he provided for the elite literati to exchange at parties in 1765 and 1766. Harunobu's prints, as well as those by numerous other artists of the period, can be damaged or disfigured by worm holes, staining, foxing, and accumulated surface dirt. The latter is especially problematic where it collects into the recesses of blind printing (*gauffrage*). The delicate nature of these prints requires extreme caution on the conservator's part, especially with the use of moisture, as both the colorants and raised surfaces could be adversely affected.

By the late eighteenth century, brocade pictures featured more opaque pigments in some parts of their compositions. It was common for lead white to be mixed with a colorant to produce an opaque, matte effect. As a result of this use of lead white, these passages often discolor and darken. A similar occurrence is found in passages where red lead was used. Further complicating the matter, discoloration of red lead was sometimes deliberately induced by the artist – it is not unusual to find this in Harunobu's prints. As has been published elsewhere, reversion of discolored and darkened passages of lead-containing media must be carried out with the utmost discretion so that an artist's aesthetic intent is not altered.[5]

Mica, a mineral composed of the silicates of aluminum and potassium, was used in much the same way as metal dust to create a lustrous surface. The mineral was ground and applied with animal glue (*nikawa*) or seaweed gel (*funori*). The mica could also be mixed with a variety of colorants to produce flat, continuous areas of colored sheen that were particularly effective as backgrounds. The artists best known for using mica in their prints include Yeishosai Choki (active 1785–1806), Kitagawa Utamaro (1735–1806), and Toshusai Sharaku (active 1794–95). Because of its large particle size, mica tends to sit on the surface of the paper, and its adhesion with the paper is often tenuous. Thus, it is greatly at risk of damage through careless handling; the mica can be lost in patches where a print has been deformed, or in lines along creases where the paper has been creased or folded, or where fingerprints and surface dirt mar its surface. It is also susceptible to mold and insect damage. Mica is extremely moisture sensitive, and any excess of moisture in conservation treatment can result in lifting and loss. All localized treatments should be undertaken from the verso. For repair of losses such as insect or worm holes, prepared sheets of mica that match the original as closely as possible in color, shape, and surface texture may be set into place from the verso of the print.

The opening of trade with the West in the nineteenth century introduced new technologies and graphic media such as photography, etching, engraving, lithography, and oil painting to Japan. These all had an influence on how artists of the late Edo period would conceive their prints. Most significantly for the purposes of this essay, importation of the more strident colors produced with aniline dyes also precipitated changes in traditional Japanese color palettes.[6] Prussian blue (Berlin blue, *bero-ai*), introduced to Japan by the early 1800s through China, was perhaps the single most important artists' color to enter Japan at the time.[7] The imported synthetic blue (ferric-ferrocyanide) was costly and rare at the turn of the nineteenth century, and was used sparingly at first. By the 1820s it was available in greater quantities, contributing to the rise of a new genre, that of pure landscape. Among the most admired prints of the period were those of landscape artist Katsushika Hokusai (1760–1849). His series *The Thirty-Six Views of Mount Fuji* (1823–30), which includes *Great Wave off Kanagawa*, illustrates the range of tone and brilliance that could be achieved with Prussian blue. It was a color that lent itself well to the *bokashi* technique of gradating color from dark to light in the sky or water of a given composition.

The fading of the many colorants used in *ukiyo-e* prints over a period spanning almost three centuries has been well documented, as has the darkening of red lead and lead white.[8] Analysis and identification of organic colorants and dyes used in these prints has been accomplished non-destructively using a 3-D fluorescence spectrophotometer.[9] The research outlined below was concerned with ascertaining what sorts of conservation problems could arise in the delicate hand-applied and printed colors described above when moisture is used in a treatment protocol to remove surface dirt, staining, backings, and in the humidification process itself. First, the organic dyes and Prussian blue potentially present in the prints were identified in mocked-up samples and in fragments from destroyed prints from the Edo period. "Known" samples were identified, and the optical changes that occurred with treatment were measured.

Colorant	15 minutes	30 minutes	45 minutes	60 minutes
Beni				
Float washed	1.53	1.68	3.42	4.35
Submerged	1.67	3.01	3.56	4.04
Ukon				
Float washed	1.37	2.00	1.74	1.06
Submerged	2.54	2.31	3.14	3.26
Sumi				
Float washed	0.46	0.50	0.61	0.55
Submerged	1.11	0.71	0.25	1.27
Bero-ai				
Float washed	0.63	0.50	0.81	0.31
Submerged	0.67	0.64	0.57	0.58

Table 4.1 Color drift measured as ΔE after float washing and submersion of Japanese print samples.

Colorant	M1 – verso only	M2 – recto & verso
Beni	1.17	1.20
Ukon	0.70	0.32
Sumi	0.47	0.32
Bero-ai	0.62	0.46

Table 4.2 Color drift measured as ΔE after misting of Japanese print samples.

Establishing the moment of a color change in the treated samples provided important information that could be used to determine more appropriate washing and humidification times for the prints used in the three case studies presented below.

Beni (red/pink), *aigami* (blue), *ukon* (yellow), and *bero-ai* (Prussian blue) were chosen for analysis because they are the four most commonly used colorants in Edo-period prints. To produce the mock-ups, each colorant was painted out on a strip of Japanese paper prepared with rice starch paste. Safflower petals from which the beni was extracted were first washed to remove the yellow dye that solubilizes before the pink and red dyes emerge. The *aigami* was supplied on a sheet of paper saturated with it; when the paper was immersed in water the dayflower blue dye solubilized from it. The *ukon* and *bero-ai* were produced by simply mixing the powdered dyes with a small quantity of water. *Sumi*, extracted by rubbing a sumi stick with a bit of water on a clean tile, was also tested. Each colorant produced a unique fluorescence spectrum when placed under the fiber-optic light source of a Hitachi fluorescence spectrophotometer F-4500 and irradiated with light ranging from 250 to 600 nm.[10]

Testing was repeated for the fragments of destroyed prints from the Edo period, which ranged in size from 12 x 5 mm to 25 x 25 mm, depending on how evenly and consistently a desired colorant appeared in a given area. Fixed locations for testing were established by drawing 3 mm circles in pencil directly on the verso of the fragments, to ensure accurate placement for each reading. Dr. Richard Holman of the Paint Research Association conducted the testing. Spectroscopic quantification of color and color variation was measured by the magnitude of change in L, the value indicating whether the sample became lighter or darker; C, the chroma value indicating whether a shift in color saturation had occurred; and H, indicating whether a change in hue had occurred. The ΔE value indicates the perceptible color change by taking C, L, and H values into account mathematically.[11] For the purposes of this study, a change in ΔE above 1.0 indicated a perceptible change in the color samples and in the actual prints later tested. Samples were float washed and submersion washed in water with a pH of 6.7 for 15, 30, 45, and 60 minutes. Misting with a light spray of water was carried out on the verso only (M1) and the recto and verso (M2) of the samples. Readings were taken immediately after misting. Humidification with Gore-Tex® positioned under the samples took place for 30, 60, 120, and 180 minutes.

After 15 minutes of either float washing or submersion washing, the *beni* began to show a drift in color. The red colorant became more yellow and lighter with a ΔE of 1.53 and 1.67 respectively. The ΔE increased as washing time increased. Likewise the *ukon* displayed ΔE values of 1.37 and 2.54 under similar conditions. The yellow colorant maintained its yellow hue, but got darker. The *sumi* did not drift as significantly, but after 60 minutes of submersion washing, it was no longer as intensely black. Surprisingly, the *bero-ai* showed more stability than the *sumi*, although its color veered towards the green (Table 4.1). With misting, only the *beni* had a ΔE greater than 1.0, and it again drifted toward the yellow spectrum. The other three colorants exhibited no significant drift in color

Colorant	30 minutes	60 minutes	120 minutes	180 minutes
Beni	1.46	0.61	2.36	1.90
Ukon	0.81	1.38	1.33	0.73
Sumi	0.40	0.38	0.24	0.89
Bero-ai	0.75	0.38	0.24	0.89

Table 4.3 Color drift measured as ΔE after Gore-Tex® humidification of Japanese print samples.

Colorant	Sample point	ΔE	Color drift significance
Aigami	A1 A2	0.73 0.49	Not significant Not significant
Beni	B1	0.30	Not significant
Ukon	T1 T2	1.31 0.94	Significant Not significant
Unidentified purple	M1 M2	1.20 1.29	Significant Significant

Table 4.4 Color drift measured as ΔE after Gore-Tex® humidification of Utamaro's *Unraveling the Threads of Desire* (Fig. 4.9) for 20 minutes.

with misting (Table 4.2). After 30 minutes of Gore-Tex® humidification, the *beni* got darker; so too did the *ukon* after 60 minutes. The *sumi* did not drift significantly after 30 minutes, nor did the *bero-ai* although it looked slightly darker (Table 4.3).

Once testing of the mock-ups was complete, tests were repeated (with some modification) for three actual prints: an example by Utamaro, an Utamaro II (active 1804–1818), and a Hokusai. The Utamaro is an erotic print entitled *Unraveling the Threads of Desire*, dating from 1799 (Fig. 4.9). Before treatment, the edges of the print were frail, and the sheet had a very pronounced central crease. All the colorants present – *ukon*, *beni*, *aigami*, and a purple derived from a mixture of *aigami* and *beni* – were moisture sensitive. A water stain that altered the purple colorant to beige is visible on the male figure's kimono and indicates how especially vulnerable this colorant is to moisture damage. Data were obtained from the print before and after treatment; a Mylar® tracing of test locations ensured that the same areas were tested each time. The areas were labeled A1, A2 for *aigami*; B1 for *beni*; T1, T2 for *ukon*; and M1, M2 for the purple colorant.

It was determined that humidification on Gore-Tex® would be the best way to relax the paper so that the crease could be eased out of the center of the sheet. The print was humidified in a sealed environment for 20 minutes rather than 30 minutes as it was feared that longer exposure would perhaps put the *aigami* at risk. Misting the sheet was rejected as a treatment option for fear that droplets of water falling on the surface of the print could adversely affect the *aigami*. In anticipation of reinforcing the crease from the verso, a 12 gsm strip of *tengujo*, 7 mm wide and the length of the crease, was water torn together with another, shorter strip of *tengujo* to be applied over the leg area, where the crease was most pronounced. The strips were applied with starch paste

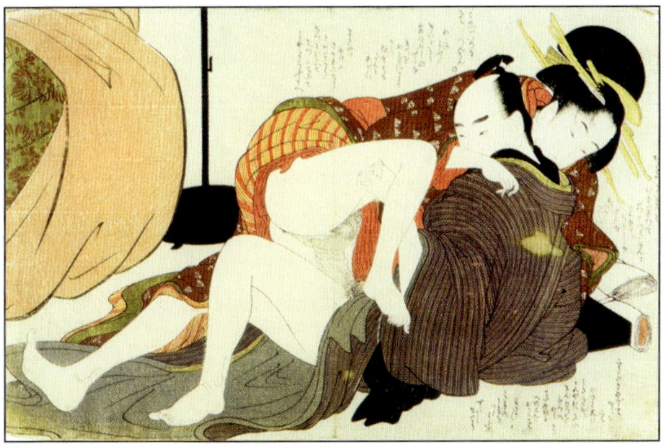

Figure 4.9 Kitagawa Utamaro (Japanese, 1735–1806), *Unraveling the Threads of Desire*, 1799 (before treatment). Woodblock print, 25.2 x 38 cm. Matsumi Salon Fine Japanese Prints. (Photograph by Ian Hessenberg, London.)

immediately after the print's removal from the humidity package. When treatment was complete, the print was placed between sheets of non-woven polyester and blotters, under minimal pressure, to dry and flatten. After conservation the deep central crease was successfully eased out of the sheet and the frail edges reinforced with Japanese tissue. Comparison of the data obtained before and after treatment reveals that the *beni* (B1) and the *aigami* (A1, A2) did not drift significantly, although the *ukon* (T1, T2) and the purple (M1, M2) did (see Table 4.4). The *ukon* got slightly darker, although its hue did not change. The purple shifted towards the yellow.

The second print to be analyzed before and after treatment was an Utamaro II entitled *First Pulling of Pine Shoots on the First Rat Day of the New Year*, dating from 1806 (Fig. 4.10). A significantly yellowed backing paper was firmly attached to the verso of the print and some color transfer from the recto was evident. The print

Colorant	Sample point	ΔE	Color drift significance
Aigami	A1	0.47	Not significant
	A2	0.78	Not significant
Beni	B1	1.71	Significant
	B2	1.41	Significant
Ukon	T1	1.79	Significant
	T2	0.74	Not significant
Unidentified green	G1	0.83	Not significant
	G2	0.36	Not significant

Detailed analysis of sample point B1 – *Beni*

	L	a	b	C	h	ΔE
Before treatment	57.28	28.97	17.01	33.60	30.42	
After treatment	56.93	27.88	17.55	32.53	32.66	
Color drift	–0.35	–1.09	0.54	–1.07	ΔH 0.29	1.71

The *beni* has become more yellow.

Detailed analysis of sample point T1 – *Ukon*

	L	a	b	C	h	ΔE
Before treatment	67.27	4.28	27.25	27.58	81.08	
After treatment	67.97	4.27	28.89	29.21	81.59	
Color drift	0.70	–0.01	1.64	1.63	ΔH 0.25	1.79

The *ukon* has become more yellow, less gray, and slightly lighter.

ΔH is the linear change in hue, and is calculated from the change in hue angle (h) together with the changes in a + b.

Table 4.5 Color drift measured as ΔE after backing removal with CMC and misting of Utamoro II's *First Pulling of Pine Shoots on the First Rat Day of the New Year* (Fig. 4.10).

includes *beni*, *ukon*, and *aigami*. The *aigami* in the background had obviously faded to a pale gray, and it was water stained on the left as well. In addition to these colorants, an unidentified green and *sumi* were also present. Once again, before treatment began a Mylar® tracing that marked test areas was made to ensure accuracy in the placement of the equipment that would take the readings.

It was decided to remove the yellowed backing with a carboxy methyl cellulose gel (CMC, 5% w/v) poultice rather than subject the print to either float washing or misting, since either of these measures would have necessitated an excessive amount of moisture and would certainly have put the colorants at risk. The CMC poultice was applied to the verso of the print in 12 cm^2 sections and weighted with glass for a few minutes to facilitate penetration of the moisture held in the poultice. In this way the backing adhesive softened so that the backing paper could be rolled back and away from the print. Using this method there was minimal risk of moisture penetrating to the recto. When necessary the verso of the print was lightly misted, as it started to dry out during backing removal. Once treatment was complete, the print was placed between sheets of non-woven polyester and blotters and pressed to dry and flatten. Comparison of the data taken before and after treatment reveals that the *beni* (B1, B2) and the *ukon* (T1) drifted significantly, so much so that a color difference was apparent to the unaided eye (see Table 4.5). *Beni* (B1) became more yellow; *ukon* (T1) became more yellow, less gray, and slightly lighter in tone.

The third print to be treated and analyzed was a Hokusai entitled *Totomi Province*, from the series *One Hundred Views of Mount Fuji* of c. 1830 (Fig. 4.11). Significant amounts of Prussian blue were identified throughout the print, so a series of Prussian blue samples was first subjected to accelerated aging and analysis for comparative purposes. Three samples were chosen: a Japanese woodblock print fragment dating from 1858 was subjected to exposure that equaled 92.14 years of exposure at 50 lux; and two contemporary Prussian blue swatches printed from a woodblock, one dark and one light, were subjected to exposure that equaled 79 years of exposure at 50 lux. The Prussian blue in the nineteenth-

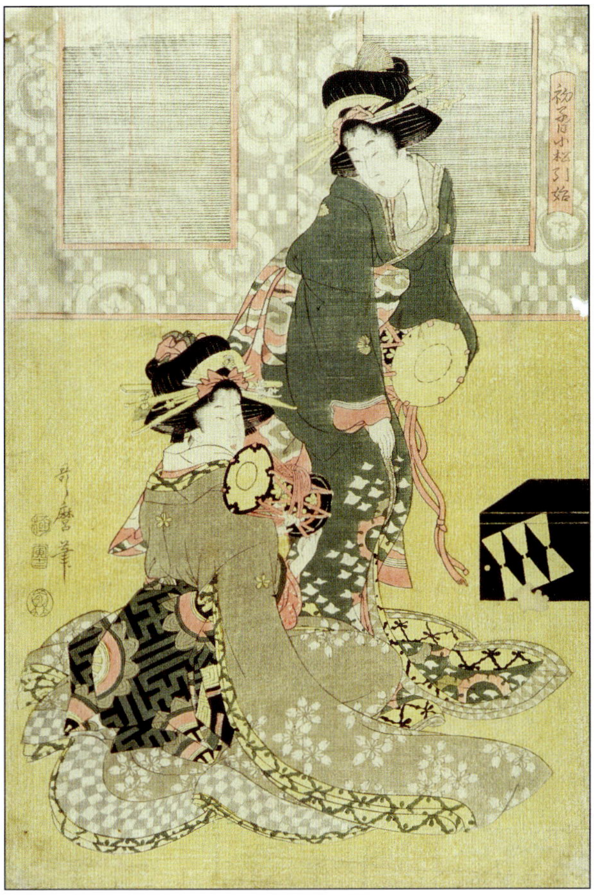

Figure 4.10 Kitagawa Utamaro II (Japanese, active 1804–1818), *First Pulling of Pine Shoots on the First Rat Day of the New Year*, 1806 (before treatment). Woodblock print, 25 x 38.5 cm. Matsumi Salon Fine Japanese Prints. (Photograph by Ian Hessenberg, London.)

century sample became fractionally lighter after exposure to accelerated aging. The Prussian blue in the contemporary samples exhibited more fading than their nineteenth-century counterpart; the light blue more than the dark blue. The hue drifted towards the green in both samples, and the chroma of the lighter blue sample became slightly grayer (see Table 4.6).

The Hokusai print had an accumulation of surface dirt as well as embedded dirt throughout. There was yellow staining in the clouds to the right and the sheet was quite yellowed overall. Eight areas were located on a Mylar® tracing for analysis before and after treatment: P1a, P1b, P2, P3, P4, and P5 represent Prussian blue areas; B1, B2 represent areas where there is a faint pinkish color present in the clouds. The print was lightly surface cleaned before submersion in water with a pH of 6.7 for 45 minutes. Since so much ingrained surface dirt was present, submersion was preferred over float washing; 45 minutes was chosen as the optimal length of time since the results of the preliminary testing on the samples indicated that submersion of the *bero-ai* sample for 45 minutes had resulted in a ΔE of only 0.57.

After 20 minutes of submersion washing, the wash water was very yellow. This water was discarded, replaced

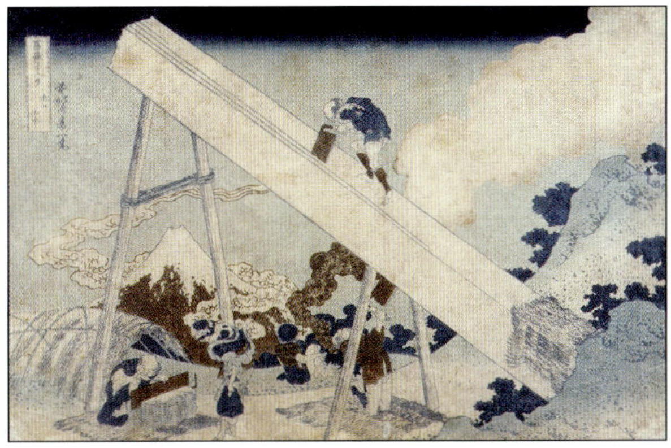

Figure 4.11 Katsushika Hokusai (Japanese, 1760–1849), *Totomi Province*, c. 1830, from the series *One Hundred Views of Mount Fuji* (before treatment). Woodblock print, 24.7 x 38 cm. Matsumi Salon Fine Japanese Prints. (Photograph by Ian Hessenberg, London.)

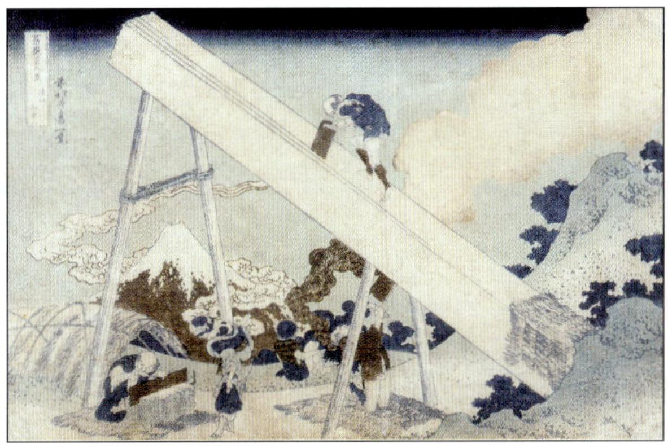

Figure 4.12 Katsushika Hokusai, *Totomi Province* (after treatment). Matsumi Salon Fine Japanese Prints. (Photograph by Ian Hessenberg, London.)

with fresh water, and the print was washed for an additional 25 minutes. After washing, the print was air dried slightly, then placed between sheets of non-woven polyester and blotters and lightly pressed to dry and flatten. The sheet was not resized as this might have interfered with the readings. Comparison of the data indicates that in some areas the Prussian blue got darker and became grayer (see Table 4.7). In every area tested, the color became bluer as a result of removing yellow degradation products from the paper in the washing process (see Fig. 4.12).

The three late Edo prints presented here as case studies show that organic and synthetic colorants alike are affected by moisture. *Aigami*, traditionally thought to be most vulnerable to moisture, stood up rather well to various conservation treatments but always showed some degree of drift – either lighter, darker, or in some cases veering towards an adjacent color spectrum altogether. In assessing the origin of color change, not only should ΔE be considered, but also the individual H, C, and L values. If a particular conservation treatment results in a ΔE of 1, this may not be indicative of an unacceptable change or damage. Rather, the change may arise only from an increase in the cleanliness (i.e. whiteness) of the

Sample A	L	a	b	C	h	ΔE
Covered	34.40	–5.13	–7.97	9.48	237.24	
Uncovered	35.62	–4.92	–8.86	10.14	240.94	
Color drift	1.22	0.21	–0.89	0.66	ΔH 0.43	1.52
Sample A – Japanese woodblock print sample from 1858						
Sample B	L	a	b	C	h	ΔE
Covered	23.86	1.08	–24.94	24.98	272.52	
Uncovered	25.74	–0.64	–24.95	24.96	268.50	
Color drift	1.88	–0.72	–0.01	–0.02	ΔH 1.72	2.55
Sample B – Dark *bero-ai* sample printed from woodblock in 1999						
Sample C	L	a	b	C	h	ΔE
Covered	37.56	–7.37	–22.37	23.56	251.79	
Uncovered	38.79	–7.64	–19.00	20.48	248.12	
Color drift	1.23	–0.27	3.37	–3.08	ΔH 1.41	3.60
Sample C – Light *bero-ai* sample printed from woodblock in 1999						
ΔH is the linear change in hue, and is calculated from the change in hue angle (h) together with the changes in a + b.						
Note that +a = redder; +b = yellower; –a = greener; –b = bluer						

Table 4.6 Fading of *bero-ai* samples subjected to accelerated light aging.

Colorant	Sample point	ΔE	Color drift significance
Bero-ai	P1a	0.79	Not significant
	P1b	1.93	Significant
	P2	2.99	Significant
	P3	1.18	Significant
	P4	1.14	Significant
	P5	1.58	Significant
Unidentified pink	B1	1.38	Significant
	B2	1.06	Significant

Detailed analysis of sample point P4 – *bero-ai*						
	L	a	b	C	h	ΔE
Before treatment	42.11	–10.10	–8.19	13.00	219.04	
After treatment	41.40	–9.47	–8.83	12.95	222.98	
Color drift	–0.71	–0.63	–0.70	–0.05	ΔH 0.89	1.14
ΔH is the linear change in hue, and is calculated from the change in hue angle (h) together with the changes in a + b.						

Table 4.7 Color drift measured as ΔE after submersion of Hokusai's *Totomi Province* (Fig. 4.12) for 45 minutes.

paper, that is the brightness value L, while leaving the underlying color characteristics C and H relatively unchanged.[12] Edo-period prints printed with organic dyes and Prussian blue not only fade, they are moisture sensitive. In light of these findings, moisture must be used judiciously. Only with due consideration of the prints' poor condition were crease removal, backing removal, and surface dirt removal using a variety of aqueous methods undertaken. Although capable of fading a print when exposed in excess, light is necessary to appreciate a Japanese print. So too is water in conservation treatment, although it must not be used to the detriment of the color harmonies or the stability of the colorants that constitute these fragile objects.

Acknowledgments

I would like to thank the following people who gave freely of their time in the preparation of this research: Bill Foster of Hitachi Nissei Sangyo Co. Ltd.; Dr. Richard Holman and Phillip Roberts of the Paint Research Association; Dr. Vincent Daniels of the British Museum Conservation Department; Matsumi Salon Fine Japanese Prints, the owner of the prints used as case studies; Ian Hessenberg for providing photography; and Kirsten Hartley.

Endnotes

1. On these early woodblock-printed Buddhist texts, see A. R. Newman and E. Ryerson, *Japanese Art: A Collector's Guide* (London: G. Bell & Sons), pp. 28–29.

2. L. R. Bickford, "The pigment story," *Impressions: Publication of the Ukiyo-e Society of America Inc.* 1, no. 7 (autumn 1982); and R. J. Gettens and G. L. Stout, *Painting Materials* (New York: Dover Publications, 1966), pp. 152–53.

3. J. Hillier, *The Japanese Print* (London: G. Bell & Sons, 1960), p. 36.

4. R. S. Keyes, *Japanese Woodblock Prints, A Catalogue of the Mary A. Ainsworth Collection, Allen Memorial Art Museum* (Oberlin, Ohio: Oberlin College, 1984), p. 254.

5. J. Walsh, B. Berrie, and M. Palmer, "The connoisseurship problem of discoloured lead pigments in Japanese woodblock prints," *IPC Conference Papers London 1997*, pp. 123–24; and V. Daniels and D. Thickett, "The reversion of blackened lead white on paper," *IPC Conference Papers Manchester 1992*, p. 109.

6. J. Meech-Pekarik, *The World of the Meiji Print* (New York: John Weatherhill, 1986), p. 40; P. de Tristan, "The analysis and conservation of aniline-dyed, nineteenth century Japanese prints," *IPC Conference Papers London 1997*, p. 116.

7. Gettens and Stout 1966 (as in note 2), pp. 149–51.

8. On fading see R. S. Keyes, R. L. Feller, M. Curran, and C. Bailie, "Japanese woodblock prints – identification of traditional organic colorants employed in Japanese prints and determination of their fading," in Keyes 1984 (as in note 4), pp. 256–63. On darkening see Walsh et al. 1997 (as in note 5), pp. 118–23.

9. S. Shimoyama and Y. Noda, "Non-destructive analysis of ukiyo-e prints," in *Dyes in History and Archaeology*, postprints of the 15th annual meeting of the Textile Research Associates (York, U.K.: 1996), pp. 27–30.

10. Data were obtained and processed using Hitachi F3D software. 3-D fluorescence spectra of known colorants were plotted as contours that expressed intensities of fluorescence in terms of individual fingerprints. These spectra and fingerprints were then used as references for comparison with unidentified colorants.

11. R. Holman, Paint Research Association, 8 Waldegrave Road, Teddington, Middlesex TW11 8LD, U.K., personal communication, 1999.

12. Ibid.

Painted Japanese Paper Screens: The Consolidation of Paint Layers on a Paper Substrate

Sandra Grantham and
Alan Cummings

Between 1995 and 1999, Ph.D. research was conducted on the conservation of painted Japanese paper sliding screens (*fusuma*) and folding screens (*byōbu*) (Grantham 1999). Within this wide-ranging, holistic study, a substantial section was devoted to the development of a consolidation procedure for the matte, porous, and friable paint layers characteristic of these screens. This paper presents a short introduction to this research, providing background information on the screens, their history, construction, and materials. There follows a discussion of the issues concerning consolidation, an account of the procedure employed, and a brief account of some of the tests used to assess the effectiveness of the procedure. While this work focussed on Japanese materials, the approach and the outcomes are relevant for any object incorporating paint layers of similar character, particularly where these are applied to a paper support.

Japanese screens formed the lightweight, movable walls in temples and large houses from the eighth-century Nara period to the nineteenth-century Meiji period. With few solid walls on which to hang paintings, these screens became a focus for the display of works of art. The wood lattice and paper screen walls could be placed or removed at will and were an ideal support for large horizontal paintings.

The rhythm of the seasons had a profound effect on life in the Far East; "experiencing the eternal circular movement of nature" was an integral part of life and work (van Gulik 1958). This sensibility, in turn, had a far-reaching effect on Chinese and Japanese art. In both countries it was established custom to harmonize the interior decoration of a house with the constant change of natural life outside. Architecture was conceived as an inseparable unit encompassing building and garden. The enclosed garden, where rocks represented mountains and stone or gravel paths were imagined mountain torrents, created a landscape in miniature in which a building was placed. The walls of the building were made to slide open, allowing the garden to enter the room. From this principle the movable Japanese screen evolved.

The construction and decoration of screen panels

The single standing screen imported from China was refined in Japan, and although *tsuitate* (single screens) were used across entrances, panels were also fitted into *shikii* (floor grooves) and *kamoi* (upper grooves) to become *fusuma* (room partitioning), or they were hinged to form free standing *byōbu* (folding screens). As a general rule, *byōbu* could be placed anywhere to section off an area of a room or veranda, and *fusuma* made up interior walls.

The frames and linings were designed to remain flat through the extreme changes of the Japanese climate. They were produced by pasting layers of thin, strong Japanese paper (usually made from *kozo* fibers) onto lightweight wooden lattice frames with a starch paste (see

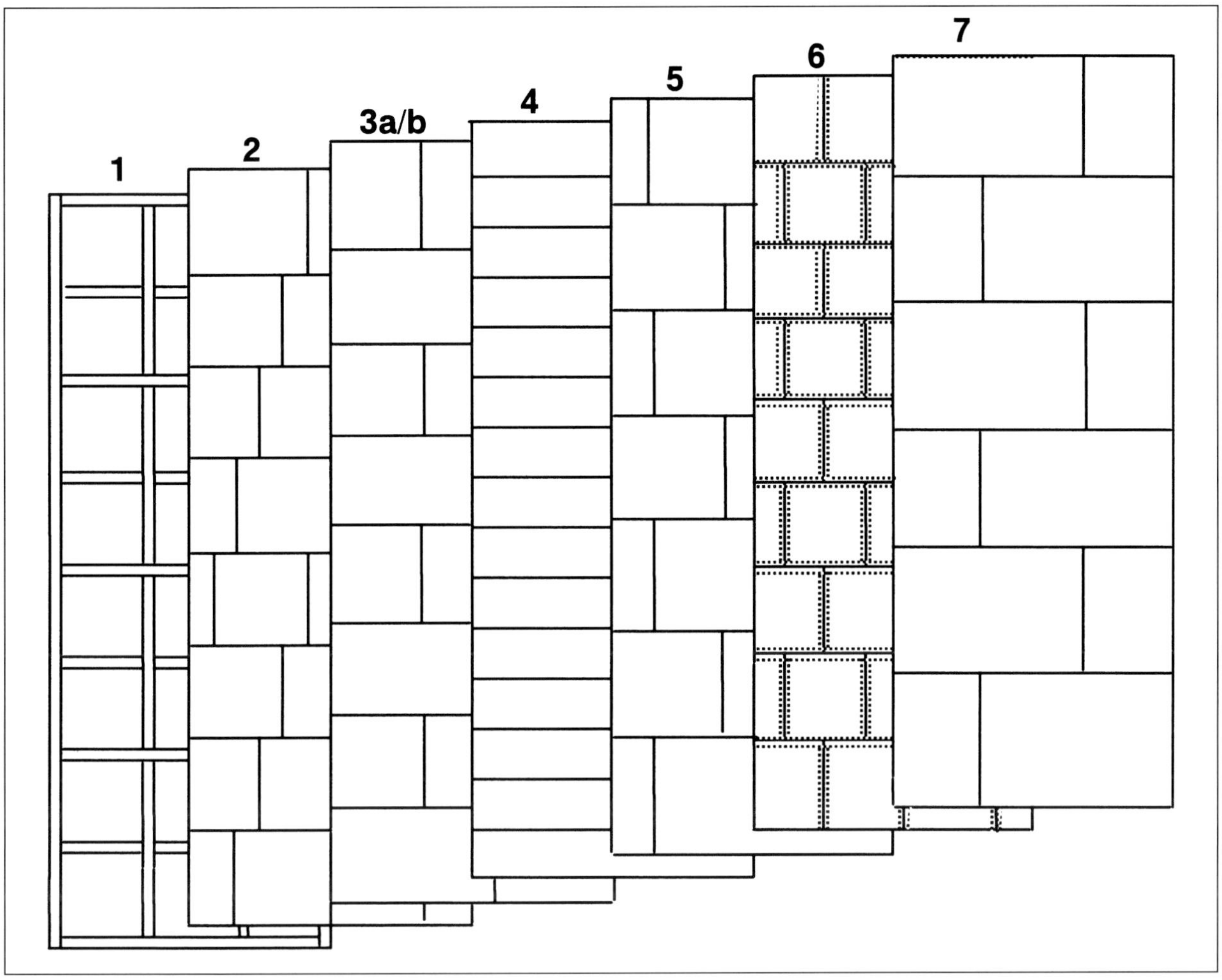

Diagram 4.1 The *Shitabari* (paper lining layers).

1. *Shitaji* – straight-grained cedarwood frame
2. *Honeshibari* – the first paper lining

3a or b. Tokyo- or Kyoto-area *Dobari* – a holding layer (different area styles)

4. *Minokake* – the cushion or shingle layer (resembling a traditional layered Japanese straw raincloak) to give a smooth, slightly padded surface
5. *Minoshibari* – this layer is similar to the *dobari*, another holding layer
6. *Ukekake, ukebari*, or *fukurobari* – the floating, bag, or pocket layer which allows the top painted layers to be removed for repair etc.
7. *Betabari* – the last layer of the prepared panel

Diagram 4.1). The paper layers were pasted onto both sides of the panels, each being allowed to dry before the next was adhered; about six lining layers were traditional, although the construction and layering varied from workshop to workshop. This construction was called the *shitabari*. Each layer had a function related to the next or previous layer, and to the overall structure (Meredith 1999). Matching linings were pasted to each side of the panel before the next was adhered to keep even tension and to prevent curvature in the panel.

Second-hand and seasoned paper was often used for the *shitabari* as it proved less reactive to humidity changes and was less expensive. Old account books were a good source of seasoned paper (see Fig. 4.13). The final layer was made with a clean, good-quality paper (usually a *mino* or *maniai*), sized with *dosa eki* – animal glue with the addition of a few *myoban* (alum crystals) to harden it. Then the painting was applied directly onto the lined panels, the methods and materials hardly changing in 600 years. Clay-filled *maniai* paper was not usually sized and was mostly used for ink paintings.

Depending on the period of Japanese history and religious proclivity in which the screens were made, their decoration could take several forms: from the cool, fairly abstract landscapes painted almost entirely in shades of gray-black ink, known as the Chinese, Kara, or Zen Literati style, to the highly decorated narratives made with a small but intense color palette, the Japanese or Yamato style. Examples of the latter style were usually enhanced with luminous backgrounds of gold or silver-leaf, and nearly always included the intense blue and green of azurite and malachite, basic copper carbonate pigments. These minerals were ground and graded into ten particle sizes by water levigation.

Paintings known as *tsukuri-e*, "built-up" pictures, were created by layering the paint to form certain colors.

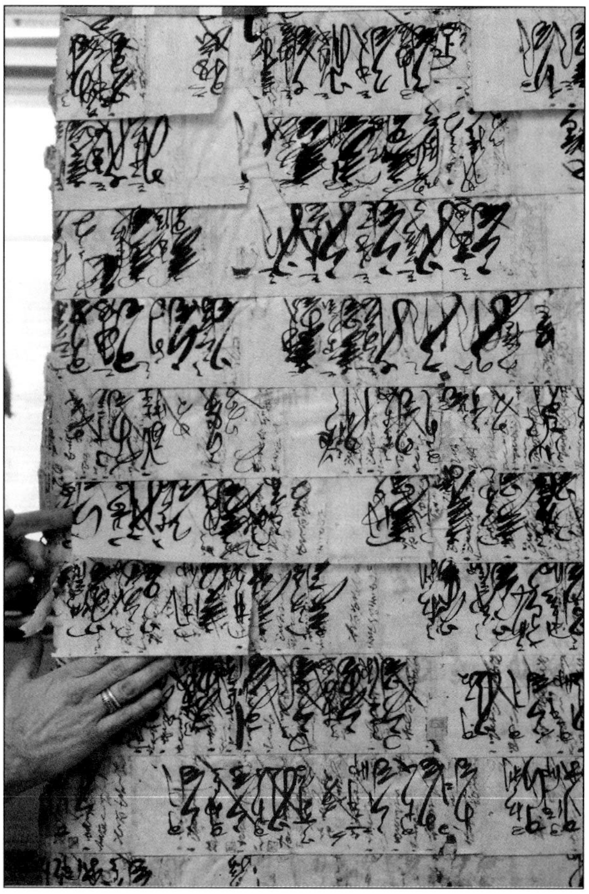

Figure 4.13 Japanese account-book papers recycled for the inner linings of screen panels.

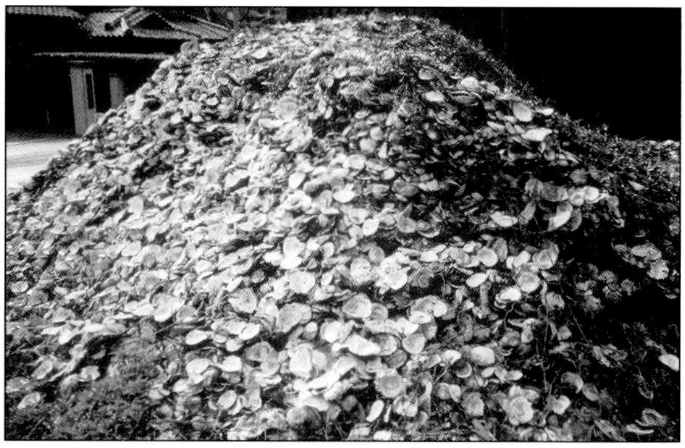

Figure 4.14 Oyster shells aging for up to 20 years outside a *gofun* manufactory at Uji, southeast of Kyoto.

The preliminary paint layers were made with white or fine-ground and pale-colored pigments mixed well with animal glue binder with no alum added. As each paint layer dried, another darker, more coarsely ground and lightly bound coat would be added until the required hue was achieved.

Many of the colors were prepared with *gofun*, a white calcium carbonate pigment made from aged ground oyster shells (see Fig. 4.14). To produce a dense, opaque, glowing white, many layers were applied and the addition of glue kept to a minimum. (It was known that aging glue turned white pigment gray.) However, this relative lack of glue makes *gofun* paint layers prone to flaking. Ground mica pigment was also used. The plate-like particles settle on each other like scales when painted out. To some extent, this was also the case with ground shell-white, and in both this leads to instability. *Gofun* was also used under gold leaf to form *moriage*, a kind of gesso patterning that often develops cracking and flaking problems.

Pictorial painting was usually limited to one side of a screen. The verso was covered with a precursor to Western wallpaper known as *karakami*. These were small sheets of decorative paper, painted with a base color and then block-printed or stenciled with patterns that would link up to make a continuous design across an entire panel or panels. These verso *karakami* papers have often deteriorated severely, partly because the pattern was usually printed with a weak *funori*-bound mica paint that detaches easily if not handled with extreme care.

In summary, several factors in the production of Japanese painted paper screens resulted in paint layers that are porous and matte, prone to flaking and powdering, physical damage, and visual change. These factors include the physical character of the pigments used, the high pigment-to-binder ratios, the complex layering structures, the hygroscopic nature of the binder and support, and the degradation of the binding medium. Primary research in Japan, gaining practical experience in the making and decoration of screens, has been pivotal in developing an understanding of the deterioration processes and appropriate conservation strategies.

Consolidation

Paint loss in Japanese screen paintings is caused either through adhesive failure between paint layer and support (sometimes between paint layers), or through cohesive failure within paint layers. Adhesive failure is likely to be manifested as flaking, while cohesive breakdown usually leads to powdering. It is common to encounter both forms of loss in different areas on the same object.

Even where paint degradation is severe, conservation can sometimes be limited to safe storage with little or no handling. However, often this is not an option, and an alternative, more interventive solution may be required. Consolidation is therefore sometimes an essential aspect of conservation treatment, and it requires careful consideration of the ethical as well as the physical problems involved. For example, darkening or saturation of paint layers as a result of consolidation may be acceptable or even desirable in some contexts but should be avoided in others.

The work on consolidation reported here began with the assumption that, while paint layers should be stabilized, change in appearance was undesirable. This meant avoiding complete filling of the pores within the paint layer with consolidant, as this would cause maximum visual change. In such circumstances, the quantity and distribution of an introduced consolidant are as critical as the refractive index in controlling visual change. Enough consolidant needs to be introduced and effectively distributed to provide mechanical integrity, but if the consolidant completely fills the spaces between pigment particles, color change will be maximized.

There are many criteria to bear in mind when selecting a consolidant, and there is room for debate about desirable properties. Space prevents a lengthy discussion of these issues. Suffice it to say that the impact on the mechanical, optical, and chemical properties of the object needs to be considered, as well as solubility of the consolidant and viscosity in solution, as these affect working properties.

There are similarly many variations in approach to consolidation that must be considered. These include: single application of concentrated consolidant solution, or multiple applications of dilute solution; the use of suction to mobilize the consolidant and limit dimensional response of the object; humidification of the object before, during, and after treatment; and application by brush, spray, mist, or immersion. Although current practice for painted paper objects is usually to apply consolidant with a fine brush to localized areas of dry, flaking, or powdery paint, this can produce several potential problems, such as tension interfaces between treated and untreated areas, detachment of paint with the brush or liquid consolidant, tide lines through the migration of soluble degradation products, and surface accumulation of consolidant.

The higher the viscosity of a consolidant solution, the less likely it is to penetrate throughout the paint layers. If viscosity is too low, the consolidant may migrate in solution away from the area to be consolidated during application. Similarly, if the consolidant is applied in a "dry" or "normal" room atmosphere, reverse migration may take place, where the consolidant is drawn back up to the drying surface. Reverse migration may be reduced by keeping some moisture content in the substrate, paint layer, and surrounding atmosphere while applying a consolidant (Hansen et al. 1993). The high surface tension and relatively high viscosity of water can cause slow initial penetration, but this can also be countered by humidification.

It was not possible in the course of this work to address all the variables that potentially affect the consolidation process – some choices had to be made. The consolidants and procedure used in the main series of tests were not, however, chosen arbitrarily. They were guided by the arguments set out above, by observations in the studio, by previous experiments with consolidation procedures carried out at Camberwell College of Art in 1994, and by a set of exploratory experiments made as part of this research. The preliminary tests were conducted using prepared facsimiles and three consolidants considered most appropriate for the treatment of Japanese screens.

Following this preliminary work, the main series of tests proceeded on the basis that most effective consolidation could be achieved by making multiple applications of low-concentration aqueous solutions, as a mist, to humidified objects, without suction, followed by slow drying. The concentration/viscosity for each consolidant needed to be such that a mist could be produced using the equipment employed (see below). This was between 1 and 3% w/v in water.

The paints on Japanese screens were usually prepared with a dilute animal glue binder. Sometimes, additions of *funori* (seaweed gel) or starch paste or both were made. For consolidation, it was considered that aqueous binders resembling the original materials in physical properties would probably change the character and appearance of the objects least. Water-soluble cellulose ethers, although not traditionally used in Japan, were included because of their similarity in properties to the traditional materials selected. However, consolidants of significantly different physical character (acrylics, PVAc, waxes, etc.) were not considered, even where these materials were available in aqueous forms.

For the tests, water-soluble ethyl hydroxyethyl cellulose (WS-EHEC) was selected for comparison, or use in combination, with the traditional Japanese binders *funori* (seaweed gel) and *nikawa* (animal-skin glue). Water-soluble cellulose ethers are known for their long-term stability, resistance to biological degradation, and pH stability. WS-EHEC has been used in conservation for some years (Feller and Wilt 1990), but as there appeared to be no published reports of its testing for

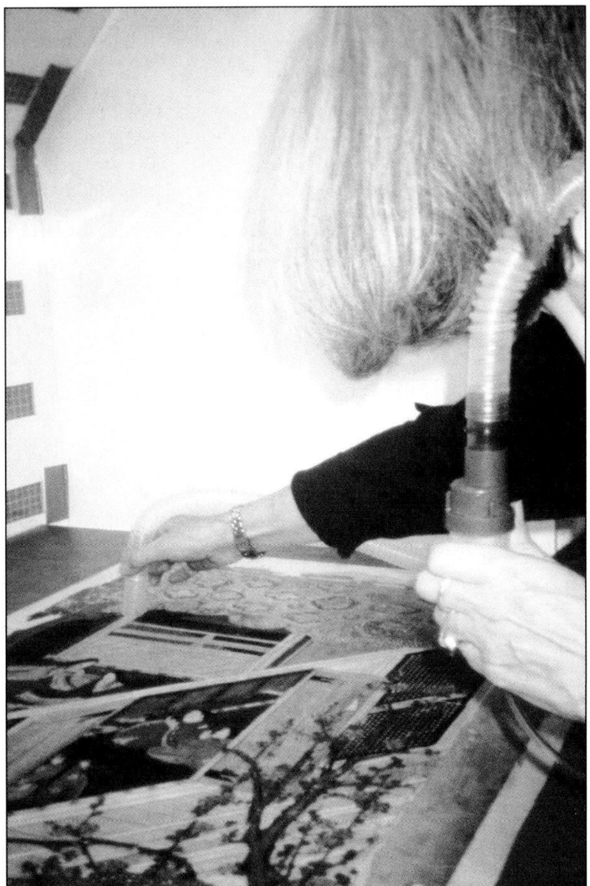

Figure 4.15 The Sidestream nebulizer in use.

1A	White – *gofun*
1B	White – mica
2	Blue – *gofun* and indigo
3	Green – *gofun*, yellow ochre, and indigo
4	Light brown – *gofun* and iron oxide
5	Brown – *gofun*, burnt iron oxide, and lamp black
6	Yellow – *gofun* and yellow ochre
7	Gray – *gofun* and lamp black

Table 4.8 Plain painted facsimiles.

Plain painted sheets (8 in all) 5 sample areas for all sheets except sheet 5 6 sample areas for sheet 5		
Consolidant application	**First aging**	**Second aging**
Control	√	√
EHEC x 6	√	√
Funori x 6	√	√
Nikawa x 6	√	√
EHEC x 3 + *Funori* x 3	√	√
Further samples for Sheet 5 – dark brown		
EHEC x 4	√	√

Table 4.9 Segments of the plain painted facsimiles, treated with various consolidants and artificially aged.

conservation, its examination was thought to be a useful exercise. It is of lower viscosity than methyl cellulose at the same concentration and more matte.

The Japanese seaweed adhesive *funori* (*Gloiopeltis furcata*) was originally chosen because of its traditional link with the materials of the objects to be treated. Moreover, in preliminary experiments *funori* had been seen to cause lightening of consolidated surfaces – a potentially useful effect where darkening of paint layers is to be avoided. A possible explanation for this is provided elsewhere (Grantham 1999).

Examination of objects and available paint samples, together with the study of painting techniques in Japan, enabled the production of facsimiles close in appearance and properties to the originals. Paints prepared in the traditional manner were applied to a selection of Japanese (mostly *kozo*) papers, 40 to 60 years old. The fresh facsimiles were artificially aged in a humid oven at 80°C, 65% RH (ISO9706 International Standard) for 15 days. The sheets were then rolled and unrolled in each direction by hand, after which they displayed a condition distinctly like the original objects. Light aging was not undertaken for reasons explained elsewhere (Grantham 1999).

Testing the facsimiles

Prior to consolidation, the aged facsimiles were placed in a humidity chamber. Mist from an ultrasonic humidifier was inserted to slowly raise the humidity level to approximately 69% RH. The humidifier was then turned off and the facsimiles left to absorb moisture for about an hour. Slow humidification reduces the likelihood of extreme dimensional changes in the objects.

The ultrasonic mister system devised by Michalski (Michalski and Dignard 1991) was first used, but with inconsistent results. The search for a more reliable device led to a Sidestream nebulizer (see Fig. 4.15). At the Royal Brompton Hospital, this device was employed to produce a fine mist of medication to be inhaled directly into the lungs of patients with severe breathing difficulties. The small plastic device has an even smaller, inner plastic component oscillated by a tiny jet of compressed air (440 ml per minute). This produces a very fine mist of whatever liquid it is charged with (65% of the droplets are said to be less than 5 microns in diameter). It worked reliably and misted quite viscous solutions.

The nebulizer was charged with measured 3 ml doses, and applications were made from the open end of a hose 22 mm (⅞ in.) in diameter. Smaller nozzles can be used, but the wider diameter helped to avoid visible wetting of the paint layer. One mode of application for the tests consisted of a slow, steady passing of the hose end approximately 15 mm (⅝ in.) above the surface, backward and forward, covering the whole area, first horizontally and then vertically. The samples were left in the humidity chamber throughout the process, with 5-minute rests between each application. Different numbers of applications were made to individual

light											*dark/yellowing*
sample 1A *gofun* – white											
a0t0	a1t2	a1t3	a1t1	**a1t0**	a2t2	a2t1	a2t0	a2t3	a2t4		
sample 1B mica – white											
a0t0	**a1t0**	a2t0	a1t2	a1t1	a1t3	a2t1	a2t2	a2t3	a2t4		
sample 2 indigo/*gofun* – blue											
a0t0	a1t1	a1t3	a1t2	**a1t0**	a2t1	a2t2	a2t3	a2t0	a2t4		
sample 3 indigo/yellow ochre/*gofun* – green											
a0t0	**a1t0**	a2t0	a1t1	a1t2	a2t1	a1t3	a2t2	a2t3	a2t4		
sample 4 iron oxide/*gofun* – light brown											
a0t0	a2t0	a2t3	**a1t0**	a1t3	a1t1	a2t1	a1t2	a2t2	a2t4		
sample 5 burnt umber/*gofun* – dark brown											
a1t1	a2t1	a1t2	a2t2	**a1t0**	a2t0	*a0t0*	a2t5	a1t5	a2t3	a1t3	a2t4
sample 6 yellow ochre/*gofun* – yellow											
a0t0	a2t0	**a1t0**	a2t1	a1t1	a2t2	a1t2	a1t3	a2t3	a2t4		
sample 7 lamp black/*gofun* – gray											
a0t0	a2t0	a1t1	a2t2	a2t1	**a1t0**	a1t2	a2t3	a1t3	a2t4		

Key: a0 = no aging a1 = first aging a2 = second aging
t0 = no treatment
t1 = 6 applications of WS-EHEC 1%
t2 = 3 applications of WS-EHEC 1% + 3 applications of *funori* 1.5%
t3 = 6 applications of *funori* 1.5%
t4 = 6 applications of *nikawa* 2%
t5 = 4 applications of WS-EHEC 1%

Table 4.10 Observer eye color tests.

sections of a facsimile, ranging from two to six. The preliminary tests had suggested that it would be worth experimenting with at most six applications of consolidant solution. Beyond this number, visible changes in the paint layers were clearly beginning to occur, and a significant improvement in adhesion was often achieved with fewer applications. After misting, the samples were allowed to dry slowly in the closed chamber until ambient conditions were re-established. This took approximately four hours.

The most important tests carried out on the prepared facsimiles were to assess changes in appearance and adhesive/cohesive strength both before and after the application of consolidant, and before and after artificial aging. Additional tests were conducted to assess changes in weight, tensile properties, flexibility (bending), and dimensions. Throughout the tests (with one or two exceptions, not reported here), the effect of applying the consolidants was assessed by comparing treated samples (unaged, aged once, and aged again after consolidant applications) with control samples (unaged, aged once, and aged again).

Color change

Two approaches were adopted. The first was to record where perceptible (to the author SG) changes had occurred and rank these changes in order by visual comparison. The second was to measure color change using a simple reproducible procedure that provides comparable, quantitative results using an internationally recognized system.

Plain painted facsimiles prepared with eight different paints were included in the tests (Table 4.8). Segments of these were treated with various consolidants and varying numbers of applications, then artificially aged (Table 4.9). Visual comparisons and color measurements were then made for each segment at each stage of the process.

Observations were made using a standard Munsell color test light box and checked again in natural north light. The swatches for each color test sheet were laid out at random, each swatch representing a different treatment. The swatches were then sorted by eye, grading from whitest or lightest to yellowest or darkest color. When all the color test sheets had been thus graded, a table was drawn up showing the sequence of swatches for each test sheet (Table 4.10).

For quantitative comparison of color change, a Minolta CR-300 chromaticity meter was used. This is a tristimulus color analyzer for measuring reflective surface color. Technical details and a discussion of the problems associated with color measurement are provided elsewhere (Grantham 1999).

Fifteen measurements were made on each section using penciled rings to relocate the measuring head as necessary. Results were recorded using the L*a*b* color space method (Minolta 1994). Although changes in L* a* and b* values were recorded for each segment, these values were also combined into a single color difference, the ΔE value, which indicates the overall change that has occurred in relation to the chosen control. The intention here was to identify where minimal change occurred following use of a consolidant and aging, regardless of the character of the change.

Adhesive/cohesive strength tests

It is difficult to quantify the tendency for paint to be lost through flaking or powdering from a paper substrate. The delicate nature of the paint layer, even when relatively stable, requires a test equally delicate. As there was no suitable standard test, a simple method was devised to assess the relative effectiveness of the consolidants in various numbers of applications in securing the paint layers. This employed Post-It™ memo block sheets (76 x 76 mm).

The adhesive strip on these sheets, covering an area 76 x 15 mm, has a non-aggressive, soft tack. The first step in the test was to prepare a strip for use by pressing it onto a sheet of tracing paper. Any loose adhesive was thereby removed, reducing the danger of offsetting adhesive onto the sample surface during the test. The sheet was then weighed. In the test itself, a brass roller, weighing 139 g, was rolled down a slight incline from the same starting point each time, over a strip which was simply laid onto a sample paint layer surface. It was then gently peeled off by hand (as consistently as possible) and weighed again. All weighing was carried out using an Oertling NA 114 analytical electronic balance, with an error of ± 0.00005 g. The testing process was carried out in a controlled atmosphere chamber, temperature 23°C, RH 55% ± 4, all samples and Post-It™ sheets having been left in the chamber for two hours to acclimatize.

The validity of this method clearly depends on a number of factors, including the consistency of "tack" in the Post-It™ sheets and the consistency of adhesion/cohesion in the paint layers of the facsimile. However, preliminary experiments convincingly demonstrated that, while there may be some deviations, clear and consistent trends were evident in the results, and the method does seem reasonably reliable. Comparing the loss of pigment from area to area on a sample does show the relative effectiveness of each number of applications for each concentration of each consolidant or combination of consolidants. A series of tests was carried out on the same selection of plain painted facsimiles as those used for the color change tests so that comparisons could more easily be made.

Results and discussion

In considering the results of the various tests, it is important to bear in mind that the test methods were devised through a quite empirical, studio-based approach, rather than rigorous scientific method with statistical analysis of errors, etc. Also, the facsimile sheets tested were extremely variable in properties at the outset (as real materials would be). Clear and consistent trends are therefore worth noting, but some anomalies are to be expected. In some experiments, only internal comparisons may be valid, i.e., the results for a single facsimile treated with different consolidants may be reasonably compared, but comparison of the results from one sheet to another are likely to mislead.

Table 4.10 presents the results of the visual color comparison of the samples treated in various ways. A key to the table indicates the status of each "swatch." This simple visual assessment of the samples indicates that, while each different color facsimile sample was made to roughly the same formula, some paint layers were visually changed by the consolidants more than others. Dark colors tended to lighten and light colors darkened, but even this is not a totally consistent observation.

This variability is a predictable consequence of differences in paint composition, surface morphology, refractive indices, degree of degradation, etc. – what would be expected in a real screen. In advocating overall application of consolidant, the aim must be to find a material that, while having varying results on different paint surfaces, does not alter any of them to an unacceptably perceptible degree. Again, the visual assessment showed that most consolidant applications (other than *nikawa*) did not change the appearance of the samples any more than artificial aging had changed the color from the original unaged control.

The most noticeable result from the simple visual examination, which is clear in Table 4.10, is that the a2t4 swatches are consistently darker/yellower than all other swatches from the same sample sheet. This is strong evidence that after aging, *nikawa* (animal-skin glue) causes the most yellowing or darkening of the consolidants tested. The only other pronounced visual effect worth noting concerns the dark brown sample, no. 5. Here, six applications of WS-EHEC were noticed at an early stage to have caused significant "whitening" of the surface (a1t3 and a2t3). For this reason, an extra treatment involving four applications was tried (a1t5, a2t5). As can be seen in the table, this caused less change in appearance. For most of the samples, applications of WS-EHEC and the mixture of WS-EHEC and *funori* seem to have produced the least perceptible changes in appearance.

In Figure 4.16, ΔE values for a set of samples is plotted on the vertical axis. The code at the foot of each column indicates the source of the data used to calculate these values, e.g. the first column on Figure 4.16 is labeled a1t1/a1t0. This means that the L*a*b* values for the once-aged (a1) swatch with treatment 1 (6 applications of WS-EHEC 1%) and the values for the once-aged (a1), untreated (t0) "control" sample have been used in the calculation of color difference. The height of the column therefore gives a quantitative idea of the change in appearance caused by this consolidation procedure for an aged plain-painted *gofun* sample

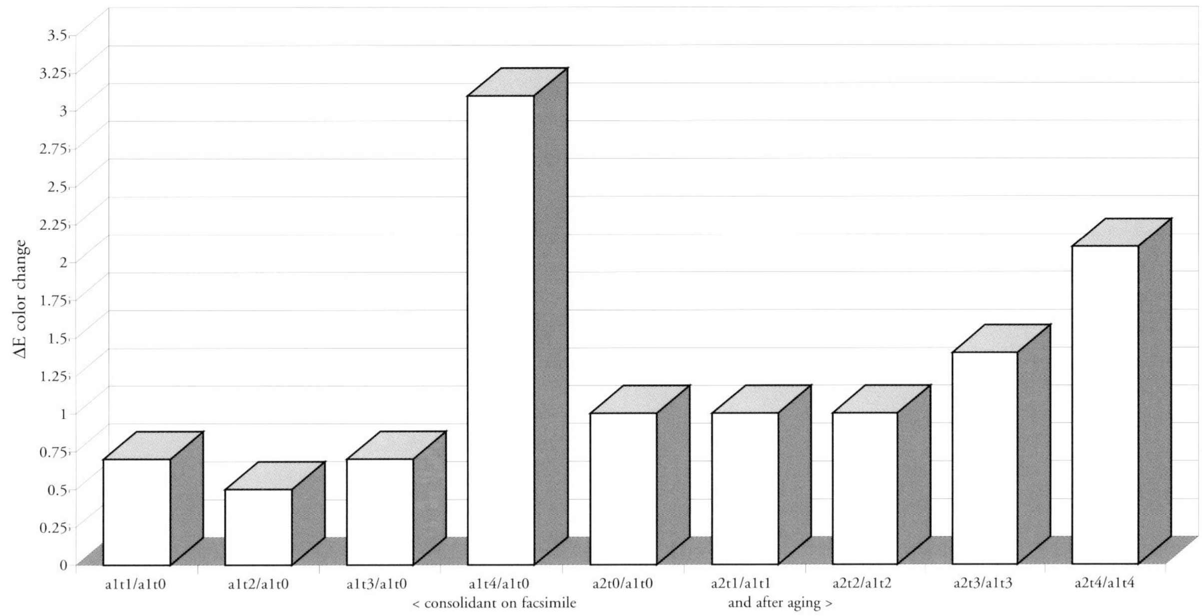

Figure 4.16 Chromameter color reading – plain painted sample 1A *gofun*.

Key

Paint layers brushed onto lightweight *kozo* paper (23 gsm) to represent those on Japanese *karakami* (decorative screen papers) and artificially aged. First aging for 15 days before consolidants applied @ 80°C, 65% RH to produce friable paint layers similar to those found on screen papers. Second aging for 15 days after consolidants applied @ 80°C, 65% RH to test aging stability of the applied consolidant. All consolidants applied after first preliminary aging and before second aging with a nebulizer in a humidity chamber.

a0t0 control with no aging and no treatment
a1t0 control with no treatment, but after preliminary aging for 15 days @ 80°C, 65% RH
a2t0 control with no treatment, but after second aging for 15 days @ 80°C, 65% RH
a1t1 6 applications of WS-EHEC 1%
a2t1 6 applications of WS-EHEC 1% after 2nd aging
a1t2 3 applications of WS-EHEC 1% + 3 applications of *funori* 1.5%
a2t2 3 applications of WS-EHEC 1% + 3 applications of *funori* 1.5% after 2nd aging
a1t3 6 applications of *funori* 1.5%
a2t3 6 applications of *funori* 1.5% after 2nd aging
a1t4 6 applications of *nikawa* 2%
a2t4 6 applications of *nikawa* 2% after 2nd aging

swatch. However, the character of this change is not indicated by the ΔE value. On the same figure, the bar labeled a2t1/a1t1 shows that a further and slightly larger change in appearance occurred when this swatch was aged for a second time.

To facilitate the analysis, a standard was adopted that defined an acceptable color change as a ΔE value of 1.0 or less.[1] Consolidation treatments that lead to ΔE values of less than 1.0 when they are made initially to the once-aged samples can be identified. A further check can be made that a ΔE value of less than 1.0 also occurs when the treated swatch is aged for a second time.

For the majority of paint layers, minimal color change was caused, both on application of consolidant and after aging, by six applications of WS-EHEC and by three applications of WS-EHEC and three applications of *funori*. Six applications of *funori* were also satisfactory in most cases. Six applications of WS-EHEC proved, however, unsatisfactory for dark browns (and may be unsatisfactory for other dark colors). As a compromise, four applications of WS-EHEC could be used across the board if this treatment adequately improves the adhesion/cohesion of the paint, as considered below. Six applications of *nikawa* caused unacceptable color change in all the samples, so if the aim is to prevent color change, it should not be used. However, all these observations need to be considered alongside the results for adhesive/cohesive tests described below.

In all cases the first aging with no consolidation treatment caused the paint surface to produce some losses (both flaking and powdering) when tested by the Post-It™ method. The second aging with no treatment produced a still greater paint loss. This is a positive indication of the usefulness of the samples and the artificial aging process as representative of real materials.

The weight differences measured were recorded in a series of tables, and a detailed analysis of the results was made. However, there is space to mention only the main observations here. For seven out of eight samples tested, six applications of WS-EHEC 1% gave the best adhesive/cohesive strength and weakened least after artificial aging. The exception, sample 1B (mica), was more effectively adhered with three applications of WS-EHEC 1% and three applications of *funori* 1.5% w/v. In conclusion, it would appear that of all the consolidation procedures tested on the samples, six applications of WS-EHEC 1% w/v generally gave the most effective consolidation both before and after aging.

Bearing in mind the study's empirical approach and the variability of the facsimiles used in the testing, conclusions must be tentative. As regards adhesion/cohesion of the paint layers typically found on a Japanese screen, optimum results can be achieved using dilute aqueous consolidants, applied in multiple overall applications as a nebulized mist to a pre-humidified material, allowing only partial drying between applications and slow drying after treatment. The pause between each misted application allows for inspection of the surface. In some areas of flaking, further conventional consolidation treatment using a brush may be necessary, but the risk of displacing particles is reduced by the previously misted application. The nebulizer used in the majority of the experiments has proved to be a highly effective, efficient, inexpensive, and straightforward device for the application of consolidants as mist.

For the consolidants and procedures tested, varied changes in appearance and adhesion were observed. However, with the exception of *nikawa*, the treatments based on WS-EHEC and *funori* caused only slight changes in appearance and significant improvements in adhesion/cohesion resulted. The best results overall were obtained using six applications of 1% WS-EHEC. However, because this number tended to cause lightening of dark paint layers, four applications could be used to good effect in such areas.

Despite the encouraging results, a more systematic and rigorous study of standardized samples would be necessary to confirm the observations made here. However, such a scientific study may not be able to take into account the variability of the objects in need of treatment and the demands to be made upon them. The judgment of the conservator will continue to be of major importance in assessing the condition of paint layers, the degree of improvement in adhesion actually required, the restrictions on change in appearance, and the most appropriate method of consolidation.

Endnote

1. This approach was recommended by Boris Pretzel of the Victoria and Albert Museum Conservation Science Department.

References

Feller, R., and M. Wilt 1990. *Evaluation of Cellulose Ethers for Conservation.* Research in Conservation series, no. 3. Marina del Rey, Calif.: The Getty Conservation Institute, pp. 94–95.

Grantham, S. 1999. "Byōbu & Fusuma: Developing an Approach to the Conservation of Japanese Screens through Historical and Technical Study and an Investigation of Current Practices". Ph.D. Diss., Royal College of Art, London.

Hansen E., R. Lowinger, and E. Sadoff 1993. "Consolidation of porous paint in a vapour-saturated atmosphere," *Journal of the American Institute of Conservation* 32, no. 1, pp. 1–14.

Meredith, P. 1999. "A new conservation lining for historic wallpapers," *Preprints from the 9th International Congress of IADA (Internationale Arbeitsgemeinschaft Der Archiv, Bibliotheks und Graphikretauratoren)*, Copenhagen, pp. 41–45.

Michalski, S., and C. Dignard 1991. "A new method of consolidating powdery paint: The ultra-sonic mist," *International Institute for Conservation – Canadian Group Vancouver Conference Papers*, pp. 4–11.

Michalski, S., and C. Dignard 1997. "Ultrasonic misting, Part I. Experiments on appearance change and improvement in bonding," *Journal of the American Institute for Conservation* 36, no. 2, pp. 109–26.

Michalski, S., C. Dignard, L. V. Handel, and D. Arnold 1994. "Ultrasonic mister: application to consolidation treatments of powdery paint on wooden artifacts," in *The Ultrasonic Mister: Starter Packet.* Ottawa: Canadian Conservation Institute.

Minolta (U.K.) Ltd. 1994. *Precise Color Communication.* Milton Keynes: Minolta (U.K.) Ltd.

van Gulik, R. H. 1958. *Chinese Pictorial Art as Viewed by the Connoisseur: Notes on the Means and Methods of Traditional Chinese Connoisseurship of Pictorial Art, Based Upon a Study of the Art of Mounting Scrolls in China and Japan.* Rome: Istituto Italiano per il Medio ed Estremo Oriente.

The Conservation of a Set of Japanese *Fusuma*

Mitsuhiro Abe and
Sydney Thomson

The painted Japanese paper screens discussed in this paper are a series of four temple sliding doors. Originally from the Tanzan Jinja Shrine at Tonomine, south of Nara, they are now in the British Museum, London. The paintings on them (c. 1612) are attributed to the artist Kano Sanraku (1559–1635), who worked near the end of the Momoyama period (1575–1615). Depicting scenes from the four seasons, the screens were painted using opaque, inorganic pigments and impasto on paper, heightened with the lavish use of gold and silver. They are mounted in the functional format known as *fusuma*, traditionally used in Japanese interiors.

The four-panel composition depicts autumn and winter, displaying flowers and birds beneath the snow-covered branches of a great pine tree (see Fig. 4.21). The outline of the design was first drawn onto smooth *maniai*, which is a clay-filled *gampi* paper, with a brush using *sumi* (black ink).[1] The artist then used expensive, coarsely ground blue and green mineral pigments such as azurite and malachite. A white paint, *gofun* (ground oyster shell), was applied thickly and proved to be very susceptible to flaking and chipping. The paintings were highlighted with gold and silver, the latter now tarnished and irreversibly blackened. At their height the bright mineral pigments and the gold and silver would have created a shimmering effect, reflecting the diffused light from *shoji* (paper windows) by day and candlelight by night, greatly enhancing the darkness of a temple room. These particular examples are painted in a broad style in

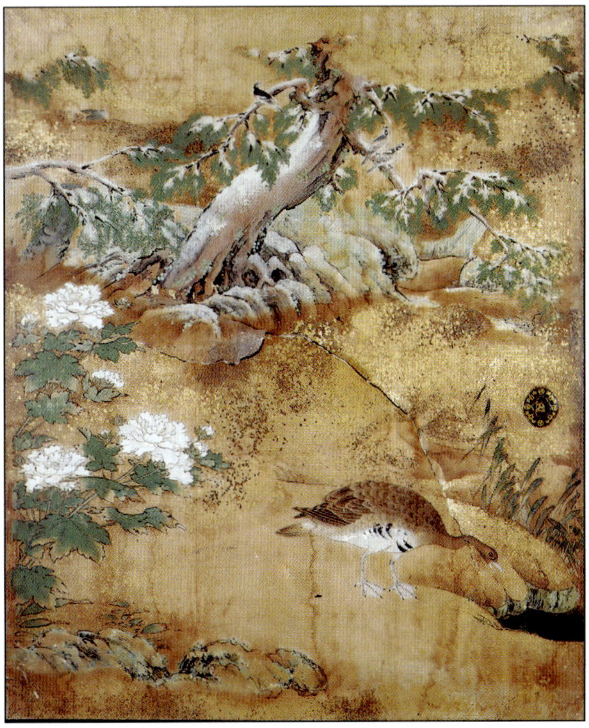

Figure 4.17 Attributed to Kano Sanraku (Japanese, 1559–1635), *Bird and Flower fusuma*, c. 1612 (third panel from right, before treatment). Mineral pigments with gold and silver on Japanese *maniai* paper, 173.5 x 140 cm. Reproduced by permission of the Trustees of the British Museum, London.

keeping with the scale of a large space. It is most probable that they were painted at the shrine, directly onto the prepared panels, each measuring approximately 173.5 x 140 cm. It is likely they were painted lying flat and bridged for access to the center, enabling the heavy pigment to bond to the paper support evenly while drying.

When these paintings were acquired for the British Museum in 1937, documentation associated them with another group of four *fusuma* paintings entitled *Seven Sages*, now in the Seattle Museum of Art. It has been suggested that they were companion pieces, and that the Seattle paintings may even have been originally placed on the other side of the British Museum's. It seems that the two sets of paintings were separated for sale in about 1935. The Seattle *Seven Sages* paintings passed through the hands of a number of dealers before finally being purchased by the Seattle Museum of Art in 1950.

Since *fusuma* are made for architectural settings, their dimensions are predetermined by the structure of the building, which in turn dictates the proportions of the rooms. The *Bird and Flower* and *Seven Sages* panels are exactly the same size and, in addition, the metal door pulls feature an identical Japanese motif that suggests a link between them. Further evidence has also been found in the under-papers: in the British Museum paintings, fragments of gilded paper that remained after the last remounting were still adhered beneath the lacquered rail. This suggests that the paintings that once formed the reverse side were heavily gilded, as are the *Seven Sages* paintings. While dismantling the British Museum paintings, a burn hole previously hidden by the backing silk was discovered on the reverse side of the second panel in the lower right. Investigations are now under way to discover whether one of the Seattle paintings displays similar damage.

The Tanzan Jinja Shrine

In May 1998 one of the authors (ST) visited the Tanzan Jinja Shrine to inform the authorities there about the British Museum's *fusuma* project, and about the paintings' affiliation with the shrine. The head priest and his assistant gave the project a formal Shinto blessing. This followed research in the Tanzan Jinja archives with the head priest and the shrine historian to look for any record or mention of the paintings.

They found that, at the beginning of the seventeenth century, a rebuilding program was under way, and new paintings were commissioned for the guest accommodation at the shrine. There is a letter dated 1612 from Chiku Rinbō Kenjō, the monk in charge of the renovation, saying that "the paintings were almost finished." This agrees with the Kano school attribution and supports the possibility that the paintings are by Kano Sanraku (like most *fusuma*, they are not signed). The temple records state that in c. 1895, due to a landslip, the buildings housing the paintings collapsed. The paintings, however, survived and were moved to another part of the shrine complex. It is important to note that this was a time when Buddhism was suffering persecution after the radical changes imposed under the new Meiji regime. Tanzan Jinja was originally a Buddhist temple, but to survive it became a Shinto shrine. It is recorded that by c. 1920 the paintings were no longer required, and the shrine sold them to a private collection.

The careful deconstruction of objects as complex as *fusuma* illustrates how important it is to be aware of the tradition from which they come. The evidence of provenance and construction can easily be overlooked or lost. On removal of the lacquered wooden rails (*fuchi*), an inscription was found on the underside of one rail that dates from the last restoration work in c. 1850; it includes the name of the carpenter who did the work and the name of a lacquer shop in Miwa close to the shrine. Additionally, the calligraphy on the recycled papers that formed the under-papers on the screen lattice frames (*shitaji*) provides a great deal of local historical information, both social and economic. While it is not unusual to find some historical evidence in Japanese paper supports, this calligraphy relates directly to "a central purchasing department" serving a number of Tonomine temples. It also includes the account books of kimono and obi pawn shops (*shichiya*), as well as accounts of travel expenses submitted to the temples. Already some fascinating details have emerged about the local history of the Tonomine area around 1800, although further translation is necessary to obtain a fuller picture drawn from this old-fashioned script. The earliest date found so far is 1798. These records seem to have been discarded by around 1850, when it is thought that they were recycled for the last remounting.

The function and construction of *fusuma*

Fusuma, almost square in shape, are functional objects used to form partitions between rooms, defining a space. They are easily removed from narrow double wooden grooves, top and bottom, which run between wooden pillars and hold the panels in place. There is a slight tolerance in the upper track to lift the panels up and out should they need to be exchanged or moved to create a larger space. This movable division of interior space is an important architectural element in Japanese temples. It is the primary function of the sliding panels to provide doorways and, when closed, to form walls with paintings to create a visual background.

The inner lattice wooden cores, made of white cedar, are much more robust and have broader cross-members than those used, for example, in folding screens (*byōbu*), because they are designed to function as sliding doors. The technique for papering the support lattice is identical to that used for *byōbu* and almost identical to that used for Japanese drying boards (*karibari*). The lattice cores are covered on both sides using the traditional system involving many layers of paper (Webber and Huxtable 1985). In the careful deconstruction of Japanese screens, it is not unusual to find recycled paper – old account books were often used. In this case, the strong and opaque paper, made from *kozo* fiber, had already been written on two or three times before, such was the economic use of paper in Japan. These old papers provided a secure support onto which the paintings were finally attached. The *fusuma* are framed with black lacquered, wooden rails known as *fuchi*. Other features include the door pulls (*hikite*), which are oval or round decorative plates of thin metal and are inserted into spaces made for them in the wooden frame.

Because the sliding panels from the Tanzan Jinja Shrine were used as well as admired, they have inevitably incurred physical damage from handling. They have suffered considerable water damage and insect attack as constituents of a wooden temple complex for more than 300 years. The building and its contents have been subjected to a huge range of climatic conditions – from hot humid summers to cold dry winters, not to mention earthquakes and monsoons. By nature, temples are dark, cold, and drafty on the inside, and the sliding rain doors and deep wooden verandas provide the only protection from the elements.

The condition of these large works of art on paper was unstable; each panel presented a variety of problems including splitting, flaking pigment, and water staining. The aim of the treatment program was to stabilize the paintings and restore them to a state in which their beauty could be appreciated. Any treatment would necessitate consolidation of the friable paint surface before removal of the paper linings. Consolidation and debacking would be followed by repair, relining, and remounting once again as *fusuma*.

The deterioration was similar throughout the four panels. Overall, however, the quality of the support paper is good. Although it may be darker in tone and has suffered some physical damage, it is still smooth and lustrous. *Maniai-shi*, a heavy, strong, and durable paper, was made specially for panel construction, the sheet size measuring 95.5 x 37.3 cm.[2] The papers are joined in such a way that the bottom edge of the top sheet overlaps the top edge of the sheet beneath it, like a roof tile. This prevents dust from being caught on the edge.

When the paintings were removed from the panels, it was noted that they were only attached around the edges, and that the center was not fully adhered to the final paper layer (*ukekake*). This created a discrepancy between the supported and unsupported areas. Changes in temperature and relative humidity (RH), occurring over time, caused tensions in the paper support that resulted in severe physical damage, such as long splits and flaking pigment (see Fig. 4.18). For some reason this mounting appears to have been rather incomplete and as such is an insubstantial method of support compared with the more traditional methods still taught today.

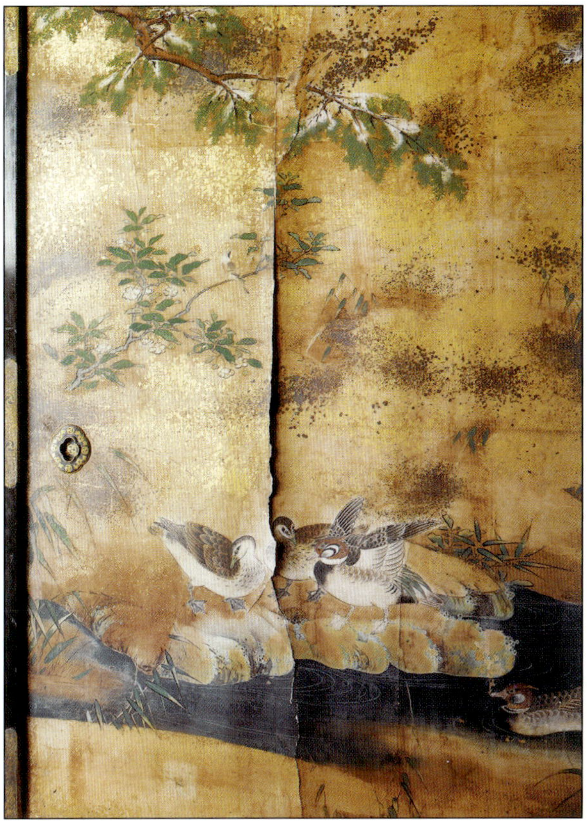

Figure 4.18 Detail of second panel from right in the series showing a large vertical split caused by tensions created by incomplete mounting.

Pigments

These paintings feature a limited number of colors. The pigments are opaque and were less finely ground than those used for paintings on silk. The binding medium was traditionally high-quality animal glue (*nikawa*). Over time this medium can lose some of its adhesive properties; the bond between individual pigment particles or between the support and painting may degrade, with a result that the slightest change in tension on the surface of the painting may unsettle the pigment layer. This was precisely the case with the pigment on these four paintings.

Pigment samples were taken for identification (Table 4.11). Analysis performed using polarized light microscopy and X-ray fluorescence (XRF) revealed the presence of malachite and azurite (both copper minerals), vermilion, and *gofun* (ground oyster-shell white; widely used in the decorative arts in Japan) (see Fig. 4.19). Three types of metal foil were used: gold, an alloy of gold and silver, and silver. The gold/silver alloy is now a reddish brown, whereas the silver alloy has tarnished and blackened irreversibly (Duncan et al. 1990). The results do not show any unexpected pigments. A high bromine concentration was detected in the *gofun*, but this is consistent with the marine origin of this pigment (crushed oyster shells), as seawater contains some bromine.

Hue	Elements by XRF	Polarized light results	Conclusion
White (flower)	Br, Cu, Ca	n close to 1.66, colorless large flat irregular particles, with acicular needles n > 1.66, which twinkled	*gofun*
White (goose)	Ca	as above	*gofun*
Red	Hg	small particles almost black in transmitted light. Isotropic, n > 1.66, 5–10µ diameter	ground vermilion
Green (rock)	Cu, Zn, As	small particles, pale green n > 1.66. Strongly birefringent, some twinkled	malachite
Green (leaf)	Cu, As	as above	malachite
Bright blue	Cu	pleochroic, birefringent, n > 1.66. Gray/blue with Chelsea filler, high relief, broken particles	azurite
Dark blue	Cu	as above, but with the addition of opaque black particles and red pigment above	azurite + vermilion + carbon black
Yellow metal foil	Au	N/A	gold
Brown foil	Au, Ag	N/A	tarnished gold alloy
Gray foil	Ag, Fe, Ca	N/A	tarnished silver
Brown tree trunk	Fe, Cu, Ca	n > 1.66 small yellow/brown particles, low relief, isotropic	iron earth
Paper (plain)	Ca, Fe, Cu	N/A	N/A
Paper (brown, near goose)	Ca, Cu	N/A	N/A

Table 4.11 Analysis of pigment samples.

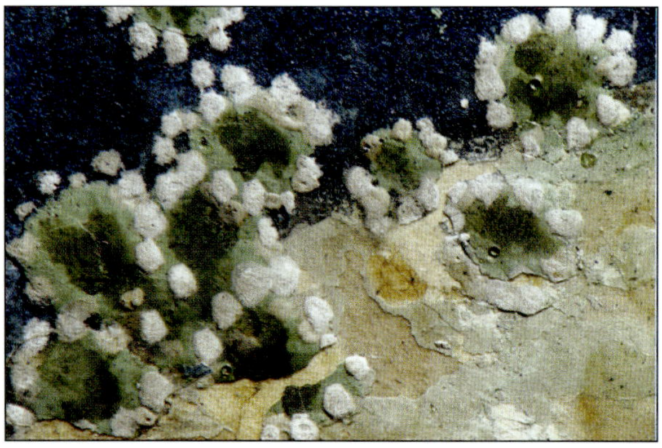

Figure 4.19 Photomacrograph of the pigment surface displaying *gofun* (ground oyster shell), malachite, and azurite.

In the depiction of rocks, the *gofun* was applied as a thick layer and overpainted with a thin glaze of pale green malachite. In these areas there is severe pigment loss where the brittle *gofun* layer has chipped off the paper surface (see Fig. 4.20). There are many other areas of similar pigment loss where *gofun* was the ground layer; for example, the vermilion maple leaves and the white birds, especially one of the geese. The goose feathers are painted in *sumi*.

The tree trunk probably derives its brown appearance from a brown pigment rather than the copper/paper degradation products noticed elsewhere on the screens. Analysis of the brown paper adjacent to the goose suggested the presence of a copper-containing pigment. Very slight traces of copper and calcium were detected in the dark brown paper substrate, but these were not significant enough to prove a theory that the change of color may have been caused by the deterioration of a copper-based pigment, now lost, which in turn may have induced the degradation of the paper substrate. This is certainly true, however, of the paper support around and directly beneath the malachite paint layers (Banik et al. 1982). On the verso it was evident that both the paper support and previous linings had deteriorated in these areas, which now appear dark brown in color. As is typical, the paper fibers had also degraded and become brittle due to the proximity of the copper-based pigment.

By contrast, beneath the azurite pigment there is a thin base layer of *gofun*. The color has increased luminosity and vibrancy, not only due to the white ground but also because of the size of the azurite pigment particles and their refractive index. The *gofun* has protected the paper support and buffered any acidic degradation from this copper-based pigment. The azurite

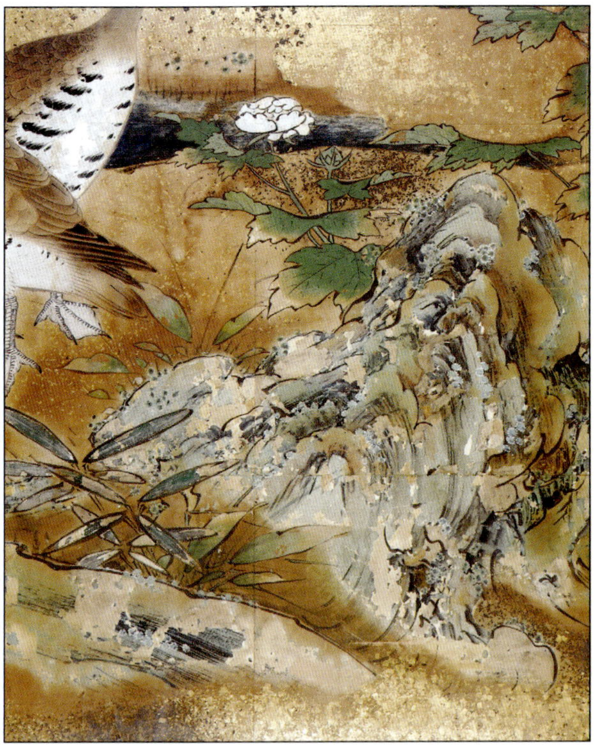

Figure 4.20 Detail of fourth panel from right in the series showing severe pigment loss where the brittle *gofun* (ground oyster shell) layer has chipped off the paper surface.

paint layer is flaking and has become discolored and "patchy" in places where old repairs and inpainting are present.

Very fine working alterations, made by the artists, came to light on close examination of the small birds in the second panel and the flying geese in the first panel. Very subtle, confident, fine-line incisions on the paper surface were observed using a magnifying lens. They were clearly intentional and contributed to alterations that are almost invisible. To understand the method of alteration, it was necessary to determine whether the incisions were made on the surface of the paper only or whether they had been cut through and repaired from the back. Knowledge of *maniai* papermaking is crucial here. *Maniai-shi*, a *gampi* paper containing clay, forms a laminated sheet known as *nagashi-zuki* during the traditional papermaking process (Hughes 1978). The artists may have removed sections of the top layer to provide a clean surface for the second attempt – a hypothesis that was confirmed by the fact that no repairs were found on the back relating to the position of the birds during lining removal.

In the second panel, it appears that five small birds were repainted. In the first panel, however, the two flying geese have been painted on top of the applied *sunago* (sprinkled gold), as flaking of the *gofun* reveals a very bright gold leaf, with underdrawing in *sumi* brush strokes. This suggests they were a later addition, for between these birds are other incised lines indicating the initial contours of a bird in flight, now removed. This area is now decoratively disguised with *sunago*. These appear to be the only changes made to the composition of the birds, and the geese painted in the fourth panel have not been changed. Here the sprinkled gold has been applied around them as though a template had been used.

Treatment of the *fusuma* paintings

As these paintings have been in a Western collection for more than 60 years, it is likely that a number of attempts have been made at consolidating the pigment. A sheen over some areas of pigment indicates a former conservation treatment, and analysis using Fourier transform infrared spectroscopy (FTIR) shows that the sample of consolidant taken from the surface of the painting was a polysaccharide similar to gum or starch.

The pigments were tested for water sensitivity, and none were found to be fugitive. However, with gilding techniques, it is not safe to assume that *nikawa* is always the adhesive used, as *funori* (seaweed gel) is also traditional. This has a much less firm bond and is water sensitive.

A thorough examination and photographic record of each panel was made. This included extensive notes on the location of tears, condition of pigments, repairs, previous inpainting, water damage, and insect attack. The treatment program involved consolidation of the unstable pigment using rabbit-skin glue size to enable safe handling during wet treatment, to reduce staining, and to remove soluble acidity.

To prepare the paintings for consolidation and cleaning, a grid method was devised using elasticated cotton thread so that the treatment could be carried out systematically over a considerable area. In order to improve the adhesion of the friable paint surface, a 1.5% solution of rabbit-skin glue was applied with a small brush to the surface of individual pigments, and small syringes with flexible needles were used to release the adhesive under the flaking pigment. These time-consuming processes were closely observed to see that each application was perfectly absorbed. Any residue or seepage would remain glossy and upon drying could cause the pigment surface to cup and flake. Absorption of consolidant by each pigment depends upon the particle size and how much of the former binder still remains. Pigments therefore have different rates of absorption and must be treated individually. The rabbit-skin glue was applied several times until each pigment ceased to absorb further application. Consolidation could be repeated if and when necessary during later stages of conservation. The consolidant was left to dry thoroughly for at least four or five weeks before carrying out any aqueous treatment to clean and remove surface dust. The extensive discoloration and obtrusive water staining was reduced by applying a mixture of alcohol and water locally. This was continued until the unity of the paintings could be fully appreciated.

The replacement of the backings began with the preparation of the worktable by covering it with thick

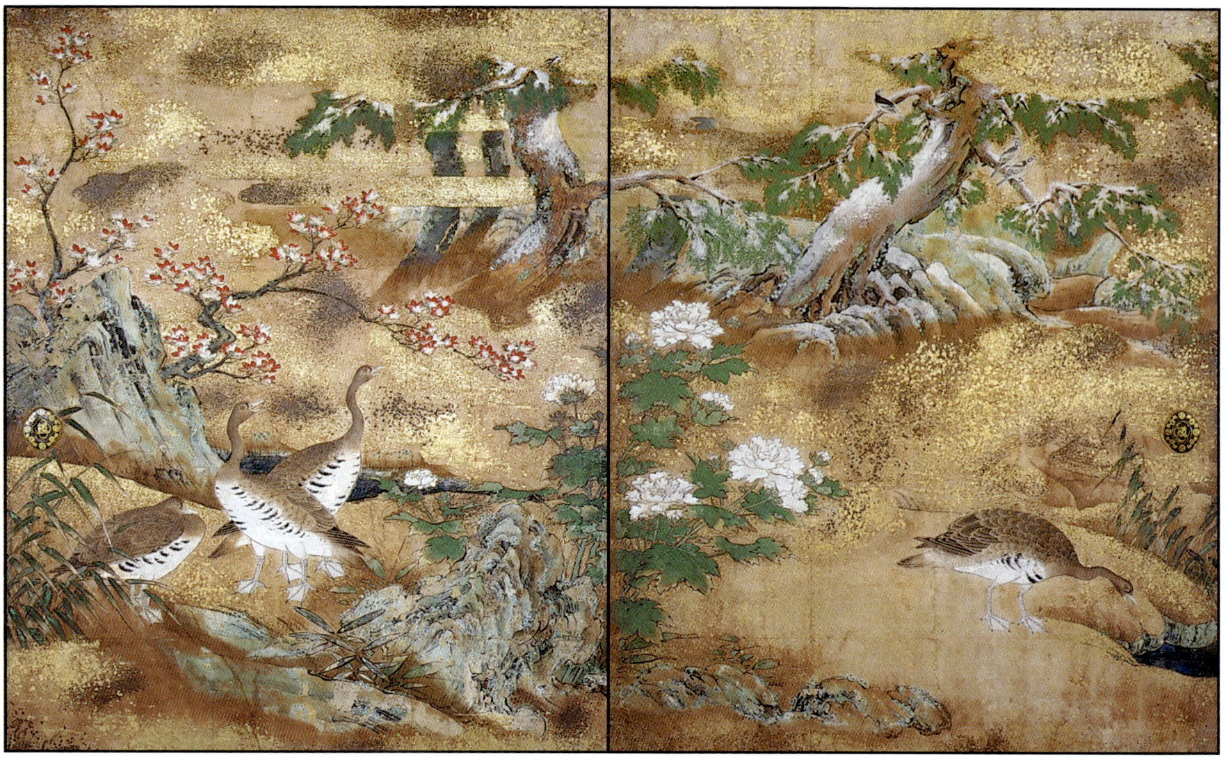

Figure 4.21 Attributed to Kano Sanraku (Japanese, 1559–1635), *Bird and Flower fusuma*, c. 1612 (after treatment). Mineral pigments with gold and silver on Japanese *maniai* paper, each panel 173.5 x 140 cm. Reproduced by permission of the Trustees of the British Museum, London (this page and opposite).

Japanese rayon paper and placing a fine rayon paper on top of that. The painting was then carefully placed face down onto this protective surface in preparation for lining removal. As previously mentioned, the last paper applied in the preparation of the panel was incompletely adhered (only around the edges) leaving the large center completely unsupported and vulnerable to damage. (The "floating" layer is traditionally designed so that paintings can be freed slowly but easily from panels using a bamboo spatula without risking stress to the painting or even to the deeper layers of paper on the panel during remounting.) To save the calligraphy on some of the lining papers, they were removed carefully by applying a little moisture using the water brush (*mizubake*). Then, all other lining papers and previous repairs made to splits in the painting support paper were removed. This required critical control of moisture and limiting the local application of water in order to minimize the expansion and subsequent contraction of the original paper support and disruption to the paint layer. The water was applied by brush to small areas (approximately 5 cm^2), and the moisture slowly penetrated through the two layers of backing. When the paste had softened, the backing paper was carefully peeled away in small pieces using tweezers and bamboo spatulas. All the tears were then realigned and strengthened with two water-torn *minogami* paper strips, the first using a paper dyed with *yasha* to give a sympathetic color in case it remained visible after the realignment of the tears.[3] The second strip was much broader than the first and lighter in weight.

When all repairs had been made on the reverse, the paintings were treated with a 3% saturated solution of $Mg(HCO_3)_2$ (magnesium hydrogen carbonate) prior to lining. This raised the pH of the paper from 4.32 in the degraded areas to 6.0. The moisture slowly penetrated through the paper substrate, and the painting was then carefully brushed out flat for lining. Until this point, aqueous treatment was kept to a minimum to reduce stress to the inflexible pigment layer. Then the first new backing of medium-weight *minogami* paper was applied using wheat starch paste. After drying, this was followed by a second lining, changing the fiber direction. The painting was allowed to dry overnight before being lightly tensioned by adding moisture to the reverse, and attaching it to the *karibari* by the paper margin (*tegami*) using wheat starch paste.

All four *fusuma* paintings are now lined and attached to new, papered, white cedar lattice supports (Fig. 4.21). New lacquered frames and the original decorative metal fittings are reattached. In preserving these great paintings for future study and display, the functional nature and traditional methods of manufacture influenced the overall aims of the treatment program – not with the intention to make such rare, beautiful, and fragile paintings into functional objects, but so that all the individual parts of the screen make-up would be given the appropriate consideration in order to achieve an integrated whole.

Acknowledgments

The Department of Conservation acknowledges with gratitude the generosity of the Art Research Foundation

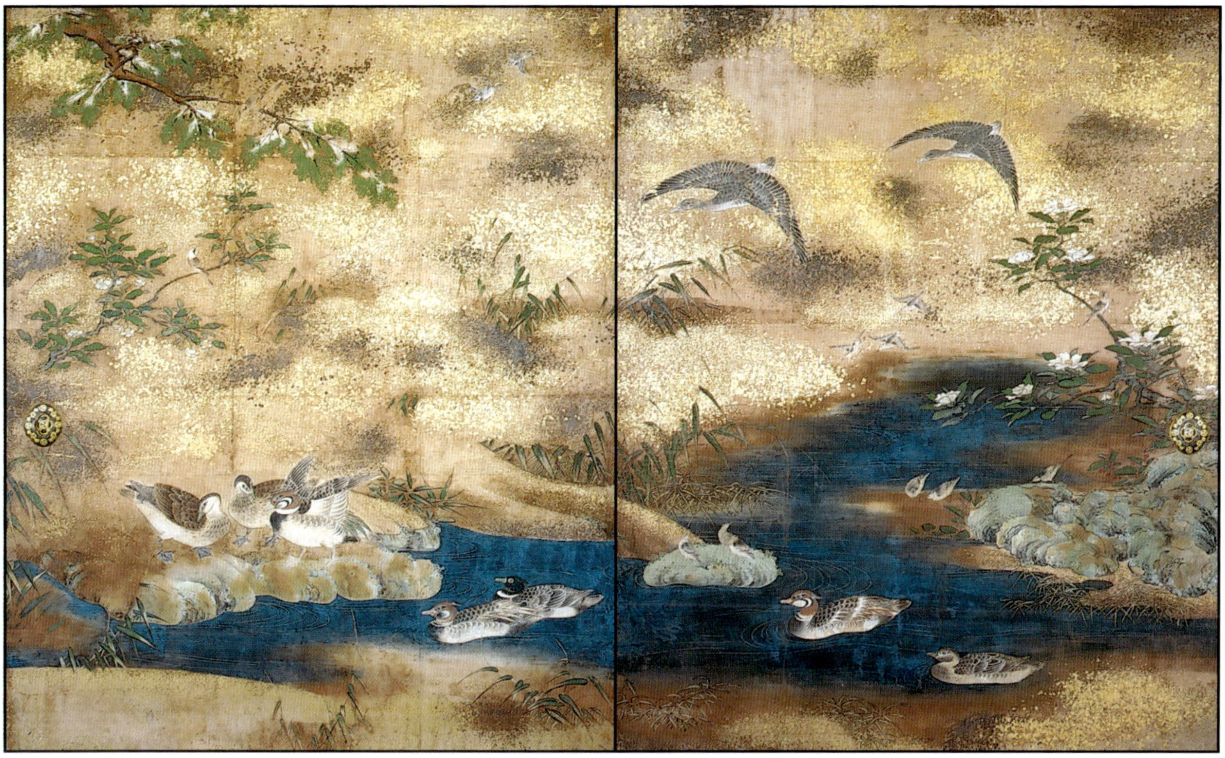

of Tokyo for helping to fund the Hirayama Fellowship, which is held by Mitsuhiro Abe at the British Museum. We are grateful to our colleagues in the Hirayama Studio for their support, and in particular to Vincent Daniels for the pigment analysis.

Endnotes

1. *Sumi* is a black ink made from soot, animal glue, and incense. It originates from China, and the soot is usually from non-resinous pine wood.

2. *Maniai-shi* is a *gampi* paper originating in the Muromachi period (1336–1573). The name is derived from the fact that it can be pasted without any joint for half the width (90 cm) of a sliding panel. Literally translated, it means "paper serving the purpose."

3. *Yasha* is an organic brown dye made from alder cones.

References

Banik, G., H. Stachelberger, and O. Wächter 1982. "Investigation of the destructive action of copper pigments on paper and consequences for conservation," in N. S. Brommelle and G. Thomson (eds), *Science and Technology in the Service of Conservation.* London: International Institute for Conservation, pp. 75–78.

Duncan, S. J., V. Daniels, and L. E. Fleming 1990. "The identification of metal foils and powders used on Japanese prints and paintings," *Restaurator* 11 (1990), pp. 244–53.

Hughes, S. 1978. *Washi, The World of Japanese Paper.* Tokyo, New York, and San Francisco: Kodansha International.

Webber, P., and M. Huxtable 1985. "Karibari – the Japanese drying board," *The Paper Conservator* 9, pp. 54–60.

Chinese Color Printing Technology and History in the *Ten Bamboo Studio Manual of Painting and Calligraphy*

M. Brigitte Yeh

> It is like a dream that I cannot ever forget – the three great publications of Chinese color printing: the *Ink Garden of Cheng Junfang*, Hu Zhengyan's *Collection of Decorated Writing Paper from the Ten Bamboo Studio*, and the *Ten Bamboo Studio Manual of Painting and Calligraphy*.[1]
> (Zheng Zhenduo, noted 20th-century scholar and bibliophile)

The final years of the Ming Dynasty (1368–1644) witnessed the apex of printing in China. It was during this time that true, registered color printing came of age in the production of the *Shizhuzhai Shuhua Pu*, or the *Ten Bamboo Studio Manual of Painting and Calligraphy*. Skillfully printed editions of this manual represented the pinnacle of the Chinese color printing technique, which was valued for its verisimilitude to subtly rendered watercolor paintings. Through the woodcut method of reproduction, the printer attempted to convey important aspects of the painter's art, such as the deployment of the brush and the nuances of ink tone.

Prior to this time, multicolored woodblock illustrations were not necessarily printed from multiple blocks. Rather, in the absence of a registration method, color prints were made by applying the various colors directly onto a single woodblock and then printing – a relatively slow and laborious process. Mastery of the problem of color registration enabled printers to produce colored multiples in great numbers for the first time. It has been reported that as many as 1,500 to 2,000 double-page sheets per day could be produced by an experienced printer using this method.[2] Although the *Ten Bamboo Studio Manual* was not the first instance of color registration, it is a virtuoso achievement of this printing technique since complete sets contain over 180 illustrations and 140 pages of text.

The printer and publisher of the manual, Hu Zhengyan (appellations Yuecong and Zigong), was born around 1582 in Anhui province. He studied the classics and medicine and later served as a minor government official until he eventually withdrew to Nanjing in order to pursue his artistic interests in seclusion. The name of Hu Zhengyan's studio and manual was reportedly derived from the cluster of bamboo trees growing outside his Nanjing home and studio. He began his artistic career as a well-respected engraver of seals before eventually earning fame as the publisher of the *Ten Bamboo Studio Manual*. In the city of Nanjing, the high scholarly taste typical of the late Ming Dynasty flourished in the tradition of printing manuals, and so it is no coincidence that the well-known *Mustard Seed Garden Manual (Jiezi yuan huazhuan)* was later (1679) published in Nanjing as well. The various volumes of Hu's *Ten Bamboo Studio Manual* were most likely produced over a period spanning eight years, from 1619 to 1627, with a publication date of 1633 given for the first completed edition.

Collected and appreciated even during Hu's lifetime, the manual has been sought after by Western museums and libraries ever since. In a preface written for the first edition, Hu Zhengyan is acknowledged thus:

> Recently he brought together a number of his wood-cuts into a repertory of writings and paintings. These are most refined and original, the whole collection being well composed and extensive ... The repertory by Yuecong [Hu Zhengyan] which is now being published may be acquired for a few coins by poets and painters who thus can take it to their home and keep it on their table together with the five Classics and the three Histories. Hu has rendered them a very great service.[3]

Approximately 100 exemplars of the *Ten Bamboo Studio Manual* are estimated to exist in collections around the world, including roughly 10 to 12 superior copies.[4] The nine copies that have been studied during the course of research for this paper include those in the Library of Congress, Washington, D.C.; The Art Institute of Chicago; the University of Chicago; the Museum of Fine Arts, Boston; the Sackler Museum at Harvard University, Cambridge, Massachusetts; the Yenching Library, also at Harvard University; the Cleveland Museum of Art; and the British Museum, London.

The manual's historic interest resides principally in the technique of color registration and in its high level of artistry. These images of pictorial printing were ultimately transmitted to Japan, where they influenced the development of the great tradition of woodblock printing. This paper introduces the history, technology, and materials of early Chinese printing as they relate to the manual. The very relevant issues of quality, dating, and the conservation concerns associated with treatment of the edition in the Library of Congress are also discussed.

The *Ten Bamboo Studio Manual* was originally intended as an elegant guide to the art of painting through its illustrations of various compositions on an assortment of traditional painting subjects. Originally bound in slender volumes using the "butterfly" format, the manual is divided into eight sections, each devoted to a particular theme: Orchids, Bamboo, Rocks, Blossoming Plum, Birds and Flowers, Round Fans, Fruit, and Miscellany. The volumes adopt the format of a double-page illustration followed by a double-page poem written in refined calligraphy (see Fig. 4.22). Most of the illustrations in the *Ten Bamboo Studio Manual* were designed by Hu; however, he also befriended numerous painters, poets, and calligraphers who contributed designs, the inclusion of which suggests that Hu collaborated with the artistic community thriving in the region at the time. One such artist-friend was the painter Gao Yang, famous for his renditions of rocks, who supplied most of the compositions for the volume on that subject. A hanging scroll entitled *Strange Rocks* by Gao Yang at The Art Institute of Chicago, which bears a

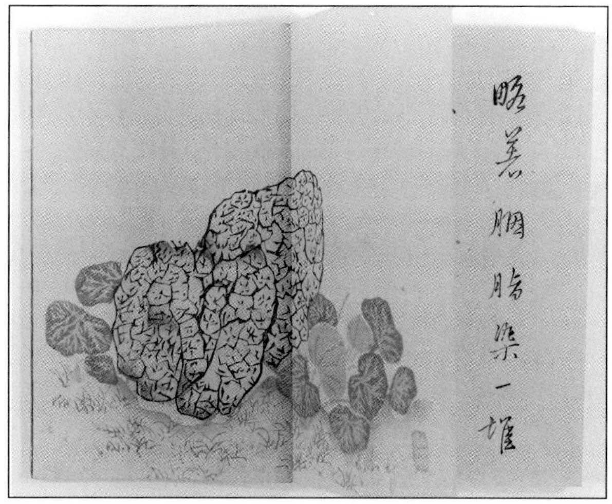

Figure 4.22 *Ten Bamboo Studio Manual of Painting and Calligraphy*, 19th-century (?) reprint. Woodblock printed book, 23.9 x 27.8 cm. The Art Institute of Chicago.

seal that matches the print, confirms the stylistic consistency.

Because it had great reference value for those beginning to paint, the *Ten Bamboo Studio Manual* has been called "the jewel of later study, the golden needle of painting study, and a treasury for calligraphers."[5] It preceded the better-known and even more explicitly pedagogical *Mustard Seed Garden Painting Manual* by approximately half a century. The *Ten Bamboo Studio Manual* includes sketches that focus on manual instruction in the volumes on Orchids and Bamboo. One example illustrates the proper way to hold a brush, while other pictures demonstrate how to render orchids under various meteorological conditions and poses: in the dew of the morning, bending under the weight of rain, or blowing to one side in the wind.

Testimony that the manual was indeed used for the intended didactic purpose may be seen in pages from an edition in the Cleveland Museum of Art (see Fig. 4.23). These pages, which show the proper sequence of brush strokes for painting orchids, display more recent ink spots that curiously follow the contour of the printed illustration. These ink stains were most likely made by a beginner who was learning how to paint by tracing over the manual's illustration. Placing a piece of semi-transparent paper over the open manual, he apparently traced the brushstrokes of the orchid leaf, and the ink bled through to the page beneath. Those who have studied Chinese calligraphy are familiar with this method of learning brush calligraphy by tracing over models. Not often, however, would one have the luxury of tracing over such a prized model.

Both the *Ten Bamboo Studio Manual* and the *Mustard Seed Garden Manual* were introduced to Japan during a time of relative sinophilia. Japanese literati painters were curious to see models of Chinese painting, especially during the period of severely restricted trade under the Tokugawa shogunate (1603–1867). The manuals had a significant impact on the development of color printing as well as painting in Japan. Although the Japanese had

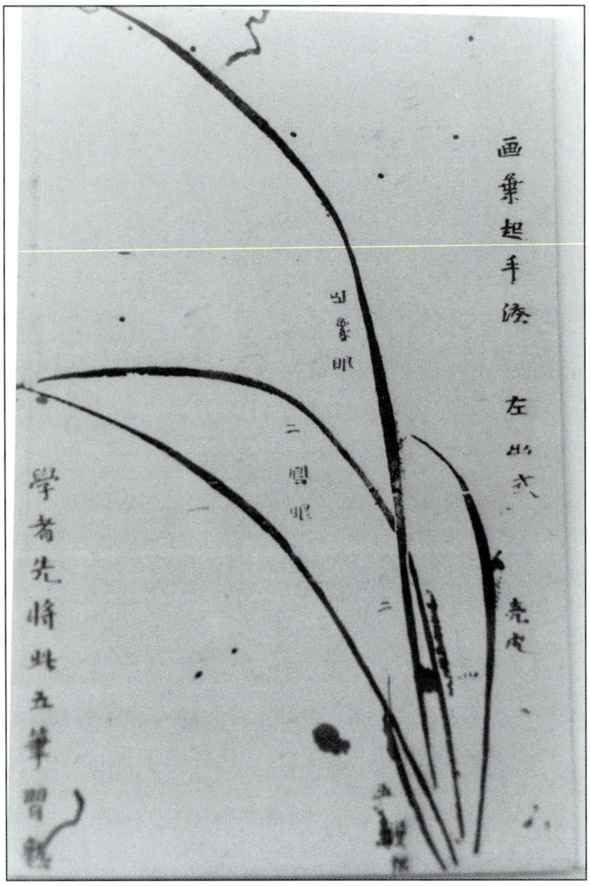

Figure 4.23 *Ten Bamboo Studio Manual of Painting and Calligraphy*, 19th-century (?) reprint. Woodblock printed book, approx. 24 x 28 cm. Cleveland Museum of Art.

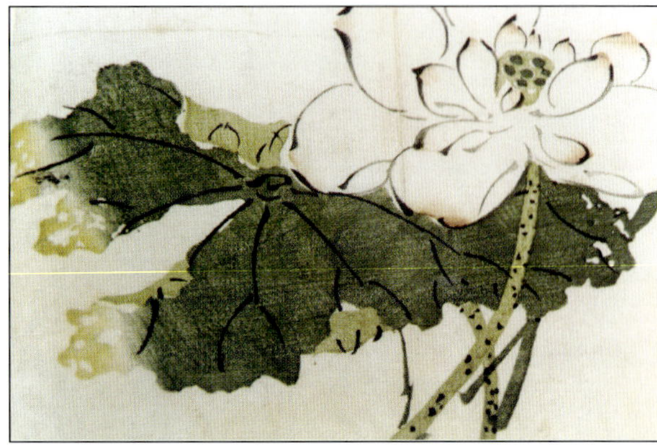

Figure 4.24 *Ten Bamboo Studio Manual of Painting and Calligraphy*, 17th century (detail showing *dan* technique). Disbound sheet from woodblock printed book. Museum of Fine Arts, Boston.

independently invented a method for color registration around 1644, color printing was still very limited until the mid-1700s, when the *kento* registration method was popularized.[6] Perhaps stimulated by the books coming from China, Japanese artists initiated a century of remarkable color woodblock printing. A floral roundel from a 1730 Japanese volume entitled *Chichi-no-On (Love for a Father)* bears an unmistakable resemblance to the fan prints found in the *Ten Bamboo Studio Manual.* Also noteworthy is the fact that the first Japanese edition of the *Mustard Seed Garden Manual*, the *Kaishien Gaden*, appeared in 1748.

Chinese printing history and technology

The ancients had devised several methods for producing multiples, such as name seals that were used like stamps and rubbings taken from engraved stone stelae. But the critical impetus for the development of printing has been attributed to the rise of Buddhism in the seventh century, with its accompanying desire to reproduce Buddhist images and texts. Although it is known that woodblock prints had been made earlier, the oldest surviving woodblock image in China is the *Diamond Sutra (Jin Gang Jing)*, dated to A.D. 868. From the twelfth century, one finds the earliest example of color printing in a woodcut depicting the legend of *The Official Dongfang Shuo Stealing the Peaches of Immortality*, printed in black, gray, and green. Another milestone in the chronology of Chinese printing history is the oldest extant example of a color-printed text, believed to be from the Yuan Dynasty, around the year 1340. Also a Buddhist text, it is the commentary to the *Diamond Sutra (Jin Gang Jing Zhu)*, employing black ink for the sutra text and red ink for the commentary and illustrations.[7]

Although not the first to employ the color printing technique, the *Ten Bamboo Studio Manual* certainly exploited it through the painterly quality of its images. It holds a place in Chinese printing history because of Hu's innovation in achieving fine color gradations using a technique called *dan*. Hu Zhengyan's second great contribution to printing, the *Shizhuzhai Jian Pu* or the *Decorated Writing Papers of the Ten Bamboo Studio*, are stationery papers bearing delicate designs. They were printed similarly to the manual, and further enhanced by one of the earliest occurrences of a *gauffrage* or embossing technique called *gonghua*.

As previously mentioned, skillfully printed editions of the manual were valued for their ability to reproduce the brush strokes, washes, and color of actual watercolor paintings (see Fig. 4.24). The new, revolutionary *dan* technique enabled the printer to reproduce the subtle gradations of color that resembled watercolor effects by applying varying grades of colors, from light to dark, to the woodblock and then either wiping the woodblock selectively so that one color fades into another or brushing on water to dilute and soften the edges of a color. This technique may be seen in images printed at the workshops of Rongbaozhai in Beijing and Duoyunxian in Shanghai, where the tradition of color woodblock printing is still continued today. The woodblock prints produced in these studios are virtually indistinguishable from original watercolor paintings.

Of profound importance to the working methods of the Chinese painter and calligrapher were the nuances of brushwork, controlled by moisture, pressure, and speed. These characteristics, in addition to the position and angle of the brush, could all be translated onto the paper. Because the woodblock prints in the *Ten Bamboo Studio Manual* were intended to imitate the subtleties of brushwork, it was necessary to conceal all visual evidence of the knife used to carve the woodblock. For instance, a slender bamboo leaf, printed darker at one end and lighter at the tip, gives the illusion that it was painted using a brush loaded with ink at the beginning of the stroke but tapering off toward the stroke's completion. Another painterly feature is the printer's effort to replicate the *fei bai*, or "flying white" technique, whereby a rather dry brush stroke is used to suggest texture, such as in the bark of a tree branch. The poems that accompany the compositions must also be considered, for here too the woodblock carver attempted to capture the elegance of a calligrapher's handwriting: lines that vary in thickness to connote the pressure and speed of the brush stroke. Because the texts were carved into single woodblocks rather than composed with individual, movable-type characters, they feature the ligatures that connect adjacent Chinese characters in the running script style of writing.

While many conservators and art historians are familiar with Japanese printing tools and techniques, relatively scant mention has been given to their Chinese counterpart and predecessor. Although Chinese woodblock printing shares many similarities with the Japanese tradition, important differences exist. For example, Chinese printers employed a number of specialized tools: the *shuabi*, a flat palm-fiber brush used for smoothing out the paper bearing the design as it is transferred to the woodblock; the *yuanmoshua*, a round horsehair brush used for inking the block; and the longpad, or *changshua*, analogous to the Japanese baren for rubbing over the back of the sheet of paper that receives the impression. Wooden hammers, metal chisels, gouges, and scrapers, along with cutting knives of various kinds, were all employed in carving the woodblock. Chinese woodcutting knives feature crescent-shaped edges, which may be distinguished from their straight-edged Japanese counterparts.

Tsuen-hsuin Tsien, an established authority on Chinese printing history, has outlined the steps involved in preparing a woodblock (generally pear wood cut along the grain) for engraving and printing.[8] These steps include: analyzing the colors in the original design, separating them, tracing the outline of each color onto thin transparent paper, and transferring the design or text to the block. Engraving is initiated with the starting cut, followed by the close cut, then chiseling out the blank spaces.

Once the blocks for each color component were carved, color registration could be achieved by printing on a table constructed especially for this purpose (see Fig. 4.25). The table is fashioned with a slit in the middle. A stack of sheets is bound under a clamp to the right side of the slit. On the left side of the table, the printing block is fixed in place with wax or pitch. The woodblock is then inked over with the *yuanmoshua*, placed on the

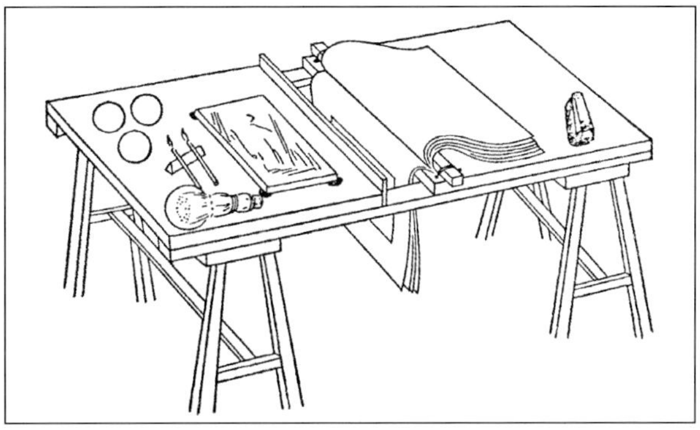

Figure 4.25 Diagram of Chinese printer's table, from Jan Tschichold, *Chinese Color Prints from the Ten Bamboo Studio* (New York: McGraw-Hill, 1970), p. 43.

table's left corner. A sheet from the stack is placed upon the inked block and brushed over with the *changshua*, featured in the right corner of the table. A narrow bar the same height as the printing block's width is attached to the left of this slit to ensure that the paper is kept flat and evenly taut during rubbing. Once the first color is laid down, the block is changed for the next, with the first five or six impressions of every stack serving to determine proper registration. The position of the printing block may be adjusted as necessary. This remarkably simple technique is still employed today in the studios of Rongbaozhai and Duoyunxian.

Materials: paper and pigments

The early development of printing in China was predicated on the availability of good writing paper that was smooth yet also durable enough to receive multiple impressions from inked woodblocks. The connoisseur Jan Tschichold offers a rather polar description to differentiate between Chinese and Japanese papers, with Chinese papers characterized as "soft, yielding and feminine, thin and extremely sensitive; the Japanese is firm and masculine, tough, strong and durable."[9] Concerning Chinese papers, he continues, "*xuanzhi* is [the type of paper] generally used for printing [since] it is comparatively absorbent and has a smooth soft surface suitable both for almost dry and for wet color."[10] *Xuanzhi* was the type of paper most frequently used by Chinese painters as well. Originally produced in the city of Xuancheng in Anhui province, it was made principally of fibers from the *qing tan* or blue sandalwood tree native to that particular locale and was prized for its whiteness. Later on, when bleaching was introduced, papermakers in other regions began to make imitation *xuan* paper using alternative fibers, such as bamboo. In regard to the paper used for the *Ten Bamboo Studio Manual*, we learn from the literature only that the printer used a *kaihua* paper that was "white like jade, printed with clear impressions, and gave a very pleasing feeling to the eye."[11]

By using polarized light microscopy and scanning electron microscopy, Jesse Munn and Anne Seibert identified the paper fibers in the copy at the Library of Congress as giant bamboo (*Bambusa arundinacea*). Rice straw and lesser additions of other, unidentified mixed fibers were also detected. Two types of giant bamboo fibers appear in the sample: primary fibers that are long and pointy, and secondary fibers that appear twisted. More importantly, the presence of bamboo in this paper may provide evidence for dating an edition, since bamboo fibers began to be used in making a type of imitation *xuan* paper sometime in the middle of the Qing Dynasty (1644–1911). In the previous dynasty, during the time of the *Ten Bamboo Studio Manual*'s creation, only pure blue sandalwood fibers were used for making *xuanzhi*.[12] These facts may corroborate the dating of the Library of Congress edition to the late eighteenth century.[13] The use of bamboo fibers certainly confirms that this particular edition was printed in Southern China, where bamboo pulp papers have exclusively been manufactured until recent times. Zhejiang province, the area around Nanjing, lies within these parameters and is particularly noted for papers made from bamboo fibers.

Underlying the efforts to conserve the edition in the Library of Congress was the need to find a compatible and archival paper for lining its embrittled pages. This edition is relatively unique because it still retains its original butterfly binding, in which the single folio images are folded down the middle and each folio section is connected by adhesive applied along the outside of the folded edges. However, it has suffered from this binding format: the center fold has become increasingly susceptible to breakage as the paper has aged. Thus, it was considered necessary to line the book pages, and it was hoped that a Chinese paper could be found as similar to the original as possible. Nineteen samples of Chinese papers gathered from mainland China and Japan were tested to judge their aging properties and general suitability. Of primary concern was the archival quality of papers presently made in China – especially since a recent visit and interview with the owner of a mill that produces hand-made paper in Zhejiang province revealed that pulp of questionable quality was being used for economic reasons.

During a research trip to mainland China, paper samples were collected from Rongbaozhai in Beijing and the arts district in Shanghai. Samples from the scroll-mounting studio at the Freer Gallery of Art and others purchased in Japan were also among the 19 samples analyzed. The papers were artificially aged in both a humid oven and in individual incubation tubes maintained at 45% relative humidity (RH) and 90°C. The effects of aging were evaluated by conducting tests for brightness, tensile strength, and pH according to TAPPI standards. The test for fold endurance was eliminated, since hand-made papers do not possess the uniformity required for this experiment. Results showed that the papers all performed equally well with the exception of the sample from the papermill in Zhejiang province. Made of imported bamboo and mixed fiber pulp, this sample performed much more poorly than the others.

Mixed with peach-tree resin or hide glue and water, the pigments used in printing the manual were largely identical to those used for painting. Naturally, when the printer sought to imitate watercolor paintings, he first reached for those same materials used to make the originals. Most of the colors were prepared by grinding and flotation; in this way different color grades could be obtained by decanting off the pigment suspensions at various stages. "As for the colors that were used in Ming dynasty print making, they were all colors that were used in traditional Chinese painting," writes Feian Yu. Feian Yu's book, *Chinese Painting Colors*, written in 1955, remains the most comprehensive source for information on Chinese pigments.[14] He identifies various pigments employed in Chinese painting, and provides a color chart with their Chinese names. One may also rely on the list of pigments furnished in the *Mustard Seed Garden Manual of Painting*, which is consistent with numerous sources on Chinese color printing.

Among the various editions of the *Ten Bamboo Studio Manual* examined for this study, some variety was found in the number of colors. There are 6 colors, including black, in the copies in the Library of Congress and the Museum of Fine Arts, Boston; 8 in The Art Institute of Chicago's; 9 in the British Museum's; and 10 in the Cleveland Museum of Art's. Preliminary examination of the Art Institute's prints revealed the identity of several pigments, including safflower (known as *beni* in its Japanese translation), which has a characteristic peach fluorescence under ultraviolet light. In China safflower has documented, widespread use as the traditional coloring agent for red. However, Feian Yu informs us that after the Opium Wars (1839–58), safflower was gradually replaced by pigments such as carmine and magenta, which were imported in great quantity.[15] This later substitution of coloring agents could one day serve as a useful fact in dating Chinese works of art.

Another pigment, lead white, was also detected among the Art Institute's prints, as may be seen in the somewhat mottled-brown appearance of the petals in the group of flowers in Figure 4.26. The mottled-brown areas represent lead white that has altered and darkened to brown lead sulfide. Slight magnification clearly shows that the area near the center fold, which was protected from exposure to air, retains the original pink color. The presence of white lead was confirmed when an area on the right-most petal was tested with hydrogen peroxide and darkening reversed. In addition, a small repair patch along the center fold was inpainted with the pink color, indicating that the repair was made prior to its darkening. Although the pink color was very likely a mixture of white and red, the possibility of darkening due to red lead was ruled out since areas printed in solid red had not darkened. Chinese artists commonly mixed one part lead white and five parts vermilion (or cinnabar) to obtain the color pink.[16] Under these conditions vermilion, composed of mercuric sulfide, may actually

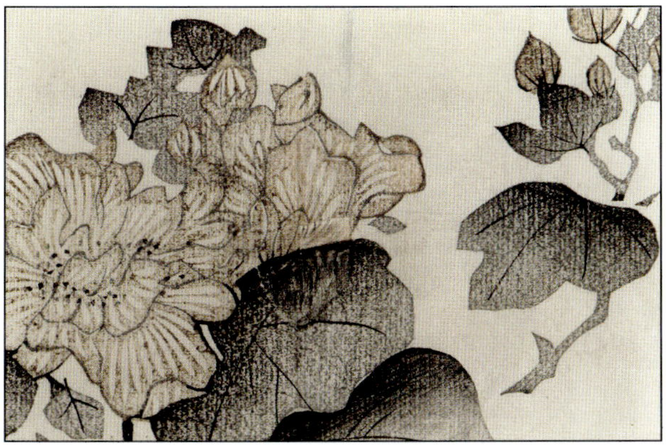

Figure 4.26 *Ten Bamboo Studio Manual of Painting and Calligraphy*, 19th-century (?) reprint. Disbound sheet from woodblock printed book, 25.7 x 29.2 cm. The Art Institute of Chicago.

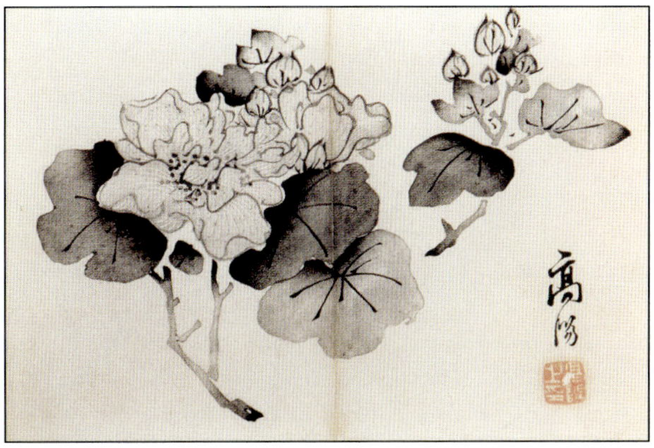

Figure 4.27 *Ten Bamboo Studio Manual of Painting and Calligraphy*, 19th-century (?) reprint. Disbound sheet from woodblock printed book, 24.7 x 28.5 cm. The British Museum, London.

serve as a source of the sulfur that converts the lead carbonate to lead sulfide.[17]

While examining the various editions of the *Ten Bamboo Studio Manual*, the color of this particular flower – it frequently appeared with black leaves and brown or gray wash for the flower petals – was exceptionally curious. The flower has been identified as the autumn hibiscus, which blooms in the fall and is distinguished by its dark leaves and dull pink flowers. In a number of cases, however, such as the British Museum's copy, the petals of the autumn hibiscus were actually printed in dark brown or gray (see Fig. 4.27). This suggests that the printer, copying from models where the lead white had already darkened, mistakenly perceived the colors of the flower petals to be brown/gray instead of pink.

Lead white's tendency to turn black over time was observed by the time of the *Ten Bamboo Studio Manual*'s printing. Methods to revert the lead white back to its original whiteness had already been proposed. One such method is recorded in the *Zhuanghuang zhi* or *Book of Mounting* by Zhou Jiazhou (d. 1660), a contemporary of the printer Hu Zhengyan:

> Melt a lump of pure white soda in water, just as if you were going to wash clothes. Then dip a new brush in this solution and therewith dab the spots that have turned black, taking good care that the soda solution does not spread over the edges of the spots treated … Paintings covered with a layer of many years will become white again after having been treated three or four times. When the pigments are again as bright as if they were newly laid on, you should finally dab the spots treated with thin tea water newly brewed, in order to remove what is left of the soda.[18]

With certain obvious exceptions, such as using thin tea water, his instructions are astonishingly similar to modern-day treatments. Not only were people familiar with lead reversion in the early seventeenth century, but there is a direct, personal connection between Hu Zhengyan, the printer of the *Ten Bamboo Studio Manual*, and the author of the *Book of Mounting*. Hu Zhengyan was known to have collaborated with Zhou Jiazhou and even kept a copy of his book in the Ten Bamboo Studio.[19]

Issues of dating

The history and dating of editions for the *Ten Bamboo Studio Manual* are quite complex and too lengthy to be included in this paper; however, it is acknowledged that a profusion of editions exist. Shortly after the Ming Dynasty came to its violent end, the Ten Bamboo Studio was burned along with the original woodblocks. Immediately following this event, new blocks were carved and printed – during Hu's lifetime, in 1643. Other known editions were printed in 1715 and 1817. It is difficult to ascertain exactly how many editions might have been made in between. Other printing shops sought to gain from making their own copies, which prompted the following advertisement prefacing the 1715 edition:

> All of the books put out by this shop on the marvels of works in ink and on copies of pictures have been most appreciated. We have sought widely for famous brushworks and have assembled a general collection. In paintings are to be found poems, in poems, pictures. They transmit the spirit and reflect nature. We have spent great sums and spared no pains. Recently, a shameless group has perceived how this shop has long carried on and acquired fame. They have fraudulently assumed this shop's name, schemed, and sought for profit. This is a hateful thing. Now we make our shop seal read Lin-t'u-yu-shu. All who patronize our salesmen should recognize our shop on the lane to the east of the T'ou-chan-men Gate of the Yung-en-ssu Temple in Chin-ling [Nanjing] and should purchase at the shop of the Ten Bamboo Studio. There can be no

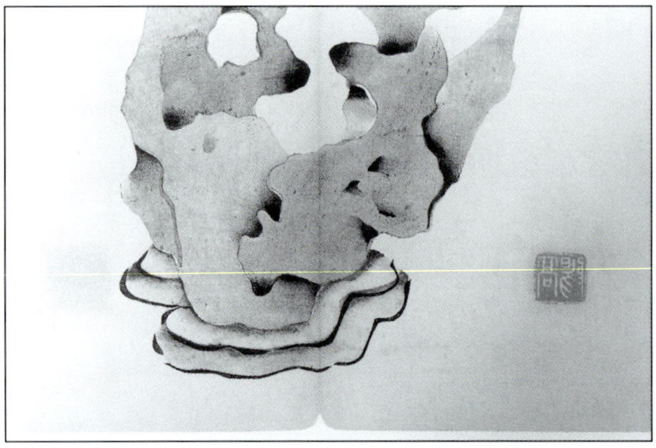

Figure 4.28 *Ten Bamboo Studio Manual of Painting and Calligraphy*, 19th-century (?) reprint. The British Museum, London.

> mistake. Respectfully written three days before the solstice of midsummer in the year yi-wei (1715).[20]

Collation of the pages varies with each copy, and images differ in each as well. Because of the manual's previously mentioned value as an image reference, it was remarkably popular, and images were reprinted many times – often from newly cut blocks or from older blocks altered by part substitutions – and not always with the same attention to color gradation and registration.[21] For example, the frame bordering the fan-shaped prints was printed with the same woodblock within each edition, and consequently these blocks wore out faster and needed to be replaced more frequently. Extensive, side-by-side comparisons are necessary to ascertain all of the nuances, and this luxury has so far only been afforded to one scholar, Robert Paine, in the 1950s.

If the end of the Ming Dynasty was the peak of color printing in China, the years following were certainly its decline. Papers used for subsequent printings varied greatly in type as well as quality. Some pages with holes were crudely patched with oversized pieces of paper and then printed upon. Pigments that were not as finely ground nor as harmonious to the composition were substituted. Later palettes are discernible. The printing, at times, grew so careless that there are several examples of pages printed upside-down, such as a page from an inferior copy in Harvard's Yenching Library. Also, the artists' chops or seal impressions, found in the lower corners of some of the designs, should exhibit halos of wax, since they would have been made from an ink paste of cinnabar and wax. Frequently, this darkened square of wax transfers to the opposite page (see Fig. 4.28). But their absence would be indicative of hastily printed later editions in which woodblocks were used to print the seals as well, since this was probably a more efficient, time-saving method for printing in great quantity.

A comparison of the extent to which the same image might vary among editions may be observed in the print of apricot branches in a vase (Fig. 4.29). The image from the Museum of Fine Arts, Boston, represents the most finely printed edition among those that I viewed. An array of this particular image would show a devolution in the quality of printing: the colors are fewer and less vibrant; the lines become less refined and less like a brush stroke. The designs become increasingly schematic, since their models were no longer actual watercolor paintings. The lower image in Figure 4.29 represents an inferior copy, which illustrates the decline in craftsmanship and artistry. "These copies, compared to the ones printed by Hu Zhengyan, have completely lost the spirit of the originals. Even for the books of the five-color *Ke le* edition, they are 10,000 miles off," wrote Zheng Zhenduo.[22] The later, inferior books may be viewed as the equivalent to modern-day paperback editions – commercial productions that printers made for profit to capitalize on the success of the *Ten Bamboo Studio Manual.*

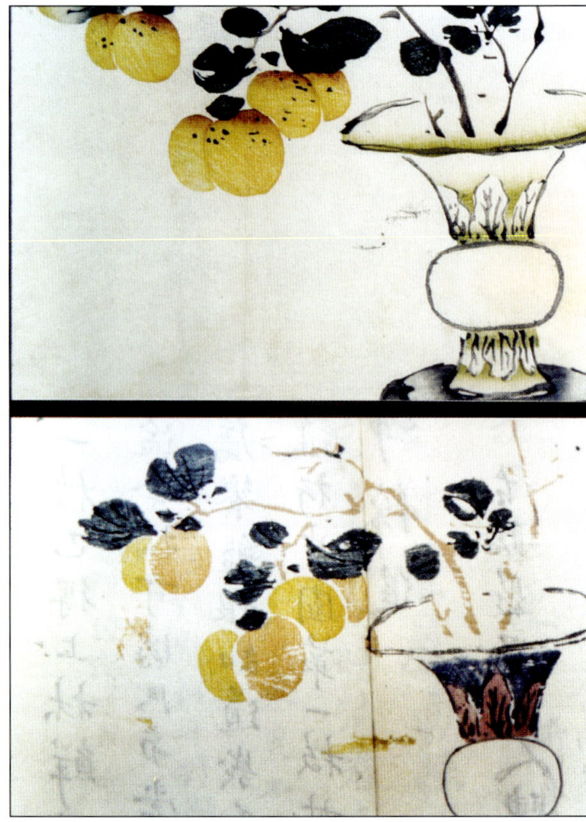

Figure 4.29 Above: *Ten Bamboo Studio Manual of Painting and Calligraphy*, 17th century. Museum of Fine Arts, Boston. Below: *Ten Bamboo Studio Manual of Painting and Calligraphy*, early 20th-century (?) reprint. Woodblock printed book. Yenching Library, Harvard University, Cambridge, Mass.

Amid this discussion of quality, the underlying questions are: can high-quality editions be equated with older, more authentic editions? Is it possible to identify the earlier copies and sort through the profusion of editions? Extensive bibliographic means have already been explored by a number of scholars. Perhaps with the aid of some analytical means, such as those presented here, a dating hierarchy may be determined eventually, but for now, it remains a topic for further research.

Endnotes

1. Jin Shen, "Hu Zhengyan yu Shi Zhu Zhai," *Duo Yun: Zhongguo Hua Yishu Zongji* 3 (1981), p. 133.

2. Tsuen-hsuin Tsien, "Technical aspects of Chinese printing," in Soren Edgren (ed.), *Chinese Rare Books in American Collections* (New York: China House Gallery, China Institute in America, 1984), p. 20.

3. Osvald Siren, *A History of Later Chinese Painting*, Vol. 2 (New York: Hacker Art Books, 1978), p. 56.

4. Discussions with Soren Edgren, spring 1998, and Jin Shen, May 1999.

5. Shen 1981 (as in note 1), pp. 133–34.

6. Jack Hillier, *The Art of the Japanese Book*, Vol. 1 (London: Philip Wilson Publishers, 1987), p. 78.

7. This example may be compared to the Mainz *Psalter*, the earliest color-printed text in the West. Produced in the year 1458, it also used a combination of black and red ink.

8. Tsien 1984 (as in note 2), pp. 18–21.

9. Jan Tschichold, *Chinese Color Prints from the Ten Bamboo Studio* (New York: McGraw-Hill, 1970), p. 41.

10. Ibid., p. 44.

11. Shen's article (as in note 1) cites the use of a type of *kaihua* paper (a paper produced in Kaihua, Jiangxi province); however, a more specific identification of the paper could not be found. Of obviously superior quality, this *kaihua* paper was manufactured and used during the Qing Dynasty for printing the palace editions at the Imperial Printing Office, according to Tsien and Shiow-jyu Lu Shaw, *The Imperial Printing of Early Ch'ing China* (Taipei: Chinese Materials Center, 1983), p. 15.

12. Renqing Liu, "Characterizing Chinese Hsuan paper," *Pulp and Paper* 82, no. 5 (1993), p. 6.

13. It has already been conjectured that the Library of Congress edition was printed after 1662 based on the substitution for a character that formed part of the personal name of the Kangxi emperor (1662–1722), as noted by Robert Paine in "The Ten Bamboo Studio: Its early editions, pictures and artists," *Archives of the Chinese Art Society of America* 5 (1951), p. 42.

14. Feian Yu, *Chinese Painting Colors: Studies of Their Preparation and Application in Traditional and Modern Times*, transl. by Jerome Silbergeld and Amy McNair (Hong Kong: Hong Kong University Press, 1988), p. 28.

15. Ibid., p. 11.

16. Ibid., p. 45.

17. Shell white, the alternative white pigment, took many days of grinding to prepare, so it is understandable that the printer may have opted to use lead white, requiring less labor. See R. H. van Gulik, *Chinese Pictorial Art as Viewed by the Connoisseur: Notes on the Means and Methods of Traditional Chinese Connoisseurship of Pictorial Art, Based Upon a Study of the Art of Mounting Scrolls in China and Japan* (Rome: Istituto Italiano per il Medio ed Estremo Oriente, 1958), p. 302.

18. Ibid.

19. Ibid., pp. 286–87.

20. Paine 1951 (as in note 13), p. 43.

21. Shen 1981 (as in note 1), pp. 133–34.

22. Ibid., p. 134.

Indian Paintings on Paper, Textile, and Mica: Conservation, Storage, and Display

Mike Wheeler, Pauline Webber, Anna Hillcoat-Imanishi, and Clair Battisson

The Victoria and Albert Museum (V&A) has a diverse collection of paintings from the Indian subcontinent, including fine examples of Mughal painting and Indian court painting as well as lesser-known examples of folk art and three-dimensional items made of paper. The supports for the paintings include paper, textile, mica, and composites of textile and paper in various combinations. This paper discusses the types of deterioration affecting these objects and the strategies devised by the paper conservation department at the V&A to prepare them for storage, display, or loan. In general, conservation treatments requiring minimal intervention are preferred. Sympathetic mounting techniques using appropriate materials, as well as improved methods of storage and display, help to ensure that the objects are seen in an appropriate setting that enhances their appearance while preserving their authenticity as museum objects.

Painting supports

The characteristics of the painting supports used by artists play a significant role in the type of deterioration that may affect these types of museum object. The smooth paper surface favored by painters of miniatures is created by burnishing. This, in combination with the surface sizing of starch, results in a compacted sheet conducive to the painting of fine details, yet it provides

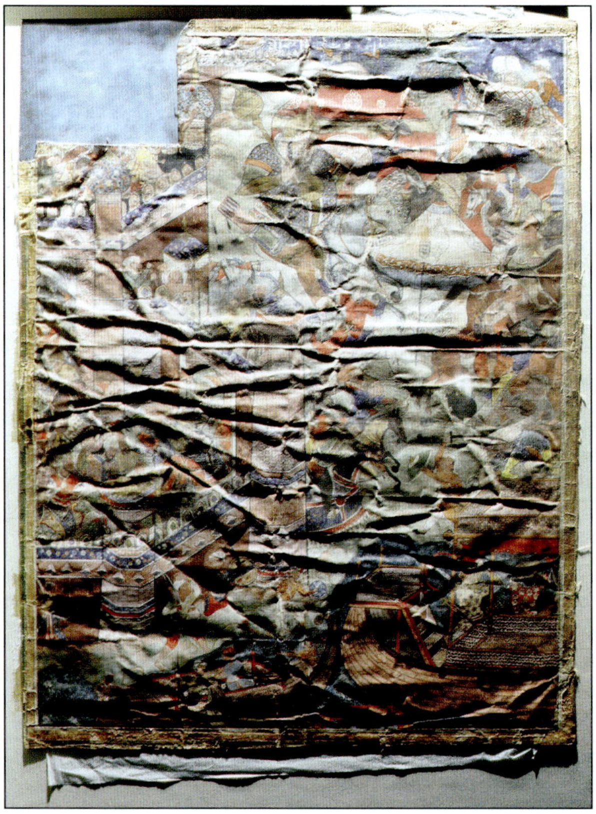

Figure 4.30 Mughal, *Hamzanama*, 1562–77. A heavily distorted page from an oversize manuscript on cloth backed with paper. By courtesy of the Board of Trustees of the Victoria and Albert Museum, London. (Photograph by M. J. Wheeler.)

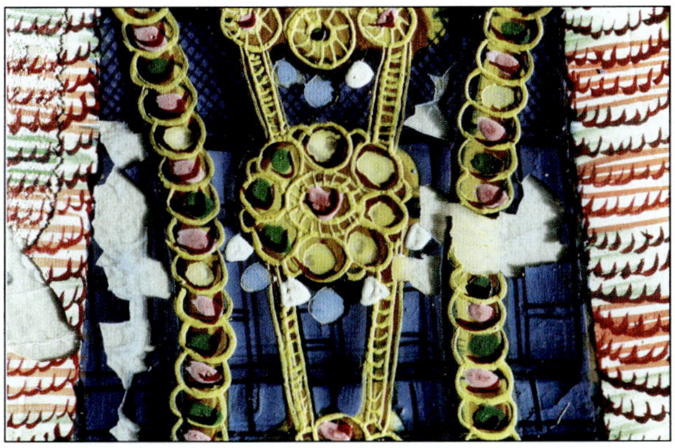

Figure 4.31 Thanjavur, South India, mica painting, 1860 (detail). Mica's smooth surface does not always provide sufficient tooth to adhere the paint successfully to the support. By courtesy of the Board of Trustees of the Victoria and Albert Museum, London. (Photograph by M. J. Wheeler.)

only a limited amount of tooth to encourage the paint to remain adhered to the paper. As a paper laminate of two or more sheets expands and contracts at one rate, the pigment layer is subjected to stresses that can cause cracking of the paint film and occasionally flaking. Many Indian paintings in the V&A's collection show signs of cockling and distortion, which are often regarded as characteristic features. However, extreme deformation can lead to abrasion and loss of pigment, especially in instances where the paint surface rubs against the mount or touches the glazing. The problem is compounded in the case of the *Hamzanama* (1562–77), a large Mughal manuscript painted on cotton cloth backed with paper. The cotton and paper expand at slightly different rates, which causes extreme distortion over time and leads to extensive loss of pigment (see Fig. 4.30).

Indian scroll paintings are often composed of many sheets of short-fibered papers that have been glued or sewn together to form a continuous roll. The papers used for these scrolls may be quite fragile, and it is common to find recycled or machine-made papers used for some modern scrolls. Such papers are easily torn by the handling they receive during storytelling sessions, and most of the examples in the V&A's collection show evidence of this type of damage along the edges of the sheets. Frequently the verso of a scroll shows signs of previous repair using recycled papers, or in some cases even documents relating to the storyteller. Some of the more robust scrolls are backed with cotton cloth, which affords them some resistance to mechanical damage but may also result in slightly decreased flexibility.

Cotton cloth is commonly used as a support for most of the thangkas in the collection. This type of support is usually more resilient than paper to the stresses imposed during rolling and unrolling, but may still be subject to deformation as a result of long periods of continuous hanging during its lifetime as a devotional object.

The V&A has a collection of approximately 700 paintings on mica from India. Mica is a mineral composed of silicates of aluminum and potassium. The

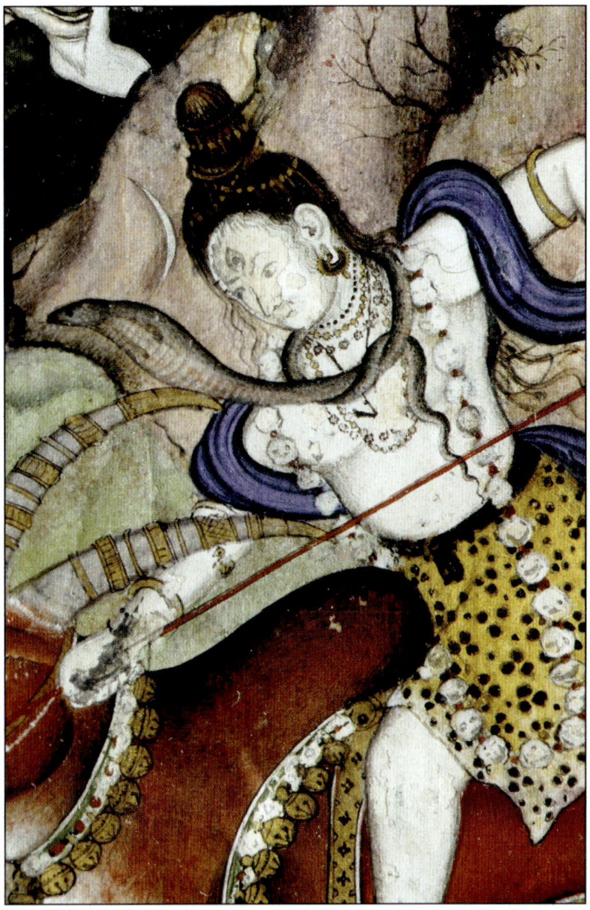

Figure 4.32 Mughal, *Harivamsa*, 1590 (detail showing areas of paint loss with underdrawing visible beneath). By courtesy of the Board of Trustees of the Victoria and Albert Museum, London. (Photograph by M. J. Wheeler.)

variety of mica used for the majority of paintings in the collection is muscovite. The use of mica as a support for paintings gained popularity in the mid-nineteenth century. Examples in the V&A's collection depict Indian flora and fauna, as well as tradespeople and religious figures. As mica is not a hygroscopic material, it does not expand and contract in response to changes in temperature and humidity. These works often suffer from problems of poor adhesion of the water-based paint to the smooth mica surface (see Fig. 4.31). Additionally, mica is prone to delamination at the edges of the support with a resulting decrease in transparency. When used as a painting support, mica may also be split into very thin laminates, which are extremely fragile and can easily be torn or broken if mishandled.

Pigments

Many of the Indian paintings have flaking or powdery pigments that require consolidation. In the case of a few Indian court miniatures, pearls and beetle wings were applied to the surface of the paper in imitation of jewels, giving the painted surface a pleasing three-dimensional appearance, but these too may be prone to abrasion and detachment from the paper surface. Paint deterioration may be caused by combinations of factors, including a lack of binding medium, poor cohesiveness among multiple layers of paint, the use of burnishing to create decorative effects, and the incompatibility of the paint media with the paper or cloth supports. The problems of pigment loss and abrasion are compounded in the case of rolled objects (thangkas and scrolls) of composite materials where the flexing of the paint film can lead to further damage. Loss of pigment may also result from flexing due to incorrect handling and, to a lesser extent, cockling of laminated paper supports as a result of fluctuations in levels of relative humidity (RH). This situation may be exacerbated when the painting is inlaid, or adhered onto another type of paper as part of the original method of manuscript production – a not uncommon practice during the Mughal era, when the ruler had teams of artists working on massive projects in several centers throughout his empire. Paintings and text of a single manuscript might be painted in different studios and only joined together at some later time. The grain direction of the papers and backings used will often vary from one painting to another. The *Harivamsa* manuscript (Mughal, 1590), for example, has several pages that are extremely distorted as a result of this type of assembly technique; the distortion has resulted in extensive paint loss (see Fig. 4.32).

The chemical stability of the pigments selected by an artist may also affect the longevity of the work. Occasionally pigments may interact with one another, causing alteration in the appearance of the painting. Lead white is prone to darkening over time as a result of exposure to atmospheric sulfur, or in certain rare instances due to cross-contamination from other pigments that contain sulfur (such as vermilion or orpiment). Silver paints composed of silver-leaf ground together with gum arabic may tarnish for the same reasons. On the Indian subcontinent, orpiment was used as a yellow pigment up until the sixteenth century, when it was largely superseded by Indian yellow. Orpiment may darken with exposure to atmospheric pollution, and sometimes decolorizes with prolonged exposure to light (Watt 1999). The mechanism by which this occurs is not completely understood, but examples can be seen in several paintings from the collections at the V&A. A page from the *Baburnama* of 1587 shows a decolorized area of orpiment on the leg of one of the horsemen. This folio also shows tarnishing of the silver paint used to embellish the horses' coats (Fig. 4.33).

In general, the relatively thick layers of opaque watercolor used to create Indian miniatures are more resistant to visible fading by light than many of the equivalents used in Western watercolor paintings. Both the good craftsmanship of the artist and the conservative palette favored by the painters help to contribute to the relative light stability of these works of art. A photograph (taken in 1933) in the V&A picture library of one of the galleries of the South Kensington Indian Museum, shows the entire *Akbarnama* series of paintings on display in galleries illuminated with high levels of natural light.

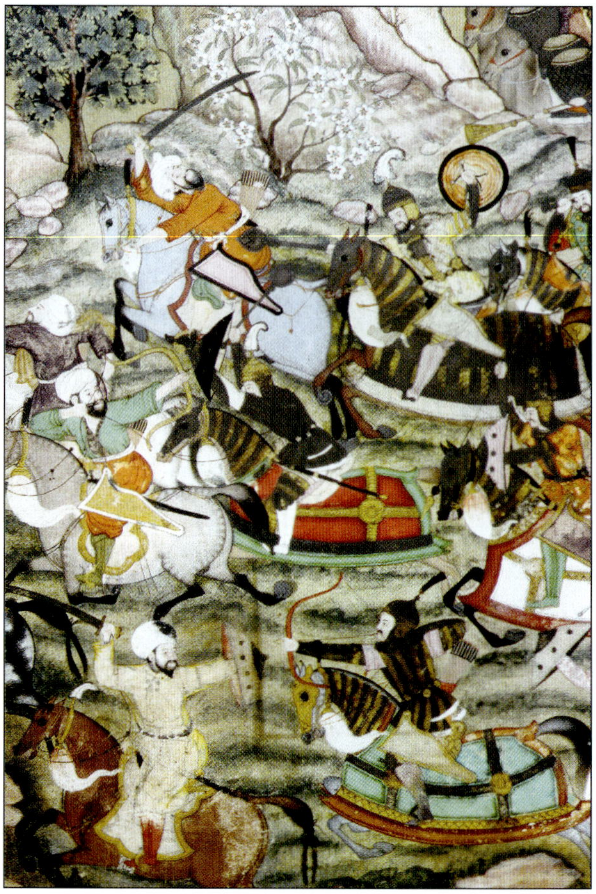

Figure 4.33 Mughal, *Baburnama*, 1587 (detail of page showing areas of tarnished silver leaf and decolorized orpiment). By courtesy of the Board of Trustees of the Victoria and Albert Museum, London. (Photograph by M. J. Wheeler.)

From records kept by the museum, it is clear that they were continuously on display for 15 or 20 years, yet there are few signs that any of the pigments have lost their intensity. This is partly because the pigments selected for use are mostly inorganic, mineral pigments applied in thick layers.

Color shifts may occur when mixtures of pigments fade preferentially. One type of green favored by Indian court painters was composed of a mixture of Indian yellow and indigo in varying proportions. Exposure to light in some cases has caused alterations in the hue and intensity of the paint, which have interfered with the original color harmonies created by the artist. The resulting appearance of the painting may be somewhat disconcerting, especially when foliage that was originally intended to appear green has faded to blue.

Conservation techniques used at the V&A

Conservation procedures devised for Indian paintings are increasingly biased toward considering the works as historical documents, not merely as painted surfaces. Treatment tends to be conservative in nature and non-invasive whenever possible. It may be confined to pigment consolidation and removal of unsightly old repairs that obscure the image or the inscriptions. Occasionally previous restoration or areas of repainting are retained, as these may constitute valuable evidence for future research.

Treatment of flaking pigment is frequently necessary on many different types of Indian paintings. After careful examination with a binocular microscope, areas of unstable paint may be consolidated with repeated local brush applications of a low-viscosity grade of methyl cellulose (0.5% solution), or ethyl hydroxyethyl cellulose (Bermocoll in a 1% solution). In the case of both of these consolidants, the viscosity is altered slightly depending on the type of application and the ambient temperature. Unfortunately in some instances the paint layers may be so poorly adhered to the surface of the paper that it is impossible to touch the surface with a brush without dislodging flakes of pigment. To cope with such situations, it is useful to first apply a very dilute skin of consolidant over the affected areas using a nebulizer. The authors have found ethyl hydroxyethyl cellulose especially suited for this purpose, and occasionally small quantities of ethanol are added to the consolidant to assist penetration. Wherever possible, the painting is gently humidified before commencing, as this appears to greatly facilitate the effectiveness of the treatment.

The nebulizer was especially useful during treatment of a miniature portrait painting of Tipu Sultan (Mysore, 1790). In this case the paint was in a highly vulnerable condition, and it was not possible to carry out local consolidation with a brush without disturbing the underlying pigment particles. Previously, there was no easy method to deal with such cases, and it was essential to anchor the paint particles without touching the surface of the miniature. Overall humidification of the miniature was not possible in this instance, nor desirable, given the very fragile, three-dimensional nature of the paint surface. Instead very small amounts of an aqueous consolidant (1% ethyl hydroxyethyl cellulose) were applied using the nebulizer while observing the effect on the painted surface with a binocular microscope (see Fig. 4.34). After each application, the surface was covered with felt and a small weight to encourage the paint flakes to re-adhere to the substrate.

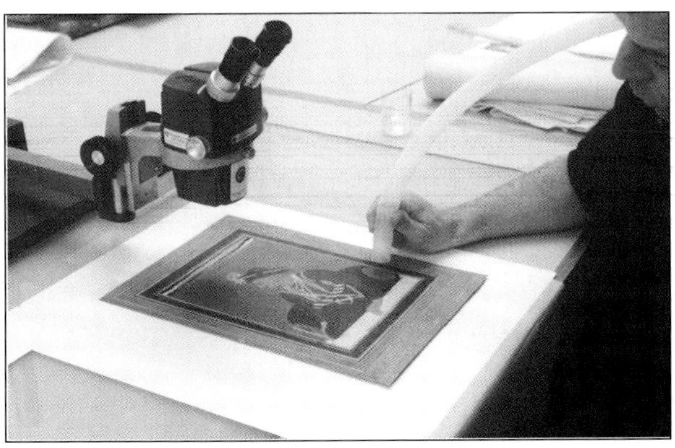

Figure 4.34 The nebulizer in use to consolidate flaking pigment on a portrait of Tipu Sultan, 1790. By courtesy of the Board of Trustees of the Victoria and Albert Museum, London. (Photograph by M. J. Wheeler.)

Old repairs can be removed with controlled local applications of moisture using water applied with a paintbrush, or with poultices of methyl cellulose. In some instances very small amounts of water vapor or steam can also be useful when trying to soften stubborn adhesives. However, with all these methods, the length of time the moisture is left in contact with the old repair is critical for safe removal. It is very easy to over-wet the repair and inadvertently skin the original paper laminate. Removal of residual adhesive can often be difficult because of the starch surface sizing used on many Indian papers. When removing any of these types of additions, roughen the surface of the repair first with fine abrasive paper and use a minimum of moisture to achieve the end result. This lessens the likelihood of skinning and localized distortion of the paper laminate.

Humidification and flattening are only carried out when absolutely necessary, as the results of this are often difficult to predict and may cause subtle alterations to the painted surface. However, if moisture is introduced it should be done gradually over a period of several hours to allow all of the layers of the paper laminate and the pigment to expand evenly. The use of Gore-Tex® is recommended for this purpose, or humidification over a bath of cold water on a nylon-mesh-covered strainer. After humidification the painting may be flattened by pressing between felts to retain the surface topography of the paint and to make allowance for the different thicknesses of the papers. Occasionally a miniature painting may require flattening on the *karibari* board, especially when the surface is particularly vulnerable to damage by pressing. In such cases the use of a strip lining of a medium-weight Japanese paper applied to the verso of the miniature with a 5% solution of Klucel G in acetone may be considered. The object is then gently humidified, attached by the strip lining to the board, and allowed to equilibrate with the ambient humidity in the studio for as long as possible.

When major losses of the paper support have occurred, they are repaired with Indian hand-made papers or European equivalents. The authors have found hand-made papers produced by Griffen Mill Papers to be especially suitable.[1] Alternatively, old Indian hand-made paper is used. Either of these types may be toned with watercolor, but no attempt is usually made to hide losses with inpainting. Frequently the replacement of missing portions of decorative borders may be sufficient to allow the miniature painting to be appreciated as a whole without the need for guesswork in trying to determine what was originally there. Wherever possible, inpainting is confined to repairs and patches. Occasionally pastel pencils (Carb Othello) are used to disguise heavily abraded areas on the image itself.

In the paper conservation department at the V&A, remedial conservation work of paintings on mica includes removal of old mounts and consolidation of pigments with local applications of a 5% solution of Paraloid B72 in acetone. Inpainting of paint losses is undertaken when they are thought to interfere with an appreciation of the image as a whole. Artists' pigments are mixed with a 5–10% solution of Paraloid B72 in a slow-evaporating organic solvent such as methyl proxitol. In some cases of excessive pigment loss, it has been proven effective to apply the retouchings to a sheet of clear polyester (125 micron Mylar®). The polyester sheet is then placed behind the object when it is mounted, which prevents the need to retouch directly onto the object. The paintings are then encapsulated in polyester envelopes prior to using transparent, self-adhesive corners to attach them into standard-size mounts.

The conservation procedures used for Indian scroll paintings are adapted from those used for the conservation of Japanese paintings; they include local repair, friction drying, and lining. Particular care is taken not to remove valuable documentary evidence from the backs of these objects, as these may assist with future dating of a painting. As machine-made papers of differing quality have been used for many of the examples in the V&A's collection, particular care has to be taken to ensure that any repair materials used have a similar coefficient of expansion as those used to make the scroll. Many of the scrolls are composed of multiple sections that are joined together by gluing or sewing and then backed with other papers or cotton cloth. The scrolls may originally have been assembled in such a way that the grain direction of the papers and cloth are not compatible.

Flattening of severely distorted scrolls is avoided whenever possible. However, when this is necessary the humidified scroll may be attached with edge-lining strips of Japanese paper to a *karibari* board. Alternatively, adaptation of a method used for flattening large objects that employs a sandwich technique (friction drying) may be effective (Webber and Huxtable 1987: 90).

Methods of display

Increasingly, emphasis is placed on using sympathetic methods of presentation to enhance the appearance and integrity of a damaged object. Missing textile borders from a thangka have been imitated by the use of a cloth-covered mount of the approximate dimensions of the original border. In the case of the *Akbarnama* (1590), a Mughal manuscript that had been separated from its original binding, a fascicule system has been devised to imitate the original bound format. In the case of the *Hamzanama* manuscript (a series of large-format paintings [600 x 800 mm] on a textile support backed with calligraphy on paper), a special mounting and storage solution was devised to allow the objects to be handled and shown safely. Individual folders were made for the objects in place of the original Solander boxes housing seven or eight items per box. These folders allow both the recto and verso of the pages to be seen without the need for a conventional window mount. Encapsulation of mica objects allows whole images to be shown without the need to apply adhesive or hinges to the objects themselves. This is also aesthetically beneficial, as most of the examples in the V&A's collection are relatively small in size.

Exhibition and loan criteria

The relatively thick applications of pigment on Indian miniatures are less prone to visible damage by light than thinly painted Western watercolors or Japanese prints. Regular rotations of Indian objects occur because exhibition space for the permanent collection is limited. The criteria for lending Indian paintings are often based more on the stability of the paint film and its perceived ability to stand up to the rigors of travel. Vibration, poor handling, and fluctuations in RH are the most likely causes of damage and loss of media. The risks involved in these and other loans of works of art on paper are assessed on an individual basis, assisted by written guidelines concerning maximum amounts of light exposure that should be received by different types of objects (Derbyshire and Ashley-Smith 1999). Extensive written condition reports together with detailed photographic records have proven essential in determining whether any damage has occurred. Inevitably, well-known works from the collection are requested repeatedly. Wherever possible, other works of a similar period or school are suggested for loan in their place.

Mounting and framing techniques

In certain cases, problems associated with flaking pigment in miniature paintings may determine the type of mounting and framing system chosen. Emphasis is placed on the use of frames with fillets and deep mounts (16-ply) to keep the pigment surface as far away from the glazing as possible. Occasionally glass is used in place of Plexiglas for in-house exhibitions when the pigment surface of a painting is thought to be particularly prone to flaking and might be adversely affected by static. Weighed against this is the obvious risk to the object in the unlikely case of breakage. Backing boards are screwed in place to minimize the shock and vibration to which an object is often subjected during the framing process. Miniature paintings are hinged into standard-size mounts and are displayed to show the entire image as well as any decorative borders. This is a slight departure from curatorial practice in the past, when decorative borders were often deliberately covered by mounts and were not considered an integral part of the object.

In the case of the *Akbarnama* series, the mounts are cut to show Persian inscriptions around the edges of the images. These folios are attached into their mounts using a fascicule system similar to that developed by Christopher Clarkson for the Bodleian Library (Lindsay and Clarkson 1994). This system was adopted because the folios were originally part of a bound volume which was disbound shortly after it was acquired by the V&A in the 1890s. In addition to providing a sympathetic format for the mounting of these miniatures, the fascicule system assists with the turning of the pages in the mounts to examine the text panels on the versos of most of the sheets. The fascicule is attached to the verso of the miniature along the left-hand edge using a thin Japanese paper (*kingami*) adhered with wheat starch paste. A thicker Japanese paper (*kozo hosakawa*) is used to attach the hinged object onto a backing paper of a third, stiffer paper (*bunkoshi*) (Wheeler and Battison 1997).[2]

The *Hamzanama* series of paintings were mounted and presented in a very different manner, in keeping with descriptions of their original use. These large paintings (600 x 800 mm) on cloth-paper laminates were intended to be held up and viewed by an audience while a storyteller read out the text written on the verso. As far as we can ascertain, they were never bound into book form, nor were they necessarily conceived as single-sheet paintings in the same manner as Indian court miniatures. The mounts devised for the *Hamzanama* paintings are intended to be as unobtrusive as possible and are similar in design to mounting systems used at the V&A for illuminated manuscript pages of parchment or vellum. The paintings are held in place with tiny Mylar® tabs that are attached to Plexiglas frames with Lascaux 360 HV acrylic adhesive and to the paintings with wheat starch paste and Japanese paper strips (Hillcoat-Imanishi et al. 1999).

Indian scroll paintings are either displayed flat in desk cases, or vertically mounted on rollers similar to those used for the display of wallpaper. Hand-scrolls may be displayed in wall cases at an angle. In both instances, the scroll is held firmly in position by Plexiglas clips at both ends. These fixings in turn are secured to the fabric-covered backing board with screws.

Storage systems

Indian miniature paintings present few problems from a storage point of view. They are presented in standard-size mounts and are stored in Solander boxes, which in turn are placed on shelves that form part of a large compact shelving unit. The exception to this rule is the *Hamzanama* series of paintings, which were previously stored in cumbersome oversize boxes that were difficult for one person to handle. As part of the rehousing scheme for these items, it was decided to make individual enclosures for each of the 27 paintings, which would allow them to be handled safely by one person, rather than in conventional window mounts. It was agreed that the design of the enclosures should be sympathetic to what was thought to be the original method of viewing, allowing the edges of the recto and verso to be seen. This system was inspired by a visit to the Museum of Applied Arts in Vienna in 1998 to view their collection of *Hamzanama* folios.

Mica paintings were previously mounted in pairs on poor-quality card with dabs of adhesive applied to the verso of the mica sheets. The cards were covered by clear plastic wrappers and stored vertically in filing cabinet drawers. Shortage of space caused over-compression of the objects, resulting in damage to the mica and the paint layers. When the paintings were rehoused, they were encapsulated before being attached into conventional mounts with transparent corners. In most cases, two or four items were included in each mount to save space. These mounts were then placed in standard-size Solander boxes, and stored horizontally on shelves.

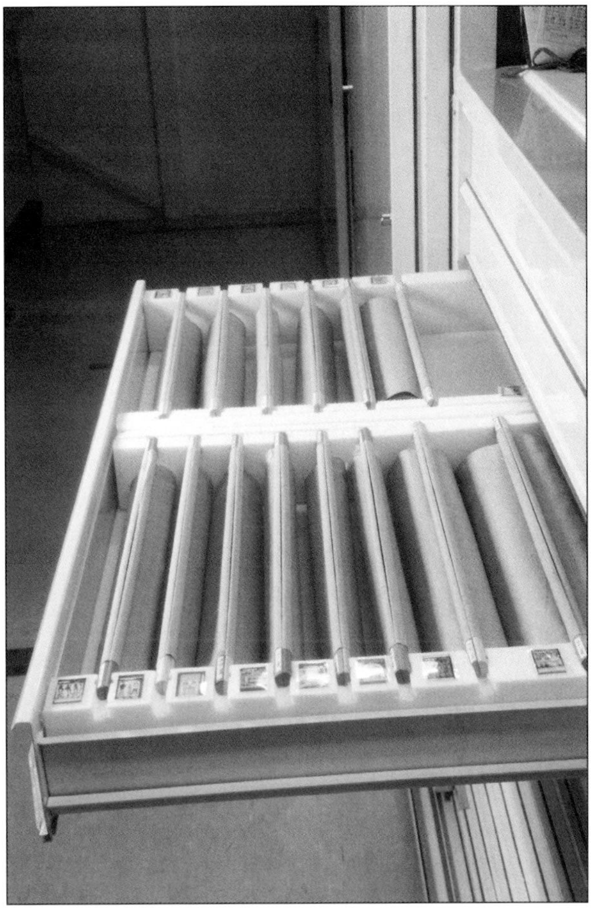

Figure 4.35 Storage system for Indian scroll paintings in the Indian Study Room at the Victoria and Albert Museum. By courtesy of the Board of Trustees of the Victoria and Albert Museum, London. (Photograph by M. J. Wheeler.)

A more specialized system was devised for long-term storage of Indian scrolls. They are suspended in standard-size cloth hammocks in drawers that are part of a compact shelving system. This concept is based on storage systems for hanging scrolls developed at the Library of Congress in Washington, D.C., and at the Museum of Ethnography in Leiden. The hammocks are made of premium-quality Irish linen that is glued to split wooden dowels using ethylvinyl acetate (EVA). The ends of these dowels fit into notches cut in thick Plastazote (ethafoam), which allow the scroll paintings to be supported evenly without touching the bottom of the drawer. To further reduce the wear and tear caused by rolling and unrolling the object, small contact-print-size color photographs are attached to the Plastazote so that it is possible to rapidly identify the scrolls and replace them in the correct position in the drawer (see Fig. 4.35).

Thangka storage has also been upgraded at the V&A. The relatively small size of the collection (250 items) allows most of the paintings to be stored flat in oversize plan chest drawers, lined with individual trays made of Plastazote. These trays allow the paintings to be lifted in and out of the drawers without exerting strain.

Many of the paintings in the Indian collection at the V&A are suffering from pigment loss and in some cases damage to the support. The structure of the supports and the pigments selected, together with the painting technique of the artist, determine the state of preservation of these unique objects. The state of preservation is further influenced by the handling and degree of use the paintings have received over time. Previously at the V&A, miniature paintings and thangkas attracted more academic interest than mica paintings or Indian scroll paintings, which were once regarded as folk curiosities and hence less worthy of study. However, as a result of recent conservation and remounting efforts, it is now possible to handle these objects more safely, which in turn allows them to be studied more easily. In recent times more attention has been focussed on the way in which the presentation of these objects affects viewers' appreciation of them as art objects. The use of sympathetic materials for display and storage help to preserve the objects' historical integrity while also enhancing their appearance.

Endnotes

1. Akbar brown and Akbar natural hand-made papers are produced by Griffen Mill Papers, Griffen Mill, Island, Ballyhaunis, County Mayo, Eire.

2. Japanese *bunkoshi* and *kingami* papers are supplied by Falkiner Fine Papers, 76 Southampton Row, London WC1B 4AR. *Kozo hosakawa* is supplied by Masumi Corporation, 4-5-2 Sugamo, Tokyo, Japan.

References

Bailey, G. A. 1997. "A palette of princes: An unpublished source on Indian pigments," *South Asian Studies* 13, pp. 45–54.

Clare, J., and F. Marsh 1979. "A dry repair method for Islamic illuminated manuscript leaves," *The Paper Conservator* 4, pp. 3–9.

Daniels, V. 1994. *A Preliminary Investigation of Princes of the House of Timur. Hummayans Garden Party.* New Delhi: Marg Publications, pp. 97–106.

Derbyshire, A., and J. Ashley-Smith 1999. *A Proposed Practical Lighting Policy for Works of Art on Paper at the V&A.* ICOM Committee for Conservation Preprints 1, pp. 38–41.

Duffy, K. I., and J. Elgar 1995. "An investigation of palette and colour notations used to create a set of Tibetan thangkas," in A. Wallert, E. Hermens and M. Peek (eds), *Historical Painting Techniques, Materials, and Studio Practice.* Preprints of a symposium held at the University of Leiden, The Netherlands, 26–29 June 1995, pp. 78–84.

Flemming, L. E. 1986. "Conservation of Oriental miniatures," *Orientations* (Aug.), pp. 51–57.

Hillcoat-Imanishi, A., P. Webber, and M. Wheeler 1999. "Conservation, mounting and storage solutions for two *Hamzanama* folios," *The Paper Conservator* 23, pp. 26–35.

Isacco, D. 1997. "The ultraviolet method of analysis: A scientific approach to the study of Indian miniatures," *Artibus Asiae* 53, Issue 3/4.

Jain, J. 1998. *Picture Showmen. Insights into the Narrative Tradition in Indian Art.* New Delhi: Marg Publications 49, no. 3.

Johnson, B. 1970. "A preliminary study of the technique of Indian miniature painting," in P. Pal (ed.), *Aspects of Indian Art.* Papers presented in a symposium at the Los Angeles County Museum of Art, pp. 139–46.

Lindsay, H., and C. Clarkson 1994. "Housing single sheet material: The development of the fascicule system at the Bodleian Library," *The Paper Conservator* 18, pp. 40–48.

Owen, A. 1997. "Discovering new layers in the structure of the *Hamzanama* illustrated folios," in Jane Eagen (ed.), *IPC Conference Papers London*, pp. 64–69.

Roy, G. 1994. "Conservation of Indian Popular Painting in Bengal." Poster displayed at the International Institute for Conservation Conference, Ottawa.

Stevens, P. 1980. "Conservation of Indian and Moghul miniatures," *IPC Conference Papers Cambridge*, pp. 157–58.

Watt, M. 1999. "Arsenic and old paint," *New Scientist* 29 May 1999, p. 20.

Webber, P., and M. Huxtable 1987. "Some adaptations of Oriental techniques and materials used in the Prints and Drawings Conservation Department at the Victoria and Albert Museum," *The Paper Conservator* 11, p. 90.

Wheeler, M., and C. Battison 1997. "The conservation and mounting of leaves from the *Akbarbama*," *V&A Conservation Journal* 24, pp. 14–17.

CHAPTER 5
THE ASSESSMENT OF FADING

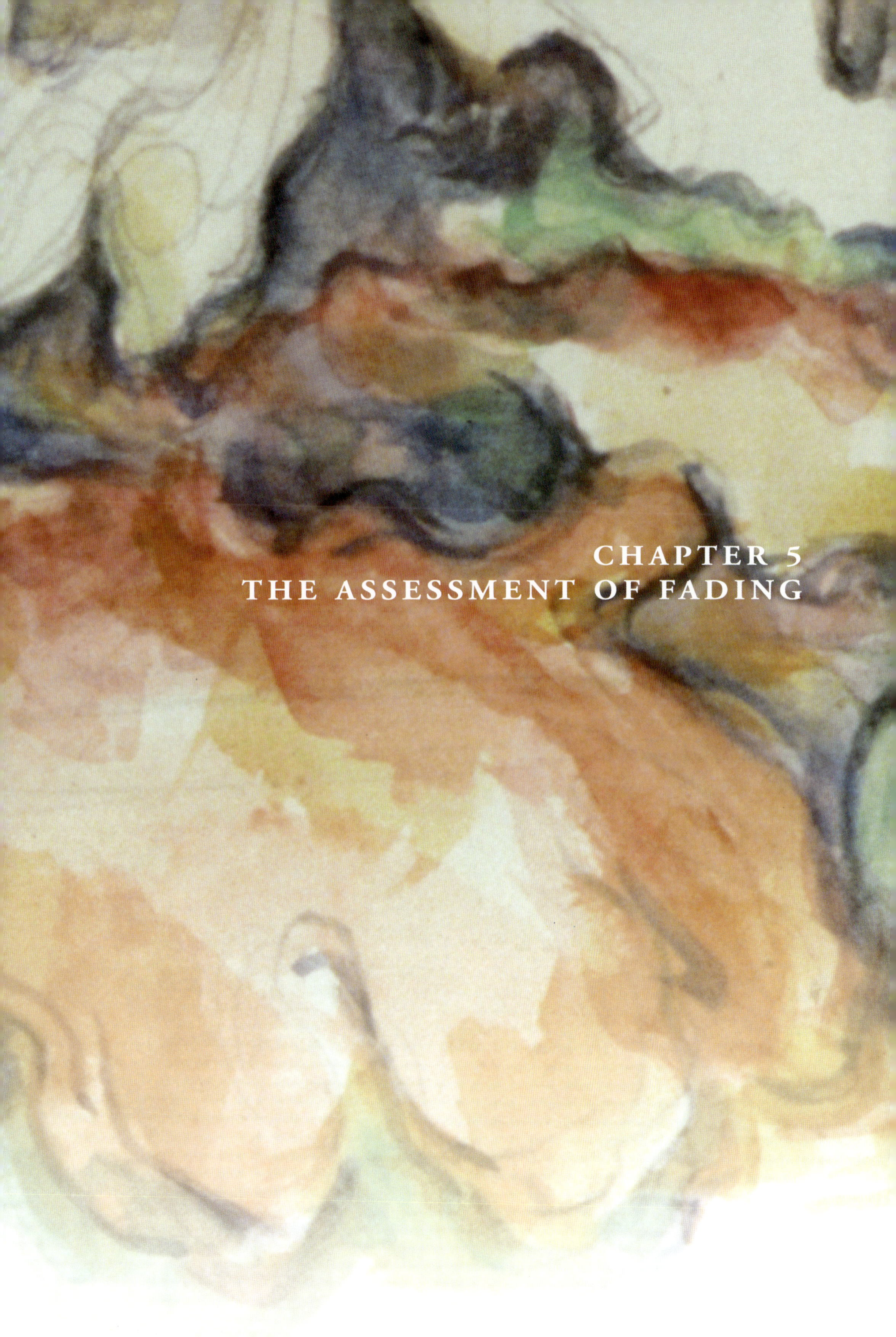

Statistics without Anesthesia: Interpreting Color Data

Roy Perkinson

Few words are more familiar and yet more misunderstood than the word "color." Because it is so familiar, it is easy to assign to it certain meanings that are inappropriate. The phrase "absolute color" is especially seductive: it is difficult to remember that the color of an object is not one of its intrinsic qualities. The word "color" encompasses three factors. The first is the physical characteristics of the object, and they are, of course, fundamentally important because electromagnetic radiation in the form of light interacts with, and is modified by, the material substance of the object. But two additional factors are inextricably linked to the real meaning of color: the characteristics of the illumination, and the nature of the viewer. These last two factors are especially relevant when considering instruments used to record color.[1]

Let's consider the first factor – the way in which a material interacts with electromagnetic radiation. A precise and fundamental way to characterize this aspect of a colored object is to record its *spectral data*: the relative intensity of electromagnetic radiation reflected from (or transmitted through) an object at each wavelength within a particular range. Spectral data reveal exactly how a source of radiation (whether within or outside a range of wavelengths visible to the eye) is changed by interaction with the object.[2] Instruments that detect and record spectral data are called spectrophotometers. A graph produced by a spectrophotometer gives important information about an object. It shows us precisely which wavelengths of radiation are reflected or transmitted by an object, independently of the kind of illumination and the characteristics of the viewer. Therefore, spectral data represent fundamental properties of the object.

Unfortunately, a graph of spectral data may be unsatisfying, especially when we are trying to determine whether the color of an art object has changed from last year to this. To achieve a more comprehensible and useful answer to such a question, it is necessary to consider the two other factors mentioned previously that are independent of the object itself. Further, if we want to have some way of quantifying color, it becomes necessary to use one or another of the various color scales or color spaces that have been developed.

In 1905 Albert H. Munsell developed one of the earliest color-ordering systems or scales, based on the characteristics of human vision. He assigned numerical values to three properties of color: hue, value, and chroma. The basis of this system is an extensive set of physical color samples, differing by equal intervals of visual perception. While the Munsell system is still in use today, because it relies on physical samples and the judgment of individual viewers it has been largely superseded by instruments that use mathematical color systems.

The basis for most of these modern, mathematical systems was officially promulgated in 1931 by the Commission Internationale de l'Eclairage (CIE). A feature of this system is the idea of a "standard observer," which was determined empirically from studies of human population. This approach sought to provide a means for converting a spectral curve into three numbers, known as tristimulus values XYZ, that identify any color. The term "tristimulus" refers to methods for communicating or generating a color using three stimuli: either additive or subtractive colorants such as RGB (red, green, blue) or CMY (cyan, magenta, yellow), or three attributes such as lightness, chroma, and hue.[3] Because these number do not, by themselves, correlate well with visual attributes, the CIE recommended further transforming them into three other numbers called chromaticity values or coordinates, Y, x, y.

Although the diagram of the 1931 CIE chromaticity coordinates will be familiar to many readers, anyone who has tried to use it to express changes in color is likely to agree that it has serious limitations.[4] For this reason, attempts have been made to transform the chromaticity coordinates into a color system that would be more "user-friendly" – one that would represent the characteristics of human vision in a more comprehensible form. The basis of these newer systems is the theory of "opponent-colors," according to which a color cannot simultaneously be both blue and yellow, nor red and green. One such system is the 1976 CIE L*a*b* color space.[5] The coordinate L* specifies how dark or light a color is, varying from 100 (pure white) to 0 (pure black). Positive or negative values of a* indicate the extent to which a color tends toward red or green; similarly, positive or negative values of b* represent tendencies toward yellow or blue. (In some respects this color space is reminiscent of the Munsell system, in that each system is envisioned in three-dimensional space, with three independent variables, one of which expresses how dark or light a color is, while taken together the other two relate, loosely speaking, to hue and intensity.) Because the CIE L*a*b* color space is relatively comprehensible, it is now widely used and has found its way into more and more conservation literature.[6]

Use of the mathematical color scales mentioned above presupposes that a color-measuring instrument is available. Although large, expensive devices for scientific measurement of color have been available for many years, within the past decade small, portable, and relatively affordable units have been developed that are appropriate for conservation studios. Instruments for gathering data on color can be divided into two groups: colorimeters and spectrophotometers. A colorimeter can be regarded as a substitute for the human eye. As a "psychophysical" instrument, a colorimeter assesses what it "sees" using a scheme that relates to the characteristics of human color vision. These devices filter light reflected from the subject into its dominant regions of red, blue, and green. Three sensors, one for each region, quantify the results. Such an instrument can help record the perceived color of an object, but it provides no information about which wavelengths of light were reflected by the object, and in what amounts, to produce the sensation of the specific color as perceived by the viewer.

A spectrophotometer, on the other hand, measures the relative intensity of electromagnetic radiation reflected from (or transmitted by) an object over a particular range of wavelengths. Some spectrophotometers are able to measure radiation that is invisible to the eye, for example that in the infrared or ultraviolet regions. Thus, a spectrophotometer assesses not only visible but also invisible light in more fundamental physical terms than a colorimeter. In addition to providing this information, the typical spectrophotometer can also present data in at least as many color systems as an ordinary colorimeter.

In 1986 I purchased our first device for recording color, the Minolta CR121 colorimeter, and have used it not only in conjunction with experiments but also to help detect and document color changes in art objects ranging from aquatints by Mary Cassatt to color photographs by Josef Karsh. This instrument is strictly a colorimeter, but subsequent to its purchase several other manufacturers also entered the market for portable color measuring instruments. For example, in addition to Minolta, X-Rite, Ocean Optics, and Gretag-MacBeth currently produce spectrophotometers that are far more versatile and useful than our original unit.

When deciding which colorimeter or spectrophotometer to purchase, it is helpful to consider the following:

1) Which category of device to choose: colorimeter or spectrophotometer? We expect to replace our existing colorimeter with a spectrophotometer so that we can record data not only in one of the color systems but also as pure spectral data.[7]
2) Do you want a completely portable unit that contains its own software and storage for data, or would it be acceptable to have to connect the unit to a computer (either desktop or laptop)? We will be choosing the latter because it is smaller and less expensive than the completely portable unit.
3) What geometry will you select for your unit: 0°/45°? Spherical? We will be continuing with 0°/45°.[8]
4) What size aperture is appropriate for the kind of measurements you will make? Our old colorimeter had a 3 mm diameter aperture, while the diameter of the new unit is 4 mm. Some makers offer more than one size, but this flexibility may be more costly.
5) What color system will you use? We use the CIE L*a*b* system.[9]
6) When making measurements in CIE L*a*b*, which CIE standard illuminant will you select? We always use the D65 illuminant, as suggested to me in conversation with Max Saltzman. This standard illuminant represents a color temperature of 6500°K, corresponding to what is described as "diffuse daylight."[10]

After obtaining an instrument, a question remains: how to use the data to discover what one really wants to know. In industry, instruments for measuring color are often used to ensure that the color of products is the same from batch to batch, or to aid in formulating colorants to match an existing sample. In the field of paper conservation, however, we are more likely to use such devices for monitoring possible fading of fugitive pigments and dyes. Or we might also want to detect color reversion in paper supports, that is, the recurrence of stains or overall darkening that might have been removed or reduced through conservation treatment. Therefore, let us turn to how to assess the measurements we have obtained from our color instruments.

Because we generally use the CIE L*a*b* color system, the result of our measurements will be three numbers: the values of L*, a*, and b*. We make several measurements in order to enhance the accuracy of the results, and the instrument automatically calculates the average values of L*, a*, and b*. How do you use these numbers to find out whether the color has changed since, say, last year? It would be wonderful if all that was necessary was to compare the average values today with those you obtained last year. Unfortunately, it's not that simple.

If several measurements have been made, something called the "standard deviation" of the measurements can be calculated. If you were ever subjected to classes in physics, this phrase may have the power to paralyze you. The professor turns and says, "Why, Mr. Perkinson, the standard deviation in your result is much too large, so you must stay after class and redo the experiment."

All of us have seen numbers that are followed by a plus and minus value, as in "83.6 ± 6.1." That numerical suffix, the plus or minus value, is the standard deviation of a group of measurements.[11] Can this statistic help answer the question about whether a color has faded? To simplify the discussion, let's look only at the value of L*, the brightness or darkness of a color. Suppose that a color was measured a year ago and the following average of the measurements of L* (I'll call it "L_{OLD}") was determined:

$$L_{OLD} = 70 \pm 4.1$$

Measurements made today provide a new *average* value:

$$L_{NEW} = 76 \pm 3.4$$

It is tempting – but incorrect – to conclude that if the plus or minus numbers of the old and new averages overlapped each other, as in this example, the two averages could then be regarded as essentially the same. To properly assess whether there is really a difference between the two, it is necessary to turn to the methods of statistics. Because I am not a statistician, I hoped that a simple, comprehensible protocol for statistical analysis could be found. With the generous help of statisticians, I developed the following two-step procedure that may spare other conservators from struggling with this same issue.

To understand this two-step method, let us look again at the phrase "standard deviation." What does it really mean? Rather than define it in the same way as statisticians – which I might garble anyway – I suggest that it should be thought of simply as an indication of how dispersed or clustered the data are. If a set of measurements all have very similar values, the standard deviation or dispersion of the data will be relatively small. On the other hand, if the measurements differ markedly one from the other, the standard deviation will be large. Furthermore, it will probably come as no surprise that comparing two sets of data could be very difficult if the measurements in one set vary enormously while those in the other set cluster neatly together.

Thus, the first step of this two-step protocol is to compare how dispersed or clustered your measurements are. This comparison is called a "test for equal variances." It may help in remembering this to think that the word "variances" suggests that what you are testing is the "variability" of your data, or, in other words, how scattered the data are in your two sets of measurements. Are you, in other words, trying to compare apples and apples, or apples and eggplants?

To conduct a test for variances, calculate the ratio of the squares of the two standard deviations and determine whether the ratio is higher than a certain number, which in this instance (with *exactly* five measurements for both sets of data) is 6.39. If, as the result of this test, you determine that the sets of data are indeed similar enough with respect to how the individual measurements are dispersed, you are in luck and you can proceed to step two. If, however, your data "flunk" the test for variances, you will need to confer with a full-fledged statistician.[12] Needless to say, I hope you never find yourself in this predicament. Let's be optimistic, therefore, and assume that your data are well behaved and that you can proceed to step two.

The second step involves the use of another statistical procedure called "Student's t test."[13] Ordinarily statisticians are happier with samples of 30 or more measurements, but collecting this much data can be very time-consuming. The t test is designed for situations where there are fewer than 30 measurements – a so-called small sample – and is therefore useful for those of us who have more to do than run a color-measuring instrument all day. Remember that this is a test you may use only if your data have passed the previous test otherwise it is inappropriate.

Discussions with statisticians led to a useful observation. One version of this formula for the t test looks like this:[14]

$$t_{n_1+n_2-2} = \frac{abs.(L_{OLD} - L_{NEW})}{\sqrt{\left(\frac{s_1^2 + s_2^2}{2}\right)\left(\frac{1}{n_1} + \frac{1}{n_2}\right)}}$$

– t is the "t factor" to be determined by this test

– abs.($L_{OLD} - L_{NEW}$) stands for the absolute value of the difference between the averages of the old and new measurements, that is, the numerical value of the difference, without respect to whether the result is positive or negative

- s_1 and s_2 are the standard deviations of the old and new measurements
- n_1 and n_2 are the number of measurements in the old and new sets

If, however, we were to make sure that the same number of measurements would *always* be made each and every time data were collected on a particular color, it turns out that this rather daunting formula can be simplified. After some experimentation in the paper laboratory, we concluded that a routine of making *five* measurements each time gave us meaningful data in a manageable amount of time. This has become our basic practice whenever we make color measurements. If all the values of n are equal to 5 in the formula above, the formula above can be simplified as follows:

$$t = 2.236 \bullet \frac{abs.(L_{OLD} - L_{NEW})}{\sqrt{s_1^2 + s_2^2}}$$

How does this help us? Insert the actual values of L_{OLD} and L_{NEW} and the respective standard deviations in this formula, and carry out the calculations. There are tables of values relating to the t test, and under these specific circumstances (n is always equal to 5), it turns out that if the result of the calculations is *greater than 2.306*, it means that there is a considerable probability (95%) that the average of the new measurements is *different* from the old one. In other words, it is very likely that L* has changed. On the other hand, if the result is *equal to or less than 2.306*, there is not likely to be a real difference. Thus, even though you determined by your measurements that L_{OLD} and L_{NEW} appear to differ, this apparent difference is probably due to the normal, expected variability inherent in the process of making measurements. In other words, the difference you observed is merely due to chance.

To see how this two-step procedure works in practice, let's work through an actual example. Some time ago, after making five measurements of a color, I obtained the following results for values of L*. (For now, we will just look at L*, but the same procedure applies to a* and b*.)

93.02, 93.29, 93.25, 93.28, 93.31

The average of these five values, L_{OLD}, is 93.23. The standard deviation of these measurements (s_1) is 0.11. Two years later I made five more measurements at exactly the same location[15] and obtained the following data:

94.05, 94.11, 94.38, 94.12, 94.36

The average of these measurements, L_{NEW}, is 94.20, and the standard deviation (s_2) is 0.15. It is clear that there appears to be a difference between L_{OLD} and L_{NEW}, but we still need to determine whether the difference is real. This apparent difference could simply be due to some of the many variables involved in making measurements, such as the expected variations that can occur whenever a limited (as opposed to infinite) number of measurements are made, or inherent non-homogeneity in the artwork.

Step one: Perform the test for equal variances. As indicated previously, divide the square of the larger standard deviation, s_2^2, by the square of the smaller standard deviation, s_1^2, as follows:

$$\frac{s_2^2}{s_1^2} = \frac{(0.15)^2}{(0.11)^2} = \frac{0.0225}{0.0225} = 1.86$$

Because the result is less than 6.39, we can proceed to the next step.

Step two: Perform the t test, using the formula indicated previously.

$$t = 2.236 \bullet \frac{abs.(L_{OLD} - L_{NEW})}{\sqrt{s_1^2 + s_2^2}}$$

$$= 2.236 \bullet \frac{abs.(93.23 - 94.20)}{\sqrt{0.021 + 0.0225}}$$

$$= 2.236 \bullet \frac{0.97}{\sqrt{0.0346}}$$

$$= 2.236 \bullet \frac{0.97}{0.186}$$

$$= 11.66$$

Because the result is greater than 2.306, we can say that there probably is a real difference between the recent measurements of L* when compared with the older measurements. In practical terms, the color measured really has gotten lighter. To cast this in proper statistical language, one would say, "The difference between the two sets of measurements of L* is statistically significant." The word "significant" in this context does not necessarily imply "interesting" or "important," as it might in ordinary conversation.[16] In statistical parlance, a "statistically significant" difference between two measured values means that there is a 95% probability that the difference is real and did not arise by chance.[17]

If you want to try this for yourself, carry out the calculations for the two sets of measurements below. Some time ago, these data about L* were collected about a certain location on a piece of paper:

95.14, 95.53, 95.34, 95.27, 95.38

Recently, this same place was measured again, obtaining the following values of L*:

95.49, 95.50, 95.61, 95.25, 95.19

After you have carried out your calculations, check your results against the figures in the endnote.[18]

The two-step procedure described above should also be applied to the other two variables in the CIE L*a*b* system. Like L*, both a* and b* are independent variables. This means that it is always possible that you could discover a statistically significant difference between, for example, L* and b*, but not a*. In other words, the lightness (or darkness) and the blue (or yellow) component might have changed, but the red (or green) component might not show a statistically significant difference. All you can do is report this fact.

Those readers who are familiar with the use of spreadsheets, such as Microsoft Excel, will agree that all of these calculations could be done much more efficiently by setting up a worksheet with the appropriate formulas and scripts. Some time ago I made such a spreadsheet for use in our laboratory. Recently, with the invaluable assistance of Roy Hudson (a volunteer consultant at the Museum of Fine Arts, Boston, and an expert in Excel), we set up an Excel template for distribution on the internet. If you would like a copy of this Excel file, direct your Web browser to one of the following two Web sites. Please note that for reasons neither I nor our Web master understand, we have noticed that after logging onto the PC site and while trying to save the file to our own computer, a cryptic dialogue box may appear that says something about "Read only ..." It seems that no matter which of the buttons is selected, the file can still be saved to the desktop (or wherever on your computer you would like to put it) and that it still opens normally once it has been saved. If all else fails, let me know and I will e-mail you a copy as an attachment.

For PC users:
URL:http://www.mfa.org/conservation/downloads/pccolor.xls

For Mac users:
URL:http://www.mfa.org/conservation/downloads/maccolor.xls[19]

Acknowledgments

I could not have done this work without the generous assistance of a number of people. I would like especially to thank Dr. Susan Milton (Radford University), Dr. Karl Hill (formerly of Wellesley College), and Roy Hudson. I am also grateful for the assistance of Phillip Getchell (Web master, Museum of Fine Arts, Boston), Dr. John McCann (formerly with Polaroid Corporation), and to Prof. Edward W. Merrill for his longstanding friendship and frequent advice.

Endnotes

1. The conference organizers asked me to provide an overview of some of the methods and instruments available for color measurement. I am by no means a color scientist, so for more complete information on the science of color measurement, the reader should consult one of the many references on color, color measurement, and the nature of human vision. An indispensable aid is F. W. Billmeyer and M. Saltzman, *Principles of Color Technology*, 2nd edn (New York: John Wiley & Sons, 1981), which is approachable by the novice and yet comprehensive. A less well-known book is Hazel Rossotti's *Colour: Why the World isn't Grey* (Princeton, N.J.: Princeton University Press, 1985). A chemist by training, but with a poet's temperament, Rossotti examines many facets of the phenomenon of color, from the physical chemistry of colored materials to the development of words for colors within various cultures (other sources of information follow in the Endnotes and References). The main purpose of this paper, however, is to share with the reader a protocol by which it can be determined whether a particular color has changed over time. To do this properly, some familiarity with statistics is required – a daunting and complex subject – and I hope that this information will spare others some of the difficulties this might entail.

2. X-Rite, Inc., *The Color Guide and Glossary* (Grandville, Mich.: X-Rite Inc., 1998), p. 47.

3. Ibid., p. 48.

4. Billmeyer and Saltzman 1981 (as in note 1), pp. 47–52.

5. For an illustration of this color space, see X-Rite, Inc., *A Guide to Understanding Color Communication* (Grandville, Mich.: X-Rite Inc., 1993), p. 11.

6. See, for example, P. Whitmore, G. R. Cass, and J. R. Druzik, "The ozone fading of traditional organic colorants on paper," *Journal of the American Institute for Conservation* 26, no. 1 (1987), pp. 45–57.

7. Billmeyer and Saltzman 1981 (as in note 1), Chapter 3, "Color and color-difference measurement," pp. 67 ff.

8. The X-Rite company has a useful document on their Web site – http://www.xrite.com/ – on the pros and cons of spherical vs. 0°/45° geometries.

9. Billmeyer and Saltzman 1981 (as in note 1), pp. 101 ff.

10. Ibid., p. 35.

11. Strictly speaking, there are two kinds of standard deviations: the standard deviation of a sample, and the standard deviation of the population. In this paper I always refer to the former. See D. Rowntree, *Statistics without Tears* (Needham Heights, Mass.: Allyn and Bacon, 1981), pp. 53 ff. If the color-measuring instrument does not produce this statistical result automatically, it can easily be calculated with many ordinary hand-held calculators or with the help of spreadsheet software, such as Microsoft Excel.

12. According to Dr. Susan Milton, if the result of the test for equal variances is greater than 6.39, a test called the "Smith-Satterthwaite t test" is appropriate for comparing the means.

13. This test was developed by William Gossett, who worked for the Guinness brewing company. Perhaps he wanted to be discreet about being employed by a brewer, so he published this technique under the pseudonym "Student."

14. This form of the t test is actually called the "pooled" t test, in which we estimate the common population variance (the square of the standard deviation) by "pooling" or combining the squares of the two standard deviations. Because in this special case the sample sizes are the same, the two standard deviations are averaged. I am grateful to Dr. Susan Milton for providing me with this information.

15. To ensure that measurements will be made in precisely the same location, we prepare a Mylar® template in which small holes are cut corresponding to each site. See Henry Wilhelm's description of this practice in his article "Monitoring the fading of color photographic prints," *Journal of the American Institute for Conservation* 21, no. 1 (1981), pp. 49–64.

16. See Rowntree 1981 (as in note 11), p. 116, on the use of the term "significant" in statistics. I am grateful to Dr. James Druzik for suggesting that it might be useful to extend one's ability to interpret the data beyond the results given by my two-step protocol. In private communication, he pointed out that a factor called the "p-statistic" (which for 95% confidence levels is 0.05 or less) can help give an idea

of how close to the 95% confidence level one might be. Here again, it would be helpful to consult a good text on statistics, or to seek the guidance of someone well versed in statistical methods.

17. If the statistical calculations are carried out at a level of 99% probability or confidence, the results can be said to be "statistically highly significant."

18. For this example, $L_{OLD} = 95.33$, and $L_{NEW} = 95.41$; the respective standard deviations $s_1 = 0.14$ and $s_2 = 0.17$. The test for equal variances gives the result of 1.47, which is less than 6.39, so the two sets of data are similar enough with regard to their variability to proceed with the t test. The result of calculating the t test is 0.81, which is less than 6.39. Therefore the data fail the t test and it is 95% probable that there is no real difference between the old and new measurements.

19. Note that the extension "xls" on the Mac version is required for some reason known only to our Web master.

References

Billmeyer, F. W., and P. J. Alessi 1981. "Assessment of color-measuring instruments," *Color Research and Application* 6, no. 4, pp. 195–202.

Comrey, A. L. 1975. *Elementary Statistics: A Problem Solving Approach.* Homewood, Ill.: The Dorsey Press.

Hunter, R. S., and R. W. Harold 1987. *The Measurement of Appearance,* 2nd edn. New York: John Wiley & Sons.

Kelly, K. L., and D. B. Judd 1976. *The Universal Color Language and Color Names Dictionary.* Washington, D.C.: National Bureau of Standards.

Popson, S. J. 1989. "A comparison of densitometers, reflectometers, and colorimeters," *TAPPI Journal* 72, no. 3, pp. 119–22.

Reedy, T. J., and C. L. Reedy 1988. *Statistical Analysis in Art Conservation Research.* Marina del Rey, Calif.: The Getty Conservation Institute.

The Use of the X-Rite Colortron II™ for Color Measurement of Watercolors

Barbara J. Mangum and
Arlen Heginbotham

The Isabella Stewart Gardner Museum was created by Isabella Stewart Gardner around the turn of the nineteenth century and consists of about 3,000 objects integrated into palatial galleries, all of which are centered around a spectacular central courtyard. Particular strengths of the collection are European paintings and works of art on paper, American paintings, Asiatic works of art, and classical sculpture. Among the several hundred works on paper are important works by Raphael, Michelangelo, Veronese, Whistler, Degas, Matisse, Zorn, and Sargent. When Gardner died in 1924, she left the collection for the education and enjoyment of the public forever, with the restriction that the collection remain intact and unchanged, and the objects remain placed in the positions she had selected for them. Although as many as half of the works of art on paper are displayed within protected cupboards, several rooms contain numerous watercolors on permanent exhibition.

In 1994 the National Endowment for the Arts awarded the museum funds to plan for the long-term monitoring of color permanency of objects in the collection. In order to understand and capture the change occurring in the art objects, a methodology was needed which would allow conservators to measure color accurately over a span of decades. Dr. Paul Whitmore, Director of the Research Center on the Materials of the Artist and Conservator at the Carnegie Mellon Research Institute, and Steven Weintraub, noted consultant in environmental conservation, were invited to the museum to consult on the development of a protocol for this project. The result was a decision to measure color over time using a reflectance spectrophotometer and to focus on works considered at highest risk. The advantage of the spectrophotometer over other color-measurement instruments such as the tristimulus meter was its ability to measure a physical property of the object rather than a relative property. That is, a spectrophotometer measures the absolute reflectance of electromagnetic radiation at a given wavelength, while a tristimulus meter measures the way a color appears to the human eye given a specific light source. Furthermore, a spectrophotometer can provide all the information of a tristimulus meter and more. Once collected, spectrophotometric data can be converted easily into tristimulus values, which can be useful for such things as expressing overall color change. By contrast, measurements made with a tristimulus meter cannot be extrapolated to provide a reflectance spectrum. In 1995 Roy Perkinson, Head of Paper Conservation at the neighboring Museum of Fine Arts, further spurred the staff to begin color measurement when he noted during a survey that our John Singer Sargent watercolors were in remarkably good condition given their permanent exhibition status.

The Colortron II™, a commercially available spectrophotometer and digital sensor, was purchased in late 1996 with support from the NEA (see Fig. 5.1). The instrument is a small, lightweight, 32-band reflectance spectrophotometer that records percent reflectance in

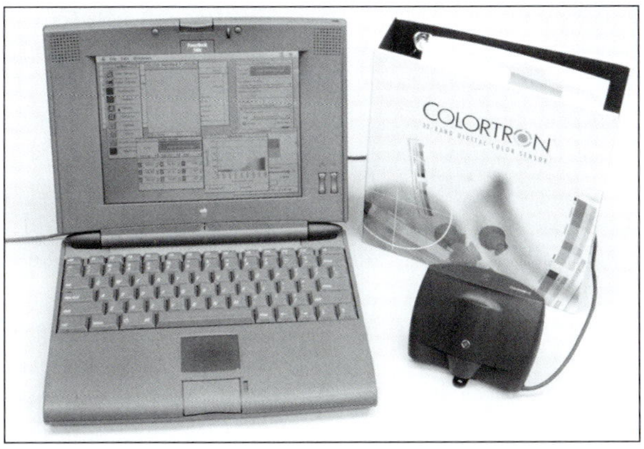

Figure 5.1 The Gardner Museum's Colortron II *User Manual* and 32-band digital spectrophotometer (right), and Macintosh Powerbook 540C (left).

10 nm bands from 390 nm to 700 nm. Designed for the desktop publishing industry, the Colortron II is a user-friendly, peripheral device for personal computers, and the results are presented through a well-designed software interface. Specific software functions or "tools" that conservators at the Gardner Museum have found very useful are the spectrum, colorimeter, and compare tools. Using the "spectrum" tool, results are plotted on a bar graph of percent reflectance vs. the 32 bands of wavelength. Unfortunately, the spectrum collected cannot be printed out using the software provided, nor can the numerical reflectance data be manipulated for statistical analysis. This shortcoming can be overcome by exporting the spectral data into a spreadsheet program such as Microsoft Excel. The "colorimeter" function translates the spectral data into any number of color space coordinates, including Commission Internationale de l'Eclairage (CIE) L*a*b* and xyY values. The "compare" function compares the L*a*b* readings between two colors and calculates the color difference value, ΔE. ΔE provides a convenient, single-variable measure of color change, which is based on how two colors *appear* to be different to a standard observer. Because the information provided by the Colortron II is captured digitally, it is possible to translate and correlate the data to color measurements made under a variety of illuminants, including standard illuminants such as D_{65} and D_{50}. This becomes important in establishing the accuracy of the data through measurement of reference standards. The instrument allows for measuring intensities up to 125% in the event that moderately fluorescent colors are scanned. However, it should be noted that many fluorescent colors, such as those that might be found in modern or contemporary works of art, could re-emit more than 125% of incident light at a given wavelength and would require other instrumentation for measurement.

Some other facts are important to know about the Colortron II. It is an instrument of 45°/0° orientation.[1] The sampling size is 3 x 3 mm. It weighs 640 g (about 6/10 lb.) and it uses the 2° standard observer (check whether this is adequate for your needs).[2] To obtain a reading it is necessary to make contact with the object, therefore this method of color measurement is not suitable for extremely fragile works of art or those with pronounced topography. The Colortron II costs approximately $1,000, making it an affordable option for smaller museums and institutions interested in undertaking a study of color change in their collections. At the Gardner Museum, the Colortron II is used in conjunction with a laptop Macintosh Powerbook 540C equipped with a color monitor; PC versions are also available (for specific system requirements visit the Web site, www.xrite.com). Although the instrument is designed to be used with batteries that recharge off the computer, it is often more convenient to use the accompanying AC adapter. In addition, a Kodak digital camera, Microsoft Excel, and Adobe Photoshop were used to perform the work described in this paper.

Investigation of color change in Sargent's *San Giuseppe di Castello*

In 1996 the authors began to develop a methodology to undertake studies of long-term color change in the museum's most vulnerable and important watercolors and works on paper. *San Giuseppe di Castello* (Fig. 5.2), a 1903 watercolor by John Singer Sargent (1857–1925), was chosen for the first study. It had been placed on permanent exhibition in the museum's Blue Room by Isabella Gardner soon after it was created. It remains there to this day and, for the purposes of this study, is considered representative of other watercolors in the collection. Readings were taken of seven areas of the painting in 1996 and again in 1999. The goal was to determine the success of the methodology after a significant lapse of time to see whether any color shifts had in fact occurred, and to present the use of the Colortron II to other conservators interested in affordable, long-term color monitoring.

Initial testing was undertaken to become comfortable with the instrument and software and to assess the degree of accuracy and repeatability of the measurements. Timed light-exposure tests were run on several Winsor &

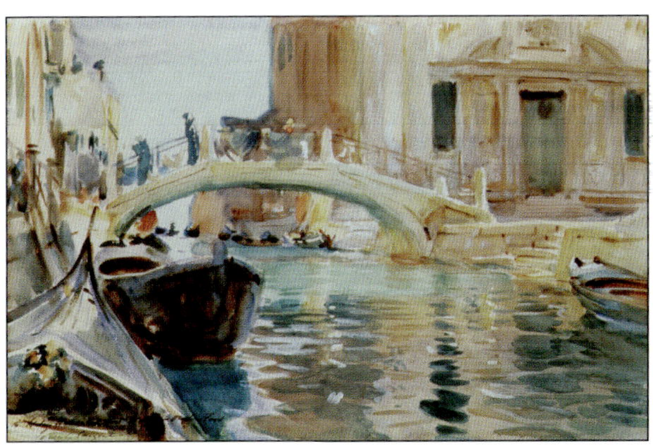

Figure 5.2 John Singer Sargent (American, 1857–1925), *San Giuseppe di Castello*, 1903. Watercolor on paper. Isabella Stewart Gardner Museum, Boston.

Newton watercolor samples painted on paper, and background information was obtained from the company to identify the coloring matter and permanence rating of the watercolors. Results conformed to expectations and confirmed the usefulness of the instrument. Other conservators might elect to run tests with even more fugitive colorants, such as those found in many felt-tipped pens.

Six different colorants used in the Sargent watercolor were selected for testing. Criteria for selection included the visual purity of color in the area, the uniformity and density of paint in the sample site and surrounds, and the ease of documenting and relocating the site. Finally, because the colorants were to some degree transparent, readings of the paper support were also taken. Documentation was accomplished by two means: 1) creation of a digital image of the artwork with an overlay of the Colortron II footprint showing the precise locations of both the sample site and the relative position of the device; and 2) written description. For easy location of the sample site and to protect the artwork from dirt that might cling to the bottom of the instrument, a Mylar® template was created. First, a tracing of the outline of the Colortron II's foot was made and positioned against a background grid of uniformly placed vertical and horizontal lines on Mylar®. A 5 x 5 mm aperture was cut into the Mylar® to allow the Colortron II to directly measure the artwork below. The 2 mil thickness of the Mylar® under the foot of the device affects the readings slightly, but it should not affect the accuracy of multiple readings provided the dimension is kept the same in all subsequent measurements. The template was positioned on the artwork with the aperture above the selected sample site and a digital photograph was taken of the template and the artwork below. By referring to this photograph during subsequent measurements, accurate repositioning of the Colortron II was found to be possible.

Calibration of the Colortron II

A reference target and "black box" accompany the instrument on purchase and are used for calibration – a process fully described in the manual and easy to do. The card is coded with the instrument's serial number and is used by the factory to set the performance characteristics of the particular instrument in question. It is therefore important to keep this card in good condition and to use it prior to measuring all other control standards and/or the artwork. An option called "Diagnostics" allows re-calibration of the colors yellow, magenta, and cyan. It is important to note that the instrument will pass diagnostics if the values read are within ΔE 4 of the values entered for the color patches in the internal software.

Accuracy and repeatability are key to the successful long-term monitoring of color. Over time, it is assumed that the electronics of the spectrophotometer will become corrupted and the light emitted by the tungsten bulbs will change in output due to deterioration of the filaments. Therefore, after the instrument is calibrated against the Colortron II reference target, readings are taken of a standard white ceramic tile. The Gardner Museum's white tile was purchased from Minolta Corporation in 1987 as a color control, and its color is documented in CIE xyY using a 2° standard observer and D_{65} standard illuminant. The difference in color readings between the two instruments is now kept in a log and reflects changes in the accuracy of the device over time. The white tile is also useful diagnostically. For example, unintelligible readings due to computer malfunction can often be seen immediately to be wrong on the computer screen.

A Graphic Communications Association (GCA) Lab-Ref™ calibration standard is also used to verify the accuracy of the Colortron II prior to each new series of readings. This reference card contains twelve color patches that have been measured using standard methods and whose accuracy is traceable to the standards of the National Institute of Standards and Technology. Verification using the Lab-Ref™ can help uncover subtle inaccuracies in the Colortron II readings even when the instrument is still passing diagnostics. An example of this occurred recently: the instrument appeared to be working accurately (passing the white tile test as well as diagnostics), yet on reading the Lab Ref™ card, errors of up to ΔE 8 were found consistently in the yellow test patch. The instrument was sent back to X-Rite, Inc., for service and a subtle misalignment of the beam was discovered. Again, it is suggested that a log of the readings be kept.[3]

Assessing repeatability

Making certain that the instrument is working properly is of obvious importance. Accuracy is one criterion that must be met by the instrumentation, but another is repeatability, or getting the same number time after time within certain acceptable limits. Factors contributing to the variability in the data include random errors and systematic errors. Random errors may arise from difficulty in characterizing events occurring on a molecular scale, or from local variations in temperature and humidity at the time of measurement, among other factors. The variability of the data can be revealed through calculation of the average and its standard deviation. For random errors, increasing the number of readings will usually improve the reliability of the data. Five readings have been recommended as a minimum.[4] Because the software does not calculate standard deviation, however, to analyze the data it is necessary to import the spectral data into a spreadsheet such as Microsoft Excel. Many such spreadsheets contain functions that will automatically calculate the standard deviation for lists of data and can also be used to graph the difference between sets of data over time.

The other type of error to consider is systematic error. Variations in color measurement may arise from a particular problem with the device or with the way different people collect the data. This is of great importance to the collection of long-term data, as many different operators and equipment are envisioned in the future. A test was devised to determine how repeatable the taking of measurements will be for conservators at the Gardner Museum in the future. Two conservators associated with the project and two others having no prior involvement were asked to locate the sample sites using the project documentation and to take readings. After two readings, they were asked to lift the instrument, reposition it, and take two more readings. This was done for a total of eight readings. Results showed that all the conservators obtained the same color reading ±1% at a 90% confidence level of standard deviation. This was considered acceptable (see Tables 5.1 and 5.2). Finally, to compare two data sets, it is important to know whether the variance in the sets is similar or not – only if they are similar can the comparison be considered meaningful.[5] The upshot is that data sets need to pass the "test for equal variances" and the "Student's t test" in order to determine that the calculated color difference is meaningful and not random.

	L*	a*	b*
Average	47.91	45.13	9.08
Standard Deviation	0.13	0.18	0.10

Table 5.1 Results of comparative readings of 1 color taken by 1 conservator 8 different times, repositioning the instrument 4 times.

	L*	a*	b*
Average	47.87	45.13	8.97
Standard Deviation	0.14	0.19	0.11

Table 5.2 Results of comparative readings of 1 color taken by 4 different conservators in a one-day period

Learning from our experience

In 1996, when this project began, the importance of verifying the accuracy of the instrument using the GCA Lab Ref card and the white tile in addition to the Colortron color reference target was not fully appreciated. Over time, the need for these further checks on accuracy has become clear, but the reliability of the original data for certain colors is now questioned. Therefore, the 1996 readings cannot be taken as a baseline for the project. They are offered here to show that the methodology is sound and leads to reasonable measurements between long periods of time (see Table 5.3). The method can now be used with confidence to include color documentation of Sargent's *San Giuseppe di Castello* and other works of art on paper. This should result in sounder exhibition policy overall and better preservation of these works of art over time.

Endnotes

1. Fred Bunting, "Measurement geometry," *Colortron™ User Manual*, 1st edn (Larkspur, Calif.: Light Source Computer Images, Inc., 1994), Chapter 8, p. 26. The two tungsten bulbs are placed at a 45° angle to the plane of the observer. This is suitable for works of art on paper and some paintings, but is not recommended for objects with a pronounced topography such as textiles. In that case, a spherical light source would be used.

2. Ibid., p. 37. This refers to a standard adopted by the CIE (Commission Internationale de l'Eclairage), to represent the color response of the average human eye. The standard proposed in 1931 is based on color-matching experiments using a 2° field of view. In 1964 a Supplementary Standard Observer was added based on a 10° field of

Sample	Date	L*	a*	b*	ΔE
Blue (bridge)	1996	64.36 ± 0.24	–4.93 ± 0.01	–15.83 ± 0.18	
	1999	65.62 ± 0.18	–6.13 ± 0.22	–14.88 ± 0.08	1.98
Paper support	1996	85.87 ± 0.14	0.46 ± 0.03	13.86 ± 0.05	
	1999	87.18 ± 0.05	0.57 ± 0.06	13.44 ± 0.27	1.38
Green (water)	1996	69.66 ± 0.07	–17.46 ± 0.05	7.17 ± 0.14	
	1999	70.07 ± 0.44	–17.74 ± 0.07	8.52 ± 0.09	1.44
Green (left side)	1996	69.73 ± 0.04	–9.52 ± 0.10	16.95 ± 0.32	
	1999	70.56 ± 0.30	–9.29 ± 0.09	16.96 ± 0.30	0.86
Wine	1996	52.81 ± 0.07	29.02 ± 0.13	28.03 ± 0.09	
	1999	53.55 ± 0.16	27.70 ± 0.25	26.22 ± 0.15	2.36
Sky blue	1996	79.74 ± 0.09	–1.72 ± 0.06	3.33 ± 0.22	
	1999	80.26 ± 0.01	–2.25 ± 0.06	2.36 ± 0.02	1.22
Sienna	1996	72.52 ± 0.09	10.09 ± 0.08	41.95 ± 0.15	
	1999	73.56 ± 0.19	10.10 ± 0.23	41.61 ± 0.38	1.10

Table 5.3 Comparative color readings of John Singer Sargent's *San Giuseppe di Castello* taken in 1996 and 1999.

view to better model the slightly different data obtained with a field of view wider than 4°. Unless otherwise specified, the CIE system assumes the 1931 2° standard observer.

3. The GCA Lab-Ref™ can be ordered from Graphic Communications Association, 100 Daingerfield Road, Alexandria, Va. 22314-2888; phone number: 703-519-8160.

4. Roy Perkinson, "Statistics without Anesthesia: Interpreting Color Data." This volume, pp. 230–35.

5. Ibid., p. 232–33.

Pursuing the Fugitive: Direct Measurement of Light Sensitivity with Micro-Fading Tests

Paul M. Whitmore

An essential element in the preservation of art objects is the protection of colors from damage due to light exposure. However, it has been extremely difficult to identify which objects have fugitive colors until fading has actually occurred, and uncertainty about fading risks has made it difficult to establish reasonable exhibition policies tailored to the needs of specific objects. Analysis of the material composition of an artifact is difficult, laborious, and often insufficient to define the details (such as pigment particle size or fading history) that can critically influence light stability. Alternatively, color monitoring programs (tracking the color changes in objects by color measurements repeated at intervals) can in principle detect very slight fading and provide warning of light sensitivity, but these efforts are also tedious, and the risk of fading damage depends on the precision and frequency of the measurements. There is a clear need for an alternative method that can rapidly assess the durability of an object's colors during exposure to light.

In response to this need, we have developed a new device – a micro-fading tester – that can perform accelerated light-aging tests rapidly and non-destructively on tiny areas of an art object. Details of the construction and performance of this instrument have been published, and the schematic design of the instrument is reproduced from that paper and shown in Figure 5.3.[1] In its essential function, this instrument is a reflectance spectrophotometer, capable of rapidly measuring the visible reflectance spectrum of very small (0.4 mm diameter) areas of an opaque colored material. The key feature of this device that allows it to serve as a fading tester is the high light intensity used for the probe beam of the reflectance measurement (about 7 million lux of visible light, provided by the focussed output of a filtered 75-watt xenon arc lamp). At these intensities, fugitive colors (Blue Wool scale #1 or #2) will fade measurably (CIE L*a*b* $\Delta E = 3$–5) within 5 minutes of exposure. Rapid spectral measurements and computation of color parameters and color differences are made possible by a photodiode array spectrophotometer controlled by a personal computer. During the test, the spectrum and color difference are updated every second, so that a "real-time" view of the course and extent of fading are provided. Thus, the fading test can be concluded after small but definite color changes have been produced, and before any perceptible change has occurred to the object. In our initial evaluation of this device, we studied the sensitivity of the measurements and precision of the fading test results. Those results (detailed in the above-mentioned paper) indicated that the instrument was capable of detecting very small color changes (about 0.1 ΔE unit), and the reproducibility of the test results on uniformly colored materials was very good.

Accuracy of test results

In evaluating the accuracy of the micro-fading test results, two questions are considered: 1) how the micro-

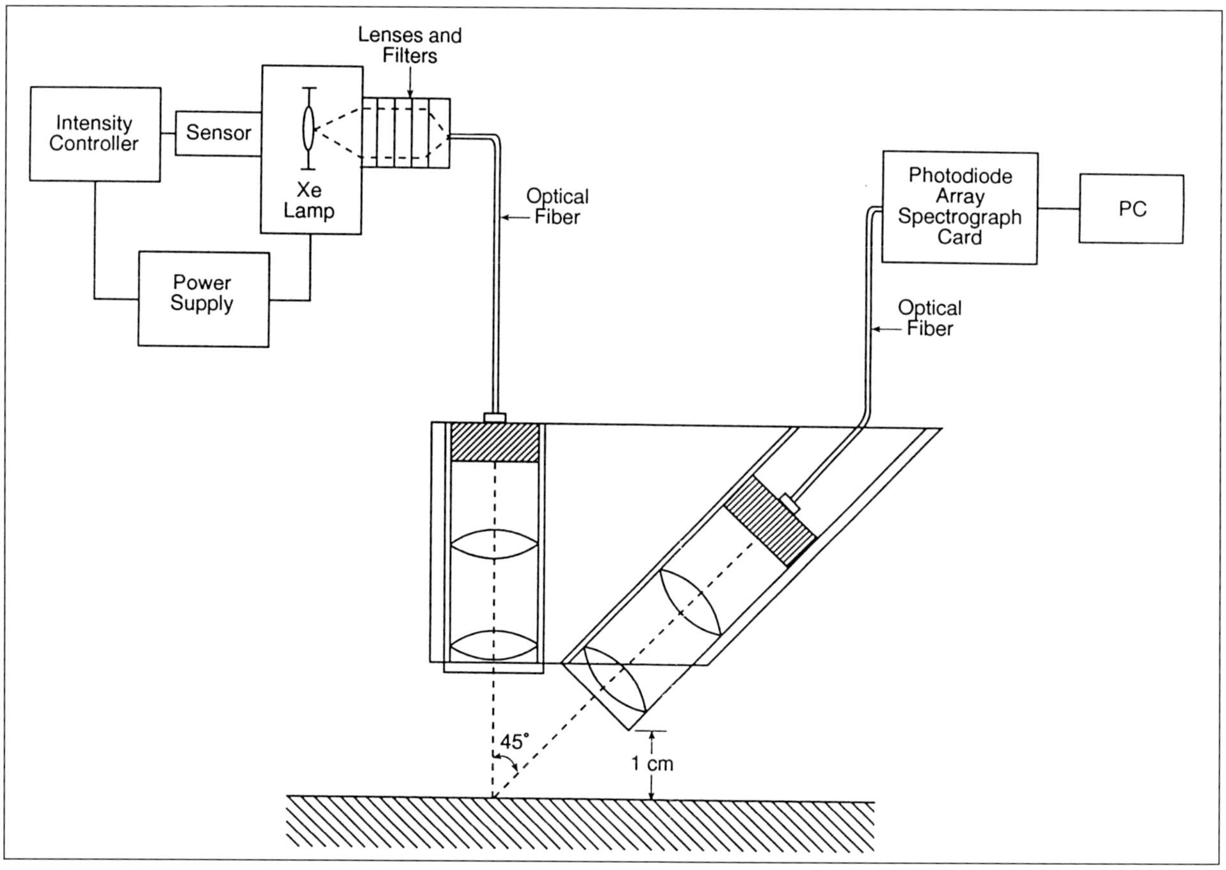

Figure 5.3 Schematic of micro-fading tester, from Whitmore et al. 1999. Reprinted with the permission of the American Institute for Conservation of Historical and Artistic Works, Washington, D.C.

fading tests compare to the conventional accelerated light-aging tests; and 2) how the results of the accelerated tests compare to actual fading upon exposure to ambient lighting in gallery settings. We have found that the accuracy of the color measurements themselves varied depending on the type of sample: smooth, opaque, uniformly colored materials were measured well (compared to a commercial reflectance spectrophotometer), but samples that did not conform to this ideal (e.g., textiles and unevenly colored or transparent samples) were only approximately described by the color measurements in the micro-fading tester. Despite these difficulties, the results produced in the micro-fading tests (the light-fastness ratings and color changes produced in the test samples) were generally in good agreement with the conventional destructive method for evaluating light-fastness (exposure to high-intensity light followed by measurement in a color-measuring instrument).

The accuracy of the micro-fading tests in predicting the future course of fading in gallery settings is less well established. To date we have very few comparisons of accelerated (in the micro-fading tester) and natural fading of materials, and that is a subject of current research. However, it is important to recognize that the accuracy of the prediction – or to put it another way, the validity of the accelerated aging test – is not an intrinsic feature of the test itself. Rather, it is a description of the performance of particular materials and objects in the test, and generalizations to describe the "validity" of the test are unlikely to be possible. There are many factors that must be considered in attempting to simulate gallery fading conditions in an accelerated fading test, and there are ways to critically examine the test results to get a better sense of their accuracy. A thorough discussion of these factors is beyond the scope of this report, but it is the subject of current study.

Safety evaluation

In assessing the safety of the micro-fading tester when using it on works of art, we examined two risks posed by the test conditions: the risk of creating a faded test area that would disfigure the object, and the risk of heat damage from exposure to the very high light intensities. Tests of the perceptibility of the faded test area indicate that the areas are imperceptibly altered until a color change of ΔE = 15–20 has been produced. This difference is much larger than the color changes usually accepted as being barely perceptible (usually on the order of ΔE = 1–2). Since we have already verified the accuracy of the color measurements, we believe that the difficulty in perceiving small color differences lies in the extremely small size of the test area: at these size scales (approximately the size of the period at the end of this sentence), much larger color changes are needed to produce a spot that the eye can pick out as having faded. Consequently, if the micro-fading tests are monitored so that they are terminated when a ΔE = 5 has been produced (a change large enough to be considered definite), the risk of this area being perceptibly faded is extremely small.

We have measured the heat generated at the surface of materials during exposure in the micro-fading tester. The worst-case heating occurs for very dark materials that absorb nearly all of the incident light. In those cases, temperatures of 45–50°C were measured, both directly (by thermocouple) and indirectly (by apparent melting damage to waxes having specified melting points). Thus, the risk of heat damage is considered small for the majority of art materials, although objects created with low-melting materials such as wax media may suffer melting damage in the small test areas. Even for those materials that may melt, however, the damage is not immediate and the fading test may yet yield important information about the material's light sensitivity before melting occurs. Just as for other analyses in which sample removal leaves a tiny defect on an object, that damage may still be judged acceptable in exchange for the knowledge that will help preserve the object.

We have also examined whether the test areas alter over time after the completion of the test, either by further fading that accentuates the faded color of the test area or through post-irradiation discoloration. In our laboratory tests, we have never been able to produce a significant post-irradiation color change on a tested material, even for those such as lignin-containing papers where such effects are well known. We believe that the photochemical changes produced in the test area are so small, particularly if the color changes are kept below ΔE = 5, that the areas are unlikely to evolve as a significantly different color from the surrounding areas. This monitoring of the tested areas on artifacts is a continuing effort, but to date we have not seen any indication that the test areas age differently enough from the original to produce a perceptible spot.

Results of field tests

More recently we have been evaluating the performance of the micro-fading tester when measuring a variety of artifact types, including easel paintings and works on paper, textiles, sculpture and ethnographic objects, and photographic materials. In general, as noted above, the micro-fading tester functions well when measuring materials that are smooth, rigid, and opaque. Measurements on artifacts having textured or irregular surfaces, such as textiles, have occasionally proven to be difficult, particularly when the artifact is so delicate that it cannot be held securely in a fixed position. With one exception, no object that has been tested in these museum visits has been damaged from the testing. The one object that was disfigured was a re-creation of an early photographic process, and this object's light sensitivity was so great that the color was completely bleached in the test area while the tester was being positioned, before the fading test had even begun. Clearly there are limits to the types of objects that will survive the test in its current configuration, and the testing of such ultra-sensitive objects would require a drastic reduction in light intensity before being considered safe. Fortunately such cases of extreme light sensitivity are likely to be recognized before a micro-fading test is attempted, and the possibility of accidental discovery of instability is unlikely.

As expected, in its current design the micro-fading tester is incapable of reliable measurements on transparent materials such as stained glass, dyed plastics, or even thick paint glazes – materials whose color is observed (and must be measured) in transmission rather than in reflection. The adaptation of this device to operate in transmission mode (monitoring loss of color from transmission measurements while being probed with high intensity light) is straightforward and the subject of current research.

Research possibilities

The main objective of the field trials was to demonstrate the utility, simplicity, and safety of this technology in assessing the preservation needs of individual objects. An additional benefit of these site visits was the stimulating discussion with colleagues who envisioned other possible applications for the device. One such interesting possibility was the prospect of accumulating evidence on the actual fading behavior of classes or types of objects, and thus putting to the test some vague but persistent "intuitions" about light stability, such as whether fading occurs steadily over time, or whether previously faded objects will tend to be more stable by having already lost the most fugitive components. During our field trials, both behaviors were observed. In some cases where faded and relatively unfaded areas of the same colorants could be tested (or in one case, where we had duplicate prints, one of which had been exhibited and faded while the other had been in storage), the faded colors tested to be just as fugitive as the unfaded. On other objects, where the fading had clearly removed most if not all of a very fugitive colorant, the object tested to be quite resistant to further fading. With such object-specific information, it might eventually be possible to construct a better, information-based "intuition" of the average behavior of groups of objects.

Another interesting idea offered during a site visit was the proposed use of the micro-fading tester as an analytical tool, essentially to discriminate between closely related materials (such as organic colorants) on the basis of their fading rates and the colors to which they tend to fade. While a considerable amount of testing would be needed to build up a database of fading behaviors, the possibility of a rapid, non-destructive analytical tool that may be free of interferences from other components in the colorant system (such as the medium) is an intriguing possibility.

Based on our use of the micro-fading tester in our laboratory and in several museums over the last few years, we remain optimistic about the promise of this technology to assess the light stability of specific artifacts. The device has proven safe and capable of fulfilling its original purpose, to detect extremely light-sensitive materials on objects before they are unwittingly subjected to inappropriate exhibition conditions. Other uses for this device – in predicting the results of particular exhibition conditions, or as an analytical tool – are intriguing future possibilities. We are continuing the development of the micro-fading tester to improve its performance and to explore some of these applications.

Endnote

1. P. M. Whitmore, X. Pan, and C. Bailie 1999. "Predicting the fading of objects: Identification of fugitive colorants through direct non-destructive lightfastness measurements," *Journal of the American Institute for Conservation* 38, no. 3, pp. 395–409.

The User's Point of View: Micro-Fading Test Results and the Shaping of Exhibition Policy

Craigen Bowen,
Barbara J. Mangum, and
Meredith Montague

At the 1998 American Institute for Conservation conference in Arlington, Virginia, Dr. Paul Whitmore, Director of the Research Center on the Materials of the Artist and Conservator at Carnegie Mellon Research Institute, gave a paper on the micro-fading tester at the paintings specialty group session.[1] It was the first talk of the day, and the attendees gradually straggled in with cups of coffee in hand. By the end of the talk, Roy Perkinson and I (CB), each of us paper conservators at the Boston Museum of Fine Arts and the Harvard University Art Museums respectively, were nearly jumping up and down with excitement at the back of the room. Dr. Whitmore had barely sat down before we assailed him with compliments about the talk and exclamations that he must come to Boston to use our museums as test sites for the device. We could think of endless works of art on which to test it and were confident that our curators would allow their works to be used as test samples. Happily, Dr. Whitmore caught onto our enthusiasm. As it turned out, Barbara Mangum had likewise been trying to get Dr. Whitmore to test his device at the Isabella Stewart Gardner Museum, and Meredith Montague, who had been struggling to determine parameters for exhibiting an important eighteenth-century American textile at the Museum of Fine Arts, was also interested in color change measurements. The coincidence of three test sites staffed with enthusiastic conservators in one city was just too much for Dr. Whitmore to refuse. Each of us, with our own institutional interests, had an agenda for the testing. Dr. Whitmore came and survived the stress of having to assemble and disassemble the device multiple times to move it between our museums.

Testing was conducted for two days at the Straus Center for Conservation at Harvard University (see Figs. 5.4 and 5.5), for one day at the Gardner Museum, and for one day at the Museum of Fine Arts. This paper discusses the results of our micro-fading tests on a disparate and sometimes seemingly random group of documents, works of art on paper, photographs, paintings, and textiles. We analyzed them for their color permanency in relation to visible light exposure. Results

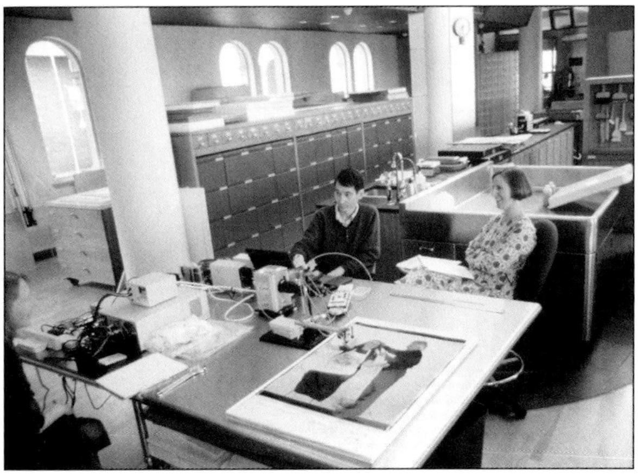

Figure 5.4 Testing at the Harvard University Art Museums, Cambridge, Mass.

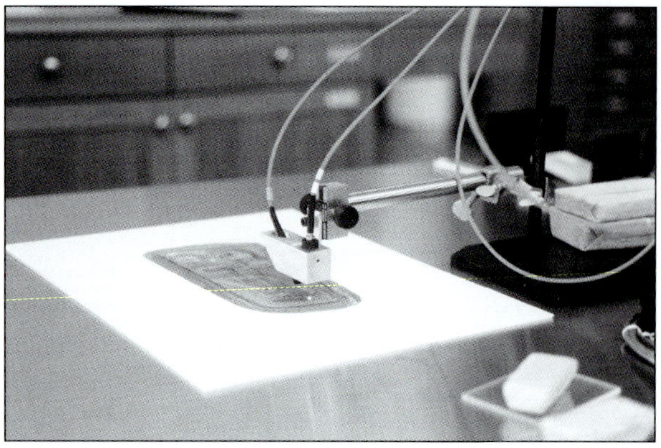

Figure 5.5 Detail of the micro-fading tester in use.

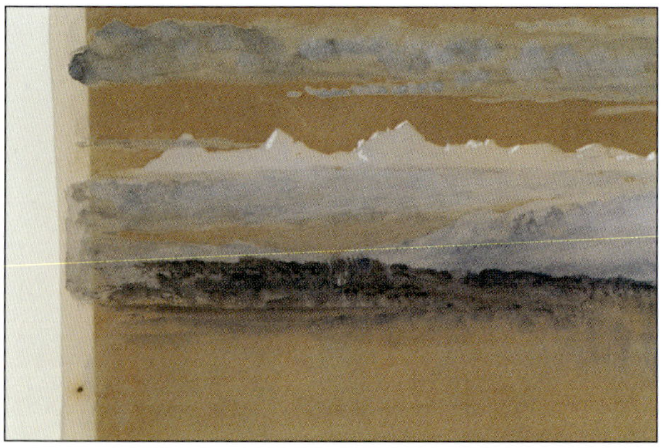

Figure 5.6 John Ruskin (British, 1819–1900), *Dawn at Neuchatel* (detail), 1866. Watercolor and white gouache on cream wove paper darkened to brown, 17.5 x 25 cm. Harvard University Art Museums, Cambridge, Mass.

confirmed that the micro-fading tester is of tremendous importance as a potential predictor of the color permanency of works of art at the level of Blue Wool #2 or less.[2] In order to understand, discuss, and compare the results of the color fading tests in the different venues, ΔE was calculated for each object after it had received a specific dose of visible light.[3] Five minutes of exposure by the xenon arc light source was equivalent to the illumination of a work for one year (no ultraviolet component) for 10 hours per day, 7 days per week at 150 lux. This is equivalent to 550,000 lux hours/year of illumination. No visible damage to an artwork was noted during testing.

The Harvard University Art Museums case study
Craigen Bowen

Twenty-one works of art chosen for micro-fading testing at the Harvard University Art Museums were tested in 43 sites. The results of the testing are given in Table 5.4 (pp. 250–52). Works were selected to test the efficacy of Dr. Whitmore's device in answering several questions regarding light exposure of works of art. The first question was whether a given work might change in appearance as a result of light exposure under museum conditions. We were also curious to know the actual sensitivity of media we had always thought to be light sensitive. Finally, we attempted to use the device as a diagnostic tool to help understand how a work might have reached its current physical state. Some cases – such as the works by Mark Rothko and John Ruskin – allowed us to consider all three questions.

In addition to extremely active exhibition schedules that include many traveling exhibitions and loans to other museums, all three museums have a policy of making their collections accessible to students, scholars, and the general public. At Harvard the study room in which the Western prints, drawings, and photographs are viewed has ambient light levels nearly six times those of a gallery. In this particular study room, watercolors, photographs, and contemporary art are often displayed to classes, so it is necessary to understand the physical properties of the thousands of pictures used there – a daunting task. Furthermore, exhibitions are becoming increasingly interdisciplinary, including objects of various kinds, both light sensitive and not, resulting in gallery light levels that sometimes are not ideal. While it is possible to record the amount of light exposure a work receives in the museum, it is cumbersome and impractical for such a large number of objects. Also, it is rarely possible to ascertain accurately the total exposure a work has received since its creation because a museum typically acquires art some time after it was made. It is not possible to know how sensitive to light a given work might be unless the chemical composition of the materials employed is known, something that cannot be determined by eye. Micro-testing would provide the knowledge of a material's light sensitivity without time-consuming and costly chemical analysis.

Certain works of art display condition problems that seem to stem from light exposure but are not clearly understood. Two drawings by John Ruskin (1819–1900) were selected for testing the diagnostic capabilities of the device. In the case of *Dawn at Neuchatel* (Fig. 5.6), the center of the sheet, exposed to light while in a window mat, is discolored from off-white to dark brown. Previous efforts to determine analytically the cause of this unusually severe discoloration were not completely successful. The micro-fading tester was focussed on a spot in the off-white margin that had been protected by a window mat. The spot was exposed for 10 minutes in the hope of inducing darkening, however, the results indicated relative stability in visible light. Since the light source in the device does not provide UV radiation, it can only be supposed that past UV exposure was the cause of damage. This, of course, suggests further development of the device to include an optional UV component to the illumination, so that it could provide information relating to a wider range of exhibition conditions. The second Ruskin drawing, *Early Byzantine Carving* (Fig. 5.7), was executed on purple paper that has faded to pale gray. An attempt to use the device to induce fading of a spot in the purple margin also was not

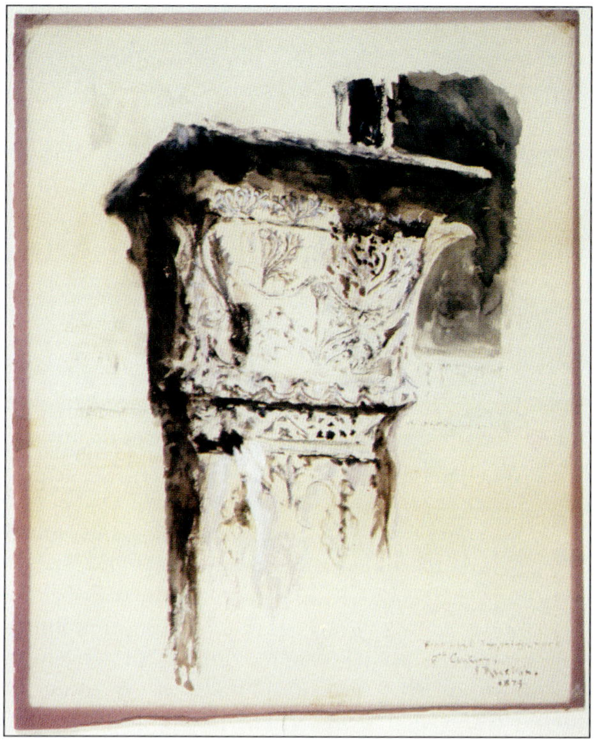

Figure 5.7 John Ruskin, *Early Byzantine Carving*, 1879. Blue-black, brown, and purplish washes and white gouache on faded purple paper, 28 x 22 cm. Harvard University Art Museums, Cambridge, Mass.

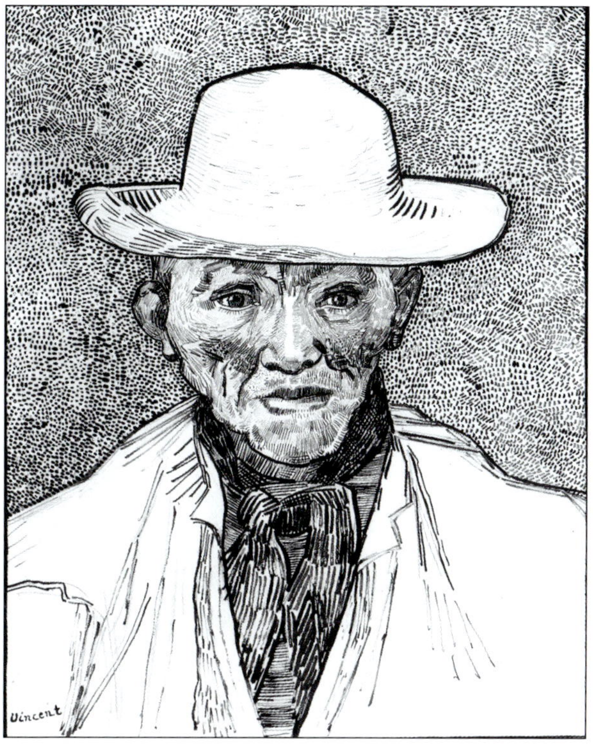

Figure 5.8 Vincent van Gogh (Dutch, 1853–1890), *Peasant of the Camargue*, 1888. Brown ink over graphite on off-white wove paper, 49.4 x 38 cm. Harvard University Art Museums, Cambridge, Mass.

effective, again presumably because UV radiation caused the damage.

Vincent van Gogh's (1853–1890) drawing *Peasant of the Camargue* (Fig. 5.8) is on excellent paper, and is thought to have been drawn with iron-gall ink, probably with a reed pen. We have always wondered whether, as a relatively "new" drawing dating to the nineteenth century, the ink might be in the process of changing from brown-black to brown. But the testing showed the ink to be very stable in visible light. The work that proved to be the most light sensitive was Alfred Jensen's drawing *Maya Structure!*, a work executed in watercolor, gouache, black ink, and blue ballpoint pen. The watercolor and gouache proved to be variably sensitive to light, with the blue less stable than Blue Wool #1, the purple equivalent to Blue Wool #2, the red more stable than Blue Wool #2, and the yellow most stable. The fact that these materials are not only sensitive to light but are variably so suggests that if this work is exposed to light it will change differentially, thereby altering its color balance.

A number of works, mostly modern and contemporary, on unusual paper supports or incorporating non-traditional media were also selected for testing with a view to protecting them from the deleterious effects of future light exposure. It can be said as a general conclusion that the micro-fading test results indicated remarkable stability in most of the artworks in the context of the museum environment with low light levels and UV filtration.

The Isabella Stewart Gardner Museum case study
Barbara J. Mangum

The Isabella Stewart Gardner Museum's collection is a permanent one, which according to the founder's will cannot be added to and is on permanent view in the position chosen by her. With the documents in the collection, however, there is some flexibility in terms of rotation. The documents are housed in bronze cases with glass tops covered by heavy textiles labeled "Visitor May Lift Cover." This keeps the documents in the dark most of the time and has proved excellent for their preservation; unfortunately, the covers also seem to act as deterrents to viewing the objects. Apparently, many visitors, after being assaulted with "PLEASE DO NOT TOUCH" signs throughout the museum, are reluctant to lift the covers to see the works they conceal. Now that the Gardner has expended considerable energy and sums of money reducing UV and visible light levels to appropriate conservation standards, interest has been expressed in removing the fabric covers during public hours.

To determine whether or not this would be appropriate, we used Dr. Whitmore's device to get an indication of how some paper-based objects would fare after six years of light exposure during normal museum operation (equal to 5 minutes of exposure using the micro-fading tester). We tested fifteen documents and

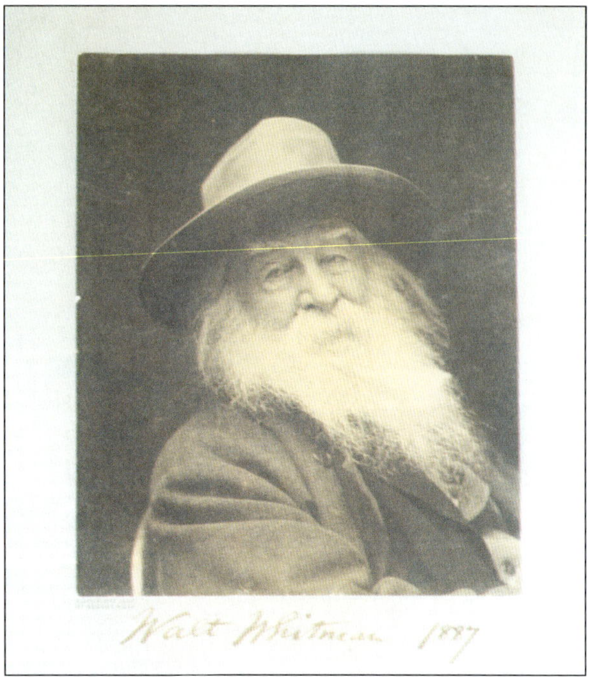

Figure 5.9 Photograph of Walt Whitman, Isabella Stewart Gardner Museum, Boston.

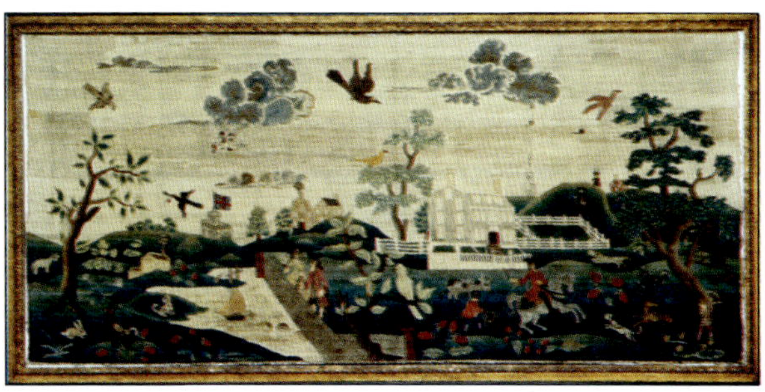

Figure 5.10 Hannah Otis (American, 1732–1801), *View of Boston Common*, c. 1750. Silk and wool embroidery on linen ground, 61.59 x 133.98 cm. Museum of Fine Arts, Boston.[5]

works of art ranging from letters by Dostoevsky, Pablo Casals, T. S. Elliot, Anna Pavlova, and Beethoven, to photographs of Walt Whitman (see Fig. 5.9) and Julia Ward Howe, and a manuscript by Washington Irving. The results are presented in Table 5.5 (pp. 253–54). In summary, all colorants tested were surprisingly more stable than Blue Wool #2, and many fall into the category of most stable. However, the results of testing the paper supports indicated that light bleaching could occur. Further monitoring would be needed to confirm these predictions over time.

Micro-fading tests of the documents did give an indication of the most fugitive qualities inherent to the individual works. It appears that parts of some of the documents should be further shaded from light, and other documents should be removed from exhibition altogether. The testing could not at this point estimate the total visual life of a work, but exposure to micro-fading tests did give an indication of fading within a decade – a fairly short period of time for an object several hundred years old. Due to the immovable and irreplaceable nature of the Gardner's collection, there is no policy of tolerance for change. At this time and until further testing can show long-term stability, the conservators recommend that the documents continue to remain covered for the foreseeable future.

The Boston Museum of Fine Arts case study
Meredith Montague

The primary goal of Dr. Whitmore's visit to the Museum of Fine Arts was to test the light sensitivity of the dyes of a recently acquired textile. This object, an eighteenth-century American embroidered picture (Fig. 5.10), had been requested for a longer period of display than normally recommended for textiles. The museum had taken steps to reduce overall light exposure by designing a case incorporating visitor-activated lighting and a data logger to record total lux-hour exposure. However, the question remained: how much light exposure would cause a visible shift in the colors of the textile? If this "color life" could be determined, how should this time be apportioned now and in the future?

Dr. Whitmore's research presented an ideal opportunity to characterize fairly quickly the sensitivity of the textile's dyes – information that was important in understanding the risk of longer display periods. Of special benefit was the ability to test areas on the face of the textile without danger of causing a visible change. Marked fading of the textile had already occurred, and these faded areas could test differently than the edges or the reverse of the textile that had been protected from light exposure.

In addition to the American embroidery, four other textiles were chosen for testing in an attempt to broaden the range of dyes and dyeing techniques that might be characterized. The results of the testing are shown in Table 5.6 (p. 255). Ten areas on the face of the American embroidery were tested, including one location along the side of the embroidery that was less faded and appeared to have been protected by the object's frame (green wool yarn, lower right edge). All of the dyes tested were more stable than Blue Wool #2, and seven of the areas showed very little change, including the area along the edge that had been protected from previous light exposure. The test results indicated greater color stability than had been expected, perhaps indicating an absence of more fugitive dye components that have already faded due to previous light exposure. The most sensitive colors were found in an eighteenth-century French silk brocade with a pink dye slightly less fugitive than Blue Wool #1, and a nineteenth-century Japanese silk woman's robe (*uchikake*) with motifs embroidered in a green yarn of sensitivity equivalent to Blue Wool #2. An orange dye in a twentieth-century French silk satin panel tested slightly less fugitive than Blue Wool #2. Three dyed camelia-fiber yarns from a Paracas textile fragment dating from before

A.D. 100 were also tested. The fragment had very little recorded previous display history and did not appear faded. The three areas tested were all significantly more stable than Blue Wool #2.

To quantify the test results on the basis of a "just noticeable fade" instead of a ΔE value, the Light Damage Slide Rule was consulted.[4] The slide rule placed a "just noticeable fade" of Blue Wool #2 at approximately 1,300,000 lux hours, roughly 8 years of continuous exposure at 50 lux. Fading of Blue Wool #1 was measured at approximately 450,000 lux hours, or 3 years of continuous exposure at 50 lux. The importance of only occasional display for objects with equivalent fading rates is immediately evident. With the micro-fading test equipment, these objects can be identified within a very short test period, without the necessity of dye or mordant analysis and without knowledge of previous display history. Objects testing more stable than Blue Wool #2 were not easily characterized with the micro-fading equipment because of the short test periods involved, and questions remain in understanding safe parameters of display for these objects. In the case of the American embroidery that had already undergone a visible color shift, a display policy can be set only after deciding what degree of further change is acceptable.

Summary

The data in the three tables is illuminating. We found, unexpectedly, that the third category, "more stable than BW #2 but less than most stable," is the most problematic category. The very light-sensitive materials, which were few in our study, are easily identified by this device. The most stable are also easily identified. Please note that we are using the phrase "most stable," not "stable," because we remain skeptical, perhaps superstitiously so, that anything is truly stable forever in light.

The problematic third category seems to fit materials not fully understood. Often the tests found a steady rise in the line charting change, but at a very low angle. We all individually found ourselves wondering what happens to those works as they are continually exposed to light. Does a visual change eventually occur such that one would limit their exposure significantly? This is definitely an area ripe for future research. Most of our sensitive works fit into this category, so it would behoove us to examine this group in depth. The inclusion of an optional UV component in the light source and longer exposure times would be helpful in illuminating this problematic category as well.

Endnotes

1. P. M. Whitmore, X. Pan, and C. Bailie 1999. "Predicting the fading of objects: Identification of fugitive colorants through direct non-destructive lightfastness measurements," *Journal of the American Institute for Conservation* 38, no. 3, pp. 395–409.

2. Garry Thomson, *The Museum Environment*, 2nd edn (London: Butterworths, 1978), p. 183: "Blue Wool standards were adopted as International Organization for Standardization Recommendation R105 and British Standard BS1006 (1961). Each card contains 8 specially prepared blue dyes on wool. They are so chosen that standard number 2 takes roughly twice as long to be perceptibly faded as standard 1, standard 3 roughly twice as long as standard 2 and so on." Note that Thomson quotes from Robert Feller's publication "Studies on photochemical deterioration," *ICOM Venice 1975*, on predicting the "intended useful life of an object" of a certain class of Blue Wool permanency. At a lux level of 450 (with some UV component), the intended useful life of a Blue Wool #1–3 object would be less than 20 years. Blue Wool #3–6 was given an intended useful lifetime of 20–100 years, and color permanency greater than Blue Wool #6 was accorded an intended useful lifetime of greater than 100 years. Thomson notes that with no UV component, the multiplying factor for extending the prediction of the intended useful life would be 6 or more.

3. ΔE is defined in the *Colortron User Manual*, 2nd edn (Larkspur, Calif.: Light Source Computer Images, 1995) as "a computed value intended to represent how 'close' two colors are to each other. The ΔE value is usually derived by computing the distance between two colors when plotted in a perceptually uniform color space (CIE L*a*b* for instance). A color difference of ΔE 1 is defined as the threshold of what is perceptible to the human eye."

4. The Light Damage Slide Rule was developed by the Canadian Conservation Institute in 1989. Readings were taken at the point when shading first completely filled the observation window. The lux-hour estimates are based on a light source with no UV content and do not consider other factors that may affect fading, such as the presence of pollutants.

5. Gift of a Friend of the Department of American Decorative Arts and Sculpture, a Supporter of the Department of American Decorative Arts and Sculpture, Barbara L. and Theodore B. Alfond, and Samuel A. Otis; and William Francis Warden Fund, Harriet Otis Cruft Fund, Otis Norcross Fund, Susan Cornelia Warren Fund, Arthur Tracy Cabot Fund, Seth K. Sweetser Fund, Edwin E. Jack Fund, Helen B. Sweeney Fund, William E. Nickerson Fund, Arthur Mason Knapp Fund, Samuel Putnam Avery Fund, Benjamin Pierce Cheney Fund, and Mary L. Smith Fund.

Object	Test Location	Less Stable than or Equal to BW #1	Less Stable than or Equal to BW #2	More Stable than BW #2 but Less Stable than Most Stable	Most Stable
Bontecou, Lee, American (b. 1931) *Untitled*, 1967 Graphite, black and red fiber-tip ink on off-white flocked paper, 51.7 x 65.9 cm HUAM 1997.18	Red			X	
	Black				X
	Plain paper, bottom right corner				X
Dehner, Dorothy, American (1901–1994) *Untitled (Nickel Drawing)*, 1954 Black ink, watercolor, and gouache on white wove paper, 26.2 x 13.1 cm HUAM 1998.77	Blue, upper right, in fourth fan segment from left				X
	Green				X
	Red, lower center at right end of oblong				X
	Yellow, upper center (yellow on top of blue)				X
Dorfman, Elsa, American (b. 1937) *My annual haircut: Diego + me. Feb 19, 1992*, 1992 Polaroid print, 86 x 56 cm HUAM P1996.6	Blue, proper left shoulder of shirt, 282 mm from top			X	
	Red, center of vertical of "E" in *Elizabeth*				X
	Black felt-tipped pen period following "1992"				X
	Plain white area, lower right corner				X
Gogh, Vincent van, Dutch (1853–1890) *Peasant of the Camargue*, 1888 Brown ink over graphite on off-white wove paper, 49.4 x 38 cm HUAM 1943.515	Two locations in top left corner: 2.7 cm down x 3.4 cm from left, and 1.5 cm down x 3.3 cm from left				X
Goldin, Nan, American (b. 1953) *Self-Portrait in Blue Bathroom London*, 1980 Cibachrome, 73.6 x 101.6 cm HUAM P1994.1	Blue				X
	Red, label on soap				X
	Yellow, upper center				X
Jensen, Alfred, American (1903–1981) *Maya Structure!*, 1965 Watercolor, gouache, black ink and blue ballpoint ink on white wove paper, 73.9 x 58.7 cm HUAM 1998.17	Black				X
	Blue	X			
	Purple		X		
	Red			X	
	Yellow				X

Table 5.4 Relative ranking of color permanency in sampled works of art at the Harvard University Art Museums.

Object	Test Location	Less Stable than or Equal to BW #1	Less Stable than or Equal to BW #2	More Stable than BW #2 but Less Stable than Most Stable	Most Stable
Jensen, Alfred, American (1903–1981) *IX*, 1965 Lithograph, 57 x 42 cm HUAM M23810	Blue, upper right				X
	Purple, lower right				X
	Red, lower left				X
	Yellow, upper right				X
Kupka, Frantisek, German (1871–1957) *Hour*, 1920 Oil on canvas, 70.3 x 70.3 cm HUAM BR62.10	Pink				X
	Purple				X
	Yellow				X
Louis, Morris, American (1912–1962) *Four-Three-Eight (Orange Column)*, 1960 Acrylic on canvas, 215.9 x 62.6 cm HUAM 1965.32	Orange, 105 cm from bottom x 8.9 cm from left				X
	Red, 118 cm from bottom x 47 cm from left				X
Rothko, Mark, American (1903–1970) *Untitled*, 1967 Acrylic on paper mounted on wood, 60.96 x 45.72 cm HUAM 1986.465	Blue, 37.5 cm from bottom x 3.8 cm from left				X
	Red, 35.5 cm from bottom x 1.6 cm from left				X
	Brown, bottom left corner				X
Rothko, Mark, American (1903–1970) *Untitled (Study for Harvard Murals)* Sides A and B Drawings HUAM 1986.629	Top right corner near edge				X
Ruskin, John, British (1819–1900) *Dawn at Neuchatel*, 1866 Watercolor and white gouache on cream wove paper darkened to brown, 17.5 x 25 cm HUAM 1919.49	Bottom left corner, 3 mm from bottom x 3 mm from left			X	
	With the UV component of the light source			X	

Table 5.4 (*continued*)

Object	Test Location	Less Stable than or Equal to BW #1	Less Stable than or Equal to BW #2	More Stable than BW #2 but Less Stable than Most Stable	Most Stable
Ruskin, John, British (1819–1900) *Early Byzantine Carving*, 1879 Blue-black, brown and purplish washes and white gouache on faded purple paper 28 x 22 cm HUAM 1924.29	Lower left margin			X	
	With UV			X	
Ruskin, John, British (1819–1900) *Tower of Fribourg*, 19th C. Black ink, watercolor, white gouache, and graphite on cream paper darkened to brown HUAM 1928.12	Upper right margin				X
Ruskin, John, British (1819–1900) *Trees and Ruins (Paestum?)*, 1854 Black chalk and white gouache on blue paper, 36.6 x 24.8 cm HUAM 1937.158	2.2 cm down x 1 cm from right				X
Talbot, William Henry, British (1800–1877) *Bonnets in a Milliner's Shop Window*, c. 1845 Varnished salt print from a calotype negative HUAM 75.1994	Upper right corner				X
Spencer album Off-white paper presumably never exposed to light, c. 18th C. HUAM S5.98	Upper right corner, 3.9 mm from right x 6.6 mm from top			X	
Blue antique laid paper c. 18th C.					X
Teal felt-tipped marker: Schwan Stabilayout 38, drawn out on ragboard several years ago and stored in the dark HUAM			X		

Table 5.4 (*continued*)

Object	Test Location	Less Stable than or Equal to BW #1	Less Stable than or Equal to BW #2	More Stable than BW #2 but Less Stable than Most Stable	Most Stable
Letter by Henry Lee Higginson Blue ink on paper Blue Room, east case	Blue ink, "b" in *before*, 7 lines down, 2nd page				X
Ogata Korin *The White Fox*, 1914 Watercolor on paper Blue Room, north case	Yellow, lower center				X
	Red, corner of mouth			X	
	Green, upper leaves				X
	Blue, lower central flower				X
Letter by Julia Ward Howe, 1890s Ink on paper Blue Room, east case	Ink, dot of "i" in *enthusiasm*				X
	Paper, to right of pencil case no.				X
Photograph of Julia Ward Howe adhered to an off-white backing board Blue Room, east case	Backing board, upper right corner				X
	Background of photograph, upper right corner				X
	Middle tone of photograph, right edge of image				X
Letter by Okakuro Kakuzo to Mrs. Gardner, 1907 Iron-gall ink (?), red, green, and lavender printer's ink on paper Blue Room, north case	Paper support				X
	Lavender, plant form				X
	Green, left-most tip of left-most leaf				X
	Red, 2nd character down on left			X	
	Black, 2nd "s" in *sensei* at bottom left				X
Letter by Sarah Orne Jewett Iron-gall ink (?) on off-white paper Blue Room, north case	Ink, "a" in *has*, lower left corner			X	
Manuscript by Washington Irving "The Conquest of Granada" Iron-gall ink (?) on white paper Blue Room	Ink, "b" in *beautiful*, 1st line, 1st page			X	

Table 5.5 Relative ranking of color permanency in sampled works of art at the Isabella Stewart Gardner Museum.

Object	Test Location	Less Stable than or Equal to BW #1	Less Stable than or Equal to BW #2	More Stable than BW #2 but Less Stable than Most Stable	Most Stable
Photograph of Walt Whitman by George C. Cox, 1887 Platinum print (?) adhered to gray backing board with iron-gall ink (?) autograph Yellow Room, north case	Photograph, sleeve at lower right corner				X
	Brown ink, dot of "i" in *Whitman*			X	
	Backing board, spot to right of "1887"				X
Letter by T. S. Eliot Black type on paper Blue Room, east case	"e" in *care,* 1st line of main body				X
Letter by Robert Benson	"o" in *lovely,* 2nd line on page with signature				X
Letter by Pablo Casals Ink on paper Yellow Room, west case	"a" in *Pablo*			X	
Letter by Dostoevski to N. N. Strakhorn, 1873 Iron-gall ink (?) on light brown paper	Ink, "z" or "y" at end of signature Paper appears to have been bleached by test			X	
Letter by Anna Pavlova, 1922 Blue type with red type return address and black ink signature on white paper Yellow Room, west case	Blue ink, comma in salutation				X
	Black ink, "c" in *sincerely*			X	
	Red ink, top of the "M" in *Mme*			X	
Letter by Beethoven, 1819 Iron-gall ink (?) on paper Yellow Room, west wall	Ink, "n" in *Beethoven*				X

Table 5.5 (*continued*)

Object	Test Location	Less Stable than or Equal to BW #1	Less Stable than or Equal to BW #2	More Stable than BW #2 but Less Stable than Most Stable	Most Stable
Hannah Otis, American (1732–1801) *View of Boston Common*, c. 1750 Silk and wool embroidery on linen ground MFA 1996.26	Blue wool yarn, lower right corner			X	
	Blue silk yarn, upper right quadrant				X
	Blue silk yarn, upper right quadrant			X	
	Green silk yarn, upper right quadrant				X
	Blue wool yarn, upper left quadrant				X
	Green wool yarn, lower right edge				X
	Blue wool yarn, lower right quadrant			X	
	Pink wool yarn, lower right quadrant				X
	Red wool yarn, lower right quadrant				X
	Yellow wool yarn, lower right quadrant				X
Japan, c. 19th C. Woman's robe (*uchikake*) Silk damask, resist and tie-dye, silk embroidery MFA 11.3860	Blue yarn, proper left sleeve, front				X
	Green yarn, proper left sleeve, front		X		
	Pink yarn, proper left body, center front			X	
	Red yarn, proper left sleeve, front				X
Peru, before A.D. 100 Mantle border fragment Stem stitch embroidery on plain weave ground camelid fiber MFA 21.2557	Purple ground yarn, upper edge near center				X
	Red embroidery yarn, upper edge near center			X	
	Yellow embroidery yarn, middle cat figure, center			X	
France, c. 1735–40 Panel, silk brocade MFA 1987.634	Blue yarn			X	
	Purple yarn			X	
	Pink yarn	X			
	Red yarn			X	
Designed by Raoul Dufy, woven by Bianchini Ferrier France, c. 1919–20 Panel, silk satin damask MFA 1988.163	Blue yarn			X	
	Orange yarn			X	
	Lavender yarn			X	

Table 5.6 Relative ranking of color permanency in sampled works of art at the Museum of Fine Arts, Boston.

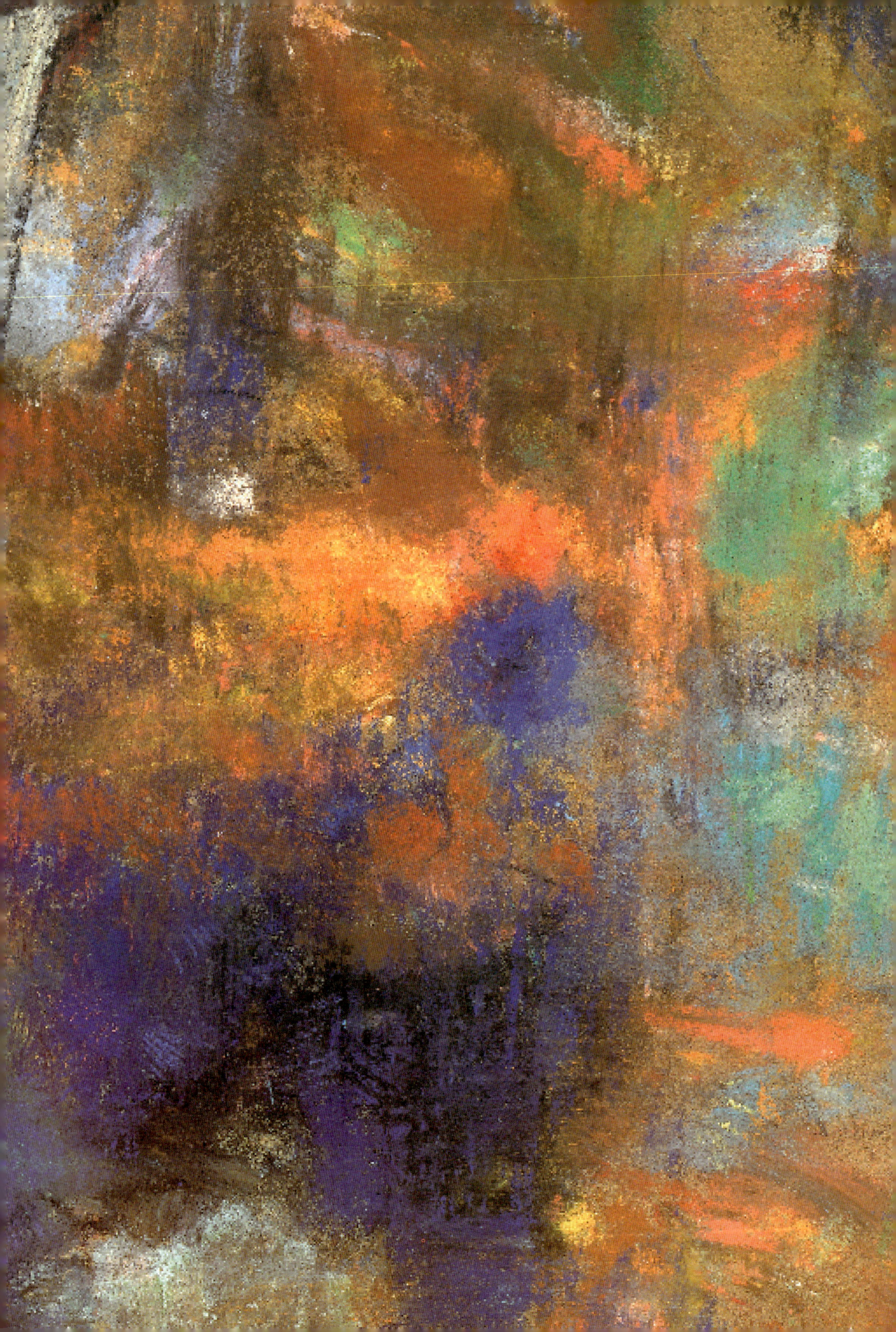

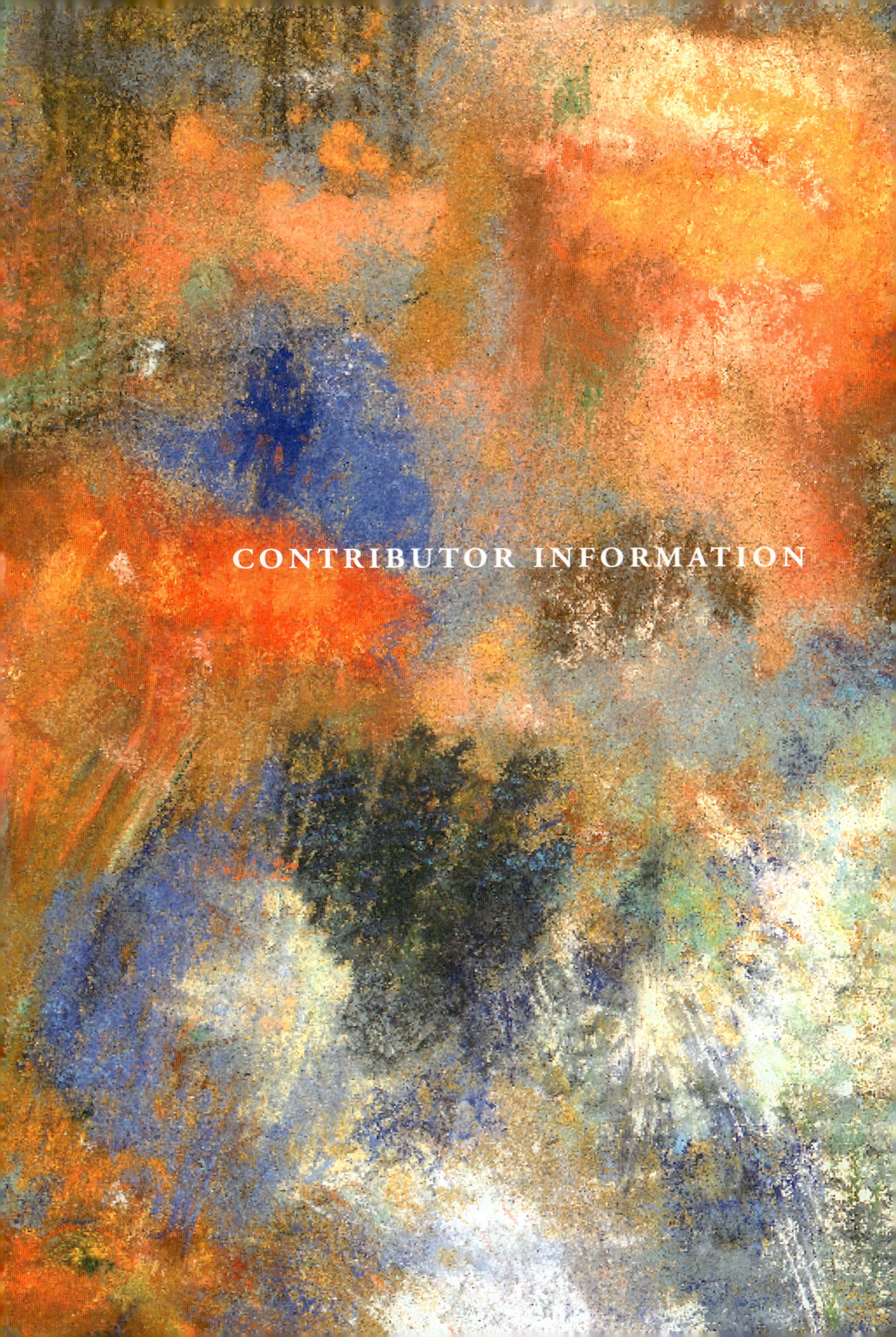

CONTRIBUTOR INFORMATION

Mitsuhiro Abe was invited to take up the post as Hirayama Fellow by the British Museum in 1998 and to oversee a five-year work program involving the conservation of *fusuma* and *byōbu*. Mr. Abe studied art restoration at the Ministry of Education of the Japanese Government over a ten-year period from 1953 to 1963. In the early 1960s he studied with Mr. Iwataro Oka, at that time Head of the Association for Conservation of National Treasures. In 1966 Mr. Abe traveled to New York where he opened the Far Eastern restoration studio at The Metropolitan Museum of Art. In 1967 he took up full-time employment there and was Master Restorer of Oriental Paintings from 1975 to 1997.

Clair Battisson studied sculpture at John Moores University, Liverpool. She graduated with a B.A. in fine arts in 1992. From 1993 to 1995 she worked as a mount cutter at the National Museums and Galleries on Merseyside. She has worked in the Conservation Mounting Department at the Victoria & Albert Museum since 1995. She has written articles on the mounting of the *Akbarnama* and on the art of paper quilling.

Maria Bedynski received an M.A. in art history from the Catholic Theological Academy in Warsaw, Poland, in 1977. That year she accepted a position in the Prints and Books Studio at the Conservation Center for National Treasures in Warsaw, where she worked for four years, first as a trainee and eventually as an assistant conservator. After arriving in Canada in 1982, she worked in private practice for two years. In 1984 she attended the National Archives of Canada in-house training program, receiving a certificate in the conservation of works on paper. Since 1988 she has been a conservator of works of art on paper, maps, and manuscripts at the National Archives of Canada.

Gabriëlle Beentjes graduated from Utrecht University in 1991 with an M.A. in history. After that she worked as a paper conservation apprentice at the Rijksmuseum and the Municipal Archives in Amsterdam. Since 1995 she has worked as a paper conservator at the Municipal Archives in Rotterdam and has also worked freelance for Art Conservation B.V. With Piet van Dalen, she has published two articles about the use of sodium borohydride in paper conservation. She has also written a booklet about the decay of scientific library collections for the European Commission of Preservation and Access.

Barbara Berrie received a B.Sc. (Honours) from the University of St. Andrews, Scotland, and a Ph.D. in chemistry from Georgetown University, Washington, D.C. From 1982 to 1984, she was a National Research Council Research Associate at the Naval Research Laboratory in Washington, D.C. She joined the Scientific Research Department of the Conservation Division of the National Gallery of Art in 1984 and is now Senior Conservation Scientist. Her current interests are in improving knowledge of artists' materials in order to help conservators choose appropriate conservation strategies and exhibition parameters.

Craigen Bowen is the Philip and Lynn Straus Conservator of Works of Art on Paper and the Deputy Director of the Straus Center for Conservation at the Harvard University Art Museums. She was educated at Smith College, where she double-majored in art and astronomy. She completed an apprenticeship in paper conservation at the Fogg Art Museum and then worked at the Williamstown Regional Art Conservation Laboratory for two years, setting up their first paper conservation studio. Following this, she worked part-time at the Northeast Document Conservation Center and privately for a year before returning to the Fogg Art Museum (now part of the Harvard University Art Museums), where she has been since 1981.

Peter Bower is a forensic paper historian and paper analyst based in London, specializing in the examination and analysis of papers for purposes of dating, attribution, authentication, and usage. He is the author of some 300 reports and consultation documents for various clients on the papers used in individual works of art, banknotes, bonds, books, and documents, and has written over 70 articles, essays, and conference papers for a range of publications and organizations. Among his most recent publications is the second volume of his research into J. M. W. Turner's use of paper: *Turner's Later Papers: A Study of the Manufacture, Selection and Use of His Drawing Papers, 1320–1651.*

Thea Burns is Helen H. Glaser Conservator for Special Collections at the Weissman Preservation Center, Harvard University. Between 1987 and 2001 she was Associate Professor and Paper Conservator in the master's degree program in art conservation at Queen's University, Kingston, Ontario. She has also worked for the Montreal Museum of Fine Arts, the Canadian Centre for Architecture, the Queen's University/City of Kingston Archives, and in private practice. She attended McGill University (B.A. 1st class honors in fine arts), Queen's University (M.A.C.), and the Center for Conservation and Technical Studies, Harvard University Art Museums. She is about to defend her Ph.D. thesis on pastel painting at the Courtauld Institute of Art, University of London (Department of Conservation and Technology).

Esther Chao graduated *summa cum laude* from the University of Arizona with a B.A. in anthropology and art history in 1996. She received an M.A. in art history and degree certificate in art conservation from the Conservation Center, Institute of Fine Arts, New York University in 2001, specializing in objects conservation. She completed her fourth-year internship at the American Museum of Natural History working on archaeological and ethnographic materials. She is currently an Andrew W. Mellon Fellow at the Smithsonian Institution's National Museum of the American Indian.

Marjorie B. Cohn received her B.A. from Mount Holyoke College and her A.M. from Radcliffe College. She is a member of the American Academy of Arts and Sciences and the Print Council of America. She has worked at the Harvard University Art Museums (earlier the Fogg Art Museum) since 1963, for 27 years as a paper conservator and also briefly as head conservator. She was acting director of the Museums in 1990–91. Since 1989 she has been Carl A. Weyerhaeuser Curator of Prints. Her major scholarly publications have been in the areas of the history of graphic techniques and the history of print collecting.

Elizabeth I. Coombs is a paper conservator in private practice in Cranston, Rhode Island, working for institutions and individuals on Western prints, drawings, and historic artifacts, and Japanese prints. She has lectured extensively on conservation topics, most recently to the Ukiyo-e Society of America. Before opening her own studio, she was Associate Paper Conservator at the Cincinnati Art Museum and the Baltimore Museum of Art. Her basic training was with paper conservator Christa Gaehde, with whom she has continued to work, and at the Center for Conservation and Technical

Studies, Harvard University Art Museums. This training was supplemented by studies with Keiko Keyes, Katsuhiko Masuda, and Jing Gao. She has published articles on paste prints and artists' mounts, and is currently writing on Japanese prints.

Alan Cummings is Head of the School of Humanities and Director of Conservation Studies at the Royal College of Art, London. In 1998 he had a Royal College of Art Professorship. Between 1991 and 1998 he was co-founder and editor of the *Victoria & Albert Conservation Journal*. In 1989 he founded the Royal College of Art/Victoria & Albert Conservation Course. Since 1974 he has also been a freelance conservator of easel paintings. He received his training in painting conservation at the Courtauld Institute of Art, London, and graduated with a 1st class degree in chemistry from Imperial College, London University.

Susan Dackerman is the Curator of Prints, Drawings, and Photographs at The Baltimore Museum of Art. In 1986 she received an A.B. in art history from Vassar College, and in 1995 received a Ph.D. in the history of art, with a specialization in Northern European Renaissance prints, from Bryn Mawr College. She is presently working on exhibitions of early printed books and sixteenth-century hand-colored prints.

Vincent Daniels studied chemistry at University College Cardiff, where he obtained a B.Sc. and Ph.D. After a period of post-doctoral research, he joined the British Museum in 1974 to work on paper conservation research with A. D. Baynes-Cope. After two years he transferred to the newly created Department of Conservation, where he worked on a wide range of conservation research problems. His research interests have focussed primarily on cellulose substrates and colorants. The author of over 100 publications, his projects have included the use of low temperature plasmas for conservation; an investigation of the Russell effect; the deterioration of lead white, vermilion, and arsenic sulfides; the study of Japanese aged paste; foxing; the wash-fastness of watercolors; the effect of water on pastels; and the degradation of black-dyed plant fibers.

Douglas W. Druick is Searle Curator of European Painting and Prince Trust Curator of Prints and Drawings at The Art Institute of Chicago. He has contributed as organizing curator and catalogue author to numerous exhibitions on nineteenth-century art, including most recently *Odilon Redon: Prince of Dreams* (1994), *Gustave Caillebotte: Urban Impressionist* (1995), and (with Peter Kort Zegers) *Van Gogh and Gauguin: The Studio of the South* (2001).

Margaret Holben Ellis is Director of the Thaw Conservation Center at the Pierpont Morgan Library and Professor of Conservation at the Conservation Center, Institute of Fine Arts, New York University. Until recently she was Sherman Fairchild Chairman of the Conservation Center. She received her B.A. in art history from Barnard College, Columbia University, and M.A. in art history and degree certificate in conservation from the Institute of Fine Arts, New York University. Recipient of the American Academy in Rome Prize in Historic Preservation and Conservation in 1994, and the Rutherford John Gettens Merit Award in 1997, she is author of numerous publications and presentations, many focussing on the materials and techniques of twentieth-century art.

Elmer Eusman finished the four-year Opleiding Restauratoren (the national conservation program) in Amsterdam in 1989. He has held internships at the Rijksmuseum in Amsterdam, the British Museum, and a three-year Andrew W. Mellon Fellowship in Paper Conservation at the National Gallery of Art in Washington, D.C., where he studied the effects of wet-dry boundaries in paper. He has worked at the Baltimore Museum of Art and at the Boijmans-van Beuningen Museum in Rotterdam. In the Netherlands he was involved in different projects concerning iron-gall ink corrosion, including the coordination of the Iron-Gall Ink Corrosion Web site. He is currently a Senior Paper Conservator at the Library of Congress, where he chairs a small research group examining various iron corrosion treatment protocols.

Gilbert L. Gignac received his B.F.A. in drawing and painting from The School of The Art Institute of Chicago in 1972, and an M.A. in Canadian art history from Concordia University in Montreal in 1992. Since 1974 he has been custodian of the art collections of the National Archives of Canada. In that capacity he participated in the planning of the art vault in the National Archives' new Preservation Center in Gatineau and was responsible for planning and overseeing the preparation and move of the collections. He has lectured and published articles on various subjects in early Canadian art and remains particularly interested in the history of early eighteenth- and nineteenth-century Canadian drawing.

Debbie Glynn received her B.A. (Honours) in conservation in 1997, and is now studying for her Ph.D. at Camberwell College of Art researching the preservation and conservation of ink jet and electrophotographic printed materials. She has presented her research project at the ICOM interim meeting in Ludwigsberg, Germany, in April 1998, and has published articles in fine-art printmaking journals in Australia and the U.K.

Sandra Grantham studied fine art at Hammersmith College of Art in the 1960s. She graduated in 1994 from the B.A. Honours Course in paper conservation at Camberwell College of Art. In 1994–95 she received a Samuel H. Kress Fellowship to work as a conservator in the art on paper section at the Pierpont Morgan Library, New York. In 1995–99 she carried out M.Phil./Ph.D. research into the conservation of Japanese screens on the Royal College of Art/Victoria & Albert joint conservation course. Her research included visits to Japan, a three-month term at Kyoto City University of Arts to study Japanese classical painting, and a trip to study with pigment and paper manufacturers and in scroll mounting studios in Tokyo and Kyoto. She is currently engaged in private practice in London.

Gloria Groom is the David and Mary Winton Green Curator of European Painting at The Art Institute of Chicago. She received her Ph.D. from the University of Texas at Austin in 1989. Between 1979 and 1981, she studied at the Université de la Sorbonne, Ecole du Louvre, and UNESCO. She came to The Art Institute of Chicago in November 1984 as a research assistant. Since then she has helped organize and write for a number of exhibitions including: *The Art of Paul Gauguin*, *Odilon Redon: Prince of Dreams*, and *Gustave Caillebotte: Urban Impressionist*. Her book, *Édouard Vuillard: Painter-Decorator*, was published in 1993. Recent exhibitions she has curated include *Renoir's Portraits: Impressions of an Age* (1996–97), *Constant Troyon and the Evolution of French Landscape Painting* (1998), and *Beyond the Easel: Decorative Painting by Bonnard, Vuillard, Denis, and Roussel, 1890–1930* (2001).

Arlen Heginbotham is Assistant Conservator of Decorative Arts and Sculpture at the J. Paul Getty Museum in Los Angeles,

California. Since taking an interest in art conservation in 1990, he has worked in several private furniture conservation studios including, most recently, Robert Mussey Associates in Boston, Massachusetts. He has also held internships and technician positions at the Yale Peabody Museum of Natural History, the Isabella Stuart Gardner Museum, and the Philadelphia Museum of Art. He received his B.A. in East Asian Studies from Stanford University and his M.A. in Art Conservation from Buffalo State College.

Gregory J. Hill is Senior Prints and Drawings Conservator for the Conservation Treatment Section, Archives Preservation Division, National Archives of Canada. Collections under the care of the print and drawings laboratory include: works of art on paper, philatelic, and caricature. Mr. Hill has an honors degree in fine art from the University of Guelph, Ontario, and a diploma in art conservation techniques from Sir Sandford Fleming College, specializing in photograph conservation. Prior to joining the staff of the National Archives in 1988, he worked for the Provincial Archives of Manitoba and in private practice.

Anna Hillcoat-Imanishi graduated from Camberwell College of Art in 1996 with a B.A. in paper conservation. She was a student in the Royal College of Art/Victoria & Albert Conservation Course specializing in the conservation of works of art on paper from India and Southeast Asia, graduating in 1998 with an M.A. She was short-listed for the Jerwood Young Conservator of the Year Award in 1998. She has been Conservator of the Smithian Herbarium, Liverpool, since June 1999.

Carlo James studied mathematics at the Sorbonne, Paris. He went to Florence in 1967 and took his master's degree in engraving at the Istituto Statale d'Arte. He was trained by Sergio Boni, head of the conservation laboratory of the Uffizi Gallery, in the conservation of prints and drawings. In 1973 he took over the conservation laboratory of the Fondation Custodia in Paris, where he still works today. From 1980 to 1986, he taught conservation of graphic art at the Institut Français de Restauration des Oeuvres d'Art. With colleagues he prepared a publication to give students a global view of conservation, the *Manuale per la Conservazione e il Restauro di Disegni e Stampe Antichi* (English transl. by Marjorie Cohn, 1997).

Richard Kendall graduated from the Courtauld Institute of Art with an M.A. in art history. Initially specializing in fourteenth-century Italian art at the Courtauld Institute, he later turned to nineteenth- and twentieth-century painting and sculpture in his career as a teacher. As Senior Lecturer in the Department of History of Art and Design at Manchester Metropolitan University, he taught art history to both academic and fine-art students, and pioneered a course that linked their two disciplines. In the mid-1980s he published a series of articles and monographs on the work and writings of the Impressionists, principally Cézanne, Degas, and Monet. These were followed by the exhibitions *Degas: Images of Women* in 1989; *Degas Landscapes* in 1993; *Degas: Beyond Impressionism* in 1996; and *Degas and the Little Dancer* in 1998.

Roger S. Keyes is Director of the Center for the Study of Japanese Prints in Cranston, Rhode Island, and a Visiting Scholar at Brown University. He is the author of over 100 books, catalogues, articles, and other publications including *The Theatrical World of Osaka Prints* (with Keiko Mizushima), *The Art of Surimono, The Male Journey in Japanese Prints,* and most recently *Creative Transformations: Japanese Prints from the Collection of William Green.* He is currently cataloguing the Japanese illustrated books at the Museum of Fine Arts, Boston, and completing a *catalogue raisonné* of Hokusai's prints. He has lectured to conservation students and co-taught two seminars for practicing paper conservators on the history, connoisseurship, and restoration of Japanese prints.

Daria Keynan is a paper conservator working in private practice in New York, specializing in the conservation of modern, contemporary, and ethnographic art on paper.

Margaret F. MacDonald is Senior Research Fellow, Centre for Whistler Studies, University of Glasgow; B.A. (Honours) Fine Art, Leeds University, 1965; D. Litt., University of Glasgow, 1997. Her publications include *James McNeill Whistler. Drawings, Pastels and Watercolours* (1995), *Beatrice Whistler, Artist and Designer* (1997), and *Whistler in Venice* (2000). She is a co-author of *The Paintings of James McNeill Whistler* (1980), and co-curator of the retrospective exhibition *James McNeill Whistler* (1995–96). Current collaborative projects include an electronic edition of the complete Whistler correspondence (11,000 letters), a *catalogue raisonné* of Whistler's etchings, and exhibitions for Whistler's bicentenary in 2003.

Anne F. Maheux received her Honors B.A. in fine art from the University of Guelph, Ontario, and an M.A. in art conservation from Queen's University, Kingston, Ontario. She is Conservator of Prints and Drawings at the National Gallery of Canada in Ottawa. Areas of research have included both practical topics which deal with conservation treatment (applications of the ultrasonic mister; treatment of oversized works on paper) and research into artists' techniques and materials. Her interest in nineteenth-century pastels stems from research on the pastels of Edgar Degas, conducted during the preparation of a major retrospective of the artist's work presented in 1988–89. Research on Giuseppe De Nittis was initiated during her term as a Fellow of the American Academy in Rome in 1996.

Barbara J. Mangum is the principal of Sculpture and Decorative Arts Conservation Services in Somerville, Massachusettes. She holds a B.A. in chemistry and art history *cum laude,* from Vanderbilt University (1977), and an M.A. from the Cooperstown Graduate Conservation Program (1983). She interned in objects conservation at the Fogg Art Museum, the Museum of Fine Arts, Boston, and the Fitzwilliam Museum, Cambridge, England. She came to the Gardner Museum as Conservator of Objects in 1986, and was Chief Conservator from 1992 until 2000. While at the Gardner she was instrumental in evaluating the museum's environment and obtaining full climate control for the collection. She oversaw the development of the conservation department, including the renovation of conservation laboratories for textiles, historic upholstery, and objects.

Christopher W. McGlinchey is the Conservation Scientist at The Museum of Modern Art, New York, and Adjunct Professor of Conservation Science at the Conservation Center, Institute of Fine Arts, New York University. He received his B.A. from Boston University and M.A. in polymer science from Polytechnic Institute, Brooklyn.

Thomas McGrath earned his Ph.D. at Harvard University in 1994. His scholarship combines a number of different interests, including Renaissance theories of art and aesthetics (particularly color theory), drawing practices, painting, and the reintegration of works of art into a period context of function and meaning. His publications include "Color in Italian Renaissance drawings: Reconciling theory and practice in Central Italy and Venice," *Apollo* 145, no. 429 (November 1997); "Federico Barocci and the history of *Pastelli* in Central Italy," *Apollo* 148, no. 441 (November 1998); and "Drawing practices and market forces in sixteenth-century Italy," in a festschrift for John Shearman (1999).

Meredith Montague is Associate Conservator of Textiles and Costumes at the Museum of Fine Arts, Boston. She received her M.A. and Certificate of Advanced Study in Conservation from the State University College at Buffalo in 1991; internship training includes work at the Los Angeles County Museum of Art, the Detroit Institute of Arts, the Textile Museum, and the Textile Conservation Workshop in South Salem, New York.

Roy Perkinson received his B.S. in physics and philosophy (1963) from the Massachusetts Institute of Technology. He worked as a physicist at Draper Laboratory, MIT, and received an M.A. in the history of art (1967) from Boston University. He carried out an apprenticeship in conservation under F. W. Dolloff, at the Museum of Fine Arts, Boston. He established the Western Regional Paper Conservation Laboratory at the Fine Arts Museums of San Francisco in 1973 and has acted as Head of Paper Conservation at the Museum of Fine Arts, Boston, since 1976. He has published his research on watermarks in nineteenth-century papers, the medium of watercolor, and the history of blue papers. Currently he is completing a translation of Schweidler's classic manual on early twentieth-century practices of paper restoration. He is a Fellow of AIC and IIC, a member of the Print Council of America (1998), and is a recipient of the Sheldon and Caroline Keck Award from AIC (1999).

Thomas Primeau is the Assistant Paper Conservator at the Baltimore Museum of Art. He received an M.A. in art history from the University of Michigan in 1990. In 1996 he completed an M.A. in paper conservation at the State University College at Buffalo. He has worked at McKay Lodge Fine Arts Conservation Laboratory, the Pierpont Morgan Library, and the Detroit Institute of Arts.

Barbara Rosenberg received her B.A. in art history from the University of Toronto (1972) and went on to get a diploma from the Camberwell School of Arts and Crafts, London (1973). After a brief internship with bookbinder William Bull at the India Office Library and Records, London (1973), she apprenticed with Robert Muma, hand bookbinder and restorer, in Toronto (1974–77). In 1978 she acquired Mr. Muma's business and started a private practice. She studied book conservation with conservator Betsy Eldridge (1986–87). While completing her M.A. in conservation at Queen's University, Kingston (1997), she interned at the McMichael Canadian Art Collection, Kleinberg, and the Art Gallery of Ontario, Toronto. She is currently in private practice as a book and paper conservator in Toronto.

Britt Salvesen is Associate Editor of Scholarly Publications at The Art Institute of Chicago. She received her Ph.D. in art history from the University of Chicago in 1997. Her area of specialization is nineteenth-century art and culture. A collaborating editor for *The Lithographs of James McNeill Whistler*, she also wrote *Impressionism and Post-Impressionism in The Art Institute of Chicago* and *Artists in Focus: Gauguin*. Most recently she was a contributor to *Van Gogh and Gauguin: The Studio of the South*, by Douglas W. Druick and Peter Kort Zegers (2001).

Marjorie Shelley is Sherman Fairchild Conservator in Charge of the Paper and Photograph Conservation Department, The Metropolitan Museum of Art. She holds an M.A. in art history from the Institute of Fine Arts, and a degree certificate from the Conservation Center, New York University. She did her internships at The Metropolitan Museum of Art, the Pierpont Morgan Library, and the British Museum. Her areas of conservation specialization have been old master drawings, pastels, and nineteenth-century American drawings. She writes and lectures frequently on the technical history of these subjects. Her essays on pastel include "American pastels of the late nineteenth and early twentieth centuries: Materials and techniques," and "Painting in crayon: The pastels of John Singleton Copley." She is the author of the Metropolitan's manual *The Care and Handling of Art Objects*.

Harriet K. Stratis is Conservator of Prints and Drawings at The Art Institute of Chicago. Her research is devoted primarily to the study of pastels, charcoals, and the drawing techniques of nineteenth-century artists including Odilon Redon, Edgar Degas, and Mary Cassatt. She received her B.A. in art history and visual arts from Barnard College, Columbia University, and an M.A. in art history and degree certificate in conservation from the Institute of Fine Arts, New York University. Most recently, she served as co-editor and is one of four authors of the definitive *catalogue raisonné* of the lithographs of James McNeill Whistler.

Yoonjoo Strumfels received her B.F.A. from Rosemont College, Pennsylvania, and her M.S. in art conservation from the University of Delaware. She is now Associate Paper Conservator at the National Gallery of Art, Washington, D.C.

Sydney Thomson gained a B.A. (Honours) degree in fine art at the University of Newcastle upon Tyne in 1983. She was awarded an M.A. in paper conservation from Gateshead Technical College in association with Newcastle upon Tyne Polytechnic, graduating in 1988. In August of the same year, she joined the Eastern Pictorial Conservation Department at the British Museum, where she began her training in the conservation and mounting of Japanese art. From April 1993 to 1994 she worked in the *Usami Shokukudo*, one of the four "National Treasure" conservation studios in Kyoto, Japan. She specialized during this time on the conservation of folding screens (*byōbu*) and hanging scrolls (*kakemono*). Since 1994 she has continued her work in the Hirayama Conservation Studio in the British Museum.

Carol Tolocka received her B.S. in chemistry from George Washington University. She was research assistant in the Scientific Research Department at the National Gallery of Art, Washington, D.C. 1997–98.

Joyce H. Townsend received her B.Sc. (Honours) in physics from the University of Glasgow in 1979. She was employed as Conservation Scientist, Glasgow Museums, 1979–87. Between 1987 and 1991, she was Conservation Scientist, and from 1991 to the present, Senior Conservation Scientist at the Tate Gallery, London. She received her Ph.D. from the Courtauld Institute of Art, University of London, in 1991. Ms. Townsend concentrates on the identification and aging of artists' materials used in the later eighteenth and nineteenth centuries, and also works with conservators who are evaluating or developing conservation processes. She has published studies on Turner, Reynolds, Whistler, Sargent, G. F. Watts, and Degas. She has been editor and technical editor of several conference proceedings, most recently *Turner's Painting Techniques in Context* (UKIC, 1995), and *Resins: Ancient and Modern* (SSCR, 1995), and a co-editor with Stephen Hackney and Rica Jones of *Paint and Purpose: A Study of Technique in British Art* (Tate Gallery, 1999).

Pamela de Tristan is in private practice in London and specializes in Japanese books and prints. She obtained a national diploma in design and an art teacher's certificate from London University. In 1970 she began dealing in Japanese woodblock prints. She studied paper conservation at Camberwell School of Arts and Crafts and graduated in 1981. She has lectured on Japanese print conservation for the Pacific Asia Museum, and the Ukiyo-e Society of America. She presented papers at the 1992 and 1997 Institute of Paper Conservation conferences.

Piet van Dalen worked in the restoration department of the Municipal Archives in Rotterdam since 1980. In 1991 he graduated from Opleiding Restauratoren, Amsterdam (M.B.O.). In the same year, together with Jan Stokmans, he founded Art Conservation B.V., a studio specializing in the restoration and conservation of art and artifacts on paper. In addition to collaboration with the Schiedam Museum, at Art Conservation B.V. he works on various collections, including that of the Nederlands Film Museum (Amsterdam), and is involved in the conservation of over 10,000 posters.

Pauline Webber studied at Canterbury College of Art and Chelsea School of Art. She obtained a 1st class honours degree and an M.A. in fine art. She has worked in the Conservation Department at the Victoria & Albert Museum for over 20 years and is presently Head of the Paper Section. She has studied in the U.S., Italy, and Taiwan. She has written frequently for *The Paper Conservator* and other conservation journals. She was a tutor for the Royal College of Art/Victoria & Albert Museum course in the conservation of historic wallpapers and large works of art on paper (1991–93). She has also been a tutor for the Royal College of Art/Victoria & Albert Museum course on Indian and Asian art on paper, the conservation of Japanese screens, and Japanese woodblock prints.

Mike Wheeler graduated from Gateshead Technical College with a diploma in paper conservation in 1984. He has worked as a paper conservator in the U.K., New Zealand, and the U.S. In 1991 he was awarded an Andrew W. Mellon Fellowship in Paper Conservation at The Metropolitan Museum of Art, New York. In 1994 he worked as a conservator of photographs at the Better Image in New York. Since 1995 he has been Senior Paper Conservator at the Victoria & Albert Museum. He has been a tutor for the Royal College of Art/Victoria & Albert conservation course, specializing in Indian and Asian art on paper. He has studied in Japan and worked on conservation projects in India.

Paul M. Whitmore received a Ph.D. in physical chemistry from the University of California at Berkeley. Following an appointment at the Environmental Quality Laboratory at Caltech studying the effects of photochemical smog on works of art, he joined the staff at the Center for Conservation and Technical Studies at the Harvard University Art Museums. Since 1988 he has been Director of the Research Center on the Materials of the Artist and Conservator at Carnegie Mellon Research Institute, where his research has been directed towards the study of the permanence of modern art and library materials.

Faye T. Wrubel has been Conservator of Paintings at The Art Institute of Chicago for over 20 years. Her responsibilities range from examination and treatment of paintings from the fourteenth through the twentieth centuries, as well as extensive exhibition work and travel. Ms. Wrubel has acted as a conservation consultant in Chicago to the Museum of Contemporary Art, the Field Museum, the Smart Museum at the University of Chicago, and the Chicago Historical Society, as well as numerous other institutions and private collections nationwide. She has received many IMS, NEA, NEH, and Samuel H. Kress grants for both surveys and treatments. In 1995 she received a USIA Grant to act as conservation consultant to museums in Brazil. She has an M.A. from the Cooperstown Graduate Conservation Program, and is a Fellow of the AIC.

M. Brigitte Yeh is a Visiting Fellow in Conservation at the Kyoto National Museum Conservation Center for Cultural Properties in Japan. She obtained a B.A. from Princeton University and an M.A. from the University of California at Berkeley in the field of East Asian art and archeology before enrolling at the Conservation Center, Institute of Fine Arts, New York University. She completed her final-year internship in the Paper Conservation Division of the Library of Congress, Washington, D.C. Between 1998 and 2000 she was an Andrew W. Mellon Fellow in Paper Conservation at The Art Institute of Chicago.

Peter Kort Zegers is Rothman Family Research Curator in the Department of Prints and Drawings at The Art Institute of Chicago. Prior to joining the staff of the Art Institute in 1986, he was Senior Conservator of Graphic Art at the National Gallery of Canada in Ottawa. Most recently he was the organizing curator (with Douglas W. Druick) of *Van Gogh and Gauguin: The Studio of the South* (2001).

Faith Zieske received a M.S. in conservation from the Winterthur/University of Delaware Art Conservation Program, specializing in works on paper and photography, and a M.F.A. from Michigan State University. Since 1983, she has worked at the Philadelphia Museum of Art, where she is presently Conservator of Works of Art on Paper. Previously she worked for the National Museum of American History, Smithsonian Institution, and held internships at the National Gallery of Art, Washington, D.C., and with private conservator Christa Gaehda. She presented techniques developed for removing aged pressure-sensitive tape from 81 Cézanne sketchbook leaves at the 1992 Institute of Paper Conservation conference in Manchester. The results of her analysis of Cézanne's watercolors were presented at the American Institute for Conservation conference and published in 1995. Her ongoing research includes characterization and fiber analysis of several papers used by Cézanne for his watercolors.

Additional Photography Credits

Page 1, Edgar Degas, *Two Dancers* (detail of Fig. 1.14).

Page 61, Paul Cézanne, *The Balcony* (detail of Fig. 2.24).

Pages 136–37, Édouard Vuillard, *Foliage: Oak Tree and Fruit Seller* (detail of Fig. 3.20).

Pages 182–83, Kitagawa Utamaro, *The Courtesan Hanaôgi Holding a Cat* (detail of Fig. 4.7).

Page 229, Paul Cézanne, *Three Skulls* (detail, faded to 20%), 1902, watercolor on paper, The Art Institute of Chicago, 1954.183.

Pages 256–57, Odilon Redon, *Flower Clouds* (detail), c. 1903, pastel on paper, The Art Institute of Chicago, 1990.165.

Figures 3.12 to 3.23, © 2001 Artists Rights Society (ARS), New York/ADAGP, Paris.

Figure 3.27, © 2002 Artists Rights Society (ARS), New York/Beeldrecht, Amsterdam.

Figure 3.30, © 2002 Andy Warhol Foundation for the Visual Arts/Artists Rights Society (ARS), New York.

Figures 3.47 and 3.48, © Gerhard Richter.

Figure 3.49, © Estate of Ben Shahn/Licensed by VAGA, New York, NY.

Figures 3.50 and 3.51, © 2002 Estate of Sam Francis/Artists Rights Society (ARS), New York.

Figures 3.52 to 3.54, © Tom Wesselmann/Licensed by VAGA, New York, NY.

Figures 3.55 and 3.56, © Dedalus Foundation, Inc./Licensed by VAGA, New York, NY.